Waste Treatment and Disposal

Waste Treatment and Disposal

PAUL T. WILLIAMS

Department of Fuel and Energy
The University of Leeds, UK

JOHN WILEY & SONS
Chichester • New York • Weinheim • Brisbane • Singapore • Toronto

National 01243 779777
International (+44) 1243 779777
e-mail (for orders and customer service enquiries):
cs-books@wiley.co.uk.
Visit our Home Page on http://www.wiley.co.uk
or http://www.wiley.com

Other Wiley Editorial Offices

John Wiley & Sons, Inc., 605 Third Avenue,
New York, NY 10158-0012, USA

WILEY-VCH Verlag GmbH, Pappelallee 3,
D-69469 Weinheim, Germany

Jacaranda Wiley Ltd, 33 Park Road, Milton,
Queensland 4064, Australia

John Wiley & Sons (Asia) Pte Ltd, 2 Clementi Loop #02-01,
Jin Xing Distripark, Singapore 129809

John Wiley & Sons (Canada) Ltd, 22 Worcester Road,
Rexdale, Ontario M9W 1L1, Canada

Library of Congress Cataloging-in-Publication Data

Williams, Paul T.
Waste treatment and disposal / by Paul T. Williams.
p. cm.
Includes bibliographical references and index.
ISBN 0-471-98166-4 (alk. paper). — ISBN 0-471-98149-4 (alk. paper)
1. Refuse and refuse disposal—Great Britain. 2. Refuse and refuse disposal—Government policy—Great Britain. I. Title.
TD789.G7W54 1998
363.72'8'0941—dc21
97–46541
CIP

British Library Cataloguing in Publication Data

A catalogue record for this book is available from the British Library

ISBN 0-471-98166-4
0-471-98149-4

Typeset in 10/12pt Times from the author's disks by Mayhew Typesetting, Rhayader, Powys
Printed and bound in Great Britain by Bookcraft (Bath) Ltd

This book is printed on acid-free paper responsibly manufactured from sustainable forestation, for which at least two trees are planted for each one used for paper production.

This book is dedicated with lots of love to
Lesley, Christopher, Simon and Nicola

Contents

Preface

This book is aimed at undergraduate and postgraduate students undertaking courses in Environmental Science and Environmental, Civil, Chemical and Energy Engineering with a component of waste treatment and disposal. It is also aimed at professional people in the waste treatment and disposal industry.

The book covers waste treatment and disposal, with particular emphasis on the UK, but with examples and comparisons with Europe, North America and the Far East throughout the text.

The first chapter is an historical introduction to waste treatment and disposal. It discusses the UK National Waste Strategy, UK waste management policy and the strategy for sustainable waste management and its implementation via policy instruments. The current options used in the UK for waste treatment and disposal are described.

Chapter 2 outlines the development of UK waste legislation. The major legislative and regulatory measures dealing with waste treatment and disposal are described, including The Environmental Protection Act 1990 and integrated pollution control, the regulation of 'Prescribed Processes' and of smaller waste incineration and combustion processes, regulation of waste disposal on land, the 'Duty of Care' and Waste Disposal Plans. The influence of the UK planning system on waste treatment and disposal are discussed. The 1995 Environment Act and the role of the Environment Agency are also described. The requirements for an Environmental Assessment for a large scale waste treatment and disposal facility are described. The economics of the different waste management options are compared and discussed.

Chapter 3 discusses the different definitions of waste. Estimates of waste arisings in the UK and the rest of the world are discussed, along with the methods used in their estimation. Various trends in waste generation and influences on them are discussed. Several categories of waste are reviewed in terms of arisings, and treatment and disposal options. The wastes described in detail are municipal solid waste, hazardous waste, sewage sludge, clinical waste, agricultural waste, and industrial and commercial waste. Other wastes described are construction and demolition waste, mines and quarry waste, power station ash, iron and steel slags and scrap tyres. The chapter ends with a discussion of the different types of waste containers, collection systems and waste transport.

Chapter 4 is concerned with waste reduction, re-use and recycling, with the emphasis on recycling. The legislative background to recycling is discussed. Industrial and commercial waste recycling and household waste recycling are reviewed in detail with examples. Examples of recycling of particular types of waste, i.e., plastics, glass, paper, metals and tyres, are presented. Economic considerations of recycling are discussed. Current examples of recycling schemes for household waste, and particular types of waste, are described.

Chapter 5 is concerned with waste landfill, the main waste disposal option in many countries throughout the world. Comparison is made with UK waste landfilling, and with that of Europe and the rest of the world. Landfill design and engineering, the various considerations for landfill design and operational practice are described. The different main types of waste which are landfilled, i.e., inert wastes and bioreactive wastes, and the processes operating within and outside the landfill are discussed. The major different landfill design types, i.e., attenuate and disperse, containment, co-disposal, entombment and the sustainable landfill (the controlled flushing bioreactor landfill), are reviewed in detail. The formation of landfill gas, landfill gas migration, and management and monitoring of landfill gas are discussed, as is landfill leachate formation, and leachate management and treatment. The final stages of landfilling of wastes, i.e., landfill capping, and landfill site completion and restoration, are described. The recovery of energy through landfill gas utilisation is reviewed in detail. The problems of old landfill sites are highlighted. The economic aspects of landfill in comparison to other waste treatment and disposal options are presented. The proposed European Directive on waste landfills and its impact on current landfill practice are discussed.

Chapter 6 is concerned with incineration, the second major option for waste treatment and disposal in many countries throughout the world. The various incineration systems are discussed. This chapter concentrates on mass burn incineration of municipal solid waste, following the process through, waste delivery, the bunker and feeding system, the furnace, and heat recovery systems. Emissions formation and control is emphasised, with discussion of the formation and control of particulate matter, heavy metals, toxic and corrosive gases, and products of incomplete combustion such as polycyclic aromatic hydrocarbons (PAH), dioxins and furans. The contaminated wastewater and contaminated bottom and flyash arising from waste incineration is described. The dispersion of emissions from chimney stacks is discussed. Energy recovery via district heating and electricity generation are described. Other types of incineration, including fluidised bed incinerators, starved air incinerators, rotary kiln incinerators, cement kilns, and liquid and gaseous waste incinerators and the types of waste incinerated in the different types is discussed. The economic aspects of waste incineration are reviewed. Current examples of different types of incinerator are described throughout.

Chapter 7 discusses other options for waste treatment and disposal. Pyrolysis of waste, the types of products formed during pyrolysis and their utilisation, and the different pyrolysis technologies are discussed. Gasification of waste, gasification technologies and utilisation of the product gas are described. Composting of waste is described, including the composting process and the different types of composter. Anaerobic digestion of waste, the degradation process, and the methods of operation and the technology for anaerobic digestion are discussed. Current examples of different types of pyrolysis, gasification, composting and anaerobic digestion are presented throughout.

The concluding chapter discusses the integration of treatment and disposal options described in the previous chapters to introduce the concept of 'integrated waste management'. The different approaches to integrated waste management are described.

Acknowledgements

Many people have helped in the preparation of this book. I would like to thank in particular, Jean Gibbs of The World Resource Foundation, Gerry Atkins of SELCHP Ltd., Hugh Moss of Dynamotive Ltd., Andrew Eeles, HMSO, and Judith Petts, Loughborough University. I would also like to thank the many copyright owners for permission to use published materials.

1

Introduction

Summary

This chapter is an historical introduction to waste treatment and disposal. It discusses the UK National Waste Strategy, UK waste management policy and the strategy for sustainable waste management and its implementation via policy instruments. The current options used in the UK for waste treatment and disposal are described. Comparison is made with waste treatment and disposal in the USA and Japan.

1.1 History of Waste Treatment and Disposal

The historical development of waste treatment and disposal has been motivated by concern for public health. The industrial revolution between 1750 and 1850 led to many people moving from rural areas to the cities, a massive expansion of the population living in towns and cities, and a consequent increase in the volume of waste arising. The increase in production of domestic waste was matched by increases in industrial waste from the burgeoning new large scale manufacturing processes. The waste generated contained a range of materials such as broken glass, rusty metal, food residue and human waste, and was dangerous to human health. In addition, it attracted flies, rats and other vermin, which in turn posed potential threats through the transfer of disease. This led to an increasing awareness of the link between public health and the environment.

To deal with this potential threat to human health, legislation was introduced on a local and national basis in many countries. For example in the UK, throughout the latter half of the nineteenth century, a series of Nuisance Removal and Disease Prevention Acts were introduced which empowered local authorities to set up teams of inspectors to deal with offensive trades and control pollution within city limits. These Acts were reinforced by the Public Health Acts of 1875 and 1936 which covered a range of measures, some of which were associated with the management and disposal of waste. The 1875 Act placed a duty on local authorities to arrange for the removal and disposal of waste. The 1936 Act introduced regulation to control the disposal of waste into water, and defined the statutory nuisance associated with any trade, business, manufacture or process which might lead to the degradation of health or of the neighbourhood (British Medical Association 1991; Clapp 1994; Reeds 1994). In the USA, early legislation included the 1795 law introduced by the Corporation of Georgetown, Washington, DC, which prohibited waste disposal on the streets and introduced the requirement for individuals to remove waste themselves or hire private contractors. By 1856, Washington had a city-wide waste collection system supported by taxes. Further, by 1915, 50% of all major US cities provided a waste collection system, and this figure had risen to 100% by 1930 (Neal and Schubel 1987; McBean et al 1995).

One of the main constituents in domestic dustbins in the late 19th century was cinders and ash from coal fires, which represented a useful source of energy. The waste also contained recyclable materials such as old crockery, paper, rags, glass, iron and brass, and was often sorted by hand by private contractors or scavengers to remove the useful items. Much household waste would also be burnt in open fires in the living room and kitchen as a 'free fuel' supplement to the use of coal. This combustible content of the waste was recognised as a potential source of cheap energy for the community as a whole, and the move away from private waste contractors to municipally organised waste collection led to an increase in incineration. Purpose-built municipal waste incinerators were introduced in the UK in the late 1870s and by 1912 there were over 300 waste incinerators in the UK, of which 76 had some form of power generation (Van Santen 1993). One of the first municipal incinerators introduced in the US was in 1885 in Allegheny, Pennsylvania (Neal and Schubel 1987). However, many of the waste incinerators were small-scale, hand-fed plants which were poorly designed and controlled and their operation was not cost-effective.

However, the growth of incineration was secondary to the main route to disposal, which was dumping, either legally or illegally. The ease of waste disposal to land and the move to centralised waste management through town or city authorities meant that this route increasingly became the preferred waste disposal option, particularly as incineration plants were difficult and expensive to maintain. As incineration plants reached the end of their operational lifetime, they tended to become scrapped in favour of landfill. The waste dumps themselves, however, were poorly managed, open tips seething with vermin and often on fire. The environmental implications of merely dumping the waste in such open sites was recognised, and increasingly waste began to be buried. Burying the waste had the advantages of reducing odours, and discouraging rats and other vermin, and consequently the sites became less dangerous to health. Through the first half of the 20th century some improvements in landfill sites were seen, with improved site planning and site management. However, this did not apply to all areas, and many municipal sites still had the minimum of engineering design and the open dump was still very common. When such sites were full, they were covered with a thin layer of soil and there was minimum regard to the effects of contaminated water leachate or landfill gas emissions from the disused site (McBean et al 1995).

Following the Second World War, waste treatment and disposal was not seen as a priority environmental issue by the general public and legislature, and little was done to regulate the disposal of waste. However, a series of incidents in the late 1960s and 1970s highlighted waste as a potential major source of environmental pollution. A series of toxic chemical waste dumping incidents led to increasing awareness of the importance of waste management and the need for a more stringent legislative control of waste. Amongst the most notorious incidents were the discovery, in 1972, of drums of toxic cyanide waste dumped indiscriminately on a site used as a children's playground near Nuneaton in the UK, the leaking of leachate and toxic vapours into a housing development at the Love Canal site, New York State, in 1977, the dumping of 3000 tonnes of arsenic and cyanide waste into a lake in Germany in 1971, and the leak of polychlorinated biphenyls (PCBs) into rice oil in Japan in 1968, the 'Yusho' incident (British Medical Association 1991) (Box 1.1).

The massive adverse publicity and public outcry led to pressure for the problem of waste disposal to be more strictly controlled by the legislature. In the UK, as a direct result of the Nuneaton cyanide dumping incident, emergency legislation was introduced in the form of The Deposit of Poisonous Waste Act, 1972. The Act made it an offence to deposit wastes which were poisonous, noxious or polluting and liable to give rise to an environmental hazard. Further legislation on waste treatment and disposal followed in 1974 with the Control of Pollution Act, which controlled waste disposal on land through a new licensing and monitoring system for waste disposal facilities. In the USA, similar land-mark legislation covering waste disposal was developed with the Resource, Conservation and Recovery Act, 1976, which initiated the separation and defining of hazardous and non-hazardous waste and the separate requirements for their disposal.

The recognition of the need for environmentally acceptable means of waste disposal following the illegal waste dumping incidents led to a revival of some interest in incineration. Between 1969 and 1981, thirty new municipal waste incinerators were constructed in the UK, with a total incineration capacity of 2.7 million tonnes. This capacity was, however, less than 10% of the municipal waste arisings in the UK each year, the remainder still being landfilled, and the majority of the plants had no form of energy recovery to offset disposal costs. The revival of municipal waste incineration was short-lived as the high capital and operational costs of incineration plant had to compete with the much lower costs of waste disposal via landfill, and local authorities with increasing constraints on monetary budgets chose the cheaper option. In addition, legislation to limit the emissions from incinerators also resulted in the closure of many incineration plants.

The current situation in the UK is that still by far the majority of waste is disposed of in landfill sites, accounting for 90% of domestic waste, 85% of commercial waste and 73% of industrial waste (Deprtment of Environment and Welsh Office 1995). The modern sites are well designed, constructed and managed, and many have energy recovery utilisation of the derived landfill gas. Whilst landfill remains the major option for waste disposal in the UK, increasing regulation has placed tighter controls on leachate and landfill gas treatment, monitoring and site aftercare, with a consequent increase in disposal costs. New developments in landfill design and operation

Box 1.1
Waste Disposal Incidents which Influenced Waste Management and Legislation

1. Love Canal, Niagara City, New York State, USA: 1977

Love Canal, Niagara City, was an unfinished canal excavated for a projected hydro-electricity project. The abandoned site was used as a dump for toxic chemical waste and more than 20 000 tonnes of waste containing over 248 different identified chemicals were deposited in the site between 1930 and 1952. Following the sale of the plot in 1953, a housing estate and school were built on the site. In 1977 foul smelling liquids and sludge seeped into the basements of houses built on the site. The dump was found to be leaking and tests revealed that the air, soil and water around the site were contaminated with a wide range of toxic chemicals, including benzene, toluene, chloroform and trichloroethylene. Several hundred houses were evacuated and the site was declared a Federal Disaster Area. There were also later reports of ill health, low growth rates for children and birth defects amongst the residents. As the actual and projected clean-up costs of the site became known, legislation in the form of the Comprehensive Environmental Response, Compensation and Liabilities Act, 1980, was introduced by Congress. This legislation placed the responsibility and cost of clean-up of contaminated waste sites back to the producers of the waste.

Source: British Medical Association, Hazardous Waste and Human Health, Oxford University Press, Oxford, 1991.

2. Cyanide Dumping, Nuneaton, Coventry, Warwickshire, UK: 1972

A series of toxic waste dumping episodes occurred in the early months of 1972, the most serious of which was the dumping of 36 drums of sodium cyanide in a disused brickworks at Nuneaton, on the outskirts of Coventry. The site was in constant use as a play area by local children. The drums were heavily corroded and contained a total of one and a half tonnes of cyanide, enough, police reported, to wipe out millions of people. Over the following weeks and months further incidents of toxic waste dumping were reported extensively in the press. Drums of hazardous waste were found in numerous unauthorised sites including a woodland area and a disused caravan site. The episodes generated outrage in the population, and emergency legislation was rushed through Parliament in a matter of weeks in the form of The Deposit of Poisonous Waste Act, 1972. The new Act introduced penalties of 5 years imprisonment and unlimited fines for the illegal dumping of waste, in solid or liquid form, which is poisonous, noxious or polluting. The basis of the legislation was the placing of responsibility for the disposal of waste on industry. Further legislation on waste treatment and disposal followed in 1974 with the Control of Pollution Act.

Source: The Times, Times Newspapers, London, 1972.

have resulted in the concept of the flushing bioreactor landfill, which recirculates the leachate to increase the rate of waste degradation. The combustion of landfill gas to produce energy in the form of electricity or power generation or district heating has now become the norm for modern landfills. Incineration has seen a decrease as a waste disposal option following the closure of many plants which cannot comply with new legislation on emission limits. However, a small number of the older plants have retro-fitted the required gas clean-up systems to meet the emissions legislation and continue operation. In addition, several large scale mass-burn municipal waste incinerators have recently been constructed, are under construction or are in the planning stage. The incineration of waste with energy recovery via either electricity generation or district heating has been developed in the 1990s to become an economic viability comparable to landfill. Many of the new wave of incinerators involve the private sector. In addition, several industrial waste, sewage sludge and clinical waste incinerator projects have been initiated during the 1980s and 1990s involving the private sector in many cases. These incinerators tend to have smaller throughputs of waste, and because of the higher costs of disposal of these types of waste are cost-effective compared with other forms of disposal. In many cases, the type of waste dictates that incineration is not only the most economic option, but also the best practicable environmental option.

1.2 The National Waste Strategy

The National Waste Strategy is a requirement of all member states of the European Union and is a document which sets out the policies in relation to the recovery and disposal of waste. The Strategy is a requirement of the EC Waste Framework Directive (75/442/EEC) amended by EC Directive 91/156/EEC. In particular, the Strategy must identify the type, quantity and origin of waste to be recovered or disposed of, the general technical requirements, any special arrangements for particular waste, and suitable disposal sites or installations. The EC Directive has been incorporated into the 1995 Environment Act as a statement by The Secretary of State for the Environment in England and Wales, but for Scotland the statement was written by the Scottish Environmental Protection Agency.

The objectives of the National Waste Strategy (Environment Act 1995; Lane and Peto 1995) include:

- ensuring that waste is recovered or disposed of without endangering human health and without using processes or methods which could harm the environment;
- establishing an integrated and adequate network of waste disposal installations, taking account of the best available technology not involving excessive costs;
- ensuring self sufficiency in waste disposal;
- encouraging the prevention or reduction of waste production and its harmfulness;
- encouraging the recovery of waste by means of recycling, re-use or reclamation and the use of waste as a source of energy.

Preparing the strategy involves a wide range of bodies including the Environment Agency, local government, industry, local planning authorities etc.

Part of the Waste Framework Directive requires the National Waste Strategy to relate to identifying suitable disposal sites or installations. This section is dealt with in the UK by planning authorities who are required to draw up land-use plans for use of land within a local authority area (Metropolitan Councils, County Councils and District Councils). These plans, the 'Development Plans' also include Waste Local Plans which define land which is preferred for the use of waste treatment and disposal facilities.

Table 1.1 *Public attitudes to the environment*

Environmental issue	'Percentage very worried'
Chemicals put into rivers and sea	63
Toxic waste: disposal and import	63
Radioactive waste	60
Sewage on beaches/bathing water	56
Oil spills at sea and oil on beaches	52
Litter and rubbish	29
Household waste disposal	22
Not enough recycling	19

Sources: [1] West Yorkshire Waste Management Plan, West Yorkshire Waste Management Authority, The Environment Agency, March 1996; [2] Survey of Public Attitudes to the Environment (1993), Department of the Environment, HMSO, London, 1994.

In addition to the requirement for the National Waste Strategy, the EC Waste Directive includes the 'self sufficiency principle', which states that Member States shall take appropriate measures to establish an integrated and adequate network of disposal installations which enable the Union as a whole to become self sufficient in waste disposal. A move towards individual member state self sufficiency is also recommended. The plan should also reflect the 'proximity principle', under which waste should be disposed of (or otherwise managed) close to the point at which it is generated. This creates a more responsible approach to the generation of wastes, and also limits pollution from transport. It is therefore expected that each region should provide sufficient facilities to treat or dispose of all the waste it produces, and such a strategy should be reflected in local development plans.

A recent survey of public attitudes to environmental issues placed waste high on the list of issues of concern (Table 1.1). Placed alongside this concern is a general public acceptance that waste requires effective treatment and disposal in a responsible and environmentally acceptable manner. However, the siting of waste treatment and disposal facilities has in many cases generated intense opposition in recent years, and recognition of what has become known as the NIMBY (not in my back yard) syndrome. The public perception in relation to waste treatment and disposal facilities has been discussed in detail by Petts and Eduljee (1994). This opposition to local siting of such facilities has implications for the implementation of waste management strategies which encompass the proximity principle, whereby waste generated in a local area is the responsibility of that area.

1.3 UK Waste Management Policy – The Strategy for Sustainable Waste Management

The themes of the National Waste Strategy have developed from the idea of 'sustainable development', which led directly from the 1992 United Nations Rio Conference on Environment and Development (the Earth Summit) and requires that society takes decisions with proper regard to their environmental impacts. The concept tries to strike a balance between two objectives, the continued economic development and achievement of higher standards of living both for today's society and for future generations, and the need to protect and enhance the environment. The economic development of society clearly has an impact on the environment since natural resources are used and by-product pollution and waste are produced in many processes. However, *sustainable* development promotes development by encouraging environmentally friendly economic activity and discouraging environmentally damaging activities. Such activities include energy efficiency measures, improved technology and techniques of management, better product design and marketing, environmentally friendly farming practices, making better use of land and buildings, improved transport efficiency and waste minimisation (Sustainable Development 1994; This Common Inheritance 1996).

The treatment and disposal of waste is one of the central themes of sustainable development. The approach of the European Union and its member states for the management of waste has developed via a series of Directives and Programmes into a strategy concerning the treatment of waste which has the key objectives of minimising the amount of waste that is produced, and minimising any risk of pollution.

The strategy for sustainable waste management in England and Wales is drawn up by the Secretary of State for the Environment and is entitled 'Making Waste Work' (Department of the Environment and Welsh Office 1995). The strategy sets out the waste management policy for England and Wales until the year 2005. The strategy is a non-statutory document and sets out the policy framework for waste management in England and Wales. The statutory requirement for a waste plan under the EC Directive will be met by a further document to be produced in 1998. The document 'Making Waste Work' takes the concept of 'sustainable development' as a strategy for waste management.

The EC strategy has been developed by the UK into the concept of a 'hierarchy of waste management' (Figure 1.1) (Sustainable Development 1994; Department of the Environment and Welsh Office 1995);

1. *Reduction*. Uppermost in the hierarchy is the strategy that waste production from industrial manufacturing processes should be reduced. Reduction of waste at source should be achieved by developing clean technologies and processes that require less material in the end products and produce less waste during manufacture. This may involve the development of new technologies or adaptations of existing processes. Other methods include the development and manufacture of longer lasting products and products which are likely to result in less waste when they are used. The manufacturing process should also avoid producing wastes which are hazardous, or reduce the toxicity of such wastes. Waste reduction has the incentive of making significant savings in raw materials, energy use and production and waste disposal costs.
2. *Re-use*. The collection and re-use of materials, for example doorstep milk delivery in the UK involves collection, cleaning and re-use of glass bottles. Tyre re-treading would also come into this category, where many truck tyres are re-treaded many times throughout their lifetime. Re-use can be commercially attractive in some circumstances. However, re-use may not be desirable in all

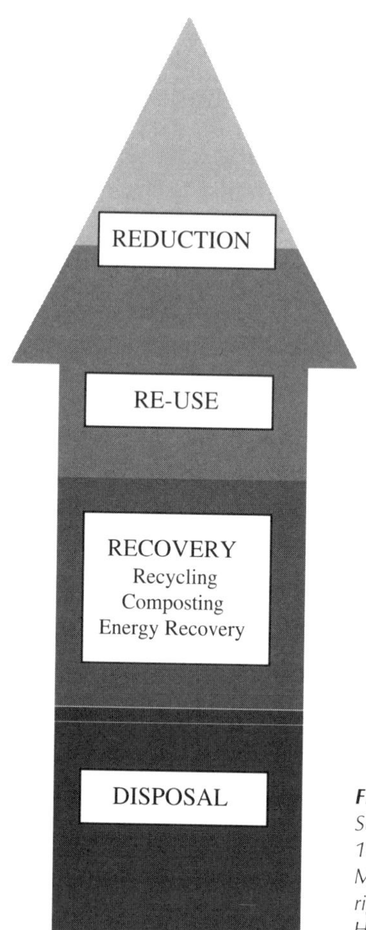

Figure 1.1 *The hierarchy of waste management. Sources: [1] Sustainable Development: The UK Strategy. HMSO, London, 1994; [2] Department of the Environment and Welsh Office, Making Waste Work, 1995. HMSO, London. Crown copyright is reproduced with the permission of the Controller of Her Majesty's Stationery Office*

cases since the environmental and economic cost of re-use in terms of energy use, cleaning, recovery, transportation etc. may outweigh the benefits.

3. *Recovery*. There are a number of different types of waste recovery.
(i) Materials recycling. The recovery of materials from waste and processing them to produce a marketable product, for example, the recycling of glass and aluminium cans is well established, with a net saving in energy costs of the recycled material compared with virgin production. The potential to recycle material from waste is high, but it may not be appropriate in all cases, for example, where the abundance of the raw material, energy consumption during collection and re-processing, or the emission of pollutants has a greater impact on the environment or is not cost-effective. Materials recycling also implies that there is a market for the recycled materials. The collection of materials from waste where there is no end market for them merely results in large surpluses of unwanted materials and also wastes additional energy with no overall environmental gain.

(ii) Composting. Decomposition of the organic fraction of waste to produce a stable product such as soil conditioners and growing material for plants. Composting is an extension of garden composting on a larger scale, and attempts have been made to use municipal solid waste for composting. However, contamination by heavy metals, glass and plastics have limited its application. Successful schemes using waste from gardens and parks have proved more acceptable, and there is research into the use of sewage sludge as a composting material.

(iii) Energy recovery. Producing energy by incinerating waste or combustion of landfill gas. Many wastes, including municipal solid waste, sewage sludge and scrap tyres, contain an organic fraction which can be burnt in an incinerator. The energy is recovered via a boiler to provide hot water for district heating of buildings or high-temperature steam for electricity generation. The incinerator installation represents a high initial capital cost, and sophisticated emissions control measures are required to clean up the flue gases. The anaerobic digestion of the putresible organic fraction of wastes such as domestic waste and sewage sludge in a landfill site produces a gas consisting mainly of methane which can be collected in a controlled, engineered way and burnt. Again the derived energy is used for either district heating or power generation.

4. *Disposal.* The disposal of waste using processes or methods that do not endanger human health and which cannot harm the environment, such as by incineration or controlled landfill without energy recovery. For newly constructed, large scale waste incinerators, some form of energy recovery is essential to ensure the economic viability of such schemes. Landfill is the predominant route for waste disposal in the UK, and throughout Europe and North America. Biological processes within the landfill ensure that over a period of time, the waste is degraded, neutralised and stabilised to form an essentially inert material. Also, the sites used for landfill are often used mineral workings which are required to be infilled, and consequently result eventually in recovered land. Methane (landfill gas) will still be produced in such sites and therefore the control of the gas emissions to prevent potential uncontrolled combustion is essential. A further major consideration for landfill disposal is the leachate, i.e. the potentially toxic liquid residue from the site which may enter the water course.

The strategy and the overall policy aim of the UK Government, encompassing sustainable development, requires that waste management practices move up the hierarchy such that waste is not merely disposed of, but should, where possible, be recovered, re-used or minimised. However, this may not be achievable in all cases, and in some cases may not be desirable. For example, some wastes are best landfilled or incinerated since the environmental and economic cost of trying to sort and decontaminate the waste to produce a useable product outweighs the benefits. Consequently, the principle of Best Practicable Environmental Option (BPEO) (see Box 2.3) should be applied to the process under integrated pollution control. In England and Wales the management of waste is concentrated towards the bottom of the hierarchy, partly due to the fact that the different waste management options do not, in all cases, fully reflect their *environmental* costs and benefits, and little information is available about the competitive benefits of waste minimisation (Department of the Environment and Welsh Office 1995).

Waste management strategy in the UK is also influenced by the mixture of public and private sector industries. The UK has a large private waste management sector, particularly for the industrial and commercial sectors. In addition, the 1990 Environmental Protection Act required Waste Disposal Authorities to divest themselves of their operations. Waste disposal is now in the hands of private companies, local authority waste disposal companies or co-owned private and local authority concerns. Disposal of waste for local authorities is the responsibility of the local Waste Disposal Authority using competitive tendering contracts. The Authority does not have to accept the lowest tendered cost where an alternative offers environmental benefits. Thereby the award of contracts by the Authority on this basis can be used to promote the policy of sustainable waste management. Waste collection is now also subject to compulsory competitive tendering, which may be through Local Authority Waste Collection Companies or the private sector.

With the aim of moving waste management policies up the hierarchy, the UK Government has set a number of indicative (not legally binding) targets for industry and waste sectors to minimise or recycle waste (Department of the Environment and Welsh Office 1995):

- to reduce the proportion of controlled (household, commercial and industrial) waste going to landfill to 60% (from 70%) by 2005;
- to recover, including materials recycling, energy recovery and composting, 40% of municipal waste by 2005;
- to reduce the waste produced by the government itself by aiming for two-thirds of Government Departments to have in place office waste minimisation targets by the end of 1996;
- to recycle or compost 25% of household waste by the year 2000. To help achieve this target, further targets are set of 40% of domestic properties with a garden to carry out composting by the year 2000, and all Waste Disposal Authorities to cost and consider the potential for centralised composting schemes, resulting in the composting of 1 million tonnes/year of organic waste. To help achieve the recycling target, 80% of households should have easy access to recycling facilities;
- to increase the use of secondary and recycled waste materials, such as construction, demolition, mining and quarrying wastes, as aggregates in the construction industry in England from 30 million tonnes/year to 55 million tonnes/year by 2005;
- to have in place a Government target before the end of 1998 for the overall reduction of waste.

The introduction of producer responsibility in the 1995 Environment Act and in response to European Union Directives has led to a number of industries setting their own re-use and recovery targets, for example recycling or recovery of 58% of packaging waste, 50% of aluminium cans, 37% of steel cans, 40% of newspapers and 65% of scrap tyres by the year 2000. Clearly the setting and achieving of targets requires accurate data on waste arisings, treatment and disposal, the provision of which is a major function of the Environment Agency.

1.4 Policy Instruments

The aims of Government strategy on sustainable waste management and the objective of moving up the waste hierarchy have led the Government to adopt a five-point policy instrument strategy to achieve those aims: a regulatory, market-based, planning, promotion and data strategy (Department of the Environment and Welsh Office 1995).

1. The regulatory strategy is based on the extensive legislative and regulatory provisions covering the management of waste.
2. The market-based strategy is centred on the use of the price mechanism in the market as a means of achieving cost-effective and economically efficient outcomes and thereby reaching the waste management goals. The Government emphasises that waste management should be carried out on a commercial and competitive basis, and that the prices of the various waste management options should reflect as far as practicable the costs of any environmental damage. The costs of the various waste management options should fall as far as possible on those responsible for the creation of the waste. The economic instrument used by the Government properly to reflect both financial and environmental costs was the landfill tax introduced in 1996. The tax, which is set at a standard rate of £7/tonne and £2/tonne for 'inactive' waste, is paid by the landfill operator, but the costs are passed on to the waste producers. Consequently, the environmental impact of landfill is better reflected in an increased cost which encourages business and consumers, in a cost-effective but non-regulatory way, to reduce the level of waste production and to dispose of less waste via landfill. However, this link is not clearly defined in the case of household waste, where the costs of waste disposal are subsumed within the local Council Tax. A further economic instrument to encourage waste recycling has been the 'recycling credit scheme'. The scheme enables recyclers to be credited via payments made by the local authority for the collection and disposal of the material they have removed and recycled from the waste stream. By recycling the waste the authorities are saving on not having to collect or dispose of waste that is recycled. In most cases, credits are paid by the Waste Disposal Authority to the Waste Collection Authority, but may also be paid to voluntary groups and business (Box 1.2).
3. The planning strategy involves the Planning Authorities, which are required to produce development plans which include provision for suitable waste disposal sites. The application of the 'proximity principle' whereby the treatment and disposal of waste should be carried out close to the point of waste production confers more responsibility on the communities which produce the waste. Planning Authorities are therefore encouraged to ensure that the development plans make adequate provision for suitable waste facilities which reflect the

Box 1.2
Landfill Tax

Landfill tax is a tax on wastes placed in landfill sites and came into force on 1 October 1996. The tax was set at £7/tonne for all wastes, except for certain specified listed wastes which are relatively 'inactive', at which the rate is set at £2/tonne. The tax is collected by Customs and Excise from landfill operators, who pass on the charges to the waste contractors and eventually to the waste producers. The rationale behind the tax was to encourage waste management to move up the hierarchy of waste management by a move away from landfill to more recycling, re-use and waste reduction. The income from Landfill Tax is expected to raise £500 million for the Exchequer.

A number of wastes are specified as 'inactive' and are subject to the lower tax regime. These include concrete, naturally occurring rocks and soil, ceramics, incinerator bottom ash, plaster, furnace slags, coal combustion ash etc. Some wastes are exempt from the Landfill Tax, including dredgings and mines and quarry waste. Other wastes may also be exempt from the tax if Customs and Excise grant a certificate of exemption. Into this category falls contaminated land where the land is removed for a construction project and therefore constitutes land reclamation. However, what waste falls into this category may not easily be defined. Definition of inactive waste is also a contentious issue where the waste falls between active and inactive or else there is contamination of inactive wastes by active waste, for example, some wastes such as roadworks waste contain high levels of bitumen.

To off-set the costs to business, a reduction in the level of employers' National Insurance Contributions has been introduced, funded by revenues from the Landfill Tax. In addition, under the Landfill Tax regime, landfill operators can claim back 90% of the sum they donate to an 'environmental body' up to an amount equivalent to 20% of their annual tax liabilities. The environmental bodies are allowed to spend the money raised on approved projects such as public amenities near landfill sites, reclamation of former landfills, educational projects and research into more sustainable waste management practices.

The introduction of the tax has seen a reduction in the level of inputs of waste to landfill, which may represent increased waste re-use, recycling and reduction. For example, some landfill operators are separating mixed wastes in tax-free zones and recycling hard core, soils and wood. However, it has also been reported that increases have been seen in some industrial wastes being used directly in land spreading on agricultural land, and increases in illegal fly tipping have occurred. This may represent waste diversion from landfill to avoid landfill tax liabilities rather than any moves up the hierarchy of waste management.

Source: ENDS (Environmental Data Services Ltd.), Bowling Green Lane, London, Report Nos. 254, 30–32, 1996; 255, 32–33, 1996; 257, 33–34, 1996; 264, 17–18 1997; 265, 18–22, 1997.

needs of the local area. The Authorities should also take account of the waste hierarchy and provide not only for waste landfill and incineration, but also for local easily accessible collection and recycling facilities. In addition, regional self-sufficiency in waste management should be a guiding principle of the Planning Authority.

4. The promotion strategy seeks to use publicity as a policy instrument to achieve sustainable waste management and thereby promote waste minimisation and good environmental waste management practices. The target groups for the

publicity are the producers of the waste in either industry or households. Publicity takes the form of a series of guides to industry on the best practice for waste minimisation in various industrial sectors, conferences, seminars, exhibitions, promotional packs, mail shots etc.

5. The data strategy is based on the key role of information and accurate data in the waste management industry. To enable suitable waste strategies to be determined, information on the sources, types and volumes of waste are produced, as well as the proportions re-used, recovered or disposed of. The National Waste Strategy produced by the Secretary of State for the Environment should also be based on sound information about the environmental and economic costs and benefits of the different waste minimisation, recovery and disposal options. The Environment Agency is a key body involved in the collection of the data.

1.5 Options for Waste Treatment and Disposal

The Government strategy of sustainable waste management is being developed through encouragement to move up the hierarchy of waste management options. To attain that objective a number of strategies have been used, including the strategy of setting targets on waste minimisation and recycling. Consequently, the economic case for each waste management option may not be the only criteria for the local authority decision makers. The concepts of sustainable waste management and also the principles of best practicable environmental option will override the minimum cost option. However, the least attractive option of landfilling may be the best practicable environmental option in some cases such as the restoration of former quarry workings by the disposal of residual, inert waste. Other guiding principles which should be taken into account in the choice of waste management option are: the proximity principle – that waste should be disposed of or dealt with close to the place where it arises; the self-sufficiency principle – that regions (and nations) accept responsibility for the wastes arising within them; the polluter pays principle – that the generator of the waste should pay for its disposal.

Table 1.2 shows the current waste options used in the UK to dispose of controlled (household, industrial and commercial) waste. In selecting a waste disposal option for a particular waste, the considerations which must be taken into account include the capital investment costs of the facility, operating costs, decommissioning and aftercare, throughput of waste and the environmental impact. These considerations are encompassed in the Best Practicable Environmental Option alongside the principle of sustainable development.

Table 1.2 shows that the dominant route for disposal of waste in the UK is landfill. However, incineration is the preferred option in some cases for low volumes of waste, for example in the hazardous waste industry. Table 1.3 shows the number of disposal licences by type in the UK. Comparison of the waste management options used in Japan and the USA are discussed in Boxes 1.3 and 1.4.

Table 1.2 *Proportion of waste landfilled/incinerated/recycled by sector (%)*

Waste type	Annual tonnage	Landfill	Incineration	Recycled/ re-used	Other
Household	20	90	5	5	0
Commercial	15	85	7.5	7.5	0
Construction and demolition	70	63	0	30	7
Other industrial	70	73	1	18	6

Other controlled wastes

Sewage sludge – Annual tonnage 35 million tonnes (wet sludge arisings, estimated on a 4% solids content) mostly landspread for agricultural use (17.5 Mt.). Also sea disposal (10.5 Mt.), incineration 3.5 (Mt.) and landfilled (3.5 Mt.).

Dredged spoils – Annual tonnage 35 million tonnes, dredged material from harbours etc., dumped at sea.

Total controlled waste arisings per annum = 245 million tonnes.

Uncontrolled wastes

Mines and quarry waste – Annual tonnage 110 million tonnes, mostly disposed of in stable above-ground tips at the production site which are then landscaped and restored.

Agricultural waste – Annual tonnage 80 million tonnes, mainly slurry and manure by-products from livestock, mostly applied to land.

Total uncontrolled waste arisings per annum = 190 million tonnes.

Total of all UK waste arisings per annum = 435 million tonnes.

Source: Department of the Environment and Welsh Office, Making Waste Work, HMSO, London, 1995.

Table 1.3 *Number of waste disposal facilities by type in the UK*

Type of disposal facility	Number of disposal licences
Landfill	4077
Civic amenity[1]	676
Transfer[2]	1745
Storage[3]	355
Treatment[4]	276
Incineration[5]	151
Scrapyards	1041
Recycling/recovery	96
Other[6]	12

[1] Waste disposal facilities open to the public.
[2] Facilities licensed for receipt, sorting, consolidation (baling and compaction), pulverisation and onward movement of waste.
[3] Facilities licensed principally for storing waste remotely from final disposal.
[4] Including physical, chemical and biological treatment and solidification.
[5] Hazardous waste, municipal, in-house, clinical and animal.
[6] Other, e.g. mineshafts.

Source: Department of the Environment, 1994. Digest of Environmental Protection and Water Statistics. HMSO, London. Crown copyright is reproduced with the permission of the Controller of Her Majesty's Stationery Office.

Box 1.3
Waste Management in Japan

Japan has a population of 124 million people and because of the highly mountainous and volcanic nature of the country, only 10% of the land is suitable for residential purposes. The shortage of land in accessible areas limits the availability of suitable landfill sites and is the driving force behind Japan's waste management policy. Policies are based on waste reduction and recycling to minimise the amount of material that is ultimately destined for landfill, and the main route for waste disposal is incineration either with or without energy recovery. The main statutory control for waste management is set out in the Waste Disposal Law 1970, which has had numerous subsequent amendments. Under the Law, waste is classified as either 'industrial' waste or 'general' (municipal) waste. The management of industrial waste, generated by business and industry, is the responsibility of the producer and is controlled by the 47 'Prefectures', which are the first tier of local government. The Prefectures authorise proposals for the construction or modification of industrial waste treatment and disposal facilities. Industrial waste production is estimated at 300 million tonnes per year and is dominated by waste from the iron and steel industry, animal waste sludge, mining and construction industries. Approximately 50% of the industrial waste undergoes intermediate treatment; recycling is also a major route for the generated waste and the remainder is landfilled. Whilst the responsibility for treatment and disposal of industrial waste is the responsibility of the producer, increasingly the waste is treated by the municipalities.

The management of general waste, which consists mainly of domestic waste and some commercial waste, is the responsibility of the second tier of local government, the 3245 municipalities, although contracting out of collection and disposal is common. Local government is very strong in Japan and completely controls domestic waste collection and disposal, and byelaws enable them to control and raise local taxes to obtain finance for waste systems. Financial aid in the form of grants, subsidies and loans are available for the construction of waste treatment facilities. Some 50 million tonnes of municipal waste is generated each year in Japan, 74.4% of which is incinerated, 20.4% landfilled and 5.2% recycled. A further 2.7 million tonnes of domestic waste is recycled privately by residents before collection. Source separation of waste by households is well established, with separation into either combustible or non-combustible material or recyclable materials such as glass, metal cans, newspapers etc. The 1991 Recycling Law seeks to promote recycling further by providing guidelines and targets for recycled material. The separated material is collected once or twice per month and taken to the materials recovery facility. The 'green' waste is recycled through 26 composting facilities throughout Japan. Waste is collected from individual households in some municipalities, but in most cases the waste is left at collection sites or stations each typically serving 10–40 residential properties. For example, in Tokyo, there are 240 000 such collection points. The collection vehicles are small, of only 1–2 tonnes capacity to negotiate the narrow streets of the densely populated towns and cities. Collections of the waste may be up to three or four times per week. There are 1873 municipal waste incinerators in Japan and incineration technology is mainly of the grate, fluidised bed or rotary kiln combustor design. The ash residue from waste incineration is taken to landfill. There are 2336 municipal landfill facilities operating in Japan and a further 1500 in the private sector. Because of the scarcity of suitable landfill sites, a new development is the construction of landfill sites offshore. Operational standards for waste treatment and disposal are similar to European practices.

Sources: [1] Waste Management in Japan. The Institute of Wastes Management, Department of Trade and Industry, Overseas Science and Technology Expert Mission Visit Report, IWM, Northampton, 1995; [2] Bonomo L. and Higginson A.E., International Overview on Solid Waste Management. Academic Press, London, 1988.

Box 1.4
Waste Management in the USA

The quantity of municipal solid waste (MSW) generated in the USA is estimated to be 210 million tonnes per year, and industrial waste generation is 400 million tonnes per year. The primary sources of solid waste legislation at the Federal or central government level are covered by the Resource, Conservation and Recovery Act 1976, amended by the Solid Waste Disposal Act Amendments 1980 and the Hazardous and Solid Waste Amendment of 1984, these Acts mainly deal with landfilling of wastes. Municipal solid waste incineration is regulated under the Federal Clean Air Acts, 1955–91. The laws are administered by the Environmental Protection Agency (EPA), but where State enforcement programs are equal to or more stringent than the Federally established regulations they may be authorised by the EPA. No Federal regulations exist for materials recovery facilities, recycling systems, compost plants or transfer stations. Regulation of such facilities is at the State level.

The generation of solid waste in recent years has more closely followed the growth in the economy rather than the population increase. The majority of MSW, about 63%, is landfilled, 17% is recycled and 16% is incinerated. There are approximately 3200 landfill sites in the US and because of more stringent regulations requiring high standards of site lining, monitoring of gas and leachate and post-closure liabilities this has led to increased costs. Consequently, the recent trend has been toward fewer, larger landfills, often located further from the source of the waste production. The majority of landfill sites are owned and operated by local government, but a significant number are privately owned. A further issue concerning landfill is that the Environmental Protection Agency (EPA) is required to enforce a law that bans hazardous waste disposal on land unless it can be proven to be less of a risk than other available technology. The reliance on landfill as the main route for waste disposal has led to capacity shortages in some areas of the US. The EPA policy is that MSW should be managed according to the hierarchy of source reduction, re-use, recycling, incineration with energy recovery and landfilling. Consequently, many states have source waste reduction programs and recycled material market development programs and have set recycling goals of up to 50%. Across the different states of the US it is estimated that there are over 500 different legislative bills dealing with recycling. Approximately 7300 residential curbside recycling programs are in operation throughout the country, with over 1400 materials recovery facilities. The major markets for recycled products are aluminium, ferrous metal and waste paper. Incineration of MSW is carried out in 156 plants, with 125 having some form of energy recovery system.

A further 400 million tonnes of industrial waste (not including mining or agricultural waste) is produced, of which industrial chemicals, agro-chemicals, waste from the iron and steel industry, electric power station waste, and plastics and resins manufacturing constitute the majority, with other sources arising from, for example, the paper, food and textile industries. The majority of the industrial wastes are managed in on-site facilities and where the waste is transported off-site, landfill is the preferred disposal route.

Sources: [1] Wiles, C.C., Municipal Solid Waste Management in the United States. National Renewable Energy Laboratory, Golden, Colorado, USA, 1994; [2] Bonomo L. and Higginson A.E., International Overview on Solid Waste Management. Academic Press, London, 1988; [3] Raymond Communications Inc., State Recycling Laws Update. Riverdale, MD, USA, 1995; [4] Hickman H.L., MSW management, state-of-practice in the US. Presented at the APPA Regulatory Compliance Seminar, Arlington Virginia, 6–7 March 1995; [5] Steuteville R., The state of garbage in America. Biocycle, April, 54-61, 1996.

Bibliography

Bonomo L. and Higginson A.E. 1988. International Overview on Solid Waste Management. Academic Press, London.

British Medical Association, 1991. Hazardous Waste and Human Health. Oxford University Press, Oxford.

Clapp B.W. 1994. An Environmental History of Britain since the Industrial Revolution. Longman, London.

Department of the Environment, 1994. Digest of Environmental Protection and Water Statistics. HMSO, London.

Department of the Environment and Welsh Office, 1995. Making Waste Work. HMSO, London.

ENDS, 1996. Report 254, 30-32, Environmental Data Services Ltd., Bowling Green Lane, London.

ENDS, 1996. Report 255, 32-33, Environmental Data Services Ltd., Bowling Green Lane, London.

ENDS, 1996. Report 257, 33-34, Environmental Data Services Ltd., Bowling Green Lane, London.

ENDS, 1997. Report 264, 17-18, Environmental Data Services Ltd., Bowling Green Lane, London.

ENDS, 1997. Report 265, 18-22, Environmental Data Services Ltd., Bowling Green Lane, London.

Environmental Assessment Guidance Booklet, 1989. Department of the Environment, HMSO, London.

Hickman H.L. 1995. MSW management, state-of-practice in the US. Presented at the APPA Regulatory Compliance Seminar, Arlington, Virginia, 6–7 March.

Holmes J. 1995. The UK Waste Management Industry 1995. Institute of Wastes Mangement, Northampton.

Lane P. and Peto M. 1995. Blackstone's Guide to The Environment Act 1995. Blackstone, London.

McBean E.A., Rovers F.A. and Farquar G.J. 1995. Solid Waste Landfill Engineering and Design. Prentice Hall, Englewood Cliffs, NJ.

Neal H.A. and Schubel J.R. 1987. Solid Waste Management and the Environment: The mounting garbage and trash crisis. Prentice-Hall, Englewood Cliffs, NJ.

Petts J. and Eduljee G. 1994. Environmental Impact Assessment for Waste Treatment and Disposal Facilities. Wiley, Chichester.

PPG23-Planning policy guidance note, 1994. Planning and pollution control. Department of the Environment, HMSO, London.

Raymond Communications, 1995. State Recycling Laws Update, 1995. Riverdale, MD.

Reeds J. 1994. Controlling the tips. The Taste Manager, 18 September.

Steuteville R. 1996. The state of garbage in America. Biocycle, April, pp. 54–61.

Survey of Public Attitudes to the Environment (1993), 1994. Department of the Environment, HMSO, London.

Sustainable Development, 1994. The UK Strategy. HMSO, London.

The Environment Act 1995. HMSO, London.

The Times, Times Newspapers, 1972.

This Common Inheritance, 1996. UK Annual Report 1996, Reporting the UK's Sustainable Strategy of 1994. HMSO, London.

Van Santen A. 1993. Wastes Management, Proceedings. 18 July.

Waste Management in Japan, 1995. The Institute of Wastes Management, Department of Trade and Industry, Overseas Science and Technology Expert Mission Visit Report. IWM, Northampton.

Waste Management Planning, 1995. Principles and Practice. Department of the Environment, HMSO, London.

West Yorkshire Waste Management Plan, 1996. West Yorkshire Waste Management Authority, The Environment Agency, March.

Wiles C.C. 1994. Municipal Solid Waste Management in the United States. National Renewable Energy Laboratory, Golden, CO.

2

Waste Management

Summary

This chapter outlines the development of UK waste legislation. The major legislative and regulatory measures dealing with waste treatment and disposal are described, including The Environmental Protection Act 1990 and integrated pollution control, the regulation of 'prescribed processes' and of smaller waste incineration and combustion processes, regulation of waste disposal on land, the 'duty of care' and waste disposal plans. The influence of the UK planning system on waste treatment and disposal are discussed. The 1995 Environment Act and the role of the Environment Agency are also described. The requirements for an Environmental Assessment for a large scale waste treatment and disposal facility are described. The economics of the different waste management options are compared and discussed.

2.1 Development of Waste Legislation

The development of waste treatment and disposal in the UK has been subject to a long history of legislative control. However, there has been no single Parliamentary Act dealing with the broad aspects of waste management. Consequently, waste treatment and disposal of waste is covered by a number of different controlling authorities. Local authorities had powers via various statutes dating back to medieval times to control waste as a public health problem. The local authority powers of waste management became established through the Public Health Acts of 1875 and 1936, the Alkali Act 1906, and the Town and Country Planning Acts of 1947 and 1971. The increasing concern for the environment and the toxic waste dumping incidents led to demands for much tighter legislative controls on waste disposal. The emergency legislation of the Deposit of Poisonous Waste Act 1972, precipitated by the Nuneaton cyanide dumping episode, was followed by the Control of Pollution Act 1974, which dealt comprehensively with domestic and industrial waste. Waste Disposal Authorities were required to prepare plans for the disposal of all household, commercial and industrial waste likely to arise in their areas. The plans would be required to include information on the types, quantities and source of waste arising in the area, the methods of disposal, the sites and equipment being provided and the cost. The Act also introduced the requirement that disposal authorities should consider reclaiming waste materials. The waste disposal operation was controlled by a comprehensive licensing system. A site licence was required for the land used as the waste disposal site, or for the incineration plant before disposal of any household, industrial or commercial waste would be allowed. The planning authorities regulation of land use impinged on waste treatment and disposal facilities through the requirement for planning permission for the land on which the facility was located; permission was required before a site licence could be issued (Control of Pollution Act 1974; British Medical Association 1991).

The late 1980s and 1990s saw further development of waste management legislation and the increasing influence of European Community legislation. Much of the legislation introduced had its source in European Directives, which set out European Community-wide objectives and standards which member states implement through individual state legislatures. For example, the 1990 Environmental Protection Act, the 1995 Environment Act, 1994 Waste Management Licensing Regulations, 1994 Transfrontier Shipment of Waste Regulations, and the 1996 Special Waste Regulations all contain measures in direct response to EC Directives (Boxes 2.1 and 2.2).

2.2 The System of Waste Management in the UK

In the UK, the management of waste is controlled through a waste legislative and planning regime which has developed over decades. The system is also subject to continuing modification by legislation from European Community Directives. Consequently, the regulatory system for waste management is covered by a number of Acts of Parliament and there is no single source which sets out waste management control in the UK.

2.2.1 The Legislative System

2.2.1.1 The Environmental Protection Act 1990 and Integrated Pollution Control

The Environmental Protection Act of 1990 set out the waste management strategy of the UK and introduced the system of Integrated Pollution Control. Part I of the 1990 Environmental Protection Act deals with prescribed processes, including waste incineration, and Part II deals with disposal of waste on land, including landfill. The basis of Integrated Pollution Control is that the environmental impacts of the process on the air, water and land environment are viewed as a whole. Previous to the 1990 Act, emissions to air, land and water were subject to separate and distinct controls. Under integrated pollution control, any emissions which unavoidably have

Box 2.1
Main Legislation Affecting Waste Management

The Control of Pollution Act 1974. Part I: Waste on Land
The Act established a new licensing and monitoring system for waste disposal facilities, and set out the duties of Waste Disposal Authorities and Waste Collection Authorities. The provisions of the Act have largely been superseded by later legislation.

The Control of Pollution (Special Waste Regulations) 1980
Sets out the definition of certain hazardous wastes as 'special wastes' with control of their movement via a pre-notification and consignment note system.

The Special Waste Regulations 1996
Updates earlier legislation on special wastes to conform with European legislation on hazardous waste. Defines the different categories of special/hazardous waste and the consignment note system which accompanies the waste from point of production to disposal.

The Controlled Waste (Registration of Carriers and Seizure of Vehicles) Regulations 1991
Requires carriers of controlled waste to be registered with the waste regulation authority.

The Control and Disposal of Waste Regulations 1988
Provided legal definitions of household, commercial and industrial wastes and established certain exemptions from the requirement for licensing. The Act has now largely been superseded by the Controlled Waste Regulations 1992 and the Waste Management Licensing Regulations 1994.

Town and Country Planning (Assessment of Environmental Effect) Regulations 1988
Implements the EC Directive (85/337/EEC) requiring environmental assessments to be made for certain prescribed processes, including incineration or chemical treatment of special waste, landfills receiving over 75 000 tonnes/year of waste and sites in environmentally sensitive areas.

Town and Country Planning Act 1990
Defines development, sets out the development control framework and establishes the requirement for Unitary Development Plans.

Planning and Compensation Act 1991
Introduced the requirement for Waste Local Plans, setting out the designated preferred use of land for waste treatment and disposal.

The Environmental Protection Act 1990. Part I: Integrated Pollution Control and Air Pollution Control by Local Authorities
Established the principle of integrated pollution control (IPC) and prescribed processes and prescribed substances. Large processes subject to IPC placed under HMIP regulation and smaller processes for air pollution control only under local authorities. Set out the authorisation system and enforcement procedures.

The Environmental Protection Act 1990. Part II: Waste on Land
Defines the duties of the Waste Regulation Authorities, Waste Disposal Authorities and Waste Collection Authorities. Established the new licensing and monitoring system for waste disposal

(continues overleaf)

Box 2.1
(continued)

on land. Prohibits unauthorised or harmful deposit or disposal of waste. Established the principle of the 'Duty of Care' on the storage, handling, transport, treatment and disposal of waste. The disposal of household waste is subjected to compulsory competitive tendering. Waste Regulation Authorities are required to produce a 'waste management plan' and Waste Collection Authorities a 'waste recycling plan'.

The Environmental Protection (Prescribed Processes and Substances) Regulations 1991 (and subsequent amended regulations)
Specifies the 'prescribed processes' of the 1990 Environmental Protection Act and designates them to either integrated pollution control under Her Majesty's Inspectorate of Pollution or air pollution control by local authorities.

The Controlled Waste Regulations 1992
Provides definitions of the controlled wastes (household, commercial and industrial waste) and prescribes certain types of litter and refuse to be treated as controlled waste.

The Environmental Information Regulations 1992
Implements EC Directive 90/313/EEC on the Freedom of Access to Information on the Environment, providing public access to environmental information held by public authorities.

Animal By-Products Order 1992
Implements EC Directive 90/667/EEC on veterinary rules for the disposal and processing of animal waste and the requirement that certain animal by-products must be disposed of at approved premises by incineration or burial.

The Waste Management Licensing Regulations 1994
Implements the EC Waste Framework Directive (91/156/EEC) on waste reduction, re-use and recycling of waste, and the use of waste as a source of energy. The regulations required the Waste Regulation Authorities to take account of the Directive in producing their 'waste management plans'. Enhances the licensing system for the treatment, keeping or disposal of controlled waste on land.

The Transfrontier Shipment of Waste Regulations 1994
Specifies the requirements of the EU Council Regulation 259/93/EEC on the supervision and control of waste within and into and out of the European Union.

The Environment Act 1995
The Act establishes the Environment Agency by combining the functions of the National Rivers Authority, Her Majesty's Inspectorate of Pollution, the Waste Regulation Authorities and Waste Disposal Authorities. Requires the Secretary of State for the Environment to produce a national waste strategy. Introduced the concept of sustainable development and the principle of 'producer responsibility' for waste.

Sources: [1] West Yorkshire Waste Management Plan, West Yorkshire Waste Regulation Authority (now Environment Agency), Wakefield, March 1996; [2] Garbutt J., Waste Management Law. A Practical Handbook. John Wiley and Sons, Chichester, 1995; [3] Lane P. and Peto M., Blackstone's Guide to the Environment Act 1995. Blackstone Press Ltd., London, 1995.

Box 2.2
The Inter-relationship of European and UK Legislation

European measures do not usually operate directly in the UK but set out standards and procedures which are then implemented by the Member States via their own legislative systems. In the UK an act of Parliament or subsidiary legislation would be required, or where the necessary legislation is already in place, a minor adjustment can be made through the relevant government department. European Community (EC) law is set down mainly in 'Framework Directives', which is commonly shortened to 'Directives', which set general standards and objectives. More detailed, subsidiary 'daughter Directives' deal with specific subjects within the Framework Directive. EC law may also be implemented by 'Regulations', but these are much rarer than Directives, since they apply directly to the UK without the need for an act of Parliament. For example, Council Regulation 259/93/EEC deals with the transfrontier shipment of waste into and out of the Community and also between member states. Enforcement of EC law is devolved to Member States but each state is answerable to the Community as a whole for the implementation of that law.

The main legislative measures introduced by the EC in relation to waste are listed below.

Council Directive/Regulation	**Area Covered**
Council Directive 89/429/EEC (1989)	Reduction of air pollution from existing municipal waste incinerators
Council Directive 89/369/EEC (1989)	Reduction of air pollution from new municipal waste incinerators
Council Directive 75/442/EEC (1975)	Establishment of waste disposal authorities, proper waste control regimes and the requirement for a waste plan, amended by Council Directive 91/156/EEC (1991)
Council Directive 94/67/EEC (1994)	Describes operational standards and emission limits for new and existing hazardous waste incinerators
Council Directive 91/156/EEC (1991)	Amends the 75/442/EEC Directive and introduces the polluter pays principle and encourages recycling
Council Directive 78/319/EEC (1978)	Toxic and dangerous waste, amended by Council Directive 91/689/EEC (1991)
Council Directive 91/689/EEC (1991)	Hazardous waste
Council Directive 85/337/EEC (1985)	Requirement of an environmental assessment for certain prescribed developments, e.g., incinerators and landfill sites
Council Directive 84/360/EEC (1984)	Control of air pollution from industrial plant requires authorisation and the uses of BATNEEC for specified incinerators
Council Directive 86/278/EEC (1986)	Control of sewage sludge to land
Council Directive 91/271/EEC (1991)	Restriction of sewage sludge disposal to sea
Council Directive 84/631/EEC (1984)	Supervision and control of transfrontier shipments of hazardous waste within the EC
Council Directive 80/68/EEC (1980)	Protection of groundwater against pollution from certain dangerous substances; may apply to landfill leachate and incinerator wastewater
Council Regulation EEC 259/93 (1988)	Supervision and control of shipments of waste, within, into and out of the EC

Source: Garbutt J., Waste Management Law, A Practical Handbook. John Wiley and Sons, Chichester, 1995.

to be released into the environment are released to the medium in which they would cause the least environmental damage. The emphasis in integrated pollution control is the control of wastes at the point of production rather than at the point of disposal, and also the integration of pollution control strategies to minimise pollution to the air, water and land (Environmental Protection Act 1990; Lane and Peto 1995; Slater 1995).

The 1990 Act defined 'controlled waste', which is the main category of waste covered by the Act. The term was introduced under the 1974 Control of Pollution Act and redefined under the 1990 Environmental Protection Act and 1994 Waste Management Licensing Regulations. Controlled waste refers to household, industrial and commercial waste. Each category is separately defined. Household waste means waste from private domestic accommodation, caravans, residential homes, universities or schools or other educational establishments, hospital premises, nursing homes, camp sites, prisons etc. Industrial waste generally includes waste from any industrial undertaking or organisation and includes industrial waste produced from the manufacture of food products, textiles, wood products, plastic products, motor vehicles, chemicals, etc. Commercial waste examples include waste from offices, hotels, shops, local authorities, markets, fairs etc.

The 1990 Act also defines the duties of the Waste Regulation Authorities, Waste Collection Authorities and Waste Disposal Authorities. The Waste Regulation Authorities, comprising the metropolitan district councils, county councils, Greater London, Greater Manchester and Merseyside, were responsible for granting and supervision of Waste Management Licences and licensed sites, and enforcement of the provisions of the Act regarding unauthorised waste operations. The Regulation Authority was also responsible for the development of a Waste Plan to identify the arrangements needed for the treatment and disposal of controlled waste, defined as household, industrial and commercial waste (Environmental Protection Act 1995).

The Waste Disposal Authority is responsible for arranging for the disposal of waste collected by the Waste Collection Authority. A payment per tonne of waste collected by the Collection Authority is made to the Disposal Authority for the disposal of the waste. In England and Wales the Waste Collection Authorities are the District Councils, in Greater London they are the London Boroughs, and in Scotland they are the Island or District Councils. Waste Collection Authorities provide receptacles for the waste of households, collect the waste in their area and deliver it to the place of disposal. The Collection Authority is also responsible, if requested, for collecting any commercial waste, but there is no obligation to collect industrial waste. The 1990 Environmental Protection Act also provides for the Collection Authority to encourage the recycling of waste by investigating arrangements for separating, baling or otherwise packaging waste for the purpose of recycling. A Recycling Plan dealing with kinds and quantities of recyclable waste is also required. Since recycling results in less waste to be disposed of, the Collection Authority is then entitled to a payment from the Disposal Authority for the net savings made. Where a third-party recycles the waste, they too are entitled to payment, although on a discretionary basis, from the Disposal Authority. Following the Local Government Act of 1988, a requirement for compulsory competitive tendering was introduced for services provided by the local authority. The local authorities set up in-house collection organisations through 'Direct Services Organisations' to bid for collection and street cleansing contracts. At the same time private sector or joint public–private sector companies entered the market for collection and street cleansing. Currently, approximately 25% of market share by value is in private sector hands (Department of the Environment and Welsh Office 1995; Waste Management Planning 1995; Davis-Coleman 1996; Energy from Waste 1996).

Under the 1990 Environmental Protection Act, Waste Disposal Authorities were required to transfer their disposal operations to private contractors or to separately organised local authority waste disposal companies (LAWDCs). Contracts for the disposal of the waste are awarded through compulsory competitive tendering such that the authority provide an efficient waste disposal service. The authority is empowered to include in the contracts environmental and public health factors as well as value for money. Waste disposal is therefore currently either through the local authority waste disposal company (LAWDC), through a joint venture company with the local authority and the private sector, or through the privatisation of the local authority waste disposal section.

The regulation of waste incineration: prescribed processes Part I of the 1990 Environmental Protection Act covers the regulation of 'prescribed processes', which includes large scale waste incineration. A process requires an 'Authorisation' for Integrated Pollution Control issued by the Environment Agency (formerly Her Majesty's Inspectorate of Pollution – HMIP). What constitutes a 'process' is defined by the Secretary of State for the Environment through his Department, and is an extensive prescribed list of processes including not only the waste disposal industry, but also the chemical industry, mineral industry, fuel and power industry etc. The prescribed processes to be controlled under Integrated Pollution Control are set out in the Environmental Protection (Prescribed Processes and Substances) Regulations 1991. The prescribed processes relevant to waste treatment and disposal are shown in Table 2.1.

The waste treatment prescribed processes are covered by guidance notes issued by the Environment Agency (Chief Inspectors Guidance Notes – referring to the Chief Inspector of Her Majesty's Inspectorate of Pollution – HMIP), which set out the standards of operation of the process, including the substances released, emission levels, release levels etc. The guidance notes do not have any statutory force, but represent the best available techniques, including management and technology standards and the emission standards associated with the technologies of the prescribed processes. The guidance notes represent the standard against which the application for an authorisation is judged. Since each process will have its own unique details of technology and operation, each application for authorisation is judged on a case-by-case basis, which allows flexibility with regard to the choice of technique in order to achieve the given standard. The final decision on a particular application would be taken on the basis of the guidance notes, the applicants case and also the public. In fact, the regulatory process is open to public scrutiny and comment, including the application, the authorisation stages and the regular returns of monitored releases to the environment and any enforcement action by the Environment Agency (Slater 1995).

Table 2.1 *Prescribed waste treatment processes regulated by The Environment Agency in the UK under the Environmental Protection Act 1990*

Process guidance note	Process
[1] Incineration and recovery processes of over 1 tonne per hour capacity	
IPR 5/1	Merchant and in-house chemical waste
IPR 5/2	Clinical waste incineration
IPR 5/3	Municipal waste incineration
IPR 5/4	Animal carcasses incineration
IPR 5/5	Burning out of metal containers
IPR 5/6	Making solid fuel from waste
IPR 5/7	Cleaning and regeneration of carbon
IPR 5/8	Recovery of organic solvents by distillation
IPR 5/9	Regeneration of ion exchange resins
IPR 5/10	Recovery of oil by distillation
IPR 5/11	Sewage sludge incineration
[2] Combustion processes with net thermal input of 3 MW or more	
IPR 1/4	Waste and recovered oil burners
IPR 1/5	Combustion of solid fuel manufactured from municipal waste
IPR 1/6	Combustion of fuel manufactured from or comprising tyres, tyre rubber or similar rubber waste
IPR 1/7	Combustion of solid fuel from or comprising poultry litter
IPR 1/8	Combustion of solid fuel which is manufactured from or comprises wood waste or straw

Integrated Pollution Control also involves a list of prescribed substances which is set out in the Environmental Protection (Prescribed Processes and Substances) Regulations 1991, and examples are shown in Table 2.2 (Prescribed Processes and Substances Regulations 1991). The release of any of the prescribed substances from the process should be prevented or minimised.

The Environmental Protection Act 1990 imposes a duty on the Environment Agency when setting the conditions for an authorisation to ensure that certain objectives are met. The conditions should ensure that the best available techniques not entailing excessive cost (BATNEEC) are used to prevent, minimise or render harmless the release of the prescribed substances into the environmental media of air, water or land. Whilst the prescribed substances are listed, the legislation also covers any substances which might cause harm if released into any environmental medium. When a process is likely to involve releases into more than one medium, which in most processes is most often the case, then the best practicable environmental option (BPEO) is achieved. That is, the releases from the process are controlled through the use of BATNEEC so as to have the least overall effect on the environment as a whole. Boxes 2.3 and 2.4 describe BPEO and BATNEEC. The authorisation grants the operator the right to operate the process under the conditions agreed by the Environment Agency, who also ensure that the conditions are complied with. Authorisations are subject to periodic review by the Environment Agency, who have the power to revise requirements in the light of developing techniques and knowledge. The Environment Agency monitors world-wide progress

Table 2.2 *List of prescribed substances and major substances requiring control – Environmental Protection (Prescribed Processes and Substances) Regulations 1991*

Prescribed substances

Release to air: prescribed substances

Oxides of sulphur and other sulphur compounds
Oxides of nitrogen and other nitrogen compounds
Oxides of carbon
Organic compounds and partial oxidation products
Metals, metalloids and their compounds
Asbestos (suspended particulate matter and fibres), glass fibres and mineral fibres
Halogens and their compounds
Phosphorus and its compounds
Particulate matter

Release to water: prescribed substances

Mercury and its compounds
Cadmium and its compounds
All isomers of hexachlorocyclohexane
All isomers of DDT
Pentachlorophenol and its derivatives
Hexachlorobenzene
Hexachlorobutadiene
Aldrin
Dieldrin
Endrin
Polychlorinated biphenyls
Dichlorvos
1,2-Dichloroethane
All isomers of trichlorobenzene
Atrazine
Simazine
Tributyltin compounds
Triphenyltin compounds
Triflualin
Fenitrothion
Azinphos-methyl
Malathion
Endosulfan

Release to land: prescribed substances

Organic solvents
Azides
Halogens and their covalent compounds
Metal carbonyls
Organo-metallic compounds
Oxidising agents
Polychlorinated dibenzofuran and any congener thereof
Polychlorinated dibenzo-p-dioxin and any other congener thereof
Polyhalogenated biphenyls, terphenyls and naphthalenes
Phosphorus

(continues overleaf)

Table 2.2 *(continued)*

Pesticides*
Alkali metals and their oxides and alkaline earth metals and their oxides

Major substances requiring control

In addition to the prevention or minimisation of the release of the prescribed substances the following substances should be considered in each application and authorisation:

Particulate matter
Carbon monoxide
Hydrogen chloride
Sulphur dioxide
Oxides of nitrogen
Lead and its compounds
Cadmium and its compounds
Mercury and its compounds
Organic chemicals (trace amounts)
Dioxins
Furans

* That is any chemical substance or preparation prepared or used for destroying any pest including those used for protecting plants or wood or other plant products from harmful organisms, regulating the growth of plants, giving protection against harmful or unwanted effects on water systems, buildings or other structures, or on manufactured products, or protecting animals against ectoparasites.

in process and technique development to ensure up-to-date knowledge of what the best available techniques might be for a particular process.

Process guidance notes The Process Guidance Notes issued by the Environment Agency are available to the public, operators and developers. The notes include details of the required standards of operation and frequency of monitoring. They also indicate the types of information required and the different options as to how the operator might meet the required standards required by BATNEEC. Box 2.5 outlines the process guidance note for the process of municipal waste incineration. The process guidance notes provide information about the process and wastes to which it applies, prescribed substances likely to be released and their release routes, BATNEEC and BPEO, techniques for pollution monitoring and control etc. The notes include the monitoring of the surrounding ambient environment such as the ambient air, river water, soil and flora, and odour.

Regulation of smaller waste incineration and combustion processes The Part I regulations of the 1990 Environmental Protection Act apply to large scale incineration of waste in addition to a wide range of other processes. Smaller incinerator operations, such as clinical, small industrial and sewage sludge incinerators of less than 1 tonne per hour throughput, or between 0.4 and 3 MW thermal input, are regulated by Local Authorities, which control only the emissions to air (Table 2.3). Guidance notes on these less significant process have also been issued by the Environment Agency. The guidance notes prescribe the emission and control limits for each specific pollutant for new and existing plant, together with information on the level of monitoring to be undertaken. Guidance is also given on material

Box 2.3
Best Practicable Environmental Option (BPEO)

Best Practicable Environmental Option (BPEO) was defined by the 12th Report of the Royal Commission on environmental pollution, 1988, as;

> *'the outcome of a systematic consultative and decision making procedure which emphasises the protection and conservation of the environment across land, air and water. The BPEO procedure establishes, for a given set of objectives, the option that provides the most benefits or least damage to the environment as a whole, at acceptable cost, in the long term as well as in the short term'*

The principle was introduced to take account of the total pollution from a process and the technical possibilities for dealing with it. BPEO is an integrated multi-media approach which applies to polluting discharges to air, water or land, and should take into account the risk of transferring pollutants from one medium to another. The option chosen requires an assessment of the costs and benefits of the appropriate measures, but does not imply that the best techniques should be applied irrespective of cost. BPEO is a requirement for individual processes subject to Integrated Pollution Control regulated by the Environment Agency, such as municipal waste incinerators. The concept is also applied in a wider context to policy and strategy planning for waste disposal and to the management of particular waste streams. The concept implies that different alternative options have been investigated before the preferred option is chosen which gives the best environmental outcome, in terms of emissions to land, air and water, at an acceptable cost. All feasible options which are both practicable and environmentally acceptable should be identified and the advantages and disadvantages to the environment analysed. Whilst the selection of the preferred option is subjective, the decision makers should be able to demonstrate that the preferred option does not involve unacceptable consequences for the environment. The strategy of sustainable waste management has re-emphasised the need for BPEO to be applied in a wider context, so that it should not be restricted to the disposal of a particular waste stream without also examining the production process to determine whether the waste can be minimised, recovered or recycled. The use of the term 'practicable' involves a number of parameters, including that the option chosen must be in accordance with current technical knowledge and must not have disproportionate financial implications for the operator. However, the best practicable option may not necessarily be the cheapest.

Source: Best Practicable Environmental Option. Royal Commission on Environmental Pollution, 12th Report, HMSO, London 1988.

handling and storage, flue gas treatment, disposal of residues, chimneys and general operation. Box 2.6 outlines the process guidance for the incineration of clinical waste.

Regulation of waste disposal on land The regulation of waste disposal on land is set out in Part II of the Environmental Protection Act 1990. In the UK the large majority of household, commercial, industrial and hazardous wastes are disposed of in landfills (Digest of Environmental Statistics 1994, 1996; Department of the Environment and Welsh Office 1995). However, landfill, whilst being the most widely used waste disposal option, is not described as a 'prescribed process' as are, for example, incineration of municipal waste and hazardous waste, and combustion of tyres. Landfill operations have been viewed as not being in the same category as

Box 2.4
Best Available Techniques Not Entailing Excessive Cost (BATNEEC)

Best Available Techniques Not Entailing Excessive Cost (BATNEEC) is a performance standard which must be applied to the Authorisation of a prescribed process under Integrated Pollution Control (IPC) regulated by the Environment Agency. The term occurs in a number of UK and EC legislative documents, including the Environmental Protection Act 1990, in guidance to the 1995 Environment Act and the EC Waste Framework Directive and Combating of Air Pollution from Industrial Plants Directive. BATNEEC is linked to the Best Practicable Environmental Option (BPEO), since once the best environmental option is chosen for a particular process, inevitably the best available techniques for minimising or eliminating pollution from the process should be chosen. What the 'best' techniques are for a particular process are given in the Process Guidance Notes for large processes subject to IPC and for smaller processes for air pollution control issued by the Environment Agency. The term 'techniques' includes not only the technology, but also the process design and process operation and management systems. In addition, the Notes include release limits for prescribed substances. The Agency is duty bound to monitor progress both in the UK and world-wide in the development of techniques for the various prescribed processes which can be used to prevent, minimise or render harmless both prescribed and other substances to air, land and water, i.e., the principles of IPC. Consequently, the guidance notes are subject to regular updates as better techniques for meeting IPC are developed. The Guidance Notes are not legal documents, but represent appropriate techniques for a particular process. Consequently, for an Authorisation to be issued by the Environment Agency the operator must demonstrate that any alternative technique used is equivalent or better than that in the Guidance Note. The cost element of BATNEEC, 'not entailing excessive cost', relates to criteria where the advantages in reducing pollution to the environment are outweighed by the disproportionate, excessive cost to the industry. The use of BATNEEC therefore requires consideration to be given to the cost of applying best available techniques in relation to both the nature of the industry sector concerned and the degree of environmental protection achieved. In applying for an Authorisation from the Environment Agency, new processes must apply BATNEEC. However, already existing processes must also be upgraded to comply with BATNEEC but over an agreed reasonable timescale. The application of BATNEEC by the Agency is therefore site-specific. The Environment Agency continue to apply BATNEEC through IPC authorisations in such a way as to ensure that the principles of sustainable development are respected.

Source: Integrated Pollution Control – A Practical Guide. Department of the Environment, HMSO, London, 1993.

an industrial process with well-defined performance standards by which the operation can be monitored. The landfill may have uncontrolled emissions to the air of landfill gas or to water from leachate, which are difficult or impossible to characterise or monitor as a 'process', particularly since the site would have a large volume and cover a large area. Therefore, the disposal of waste on land is covered by separate regulations and requires a 'Waste Management Licence', which is issued and regulated by the Environment Agency. To avoid any regulatory overlap, a waste treatment and disposal facility requiring an authorisation does not also require a waste management licence. However, the solid residues arising from the prescribed processes will most probably be disposed of on land and hence would require to be placed in a regulated site.

The detail of the licensing system is set out in the Waste Management Licensing Regulations 1994. The licence is granted by the Environment Agency and authorises

Box 2.5
Process Guidance Note IPR 5/3 for Municipal Waste Incineration

The note covers municipal waste incinerators of greater than 1 tonne/hour capacity regulated under Integrated Pollution Control (IPC) by the Environment Agency; smaller units are dealt with by local authorities. The note covers municipal waste incineration as a 'prescribed process' under the Environmental Protection Act 1990 and therefore operation of the process requires an Authorisation from the Agency for operating the incinerator. BATNEEC and BPEO are paramount objectives to be demonstrated in the application for the Authorisation. The note lists the statutory consultees, which include the Health and Safety Executive and the Nature Conservancy Council. The key area of the note is the list of concentration limits for releases of pollutants into air.

Total particulate matter	30 mg/m^3	Volatile organic matter	20 mg/m^3
Sulphur dioxide	300 mg/m^3	Nitrogen oxides	350 mg/m^3
Hydrogen chloride	30 mg/m^3	Hydrogen fluoride	2 mg/m^3
Dioxins (toxic equivalent)	1 ng/m^3	Cadmium	0.1 mg/m^3
Mercury	0.1 mg/m^3	(As, Cr, Cu, Pb, Ni, Sn)	1 mg/m^3
Combustion control	850 °C for 2s with 6% oxygen		

Also specified are the monitoring requirements for the pollutants, for example continuous monitoring is required for particulate matter and hydrogen chloride, quarterly monitoring for heavy metals, hydrogen fluoride, nitrogen oxides, volatile organic matter and sulphur dioxide and annual monitoring for dioxins. The note also recommends the different types of instruments which can be used to carry out the monitoring. For releases of pollutants to water the note recommends that recycling and re-use is preferable to release. However, the note also includes a list of 16 'prescribed substances' with release limits and includes for example, polychlorinated biphenyls (PCBs), cadmium, mercury, hexachlorobenzene, trifluralin, and pentachlorophenol. The release of waste ash to land would be via disposal in a landfill site and consequently would be covered by regulation under Part II of the 1990 Environmental Protection Act and the Waste Management Licensing Regulations 1994. The main types of plant used in incineration are described with their advantages and disadvantages. The full incineration process is described from the handling of the waste, storage in the waste bunker, feeding of the waste to the combustion chamber to techniques for minimising pollution by way of using the 'best practice' of plant operation, for example, maintaining good combustion conditions with a high temperature above 850 °C, a 2 second residence time and with excess oxygen to destroy any dioxins which might be present. For control of pollutants before emission of the gases to air the recommended systems for cleaning up of the flue gases are described. For example, the various spray drying and wet or dry scrubbing systems for the removal of acid gases and the different types of particulate removal systems are described with their advantages and disadvantages. Also included is the requirement for environmental monitoring, which may include ambient air, river water, soil and flora monitoring.

Source: Chief Inspector's Guidance to Inspectors, Process Guidance Note IPR 5/3, Waste Disposal and Recycling, Municipal Waste Incineration. Department of the Environment, HMSO, London, 1995.

the disposal of waste on land – the treatment, keeping or disposal of controlled waste (household, industrial and commercial waste) in or on specified land. The submission for a licence involves setting standards and operational criteria and showing how these criteria will be realised by operational practices. In effect a working plan of how the facility will be prepared and operated normally forms part of the licence. In the same way that the prescribed processes for waste treatment and

Table 2.3 Prescribed waste treatment processes subject to local air pollution control (less than 1 tonne per hour or between 0.4 and 3 MW thermal input; 'Part B' processes)

Guidance note	Process
PG 1/1	Waste oil burners
PG 1/2	Waste or recovered oil burners
PG 1/6	Tyre and rubber combustion processes
PG 1/7	Straw combustion processes
PG 1/8	Wood combustion processes
PG 1/9	Poultry litter combustion
PG 5/1	Clinical waste incineration
PG 5/2	Crematoria
PG 5/3	Animal carcass incineration
PG 5/4	Waste incineration
PG 5/5	Sewage sludge incineration

disposal are placed on a public register, so are the waste management licences. Guidance on the considerations relevant to licences and operational criteria are provided by the Department of the Environment as Waste Management Paper No. 4 – The Licensing of Waste Facilities (1988). The policy behind the licensing system is based on site-specific criteria, and consequently the guidance is non-prescriptive. Operational and environmental monitoring requirements are part of all licences and include both the operational lifetime and the post-closure phase of the facility. The monitoring system set out in the licence would include baseline water quality, geological and hydrogeological data before the start of operations, and a minimum number of sampling sites for surface water, groundwater, leachate and landfill gas. In addition, many site operators monitor both on-site and off-site for litter, dust, noise and odour, but this may not be on a regular basis. Post-closure monitoring would be carried out until the site reached completion. Then the Environment Agency grants a 'certificate of completion' and the licence is surrendered. Box 2.7 outlines the conditions for a landfill site licence.

Duty of care The Environmental Protection Act 1990 also introduced the concept of the 'duty of care'. The duty seeks to ensure the safe storage, handling and transport of waste by authorised people and to authorised sites for commercial and industrial waste. The duty also places a responsibility on the producer of the waste, who must ensure that the waste is transferred for suitable treatment and disposal by an authorised operator. In addition, those involved in the storage, transport, treatment and disposal of the waste must take reasonable measures to prevent pollution to the environment or harm to human health. Also, the making and keeping of records as to the nature of the waste are required. The duty is legally enforceable.

Waste disposal plans and waste recycling plans Under the Environmental Protection Act, 1990, the Waste Regulation Authorities (now the Environment Agency,

Box 2.6
Guidance Note 5/1 for Clinical Waste Incineration

The note covers clinical waste incinerators of less than 1 tonne/hour capacity as a guide to local enforcing authorities and covers only the emissions to air. The regulations are issued by the Environment Agency in order to achieve the objectives of the Environmental Protection Act 1990, and the different processes which are subject to such local air pollution control are listed in Part B of the Prescribed Processes and Substances Regulations 1991. Clinical waste is defined in detail and includes human or animal tissue, blood or other body fluids, excretions, drugs or other pharmaceutical products, swabs and syringes. In addition, the definition includes waste arising from medical, nursing, dental, veterinary, pharmaceutical or similar practice. Clinical waste also includes the packaging associated with the waste. The design of incinerator plant used is recommended as the two-stage incineration system, consisting of the primary stage chamber with a burner fitted to ensure controlled incineration and a second stage combustion zone where temperatures should be maintained at not less than 1000 °C with a gas residence time of a minimum of 2 seconds and with sufficient combustion air to ensure not less than 6% concentration of oxygen. These conditions are specified to ensure complete combustion of organic material, including dioxins. The core part of the note are the guidance on concentration limits of emissions to air.

Total particulate matter	30 mg/m^3	Volatile organic matter	20 mg/m^3
Sulphur dioxide	300 mg/m^3	Carbon monoxide	50 mg/m^3*
Hydrogen chloride	30 mg/m^3	Hydrogen fluoride	2 mg/m^3
Dioxins (toxic equivalent)	1 ng/m^3	Cadmium	0.1 mg/m^3
Mercury	0.1 mg/m^3	(As, Cr, Cu, Pb, Ni, Sn)	1 mg/m^3
Combustion control	1000 °C for 2s with 6% oxygen in the second stage.		

* Daily average

In addition, all emissions should be free from visible smoke, colourless and free from odour and persistent fume. To meet the limits on emissions, the general principles of BATNEEC are applied. The monitoring, sampling and measurement of emissions required are also set out in the note. For example, particulate matter, oxygen and carbon monoxide should be continuously recorded, cadmium, mercury and other heavy metals should be tested every six months and dioxins monitored at least once a year. The instruments recommended for the monitoring of emissions are recommended with the preferred and alternative methods listed. The preferred methods would in most cases be the standard USA Environmental Protection Agency methods which are continuously being adopted as the standard methods for sampling and analysis world-wide. The note also sets out the preferred location in the plant where the samples should be taken, the minimum volume of gas to be sampled and the sampling time. The general operation, management systems and the procedures for dealing with plant malfunctions are also set out as recommended practices in the note. Finally, the note also includes details of the chimney stack required to disperse the air pollutants from the incinerator, including the velocity of the gases up the chimney stack and its height.

Source: Guidance Note 5/1, Secretary of State's Guidance – Clinical Waste Incineration Processes under 1 tonne an hour. Department of the Environment, HMSO, London, 1995.

following the 1995 Environment Act) are required to prepare Waste Disposal Plans for the county or metropolitan area. The Waste Disposal Plan includes provisions about what arrangements are needed for the treatment and disposal of household, industrial and commercial waste. In addition, the types and quantities of the waste arising in the area, the costs and benefits to the environment, estimated costs of the

Box 2.7
Licence Conditions for a Landfill Site Licence

The licensing of waste disposal on land was introduced in the Environmental Protection Act 1990 and the licensing system set out in the Waste Management and Licensing Regulations 1994. The facility would also be subject to an Environmental Assessment and would require planning permission. The licences are issued by the Environment Agency. The waste management licence must include details of the types and quantities of the waste, the technical requirements which must be met to operate the site, the security precautions, details of the disposal site and the treatment methods. An assessment of the physical environment of the site including topography, geology and hydrogeology should also be included, which determine the degree of engineering and site preparation that would be acceptable before granting a licence. The application for the licence should also include 'The Working Plan', which is the detailed operational plan for the site. The plan would include specifications of the engineering work to control the pollution of surface water and ground water by leachate, water balance calculations, provision of gates and fencing, weighbridge, pollution control equipment and litter screens, and measures to monitor and manage landfill gas. The ways activities are to be carried out are also included in the plan, and would include hours of operation, types and amounts of waste, staff qualifications and experience, manning levels and the processes and methods to be used on site. The principal concerns of the waste management licensing system are the generation and possible migration of polluting leachate and evolution and migration of landfill gases from the site. The leachate management system, including containment, monitoring, collection, removal and disposal or treatment at both the operational and post-closure stages, should therefore be specified in the licence. The monitoring and management of landfill gas before, during and after operation is a requirement of the licence. Monitoring of leachate and landfill gas is required at either weekly, monthly, quarterly or yearly intervals and covers a wide range of parameters, including, for example,

Landfill gas: methane, carbon dioxide, oxygen, temperature, meteorological data.
Leachate: pH, temperature, electrical conductivity, chloride, biochemical oxygen demand, chemical oxygen demand, total alkalinity, total organic carbon, sulphate, total oxidised nitrogen, calcium, magnesium, potassium and sodium and heavy metals including iron, manganese, cadmium, lead, chromium, copper, nickel and zinc.

At the end of the operational lifetime of the site, landfill, capping and restoration procedures are set out as a requirement of the licence and involve a long-term site aftercare commitment. A great deal of the licensing procedure is taken up with ensuring that the licence holder is a 'fit and proper person', and the criteria on which such a decision can be made. The criteria include whether the licence applicant has a conviction for a relevant offence, can the holder demonstrate technical competence in the management of the site, and has financial provision been made to provide for the full costs of the operation, including environmental precautions required by the licence conditions. In addition, the licensee of the landfill facility must carry out environmental monitoring before, during and after operation of the site to determine the environmental effects of the facility.

Source: Licensing of Waste Management Facilities. Department of the Environment, HMSO, London, 1996.

methods of waste treatment and disposal, and what provisions can be made for recycling the waste are also outlined. The Waste Disposal Plans are not concerned with '*planning permission*' for the use of land, but only that suitable provision can be made to deal with all the waste arising in a region. Planning controls over the use of land for waste treatment and disposal facilities, as with any other use of land, are dealt with under separate legislation. Specifically for waste treatment and disposal facilities, the Town and Country Planning Act 1990 and the Planning and Compensation Act 1991 are most relevant.

With the absorption of the Waste Regulation Authorities into the Environment Agency following the 1995 Environment Act, the data on waste arisings and treatment and disposal at the regional level, collected for the Waste Disposal Plans under the 1990 Environmental Protection Act, will be incorporated into the National Waste Strategy. It is also envisaged that regional plans regarding waste management will still be produced, but with each region planning in relation to a common strategy for the longer term objectives for waste management (Turvey 1996).

A Recycling Plan is also a requirement of the Environmental Protection Act 1990 and is prepared by the Waste Collection Authorities. The Waste Collection Authority is required to identify what arrangements are appropriate for dealing with the waste by separating, baling or otherwise packaging for recycling purposes. The plan details the type and quantities of waste collected, and information on recycling of the wastes and any plant and facilities provided.

2.2.1.2 *The Environment Act 1995: The Environment Agency*

The 1995 Environment Act established the Environment Agency to further the aims of integrated pollution control introduced in the 1990 Environmental Protection Act. The Agency came into operation in April 1996. The Agency takes over the responsibilities of the Waste Regulation Authorities for the licensing for all waste disposal, storage, transfer, and treatment plants and landfill sites. In addition, it takes over the responsibilities of Her Majesty's Inspectorate of Pollution (HMIP) for prescribed processes such as incinerators and the National Rivers Authority (NRA). The Agency was established to minimise the overlap and potential conflict of the different regulatory bodies and have a single 'one-stop-shop' Agency dealing with industry. For example, the disposal of waste, depending on the disposal method chosen, has the potential to pollute the air, land and water, each of which may be dealt with by different authorities. Instead, the aim was to produce a fully integrated, effective and multi-media approach to pollution control. The central role of the Agency would also allow a coordinated approach to environmental monitoring and the production of environmental statistics, and a more consistent and coherent approach to environmental protection. In Scotland, the 1995 Act established the Scottish Environmental Protection Agency, which took over the responsibilities of Her Majesty's Industrial Pollution Inspectorate, river purification authorities, waste regulation authorities and the air pollution controls formerly the responsibility of the local authorities (Environment Act 1995).

The main responsibilities of the Agency (Lane and Peto 1995) include:

- applying standards by giving authorisations, licences and consents for emissions, discharges and disposals to air, water and land;
- monitoring compliance and enforcement;

- regulating the import, export and movement of waste;
- assessing national waste disposal need and priorities;
- monitoring environmental conditions, publishing statistics and commissioning research;
- acting as a statutory consultee (e.g., in town and country planning matters);
- providing authoritative and independent advice to government and advice and guidance to industry on best environmental practice.

The Environment Agency has a regional structure with separate regulation (pollution control) and operation (e.g., connection with water resources and flood defence) functions and a single policy-making directorate.

The 1995 Environment Act does not significantly alter the provisions of the 1990 Environmental Protection Act in terms of BATNEEC, BPEO and the core principle of integrated pollution control as previously operated by Her Majesty's Inspectorate of Pollution in relation to waste treatment and disposal by prescribed processes such as municipal waste incineration, waste tyre combustion, clinical waste incineration etc. These principles are extended by the concept of 'sustainable development', which has been defined as 'development that meets the needs of the present without compromising the ability of future generations to meet their own needs' (Sustainable Development 1994). Guidance to the Agency by the Department of the Environment describes BPEO as central to integrated pollution control and sustainable development. The most sustainable form of development is that which achieves the optimum distribution of any pollutants to the three media of air, water and land according to the ability of those media to accept such pollutants without, for example, exceeding critical loads. By requiring an assessment of the BPEO where emission from an industrial process are likely to affect more than one medium, integrated pollution control seeks to ensure that a sustainable outcome for the environment as a whole will be achieved (Waste Management Planning 1995). The Agency apply the concepts of BATNEEC and BPEO through the issue of Authorisations based on Integrated Pollution Control (IPC) whilst having due regard to sustainable development, thereby ensuring that the levels of pollution control achieved from industrial processes achieve a sustainable balance between society's interests in the environment and industry, in both the short and the long term (Lane and Peto 1995).

The waste management licensing system for the disposal of waste on land, set out in the 1990 Environmental Protection Act and Waste Management Licensing Regulations 1994, is transferred to the Environment Agency under the 1995 Environment Act. The advantages of an integrated Agency can be seen in that the bringing together of the National Rivers Authority (NRA) with the waste management licensing system integrates the possible danger of leachate causing pollution to the water system with the danger of deposit of waste on land. Previously the Waste Regulation Authority was only required to consult the NRA in its position of being a 'statutory consultee' before granting a licence. However, the Agency, as before, must consult the appropriate planning authority before issuing a licence. In addition, the Agency is also required to consult the Health and Safety Executive and the appropriate nature conservation body.

An important policy instrument introduced by the 1995 Act was the principle of 'producer responsibility'. The policy of producer responsibility seeks to encourage the producer of a product to assume an increased share of the responsibility for dealing with the waste that arises from the product. The aim of the policy is to increase the re-use or recycling of waste materials and to reduce the production of waste. To meet the policy objectives of producer responsibility, the 1995 Act allows regulations to be introduced setting out the legal obligation for producer responsibility and setting targets for re-use, recycling or recovery that will need to be met. Industrial sectors or identifiable waste streams have been asked to set recovery targets as part of the producer responsibility initiative. For example, in response to the 1994 EC Directive on Packaging and Packaging Waste, adopted by the UK Government, the packaging industry set a target in 1994 of 58% recovery by the year 2000, 50% by recycling and 8% by energy recovery. Recovery targets alone are not the only overriding criteria, in that significant environmental or economic benefits, as against the costs of re-use or recycling, should be taken into consideration. In addition, implementation of the targets should also be fair across the industry from small to large companies.

2.2.1.3 The Special Waste Regulations 1996

Special wastes are wastes which are regarded as hazardous to human health or the environment and which require special treatment and disposal. The Special Waste Regulations 1996 replace the Control of Pollution (Special Waste) Regulations of 1980. The term 'special waste' was introduced in the 1974 Control of Pollution Act to cover wastes which are particularly hazardous. The 1996 regulations update the 1980 regulations to conform with European legislation and to unify the definition of hazardous waste throughout the European Community (Council Directive 91/689/EEC, 1991). Early legislation described wastes which were highly flammable, a prescription-only medicine or 'dangerous to life'. This risk-based approach to hazardous wastes has been replaced by the EC approach which categories a wide range of wastes as hazardous. The definition of what constitutes 'special waste', or hazardous waste, is found in the 1996 regulations, with a full list of over 200 different types of waste, each with its own six-digit code number, classified into different industrial processes, each with their own code and subcodes. Table 2.4 shows examples of different types of special waste (The Special Waste Regulations, 1996). The full list is reproduced from the list of hazardous wastes listed in Council Decision 94/904/EC (1994), which establishes the EC list of hazardous wastes in connection with the EC Directive on Hazardous Wastes (Council Directive 91/689/EEC, 1991). In addition, any waste which has certain hazardous properties, as defined in the regulations and summarised in Table 2.5, is also termed a special waste (The Special Waste Regulations 1996). Wastes with certain threshold concentrations of hazardous properties also fall into the definition of special waste under the 1996 Regulations. The Regulations do not apply to domestic household waste. The redefining of special wastes means that some wastes which under the UK definition were not special may now become so. The properties of waste (shown in Table 2.5), which define the waste as special are wide-ranging. For example, hazard

Table 2.4 *Examples of hazardous waste codes (The Special Waste Regulations 1996)*

Waste code	Description
06	WASTES FROM INORGANIC CHEMICAL PROCESSES
0601	Waste Acidic Solutions
060101	Sulphuric acid and sulphurous acid
060102	Hydrochloric acid
060199	Wastes not otherwise specified
0602	Alkaline Solutions
060201	Calcium hydroxide
060203	Ammonia
060299	Wastes not otherwise specified
07	WASTES FROM ORGANIC CHEMICAL PROCESSES
0701	Waste from the Manufacture, Formulation, Supply and Use of Basic Organic Chemicals
070101	Aqueous washing liquids and mother liquors
070102	Organic halogenated solvents, washing liquids and mother liquors
070107	Halogenated still bottoms and reaction residues
0702	Waste from the Manufacture, Formulation, Supply and Use of Plastics, Synthetic Rubber and Man Made Fibres
070201	Aqueous washing liquids and mother liquors
070202	Organic halogenated solvents, washing liquids and mother liquors
070207	Halogenated still bottoms and reaction residues
13	OIL WASTES
1301	Waste Hydraulic Oils and Brake Fluids
130101	Hydraulic oils containing PCBs or PCTs
130104	Chlorinated emulsions
130105	Non-chlorinated emulsions
1304	Bilge oils
130401	Bilge oils from inland navigation
130402	Bilge oils from jetty sewers
16	WASTES NOT OTHERWISE SPECIFIED IN THE CATALOGUE
1606	Batteries and Accumulators
160601	Lead batteries
160602	Ni–Cd batteries
160603	Mercury dry cells
19	WASTES FROM WASTE TREATMENT FACILITIES, OFF-SITE WASTEWATER TREATMENT PLANTS AND THE WATER INDUSTRY
1901	Wastes from Incineration or Pyrolysis of Municipal and Similar Commercial, Industrial and Institutional Wastes
190103	Fly ash
190104	Boiler dust
190105	Filter cake from gas treatment
190106	Aqueous liquid waste from gas treatment and other liquid wastes

Source: The Special Waste Regulations 1996. HMSO, London, 1996.

Table 2.5 *Properties of wastes which make them hazardous (special)*

Category	Definition
H1	Explosive
H2	Oxidising
H3-A	Highly flammable
H3-B	Flammable
H4	Irritant
H5	Harmful (involving limited health risks)
H6	Toxic (involving serious, acute or chronic health risks and even death)
H7	Carcinogenic
H8	Corrosive (may destroy living tissue)
H9	Infectious (known or believed to cause disease)
H10	Teratogenic (may induce non-heriditary congenital malformations)
H11	Mutagenic (may induce hereditary genetic defects)
H12	Wastes containing materials which release toxic or very toxic gases in contact with water, air or acid
H13	Wastes containing material capable after disposal of yielding another substance with properties which are hazardous
H14	Ecotoxic (may present immediate or delayed risks for one or more sectors of the environment)

Source: The Special Waste Regulations 1996. HMSO, London, 1996.

H13 is defined as wastes containing material capable, after disposal, of yielding another substance with properties which are hazardous. Under this category, leachate is given as an example, and many other wastes could be classified in the same way. Similarly, category H14-Ecotoxic defines wastes which may present immediate or delayed risks for one or more sectors of the environment. Again, many wastes could be classified as special under this definition.

The basis of the Special Waste Regulations for hazardous waste are the consignment note system whereby special wastes are accompanied by consignment notes which identify exactly the type of waste, the process from which the waste was produced, the characteristics of the waste, its quantity and type, information about containers for the waste etc. Whilst the Regulations do not specify the packaging and handling of the waste, 'the duty of care' in the description, transport, and use of authorised carriers and waste managers would apply. The consignment note accompanies the waste from the point of production to its safe and final disposal, and includes storage, treatment and recycling. The Regulator of the consignment note system is the Environment Agency (England and Wales) and the Scottish Environment Protection Agency.

2.2.2 The Planning System

A waste treatment and disposal facility, whether it be an incinerator, landfill site, waste transfer station, or materials recycling facility, represents a major use of land. Planning permission is a prerequisite to the development of such facilities so as to ensure that consideration has been given to the impact on land use of the operation of the facility. These considerations include matters such as access, effects on the environmental and visual amenity of the locality and surrounding land uses. They also include the likely nature and duration of the development, hours of operation and relation of the policies and proposals in relation to the authorities overall development plans. Consequently, there needs to be control over the location of such facilities. Control is exercised through the planning systems of the local authorities. Strategic and forward planning of land use are the responsibility of the Metropolitan authorities, District and County councils, and are set out in the Town and Country Planning Act (1990) and the Planning and Compensation Act (1991). Non-Metropolitan areas produce a Structure Plan by the County Councils which covers the strategic development of the County as a whole, and the District Councils produce Local Plans which deal with detailed planning policies and specific proposals at the local level and which conform to the overall Structure Plan. In Metropolitan areas the Unitary Development Plan covers the strategic and detailed policies of land-use planning in one document. Strategic planning must be prepared by the planning authorities to cover the future development of land in the area, which would include, for example, housing, industry, commercial development, designated green belt areas etc. However, designation of the preferred use of an area does not mean that planning permission will be granted for development. Both the non-Metropolitan areas and the Metropolitan areas are also required to produce a Waste Local Plan dealing specifically with the treatment and disposal of waste. The requirement for the Waste Local Plan was introduced under the Planning and Compensation Act 1991, to give more detailed consideration of the preferred location for waste treatment and disposal facilities in the planning. In some cases the Waste Local Plan is also produced in conjunction with a Minerals Local Plan, which covers the mineral extractions of a region, since, for example, landfilling of waste is strongly linked with infilling of quarries caused by mineral extraction. The main policies and proposals for land use for waste treatment and disposal facilities in the area are set out in the Waste Local Plan. These should take into account the aim of self-sufficiency in waste management facilities for a region, and the land-use and transport requirements of the facility. The policies in relation to waste minimisation and recycling, opportunities for energy recovery from waste, and treatment methods prior to disposal are also to be taken into account in the formulation of the Plan. The identification of spare capacity in existing storage, treatment and disposal sites and the identification of possible new areas are also included. The Plan should include the criteria against which applications for waste management developments are considered, and the criteria for improvement in the environmental acceptability of recycling and other waste management operations (PPG23, 1994; Waste Management Planning 1995; Energy from Waste 1996).

Strategic development planning under the Structure and Local Plans produced by the County and District Councils, or the Unitary Development Plans produced at the Metropolitan Councils, and the Waste Local Plans form the core of the planning procedures regarding land use for waste treatment and disposal facilities in the UK. However, they will be linked with the Waste Disposal Plan required under the Environmental Protection Act 1990 and drawn up by the Waste Regulation Authority (Environment Agency), since that Plan will set out the types and quantities of waste arising in the area and the waste management facilities available to deal with that waste. In addition, the Waste Recycling Plan drawn up by the waste Collection Authority will also be taken into account, since that also addresses the requirements of plant and facilities for the recycling of waste.

2.2.3 Environmental Assessment

Environmental Assessments are a requirement under the EC Directive on the Assessment of the Effects of Certain Public and Private Projects on the Environment, 85/337/EEC, for certain waste treatment and disposal projects. Such projects include large scale energy from municipal waste incinerators, clinical and hazardous waste incinerators and large scale (greater than 75 000 tonnes/year) landfill sites (Environmental Assessment Guidance Booklet 1989). Not all waste treatment and disposal projects require an assessment, only those which may have a significant impact on the environment. What constitutes a 'significant' impact is set out in Guidance to the Environmental Assessment Regulations of the Department of the Environment. For example, this might be where the project is of large scale or destined for a site of special scientific interest, or where the project is likely to give rise to significant pollution. The UK body which implements the EC Directive would be the planning authority which would take into account the assessment, which is presented in the form of an Environmental Statement with the planning application for the waste management facility.

For a large scale waste treatment and disposal project such as a municipal waste incinerator or landfill site, the environmental assessment would include assessments of a wide range of criteria (Energy from Waste 1996; Petts and Eduljee 1994). Some examples are given below.

Visual impact Consideration of the visual impact of a large and prominent industrial plant or landfill site upon the existing landscape and visual amenities.

Air emissions Assessment of incinerators covers existing air quality, concentration, volume and dispersion characteristics of pollutant gases, ground-level concentrations, considerations of local topography and meteorology, and comparison with legislative and guidance limits. Assessment of landfill sites covers fugitive emissions of landfill gas from the site, dispersion of the gases, odour problems, increase in ambient concentration, explosion risk, and comparison with legislative and guidance limits.

Water discharges Assessment of incinerators also includes treatment and disposal options for scrubber liquor and cooling water. Assessment of landfill sites includes treatment and disposal options for leachate, and effects on downstream treatment works and water resources.

Ash discharges Assessment of incinerators must always cover the treatment and disposal options for bottom ash and fly ash.

Human health Concerns about human health include impacts and pathways of exposure to the pollutant emissions, and ingestion via the food chain, water or inhalation. An estimation of the hazards and risks must be made.

Fauna and flora The impacts of emissions on local fauna and flora and loss of habitat, particularly for sites of special scientific interest, must be assessed.

Site operations Management controls must be assessed and an analysis of the risks associated with the plant operation and the consequences of operational failure must be made. The impact of plant operation noise should also be considered.

Traffic The number of heavy goods vehicles and other vehicle movements must be assessed, as well as their impacts on the existing road network and traffic flows. The noise from increased traffic, accident statistics and routing considerations should also be included.

Socio-economic impacts The effects of the project on adjoining residents and existing industries must be considered, including economic benefits such as employment and investment.

Land-use and cultural heritage The compatibility of the project with existing and proposed adjacent land use and conformity with local development plans must also be assessed.

The environmental assessment must identify, describe and assess the direct and indirect effects of the project on human beings, fauna, flora, soil, water, air, climate and landscape, material assets and the cultural heritage. The assessment is carried out by a project team of experts for the developer, and would include a description of the project, comprising information on the site, design, size and scale and the main characteristics of the process. In addition, information about likely environmental effects, including expected residues and emissions, is required as well as a description of the proposed measures to prevent, reduce or offset any adverse effects on the environment. The assessment would include an appraisal of any alternative sites and processes considered, baseline surveys of the site and surrounding area, a review of the options and proposed methods to prevent or minimise any environmental effects, and a prediction and evaluation of the environmental impacts. The assessment procedure involves full disclosure of information and consultation with the public. Clearly the environmental assessment can be a complex, difficult, time-consuming and expensive task. It has been estimated that the assessment can cost up to 5% of the total capital costs of the project, although more typically it is around 1%, and a timescale of one year is likely (Energy from Waste 1996).

Figures 2.1 and 2.2 show the sources of impacts and effects on the environment for an incinerator and a landfill site (Petts and Eduljee 1994).

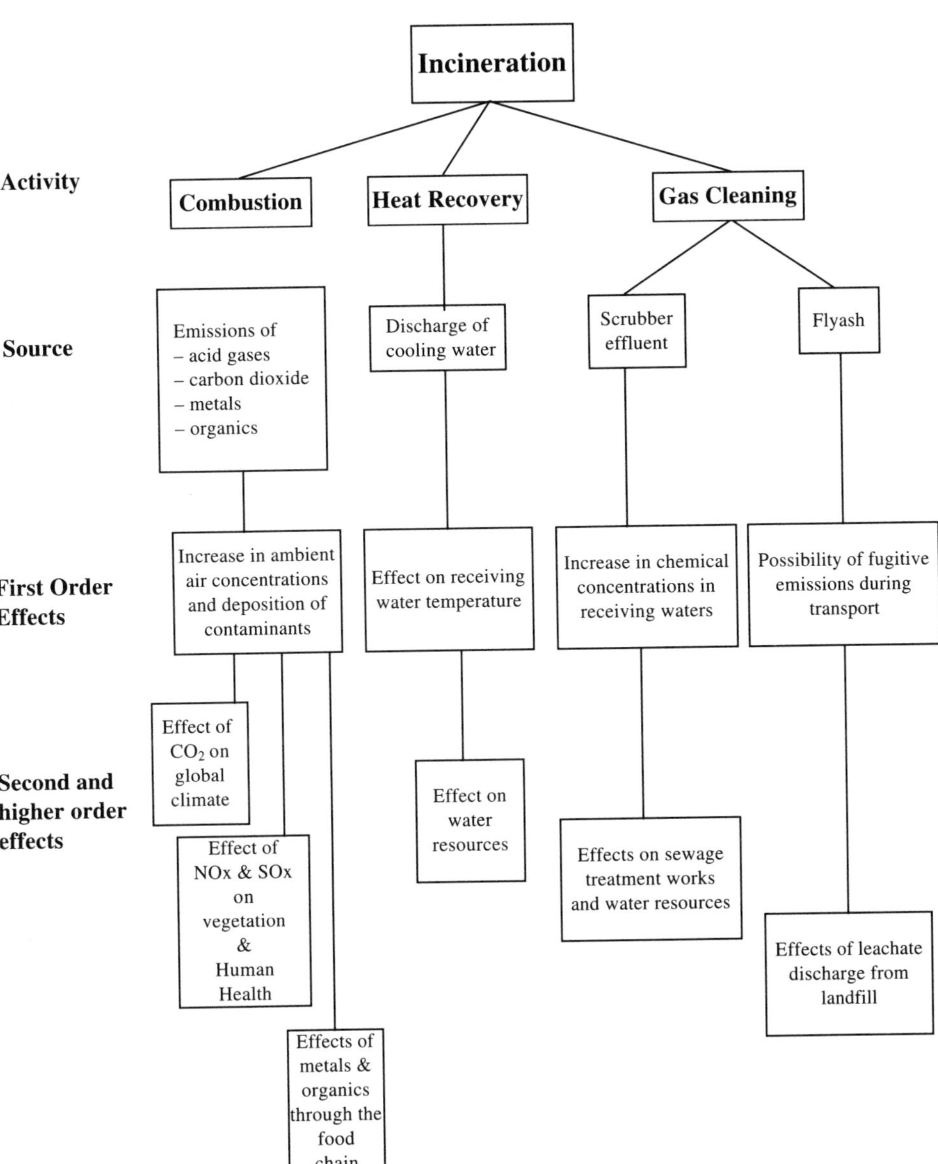

Figure 2.1 *Incineration sources of impacts and effects on the environment. Source: Petts J. and Eduljee G., Environmental Impact Assessment for Waste Treatment and Disposal Facilities. Wiley, Chichester, 1994. Reproduced by permission from John Wiley and Sons Limited.*

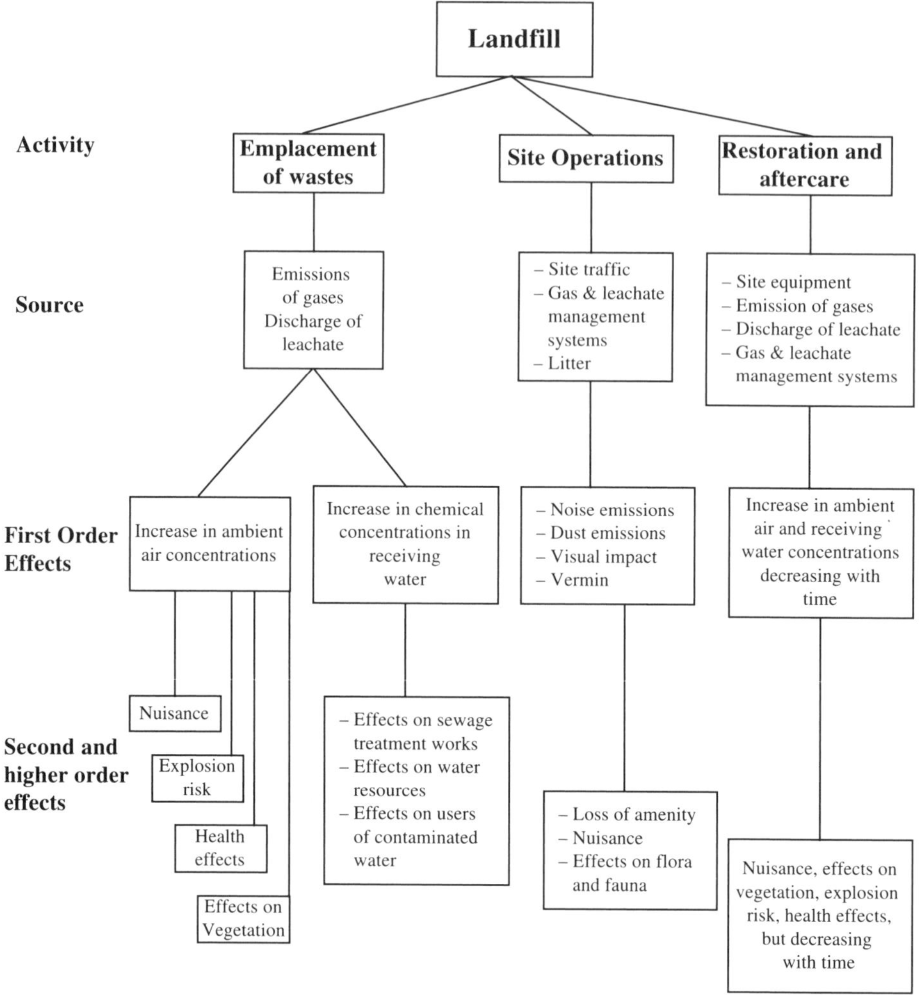

Figure 2.2 *Landfill sources of impacts and effects on the environment. Source: Petts J. and Eduljee G., Environmental Impact Assessment for Waste Treatment and Disposal Facilities. Wiley, Chichester, 1994. Reproduced by permission from John Wiley and Sons Limited.*

2.3 The Economics of Waste Management

The UK generates a total annual tonnage of waste of approximately 435 million tonnes, and the treatment and disposal of this waste represent a major proportion of UK costs (Department of the Environment and Welsh Office 1995). The dominant disposal option in the UK is landfill, which is due to the abundance of low-cost landfill sites. This is partly the result of the distinctive geological and hydrogeological environment of many sites in the UK which have natural impervious clay barriers. The barrier helps contain potentially polluting liquids and leachate and consequently markedly reduces the engineering costs at such sites. In addition, many sites are used quarries, and the disposal of waste by backfilling has been actively encouraged as a means of land recovery. However, there is increasing awareness that waste treatment and disposal options should reflect their full environmental and economic costs, a view fully endorsed by the Government's strategy on sustainable waste management (Department of the Environment and Welsh Office 1995). Consequently to the capital and operational costs of landfill (as also to other waste management options), the external costs to the environment should be added. In addition, implicit in the hierarchy of waste management is that disposal of waste by landfill is the least desirable option.

Table 2.6 shows approximate costs and charges of waste collection, transfer, treatment and disposal for a variety of wastes (Royal Commission on Environmental Pollution 1993; Holmes 1995; West Yorkshire Waste Management Plan 1996). The costs will always be approximate since they are dependent on location, scarcity of landfill capacity, the geology and hydrogeology of the landfill site, the types of waste that can be landfilled or incinerated at the facility, size of facility and waste throughput, etc. In addition, the availability of data is often restricted by the need to maintain commercial confidence. In general, the costs and charges are representative of the environmental hazards associated with the different types of waste, from low disposal costs of inert waste into landfill to very expensive chemical waste incineration. Wastes such as chemical sludges, chlorinated solvents, scrap tyres, clinical waste etc. are difficult and consequently more expensive to dispose of compared with domestic and commercial waste. Increasing the throughput of such wastes would naturally increase the revenue to the facility, but such wastes can only be accepted if there is an authorisation or licence to take them.

Incineration is capital-intensive and has higher operating costs than landfill. However, offset against this fact is the greater income from the sale of heat, power or steam. Capital costs for incinerators have increased in recent years to meet EC emissions limits and represent a major cost of the overall plant. It is often difficult to assess the cost in general terms of incinerators, particularly large scale municipal incinerators where the costs are site-dependent. For industrial waste incineration, costs will depend on the calorific value of the waste, the amount of gas clean up, the throughput, the requirement for the energy generated etc. Hazardous waste disposal costs in particular tend to be high due to the greater technical requirements of the plant. Landfill costs are also expected to increase as the implementation of the proposed EC Landfill Directive will mean more stringent gas and leachate collection, treatment, and monitoring and aftercare requirements. It should also be realised that the ash residue from incineration, which typically represents about 30% by weight of the original waste, also requires disposal by landfilling. The influence of energy recovery on both landfill and incineration costs can be seen from Table 2.6 to reduce significantly the costs of disposal. Municipal solid waste has a significant energy content (calorific value), and the combustion of the waste in an incinerator

Table 2.6 *Waste management costs*

Waste management process	Approximate cost or charge
Collection and disposal	
Local Authority waste collection	£19.90 per tonne (1986/87)
Net cost per household for refuse collection in West Yorkshire	£21.66–£27.37 per household (1993/94)
Net cost of disposal of collected household waste in West Yorkshire	£16.74–£20.39 per tonne (1993/94)
Landfill	
Landfill gate charge: non-specialised household/commercial/industrial	£10.00–£15.00 per tonne (1995)
Landfill gate charge: inert waste	£1.50–£2.00 per tonne (1995) (may be accepted free of charge for landfill engineering purposes)
Landfill gate charge: sludges-co-disposal	£18–£20 per tonne (1994)
Landfill gate charge: sewage sludge	£20 per tonne (1994)
Landfill gate charge: food waste	£20 per tonne (1994)
Landfill gate charge: contaminated soils	£15–£20 per tonne (1994)
Landfill gate charge: scrap metal fragmentiser waste	£8–£10 per tonne (1994)
Landfill gate charge: bonded asbestos	£45 per tonne (1994)
Waste tyre shredding and landfill	£125 per tonne (1994)
Landfill gate charge: incineration plant gas cleaning residue	£20 per tonne (1994)
Incineration	
Incineration: municipal solid waste for a 200 000 tonne per annum incinerator	£22–£37 per tonne (1993) (depending on site-specific factors)
Incineration with energy recovery for a 200 000 tonne per annum incinerator	£21–£29 per tonne (1993) (depending on electricity price; NFFO)
Merchant hazardous waste incinerator	£120–£150 per tonne (1993)
Clinical waste incineration	£240–£300 per tonne (1995)
Sewage sludge incineration for a 18 000 tonne per annum incinerator	£40–£70 per tonne (1989)
Animal carcass incineration	£170–£250 per carcase (1994)
Polychlorinated biphenyl waste incineration	£1000–£1500 per tonne (1994)
Chlorinated waste solvent incineration	£120 per drum (1994)
Non-chlorinated waste solvent incineration	£60 per drum (1994)
Waste transport/transfer	
Waste transport: skip wagon	£1.71 per mile (1995)
Waste transport: roll-on-off	£2.36 per mile (1995)
Waste transfer for municipal solid waste	£7.00–£10.00 per tonne (1995)
Investment costs for a 200 000 tpa facility (landfill versus incineration)	
Landfill development: MSW	£5.02M (excludes site acquisition costs)
Landfill site acquisition	£8.00M (1993)
Landfill operation	£802 000 per annum (1993)
Landfill restoration	£2.47M (1993)
Landfill aftercare	£1.60M (1993)
Landfill development: inert waste (200 000 m^3 site)	£75 000–£150 000 (excludes site acquisition costs)

Table 2.6 (*continued*)

Waste management process	Approximate cost or charge
Incinerator capital and development costs	£38.85M (1993) (excludes site acquisition cost)
Incinerator site acquisition	£500 000 (1993)
Incinerator operational costs	£3.46M per annum (1993)
Incinerator site decommissioning	£3.25M (1993)
Additional capital cost for energy recovery	£8.00M
Additional operating cost for energy recovery	£80 000 per annum (1993)
Water treatment/sewage works	
Landfill leachate treatment	£15–£20 per tonne (1994)
Food industry waste treatment	£12–£16 per tonne (1994)
Landfill tax	
Standard rate	£7.00 per tonne (1996)
Inert waste	£2.00 per tonne (1996)

Sources: [1] West Yorkshire Waste Management Plan, West Yorkshire WRA, Wakefield, 1996; [2] Holmes J., The UK Waste Management Industry 1995. Institute of Wastes Management, Northampton, 1995; [3] Royal Commission on Environmental Pollution. 17th Report, Incineration of Waste. HMSO, London, 1993.

can be used to generate electricity or the heat can be used for district heating. Similarly combustible methane generated from the anaerobic decomposition of the waste in a landfill site can be also be used to generate power or heat. Selling the energy produces income for the facility and reduces costs, particularly if Non-Fossil Fuel Obligation (NFFO) contracts have been secured with the Regional Electricity Companies. Other wastes such as clinical waste, dried sewage sludge and scrap tyres have high calorific values and incinerators with energy recovery are available. A significant influence on disposal costs, for landfill in particular, is transport costs. Increasingly, suitable landfill sites close to the point of waste generation are becoming scarcer and in some cases waste is being transported over long distances to the landfill site. Waste management facilities such as incinerators, and recycling and composting plants which can be sited within city limits clearly have a distinct advantage in such cases. In addition, pricing is dependent on the manufacturers overall involvement in the project, be it solely as an equipment supplier or as a major shareholder in the plant-operating company. Also, as the scale of operation is increased for either landfill or incineration, the gate fee becomes reduced. For example, a landfill site or incineration plant of 300 000 tonnes per annum throughput has an estimated gate fee (cost/tonne) of almost half that of a 100 000 tonne per annum facility. To support such large facilities with their associated large capital investments a long-term guaranteed supply of waste is required, and where energy recovery is employed a guaranteed long-term end market for the energy is also needed.

The introduction of the landfill tax has resulted in the costs of landfilling better reflecting the environmental impact of landfilling of waste. In addition, it has encouraged industry and consumers to produce less waste and provided an incentive to re-use or recycle waste (ENDS 1997). However, there have also been increases in tax avoidance by increases in illegal dumping of wastes (ENDS 1997). The tax has also brought landfill costs to similar levels to those found in the rest of the European Community. For example, Denmark has a landfill tax rate of £20/tonne for all types of waste, Germany between £10 and £41/tonne depending on the type of waste, and France from £2.50 to £8/tonne depending on the type of waste (ENDS 1995). The landfill tax is consistent with the Government's strategy to reduce the amount of controlled waste disposed of in landfill to 60% by the year 2005.

The UK has a large private sector waste management industry, particularly for commercial and industrial waste, with over 3000 companies involved and responsible for approximately 66% of the controlled waste market in the UK (Holmes 1995). For example, the difficult and expensive collection, treatment and disposal of industrial waste sludges and liquids is dealt with almost wholly by the private sector. In addition, in recent years, there has been the development of a market mechanism for local authority arrangements for collecting and disposing of waste. Collection of waste is now subject to compulsory competitive tendering, resulting in a significant proportion of waste being collected by private contractors. Disposal of waste is also largely operated by private companies or by Local Authority Waste Disposal Companies (LAWDCs) as separate entities. Consequently, the competitive element in waste management has influenced the choice of waste treatment and disposal methods, but has also made costs more transparent and in many cases has driven down the costs.

The Non-Fossil Fuel Obligation (NFFO) A further incentive introduced by the UK Government for the development of alternative waste management options, and thereby moving up the hierarchy of waste management, is financial support via the Non-Fossil Fuel Obligation (NFFO) and the associated fossil fuel levy. The Electricity Act of 1989 makes provision for the Secretary of State for Trade and Industry to place a Non-Fossil Fuel Obligation (NFFO) on any or all of the Regional Electricity Companies (RECs). The obligation is a requirement on the RECs to take a certain proportion of electricity generated from non-fossil fuel sources such as waste, but also including other sources such as wind, wave and hydro-power. The obligation in the context of waste applies to the recovery of energy from waste via, for example, incineration, gasification, pyrolysis or landfill gas combustion via electricity generation or combined heat and power schemes. The electricity prices paid by the RECs under the NFFO are higher than those from coal, oil and gas generated electricity. The fossil fuel levy is placed on the fossil fuel power generators as a percentage of their electricity sales revenue to compensate the RECs for the higher electricity prices paid under the NFFO. The scheme is designed as a market enablement measure to provide an initial guaranteed market in the expectation that once established, the technologies will become economically viable without further support. Once a particular technology becomes established, then the premium prices obtained above the price paid in the electricity pool will converge and the technology will become competitive with fossil fuel generated forms of electricity.

Table 2.7 *Declared net capacity and number of contracts for the Non-Fossil Fuel Obligation (NFFO)*

NFFO order	Year	Declared net capacity (contracted capacity) (MW)	Number of contracts
NFFO-1	1990	152	75
NFFO-2	1991	472	122
NFFO-3	1994	627	141
NFFO-4	1997	843	195

Source: ENDS Report 265, Environmental Data Services Ltd., Bowling Green Lane, London, February, 1997.

An NFFO specifies a period for which the RECs must make available a defined capacity of non-fossil generating plant. The NFFOs have been applied to all RECs in a fixed proportion, so that the RECs can satisfy their NFFOs through joint contracts with generators. Such contracts are held by the Non-Fossil Purchasing Agency (NFPA), which is a wholly owned agent of the RECs. Thus a generator who contracts with the NFPA is effectively contracting part of his capacity to each of the 12 RECs and contributing to the NFFO for each REC.

There have been four tranches of NFFO contracts with the number of contracts and declared net capacity shown in Table 2.7 (ENDS 1997). The declared net capacity is the contracted electricity capacity. The first three NFFO tranches had a declared net capacity totalling 1251 MW. However, not all of the projects came to fruition, and by 1997 the total actual capacity of NFFO-generated electricity was 428 MW. The UK Government have an aim of a total of 1500 MW of electricity from non-fossil fuel sources by the year 2000.

The prices paid to generators under the early NFFO tranches were determined not on a case-by-case basis, but by setting strike prices for each technology band, and contracts were for relatively short periods of time. The third and fourth tranches have shown a significant change from the previous NFFO tranches in that the order covered the period up to the year 2014, allowing contracts to be signed for up to 15 years duration. In addition, rather than all bidders receiving the strike price, they only received the actual price they bid. The 4th tranche of NFFO announced in 1997 has a declared net capacity of 843 MW, with 195 projects and contracts covering the period from 1997 to 2016 (Table 2.8) (ENDS 1997). The order will run for almost 20 years, allowing a 5-year construction and commissioning period followed by the NFFO premium payments for the electricity for a further period of up to 15 years. The 5th NFFO tranche is due in 1998. The NFFO contracts have shown a decrease in the premium price paid for the electricity, realising the UK Government's desire for the prices of fossil fuel and non-fossil fuel forms of electricity to converge and thereby for both to become competitive in their own right.

Project financing Large scale waste treatment and disposal projects, whether mass-burn municipal solid waste incineration or landfill, or hazardous waste treatment

Table 2.8 *Fourth NFFO Renewables Order – 1997*

Technology	Declared net capacity (contracted capacity) (MW)	Number of projects	Average electricity price (pence/kWh)
Landfill gas	173.7	70	3.01
Waste-fired combined heat and power	115.3	10	3.23
Waste-fired fluidised bed combustion	126.0	6	2.75
Small scale hydroelectricity	13.3	31	4.25
Wind energy >0.768 MW	330.4	48	3.53
Wind energy <0.768 MW	10.4	17	4.57
Anaerobic digestion of agricultural wastes	6.6	6	5.17
Biomass gasification or pyrolysis	67.4	7	5.51
Total	843.1	195	3.46

Source: ENDS Report 265, Environmental Data Services Ltd., Bowling Green Lane, London, February, 1997.

facilities, require a large scale investment. Local authorities often do not have the large financial inputs required, which may represent tens of millions of pounds investment. Increasingly such projects are developed as joint ventures between the Local Authority and the private sector. Private sector finance, particularly for the larger projects, would require bank loans, secured against on-going company profits and assets. The risks for such a loan lie with the borrowing company (Chappell 1995).

An alternative approach is 'project financing', where the risk to the investor is based on the success of the project. Repayment of the loan comes from the financial success of the project. Where the project fails there is no recourse to recover the debt from the company. Contracting under such financial arrangements are complex, since each investor requires clearly defined risks and returns from their investment. Such projects are financed from various sources of funding. For example, a typical capital structure for a financing project is shown in Figure 2.3 (Chappell 1995). The senior debt is usually the major portion of the total cost and consists of a bank loan. Subordinate debts represent additional funds which bridge the gap between the large bank loan and the direct investing participants. The equity part of the finance would come from shareholders investments from the direct equity participants in the project such as the fuel or waste supplier, i.e. the local authority, the energy purchaser, equipment supplier and project developer. The size of the bank loan will be determined by the risk involved in the project. Energy from waste schemes such as incineration or landfill gas projects with guaranteed sources of waste and markets for the generated energy would be liable for such loans, whereas untried technologies would find financing difficult (Chappell 1995).

Each member of the project financing team will have different objectives for the project, for example the local authority requires a waste treatment plant to dispose of its waste, the equipment supplier wishes to develop its business, the banks would merely require a return on investment. Therefore detailed contract arrangements are required to apportion suitable risk and consequently return on the investment.

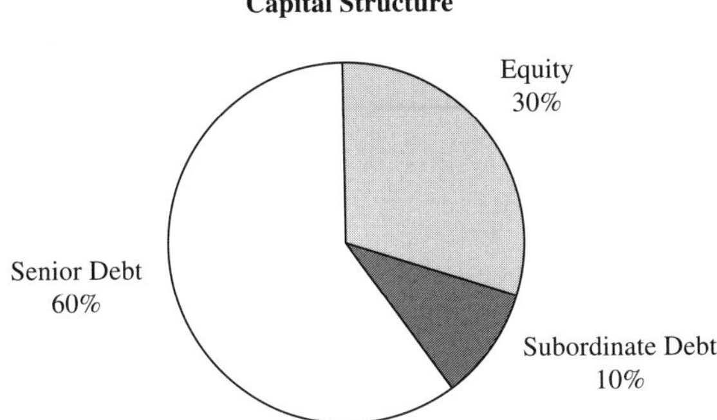

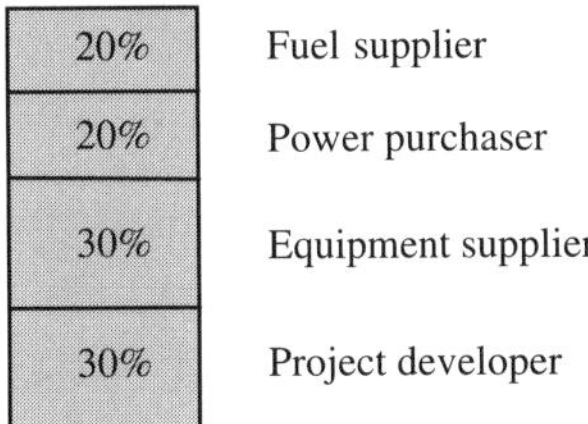

Figure 2.3 *Typical capital structure for project financing. Source: Adapted from Chappell P. 1995. Drawing Board to Reality. Energy Technology Support Unit (ETSU), Harwell.*

Bibliography

British Medical Association, 1991. Hazardous Waste and Human Health. Oxford University Press, Oxford.

Chappell P. 1995. Drawing Board to Reality. Energy Technology Support Unit (ETSU), Harwell.

Clapp B.W. 1994. An Environmental History of Britain since the Industrial Revolution. Longman, London.

Control of Pollution Act, 1974. HMSO, London.

Department of the Environment and Welsh Office, 1995. Making Waste Work: A strategy for sustainable waste management in England and Wales. HMSO, London.

Davis-Coleman C. 1996. Biting the Bullet. The Waste Manager, 27 June.

Digest of Environmental Statistics, 1994. No. 16. Department of the Environment, HMSO, London.

Digest of Environmental Statistics, 1996. No. 18. Department of the Environment, HMSO, London.

ENDS, 1995. Report 246, Environmental Data Services Ltd., Bowling Green Lane, London, July.

ENDS, 1997. Report 265, Environmental Data Services Ltd., Bowling Green Lane, London, February.

Energy from Waste, 1996. Best Practice Guide. Department of the Environment, HMSO, London.

Environmental Assessment Guidance Booklet, 1989. Department of the Environment, HMSO, London.

Garbutt J. 1995. Waste Management Law. A Practical Handbook. Wiley, Chichester.

Holmes J. 1995. The UK Waste Management Industry. Institute of Wastes Management, Northampton.

Lane P. and Peto M. 1995. Blackstone's Guide to The Environment Act 1995. Blackstone Press, London.

McBean, E.A., Rovers F.A. and Farquar G.J. 1995. Solid Waste Landfill Engineering and Design. Prentice Hall, Englewood Cliffs, NJ.

Neal H.A. and Schubel J.R. 1989. Solid Waste Management and the Environment: The mounting garbage and trash crisis. Prentice-Hall, Englewood Cliffs, NJ.

Petts J. and Eduljee G. 1994. Environmental Impact Assessment for Waste Treatment and Disposal Facilities. Wiley, Chichester.

Planning and Compensation Act, 1991. HMSO, London.

PPG23, 1994. Planning Policy Guidance Note: Planning and Pollution Control. Department of the Environment, HMSO, London.

Prescribed Processes and Substances Regulations, 1991. HMSO, London.

Reeds J. 1994. Controlling the Tips. The Waste Manager, 18 September.

Royal Commission on Environmental Pollution, 1993. 17th Report. Incineration of Waste. HMSO, London.

Slater D. 1995. Integrated pollution control and waste minimisation. In: Hester R.E. and Harrison, R.M. (Eds.) Waste Treatment and Disposal. The Royal Society of Chemistry, Cambridge.

Sustainable Development, 1994. The UK Strategy, HMSO, London.

The Environment Act 1995. HMSO, London.

The Environmental Protection Act, 1990. HMSO, London.

This Common Inheritance, 1996. UK Annual Report 1996. Reporting the UK's Sustainable strategy of 1994. HMSO, London.

Town and Country Planning Act, 1990. HMSO, London.

Turvey R. 1996. Waste regulation and the Environment Agency. Proceedings of the Institution of Civil Engineers. Municipal Engineer, 115, 105–117.

Van Santen A. 1993. Wastes Management Proceedings. 18 July.

Waste Management Planning, 1995. Principles and Practice. Department of the Environment, HMSO, London.

West Yorkshire Waste Management Plan, 1996. West Yorkshire Waste Regulation Authority (now Environment Agency), Wakefield, March.

3
Waste

Summary

This chapter discusses the different definitions of waste. Estimates of waste arisings in the UK and the rest of the world are discussed, as well as the methods used in their estimation. Various trends in waste generation and influences on them are discussed. Several categories of waste are discussed in terms of arisings, and treatment and disposal options. The wastes described in detail are municipal solid waste, hazardous waste, sewage sludge, clinical waste, agricultural waste, and industrial and commercial waste. Other wastes described are construction and demolition waste, mines and quarry waste, power station ash, iron and steel slags and scrap tyres. The chapter ends with a discussion of the different types of waste containers, collection systems and waste transport.

3.1 Definitions of Waste

The definition of waste can be very subjective; what represents waste to one person may represent a valuable resource to another. However, waste must have a strict legal definition to comply with both United Kingdom and European Community law. The regulations covering waste use a variety of terms to describe the different types of waste, including 'controlled', 'household,' 'industrial', 'commercial', 'special', 'active', 'inactive' etc. In such cases, strict definitions of waste have financial and legal implications for business, local authorities and Government.

In addition, for the requirement of a legal definition of waste, agreement on definitions and classifications of waste are required for the accurate formulation of local, regional and national waste management planning. In the UK, the Environment Agency is responsible for the collection of data on waste which is used by local authority planning departments in the preparation of their Waste Local Plans and for regional waste management planning. Waste data in the form of types and quantities of wastes are required at National level by the Secretary of State for the Environment to fulfil obligations for a National Waste Strategy under the EC Waste Framework Directive. Consequently, the interaction of waste planning authorities, including local planning, with the Environment Agency and also with the Waste Collection Authorities and Waste Disposal Authorities is desirable. This ensures that common terms and methods are used in the collection of waste data.

The UK National Waste Strategy and the European Waste Framework Directive have at their centre the hierarchy of waste management, and the basis of the strategy is to encourage movement up the hierarchy and thereby increase the levels of waste reduction, re-use and recycling. A key element of the strategy is the compilation of accurate data on waste arisings. This enables recycling targets to be set and responses measured, and the diversion of different waste types from disposal to be monitored.

Increasing regulation of the waste management industry means that accurate definitions of the different types of waste are required for licensing of waste management facilities, and also for ensuring that the obligations of the 'duty of care' regarding wastes are complied with in transferring waste loads between people. Accurate information on the source and composition of wastes also enables strategies for waste minimisation, re-use and recycling to be developed. The implementation of the Landfill Tax in 1996 requires accurate classification of wastes into inactive and other types so that the different waste types which are subject to taxation can be recognised.

The classification of waste is difficult since there can be great variation in composition between different loads of waste, but it is necessary for consistency in the description of the waste wherever the waste has arisen and whoever has described it. In addition to accurate definitions of waste, reporting methods are also required to be uniform. For example, waste arisings are notoriously inaccurate due to different methods of data collection and quantification, different dates of collection, and different reporting methods and administrative systems.

It is clear from several standpoints that accurate definitions and classifications of waste are required. However, by its very nature waste is a heterogeneous material and difficult to describe, define and classify. In many instances the waste will be a mixture of different types, or may be on the border between two categories. Also, as will be shown later, waste can vary in composition on a daily, monthly and seasonal basis, or from location to location. Consequently, when legal or financial outcomes are dependent on the definition of waste, for example for waste arising from a company, then the area is fraught with potential problems. Examples can be seen with the definitions of 'special waste' and 'inactive waste' associated with the Special Waste Regulations 1996 and the Landfill Tax introduced in 1996. Under the Special Waste Regulations, the UK has adopted a list of EC hazardous wastes which are to be described as 'special'. In addition, a special waste may be defined as one possessing materials with certain properties. The designation of waste as hazardous has financial and legal implications for a company. However, how much of a certain substance within the waste would make it hazardous is a contentious point. The Landfill Tax introduced in 1996 defines 'inactive waste' as qualifying for the lower tax band of £2/tonne rather than £7/tonne. However, if contamination of inactive waste with active waste occurs, the level of contamination before the waste is designated to the higher tax band again becomes contentious. In addition, certain wastes are exempt from the tax, and disagreements concerning the classification of the waste can become contentious.

This chapter presents the definitions and classification of waste as well as the problems of estimating waste arisings. The major categories of defined waste are discussed in terms of their arisings, treatment and disposal.

3.1.1 The UK Definition of 'Waste'

The definition of 'waste' in the UK derives from the EC Waste Framework Directive (75/442/EEC as amended by 91/156/EEC). The 1995 Environment Act defines waste as 'any substance or object which the holder discards or intends to discard'. A 'holder' means the producer of the waste or the person who is in possession of it, and 'producer' means any person whose activities produce waste, or any person who carries out pre-processing, mixing or other operations resulting in a change in the nature or composition of this waste. The 1995 Environment Act sets out different categories of waste to which the definition of waste applies (Environment Act 1995).

1. Production or consumption residues not otherwise specified below.
2. Off-specification products.
3. Products whose date for appropriate use has expired.
4. Materials spilled, lost or having undergone other mishap, including any materials, equipment etc. contaminated as a result of the mishap.
5. Materials contaminated or soiled as a result of planned actions (e.g. residues from cleaning operations, packing materials, containers etc.).
6. Unusable parts (e.g. reject batteries, exhausted catalysts etc.).
7. Substances which no longer perform satisfactorily (e.g. contaminated acids, contaminated solvents, exhausted tempering salts etc.).
8. Residues of industrial processes (e.g. slags, still bottoms etc.).
9. Residues from pollution abatement processes (e.g. scrubber sludges, baghouse dusts, spent filters etc.).
10. Machining or finishing residues (e.g. lathe turnings, mill scales etc.).
11. Residues from raw materials extraction and processing (e.g. mining residues, oil field slops etc.).
12. Adulterated materials (e.g. oils contaminated with PCBs etc.).
13. Any materials, substances or products whose use has been banned by law.
14. Products for which the holder has no further use (e.g. agricultural, household, office, commercial and shop discards etc.).
15. Contaminated materials, substances or products resulting from remedial action with respect to land.
16. Any materials, substances or products which are not contained in the above categories.

The waste definitions of particular concern in the UK are shown in Box 3.1. The majority of regulations concern 'Controlled Wastes' which are household, industrial

Box 3.1
Waste Definitions

Controlled Waste
Household, industrial and commercial waste. Sewage sludge disposed of to landfill and by incineration is controlled waste, but disposal at sea and spreading on agricultural land is regulated separately. Controlled waste is defined in the Control of Pollution Act 1974, Environmental Protection Act 1990, Waste Management Licensing Regulations 1994 and The Controlled Waste Regulations 1992.

Household Waste
Household waste means waste from private domestic accommodation, caravans, residential homes, universities or schools or other educational establishment, hospital premises and nursing homes. Other categories covered under the definition are private garages, moored vessels, camp sites, prisons, meeting halls and royal palaces.

Industrial Waste
Waste from factory premises where 'factory' is described within the Factories Act of 1961, premises used for public transport services by land, water or air, premises used for the supply of gas, water, electricity or sewerage services, postal or telecommunications services. Examples include industrial waste producers from the manufacture of food products, textiles, wood products, plastic products, motor vehicles, chemicals etc. Generally the definition includes waste from any industrial undertaking or organisation.

Commercial Waste
Waste from premises used wholly or mainly for the purposes of a trade or business or the purposes of sport, recreation or entertainment. Excluded from the commercial waste category are household and industrial waste, mine and quarry waste, and waste from agricultural premises. Examples include waste from offices, hotels, shops, local authorities, markets, fairs etc.

Clinical Waste
Clinical waste is defined, and includes any waste which consists wholly or partly of human or animal tissue, blood or other body fluids, excretions, drugs or other pharmaceutical products, swabs or dressings, or syringes, needles or other sharp instruments, being waste which unless rendered safe may prove hazardous to any person coming into contact with it. Also any other waste arising from medical, nursing, dental, veterinary, pharmaceutical or similar practice, investigation, treatment, care, teaching or research, or the collection of blood for transfusion, being waste which may cause infection to any person coming into contact with it. Clinical waste also includes the packaging associated with the waste. Clinical waste is a controlled waste and certain wastes might be segregated and separately classified as either household, industrial or commercial depending on the source within the health sector.

Special Waste
Special waste is controlled waste of any kind that is or may be so dangerous or difficult to treat, keep or dispose of that special provision is required for dealing with it. Such waste is defined as containing substances which are dangerous to life. A list of substances is set out in the legislation (Control of Pollution (Special Wastes) Regulations, 1980). The properties of such wastes are defined by physical, chemical and toxicological properties such as corrosiveness, carcinogenicity, toxicity and flammability. The wastes are subject to a consignment note system which records the production to final safe disposal. Adoption of the EC Hazardous Waste Directive (91/689/EEC) will replace the term 'special' with 'hazardous'.

Box 3.1
(continued)

Uncontrolled Waste
Wastes which are not controlled, i.e., household, industrial and commercial. This includes, agricultural waste and mine and quarry waste which are produced in large tonnages but are subject to separate legislation. Other uncontrolled wastes also subject to separate legislative control are radioactive waste and explosive waste.

Inert Waste
Uncontaminated earth and excavation waste which might include, for example, bricks, concrete, stone, building sand and gravel, ceramic materials, slates etc.

Hazardous Waste
Adoption by the UK of the definitions of hazardous waste contained in the EC Hazardous Waste Directive (91/689/EEC) will replace the term 'special' with 'hazardous'. The Directive expands the list of UK substances which if found in the waste make it hazardous, and the list of generic wastes which if they contain certain properties render them hazardous, such as corrosive, toxic, reactive, carcinogenic, infectious, irritant or harmful to human health, and also which may be toxic to the environment.

Municipal Solid Waste
A common term used for waste collected and disposed of by or on behalf of a local authority. In general the waste consists of mainly household and commercial waste. It may also include waste derived from civic amenity waste collection/disposal sites by the general public, street sweepings, gully emptying wastes and construction and demolition waste from local authority sources.

and commercial wastes. Controlled wastes are covered by the Control of Pollution Act 1974, the Environmental Protection Act 1990, the Controlled Waste Regulations 1992, and the Environmental Protection Act 1995. Within the controlled wastes category are clinical wastes, which are largely defined as industrial waste, but might also include categories of commercial and household waste depending on the source. In addition, hazardous wastes which might also include hazardous clinical waste are included as controlled wastes since they arise in most cases from an industrial source. However, hazardous wastes are subject to particular regulations (The Special Waste Regulations 1996) because of the hazardous nature of the waste or the properties of the materials in the waste. Uncontrolled wastes are wastes which fall outside the Controlled Waste categories of household, industrial and commercial waste. This includes agricultural waste and mine and quarry waste, radioactive waste and explosive waste, which are all subject to separate legislation.

3.1.2 The National Waste Classification System

Classifications of waste can be based on a number of different criteria. For example, Porteous (1992) cites several examples:

(i) origin, e.g., clinical wastes, household or urban solid wastes, industrial wastes, nuclear wastes, agriculture;
(ii) form, e.g., liquid, solid, gaseous, slurries, powders;
(iii) properties, e.g., toxic, reactive, acidic, alkaline, inert, volatile, carcinogenic;
(iv) legal definition, e.g., special, controlled, household, industrial, commercial.

It is not surprising, therefore, that waste-related statistics are notoriously difficult to compare from country to country, or even region to region, where different classification systems may be used. There are a number of problems in the classification of waste (Petts and Eduljee 1994) which primarily arise due to:

- different definitions of general categories of waste;
- different definitions of types of waste;
- different methods for data collection and quantification of waste arisings;
- lack of accurate baseline data, and different dates of collection;
- different methods of reporting and different administrative systems;
- poor or patchy responses to notification procedures.

However, the increasing environmental concern about the treatment and disposal of waste, and the desire of the UK and European legislature to encourage the reduction, re-use and recycling of waste, have led to the requirement for the classification of waste and the accurate recording of waste arisings, treatment and disposal. In addition, The Waste Framework Directive (75/442/EEC as amended by 91/156/EEC) requires that The UK National Waste Strategy includes an assessment of the type, quantity and origin of waste to be recovered or disposed of. Historically, the UK has had an extreme lack of detailed and accurate statistics in this area. Therefore, to comply with the requirements of The National Waste Strategy, the UK Department of Environment has proposed a National Waste Classification System (National Waste Classification System 1995; ENDS 1995). The system is in the proposal stage, and even if fully adopted its use would only be voluntary unless backed by an Act of Parliament. The lack of statutory enforcement may limit the widespread use of the system by waste producers in industry and commerce, waste carriers, and waste treatment and disposal operators. If the system is not adopted on a wide scale this would have major implications for the success of the National Waste Strategy.

The National Waste Classification System proposed in 1995 applies to controlled wastes, i.e., household, industrial and commercial wastes, and sets out a listing of the different types of waste and the criteria for inclusion in each category. The system enables the composition of the waste to be classified, but also allows for identification of the source of the waste. The system is designed so that the amount of detail required is proportional to the environmental significance of the waste, including recycling potential. The system also contains details of the producing industry and the waste production process, as well as the waste composition. Contaminated wastes and mixed wastes can also be classified under the system. The scheme proposed is designed to allow the aggregation and disaggregation of waste generation and disposal data on a local, regional and national level. The system is based on classification using six basic fields (boxes) in which the waste can be

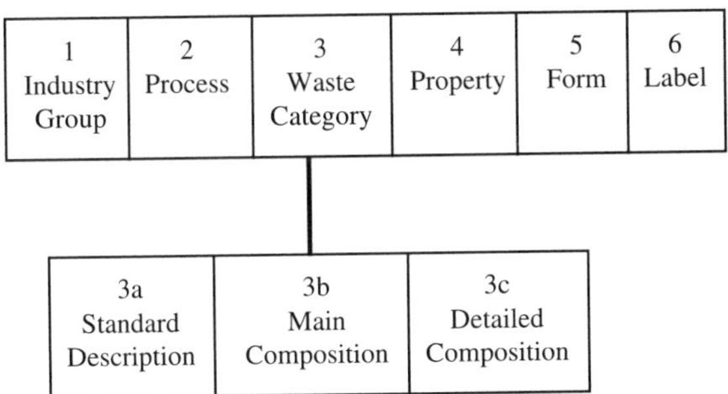

Figure 3.1 *The proposed National Waste Classification System. Source: Development of a National Waste Classification Scheme, Stage 2: A system for classifying wastes. Consultation Draft, Department of the Environment, The Welsh Office, The Scottish Office and the Department of the Environment for Northern Ireland, HMSO, London, 1995. Crown copyright is reproduced with the permission of the Controller of Her Majesty's Stationery Office*

categorised (Figure 3.1) (Development of a National Waste Classification Scheme 1995). The six fields cover the following areas.

The industry producing the waste The determination of the industry group is via a classification system (Standard Industrial Classification of Economic Activities 1992) (SIC 92) which categorises all the economic activities in the UK into classes and subclasses and designates a code number to each. Examples of codes for certain industrial and commercial sectors are shown in Table 3.1 (Standard Industrial Classification of Economic Activities 1992).

The process producing the waste The information on the process which generated the waste is important since generic research into the methods of waste minimisation can be then transferred between industrial and commercial processes. At present there is no standard process listing in the UK or Europe, but a list of processes is to be drawn up as the classification system develops.

The waste category The key area of the classification system is the waste description and detailed composition, which is based on a set of predefined descriptors at three category levels. The concept behind the category listings is the spectrum of hazard associated with each waste category, from inactive waste, defined as Category 1, through to radioactive waste and explosive waste at Categories 14 and 15, respectively. The first-level waste categories are:

1. Inactive
2. Low activity
3. Biodegradable
4. Scrap
5. Contaminated general wastes
6. Health care wastes
7. Asbestos
8. Oily waste
9. Solvents and CFCs
10. Generic types of inorganic chemical waste
11. Inorganic chemical wastes
12. Waste organic chemicals
13. Generic organic chemical waste
14. Radioactive waste
15. Explosives

Table 3.1 Examples of SIC 92 divisions for classifying industrial and commercial activities in the UK

SIC 92Code	Description
15	Manufacture of food products and beverages
16	Manufacture of tobacco products
17	Manufacture of textiles
21	Manufacture of pulp, paper and paper products
23	Manufacture of coke, refined petroleum products and nuclear fuel
24	Manufacture of chemicals and chemical products
25	Manufacture of rubber and plastic products
30	Manufacture of office machinery and computers
34	Manufacture of motor vehicles, trailers and semi-trailers
40	Electricity, gas, steam and hot water supply
64	Post and telecommunications
52	Retail trade, retail of personal and household goods
55	Hotels and restaurants
80	Education
92	Recreational, cultural and sporting activities

Source: Standard Industrial Classification of Economic Activities. HMSO, London, 1992.

Radioactive waste and waste explosives are not controlled wastes but are included for completeness.

The first-level categories are subdivided into compositional categories, for example Table 3.2 shows the subdivisions of the general household, commercial and industrial waste category, and Table 3.3 shows the subdivisions of the organic chemical wastes category (Development of a National Waste Classification Scheme 1995).

The properties of the waste The properties of the waste are described by a series of descriptors mainly relating to its potential to pollute. A waste may have more than one descriptor. The descriptive codes are:

H1 Explosive
H2 Oxidising
H3A Highly flammable <21 °C flash point
H3B Flammable 21–55 °C flash point
H4 Irritant
H5 Harmful
H6 Toxic
H7 Carcinogenic
H8 Corrosive
H9 Infectious
H10 Teratogenic (causes birth defects)
H11 Mutagenic
H12 Reacts to emit dangerous gas
H13 Reacts to form H1–H12 substances
H14 Ecotoxic (toxic to the environment)
P1 Odorous
P2 Sharp
P3 Gas-containerised
P4 Gas-pressurised
P5 Dust/powder
P6 Finely divided metal

Table 3.2 *Subdivisions of the general household, commercial and industrial waste category*

2.2 General household, commercial and industrial waste
- 2.2.1 Mixed household waste
- 2.2.2 Mixed commercial waste
- 2.2.3 Mixtures of household and commercial waste
- 2.2.4 General industrial waste or any mixtures of industrial with household waste, commercial waste or both
- 2.2.5 Mixed/unidentified textiles
- 2.2.6 Laminates

2.3 Separate components of general waste
- 2.3.1 Paper and cardboard
- 2.3.2 Coated paper and cardboard
- 2.3.3 Plastics
- 2.3.4 Glass-reinforced plastic
- 2.3.5 Rubber and tyres
- 2.3.6 Synthetic textiles
- 2.3.7 Wood
- 2.3.8 Treated timber

2.5 Vegetable matter
- 2.5.1 Vegetable fibres
- 2.5.2 Sawdust shavings and wood pulp
- 2.5.3 Vegetation and vegetable waste
- 2.5.4 Bark
- 2.5.5 Mixed garden or site clearance waste
- 2.5.6 Waste vegetable food
- 2.5.7 Compost

2.6 Animal matter
- 2.6.1 Leather
- 2.6.2 Animal fibres
- 2.6.3 Waste food – animal or mixed
- 2.6.4 Animal parts
- 2.6.5 Excreta

2.7 Animal or vegetable oil, fat, wax, grease
- 2.7.1 Vegetable oils, fats etc.
- 2.7.2 Animal fats etc.
- 2.7.3 Soap
- 2.7.4 Animal glue

2.8 Sewage
- 2.8.1 Sewage
- 2.8.2 Sewage sludge
- 2.8.3 Sewage screenings
- 2.8.4 Septic tank waste
- 2.8.5 Cesspool contents

2.9 General waste landfill leachate

2.10 Residues of fermentation and other microbiological processes

Source: Development of a national waste classification scheme. Stage 2: A system for classifying wastes. Consultation Draft, Department of the Environment, The Welsh Office, The Scottish Office and Department of the Environment for Northern Ireland, HMSO, 1995. Crown copyright is reproduced with the permission of the Controller of Her Majesty's Stationery Office.

Table 3.3 *Subdivisions of the organic chemical wastes category*

Category 10: Organic chemical wastes

- 10.1 Paints, resins and adhesives
  - 10.1.1 Resins
  - 10.1.2 Adhesives and sealants
  - 10.1.3 Paint
  - 10.1.4 Varnish and lacquers
- 10.2 Inks and dyestuffs
  - 10.2.1 Inks
  - 10.2.2 Dyestuffs and pigments
- 10.3 Cosmetics
- 10.4 Surfactants and chelating agents
  - 10.4.1 Detergents and surfactants
  - 10.4.2 Chelating agents
- 10.5 Monomers and precursors
- 10.6 Tarry wastes
  - 10.6.1 Distillation residues
  - 10.6.2 Solvent and other waste distillation residues
  - 10.6.3 Coal tars
  - 10.6.4 Acid tars
  - 10.6.5 Other tars
- 10.7 Pharmaceuticals
- 10.8 Pesticides
  - 10.8.1 Organometallics
  - 10.8.2 Other pesticides
- 10.9 Organic chemical process wastes
- 10.10 Additional codes
  - 10.10.1 Polynuclear aromatic hydrocarbons (PAHs)
  - 10.10.2 PCBs and analogues
  - 10.10.3 Dioxins and furans
  - 10.10.4 Other halogenated waste
- 10.11 Carbon wastes

Source: Development of a national waste classification scheme. Stage 2: A system for classifying wastes. Consultation Draft, Department of the Environment, The Welsh Office, The Scottish Office and Department of the Environment for Northern Ireland, HMSO, 1995. Crown copyright is reproduced with the permission of the Controller of Her Majesty's Stationery Office.

The physical form of the waste The physical form of the waste, whether solid, liquid or sludge etc., is described below.

F1 Solid – composite of materials
F2 Solid – mixed materials
F3 Solid – bulky
F4 Liquid – containerised
F5 Liquid – bulk

For Textiles
F8 Clean textiles
F9 Dry textiles
For Metals
F10 Metal rod

F6 Sludge/slurry – in a solid container
F7 Sludge/slurry – bulk

F11 Metal swarf
F12 Metal wire
For Plastic
F13 Plastic bottles
F14 Plastic film
F15 Plastic – rigid

The information label A label incorporating a number of additional factors to describe the waste must be available. These factors are: S, special waste; H, hazardous waste; P, packaging waste; C, clinical waste.

3.2 Waste Arisings in the UK

Table 3.4 shows the estimated arisings of controlled and uncontrolled waste (see Box 3.1) in the UK (Department of the Environment and Welsh Office 1995). Comparison of the estimated waste arisings in the UK with those from other countries are shown in Table 3.5 (Petts and Eduljee 1994). Difficulties arise in direct comparison of waste generation data due to the different waste classification systems, estimating methods and paucity of data available from country to country. This is also true of estimations from individual countries, where estimates for each waste producing sector can vary markedly depending on the estimator. The waste arisings shown in Table 3.5 are clearly linked to population size, with large population countries such as the USA, Germany, France, Japan and Italy having the highest arisings of municipal waste. In addition, the waste arisings of each sector are linked to the industrial development and mixture of agriculture and industry in each country.

3.2.1 Estimating Waste Arisings

Accurate data concerning estimates of present and future production and composition of different types of waste are essential for long-term efficient and economical waste management planning. The estimates are used by both planners and waste treatment and disposal engineers to determine the type, size, design and location of waste treatment and disposal facilities. The information is also used to determine the associated transport infrastructure and personnel requirements. Accurate waste arisings data are also required to meet national and international legal and policy obligations. The UK and EC strategies for increased re-use and recycling of waste requires that the composition of the waste is characterised so that potentially suitable materials can be identified. The UK strategy of targets for waste minimisation, re-use and recycling also require accurate data to determine if and when targets are met. A key element of the UK National Waste Strategy is the gathering of waste arising statistics by the Environment Agency on a regional basis which will be used at local and regional level for planning future waste management facilities and at national level to implement the Waste Strategy. The estimate of

Table 3.4 *Waste arisings in the UK*

Waste type	Annual tonnage (million tonnes)
Controlled wastes	
Household	20
Commercial	15
Construction and demolition	70
Other industrial	70
Sewage sludge	35[1]
Dredged spoils	35[2]
Total controlled waste arisings per annum	245
Uncontrolled wastes	
Mines and quarry waste	110
Agricultural waste	80
Total uncontrolled waste arisings per annum	190
Total of all UK waste arisings per annum	435

[1] Annual tonnage based on wet sludge arisings, estimated on a 4% solids content.
[2] Dredged material from harbours etc.

Source: Department of the Environment and Welsh Office, Making Waste Work. HMSO, London, 1995.

Table 3.5 *Estimated waste arisings in selected countries (million tonnes)*

Country	Municipal	Industrial	Agricultural	Mining	Demolition	Sewage	Hazardous
Belgium	3.5	27.0	53.0	7.1	0.7	0.7	0.9
Denmark	2.4	2.4	–	–	1.5	1.3	0.1
France	17.0	50.0	400.0	10.0	–	0.6	3.0
Germany	19.5	61.0	–	9.5	12.0	1.7	6.0
Greece	3.1	4.3	0.09	3.9	–	–	0.4
Ireland	1.1	1.6	22	1.9	0.2	0.6	0.02
Italy	17.3	40.0	30.0	57.0	34.0	3.5	3.8
Japan	48.0	312.0	63.0	26.0	58.0	2	6.6
Luxembourg	0.17	1.3	–	–	4.0	0.02	0.004
Netherlands	6.9	6.7	86.0	0.1	7.7	0.3	1.5
Portugal	2.4	0.7	0.2	3.9	–	–	0.16
Spain	12.5	5.1	45.0	18.0	–	10	1.7
USA	209.0	760.0	150.0	14.0	32.0	10	275.0[1]

[1] Includes wastewaters.

Source: Petts J. and Eduljee G., Environmental Impact Assessment for Waste Treatment and Disposal Facilities. Wiley, Chichester, 1994. Reproduced by permission from John Wiley & Sons Limited.

waste arisings is required on a statutory basis for controlled wastes, but uncontrolled waste data would also be required to allow for sufficient capacity for waste treatment and disposal for waste as a whole in the region sampled. The requirement under the Strategy is for an annual investigation of waste arisings, waste movements and waste management operations.

More detailed compositional analysis is required in some cases for the design of certain treatment facilities. For example, a knowledge of the chemical composition of waste, including the presence of metals, would be important for composting facilities, and the composition of waste in terms of energy content, volatile ash and moisture content would be important for the design of an incinerator. The physical particle size of the waste may also be an important characteristic, for example, in sizing feeder hoppers for incinerators and the design of screening systems to segregate waste.

Statistical data on the quantification of waste are usually by weight, although sometimes it may be more appropriate to report the data as volume. For example, plastic bottles for recycling are often reported as volume rather than weight. Classification and quantification of industrial waste may be on an industry source basis, such as the chemicals, food and petroleum industries, or generic terms such as waste solvents and oils from a number of industrial sources. Analysis of the composition of waste may also be based on the source of the waste, so that industrial waste might be analysed in terms of chemical composition, and clinical and household waste might be either by material types, such as glass, paper, metal etc., or by product types such as glass containers, tins, magazines etc. In addition to weight and composition, the energy value, moisture content, volatile content and elemental composition may also be found by a series of standard tests.

Whilst detailed classification, quantification and compositional analyses are clearly desirable, it is another matter to obtain accurate data from the producers of the waste. In the UK there is no requirement for waste producers to record waste statistics. By its inherent nature waste is heterogeneous, and the components of the waste can vary widely, so a large number of samples is required for statistical accuracy. However, for a large waste source population this may not be possible, and therefore representative samples with their consequent errors are used. Wastes may also be contaminated with other wastes, not necessarily due to the process of the organisation.

There are, however, many problems associated with waste arisings data. Waste, particularly household waste, can vary on a daily, monthly and seasonal basis, on the size of the population, the type of housing stock etc. A further factor involved in the accuracy of waste arisings data is that variations in the amount of waste collected can depend on the method of waste collection and the size of waste container. This is particularly true for household waste (Pescod 1991–93). Therefore, it should always be borne in mind that waste is a very variable and heterogeneous material.

Two approaches are generally adopted for estimating waste quantity and composition, either questionnaires to the producers of the waste, or direct waste analysis of the waste stream either at the point of waste production or at the waste treatment facility (Waste Management Planning 1995; Yu and Maclaren 1995). The questionnaire is distributed to a range of companies, with questions concerning the quantities of waste generated and the composition in relation to a predetermined list of product-based or material-based categories. Other questions which might be included are questions on the seasonality of the waste generated and any waste re-use or recycling schemes in operation. However, in many cases companies do not keep records of the amount of waste that their company generates, let alone the composition of that waste. Consequently data acquisition systems that rely on questionnaires may not be reliable because the data is not available, and reporting of both weight and composition may be estimated by visual observation of the waste stream, which is notoriously inaccurate. In addition, if there is no obligation to reply to the questionnaire then the response rate may be low due to time constraints, apathy or confidentiality. Non-response to the questionnaire may induce a bias so that responding companies may be seen to have better waste management and reporting systems than non-respondents.

In the UK, the Department of the Environment's guide to best practice in Waste Management Planning (1995) sets out requirements and methodologies for estimating waste arisings. The system recommended for commercial and industrial waste is based mainly on the questionnaire system, with questionnaires sent by post or via personal visits to the company to a sample selected from the industrial and commercial sectors represented in the survey area. To mitigate against possible errors in estimating, a dual approach to collection of data is recommended, by estimating waste arisings from the source and from the waste management facilities which receive the waste. The questionnaire, which may be up to 10 pages in length, covers such areas as the type and quantity of waste arising, the waste collection/disposal method used by the organisation, the types of container used for the waste, general categories and sub-categories of the waste, description of the process generating the waste, detailed description of the waste in terms of the percentage components in the waste, the physical form of the waste, proportion of packaging, transport method, location of the disposal site, weight of the waste etc. The waste data from the producers of the waste should then be matched against the waste received at waste treatment and disposal facilities to ensure the accuracy of the survey data. The waste treatment and disposal facilities would include incinerators, landfill sites, composting facilities, recycling centres and transfer stations. The data collected from the facility would include the classes and types of waste, quantities and recycled output, and would also include data on pollution control measures, energy recovery, remaining useful life of the facility etc. (Waste Management Planning 1995).

The alternative to the questionnaire system of obtaining waste data is the direct waste analysis method, which involves the direct examination of the waste stream. The waste, usually in 90–180 kg samples, is taken directly from the stream and sorted by hand into a set of characteristic material or product categories and weighed (Yu and Maclaren 1995). The direct waste analysis system also has disadvantages. The cost of the method is high because it is so labour-intensive and may be more than ten times the cost of a questionnaire. The direct analysis method also represents a sample typically taken during one day, which may not be representative of the waste stream on an annual basis.

Figure 3.2 compares the accuracy of the questionnaire and direct waste analysis methods for determining the quantities of industrial, commercial and institutional

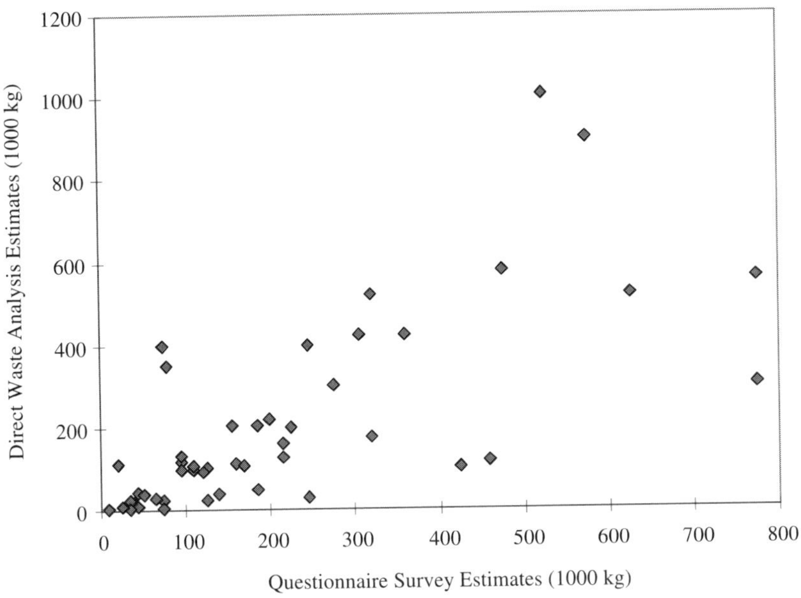

Figure 3.2 *Comparison of waste quantity estimates using the 'direct waste analysis' and 'questionnaire' methods. Source: Yu C.-C. and McLaren V., Waste Management and Research, 13, 343–361, 1995. Reproduced by permission from ISWA – The International Solid Waste Association*

waste generated in Toronto, Canada (Yu and Maclaren 1995). A good correlation is seen between the estimated weight of waste using the two methods. However, Table 3.6 shows a rather poorer comparison for the composition of the waste using the two methods, which is particularly different for paper and paperboard, plastics, textiles and food.

3.2.1.1 *Estimating Waste Generation Multipliers*

Waste generation multipliers are used for estimating waste from all the sources of waste in the region. For example, household waste generation multipliers may be based on the size of population, industrial and commercial waste generation multipliers may be based on the number of employees, and in the case of agricultural waste the multiplier might be acreage or number of livestock. For example, Table 3.7 shows typical waste generation factors based on the type of generator (Warmer Bulletin 49, 1996). The correct multiplier to use in predicting future waste arisings has major implications for planning, particularly if there is some doubt as to the correct multiplier to use.

In an attempt to obtain more accurate waste generation multipliers, some surveys have attempted to take into account a wide range of factors, for example, the size of each local population in a region, the type and age of residence occupied, the seasons of the year, the mixture and type of businesses in the area, and also

Table 3.6 *Comparison of waste composition estimates by direct waste analysis and by questionnaire (weight %)*

Waste type	Direct waste analysis	Questionnaire
Paper	24.7	33.2
Paperboard	22.3	9.0
Ferrous metal	5.9	3.3
Non-ferrous metal	0.9	0.7
Plastics	13.3	6.9
Glass	2.8	8.4
Rubber	0.4	0.5
Leather	0.0	0.0
Textiles	4.5	0.7
Wood	7.5	10.3
Vegetation	1.4	0.4
Fines	0.3	2.2
Special wastes	0.6	0.7
Construction materials	4.6	2.2
Food	10.7	20.9

Source: Yu C.-C. and McLaren V., Waste Management and Research, 13, 343–361, 1995. Reproduced with the permission of ISWA – International Solid Waste Association.

Table 3.7 *Typical US waste generation rates by type of generator*

Waste generation sector	Average	Units
Single family residential	1.22	kg/person/day
Apartments	1.14	kg/person/day
Offices	1.09	kg/employee/day
Eating and drinking establishment	6.77	kg/employee/day
Wholesale and retail trade[1]	0.009	kg/$ sales
Food stores	0.015	kg/$ sales
Educational facilities	0.23	kg/student/day

[1] Except food stores

Source: Warmer Bulletin, 49, Journal of the World Resource Foundation, High Street, Tonbridge, Kent, UK, May 1996. Reproduced by permission from the World Resource Foundation.

economic data such as industrial output, number of employees, company turnover etc. (Rhyner and Green 1988; Warmer Bulletin 49 1996).

3.2.1.2 Household Waste Generation Multipliers

Household waste generation multipliers have shown a wide variation depending on the source of the survey, for example, estimates of the arisings of household waste in the US have varied between 1.08 and 1.22 kg/person/day (Rhyner and Green 1988). It has also been shown that more accurate estimates can be produced for household waste using waste generation multipliers based on the population size of the local

Table 3.8 Comparison of household waste generation multipliers based on the population size of the community

Population	Waste generation multiplier (kg/person/day)
< 2500	0.91
2500–10 000	1.22
10 000–30 000	1.45
> 30 000	1.63

Source: Rhyner C.R. and Green B.D., *Waste Management and Research*, 6, 329–338, 1988.

community (Wisconsin, USA) (Table 3.8). Smaller communities produce a lower waste generation/person/day than larger populated communities.

Within the UK, the Wastes Technical Division of the Department of the Environment is undertaking a major analysis of household waste in terms of weight, composition and chemical analysis under the National Household Waste Analysis Project (1994, 1995). The Project has tried to produce more accurate waste multipliers and has included socio-economic classifications of households and waste arisings. Whilst there will clearly be differences in the waste amounts collected from multi-story flats compared to houses and bungalows etc., this approach does not distinguish between differences in family size, and socio-economic and other factors which are included in the UK Project. The system used in the National Analysis is a classification of all the households in the UK into demographic, housing and socio-economic units. Each unit consists of about 450–550 persons, or about 160–180 households. The classification system is a computer-based system called ACORN (A Classification of Residential Neighbourhoods). The ACORN groups are:

A Agricultural area
B Modern family housing, higher incomes
C Older housing of intermediate status
D Older terraced housing
E Council estates – Category I, e.g., older estates with well-off older workers, recent estates and better council estates)
F Council estates – Category II, e.g., low-rise estates in industrial areas and inter-war estates with older people)
G Council estates – Category III, e.g., new estates in inner cities, estates with overcrowding and estates with high unemployment)
H Mixed inner metropolitan areas
I High-status non-family areas
J Affluent suburban housing
K Better-off retirement areas

Each group is further subdivided into 38 subgroups. The classification of each group is based on such factors as unemployment, number of cars, age and number of the residents, ownership, number of rooms, toilet facilities, employment type etc.

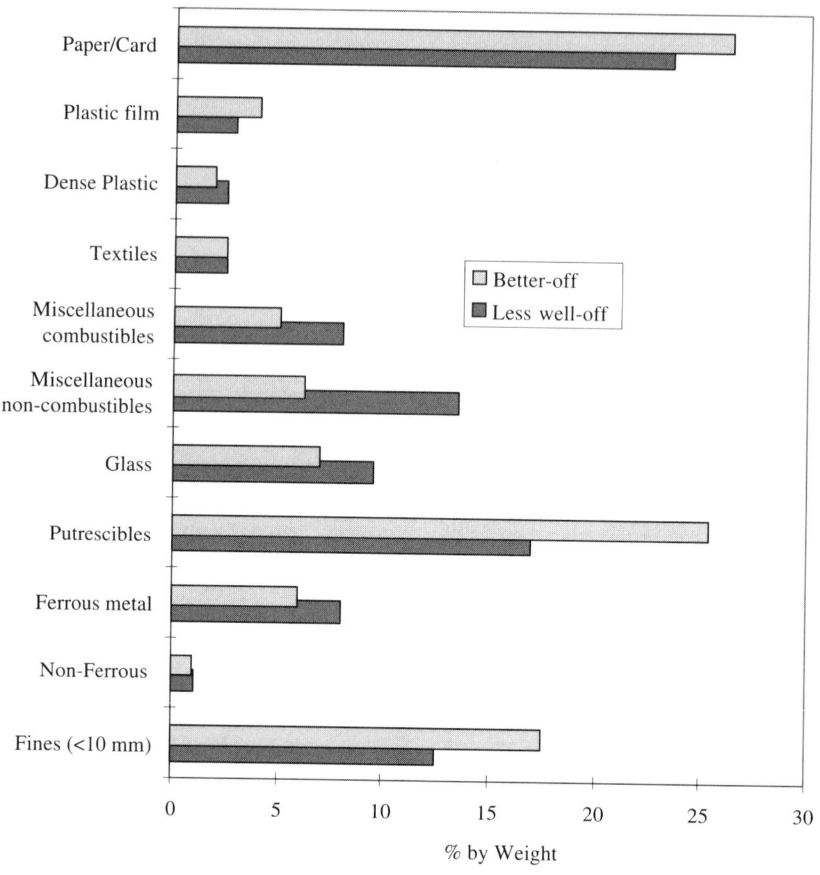

Figure 3.3 *Comparison of waste compositional analysis for 'better off' and 'less well off' ACORN household groups. Source: The Analysis and Prediction of Household Waste Arisings. Report No. CWM/037/91: The Technical Aspects of Controlled Waste Management. Wastes Technical Division, Department of the Environment, HMSO, London, 1991. Crown copyright is reproduced with the permission of the Controller of Her Majesty's Stationery Office*

The ACORN coded street listings can be obtained for any local authority area. In addition to the ACORN classification, a detailed count of the number of houses in each category is also made (Analysis and Prediction of Household Waste Arisings 1991; National Household Waste Analysis Project 1994, 1995). The results show that waste output does vary significantly between ACORN groups. However, the amount in each group also varies between different authorities. The results of the UK National Household Waste Analysis Project have shown that in many cases the better-off houses, for example in ACORN groups B, E, I and J, produce more waste compared with the less well-off groups such as D, F and G (Figure 3.3), with an average of 15–17 kg/household/week compared with 10–11 kg/household per week. The differences in waste amount produced are mainly due to differences in composition. Figure 3.3 shows that waste from the better-off households (ACORN

Table 3.9 *Waste factor variation between different local authorities for different ACORN household groups*

ACORN group	Waste factors (kg/household/week)	
	Range	Average
A	17.9–32.4	22.1
B	9.6–21.5	14.3
C	9.0–14.8	11.8
D	8.8–20.4	12.6
E	10.7–23.3	14.6
F	7.2–16.7	12.2
G	7.2–17.6	13.6
H	5.0–12.3	9.8
I	7.7–27.1	13.1
J	5.4–20.7	14.2
K	5.4–16.6	11.1

Source: The Analysis and Prediction of Household Waste Arisings. Controlled Waste Management Report, CWM/037/91, Department of the Environment, HMSO, London, 1991.

Group B) contained more putrescible and garden wastes and also more paper, plastic film and glass than that from the less well-off households (ACORN Group C).

Waste factors can be calculated for each ACORN Group, which allows the prediction of the likely waste arisings from a particular area or for the authority as a whole. However, variations can occur between different authorities for each ACORN Group, and also other factors can distort the waste factor, such as the method of waste collection. Local authorities who use a traditional bin or wheeled bin system which restrict the waste collected to that in the dustbin container tend to collect less waste than those using unrestricted waste amounts such as the plastic sack system. In addition, authorities using the larger wheeled bins with their larger capacity tend to generate more waste than the traditional smaller bins (National Household Waste Analysis Project 1994). In some cases authorities moving to a wheeled bin system have experienced increases in the waste generated of up to 30% (Pescod 1991–93). Also, some authorities collect garden waste whilst others have separate collection arrangements. The policy of the authority regarding recycling, the number and frequency of recycling sites and public waste disposal sites etc., can also influence the waste factors between ACORN Groups and between local authorities. Table 3.9 shows the range of waste factors between different ACORN Groups between different local authorities, and illustrates that waste factor calculations, particularly for households, is very difficult.

It is not only the contents of the dustbin that should be analysed to obtain an accurate picture of waste generation, since the weight and composition will be influenced by how much recycling is carried out by the residents of the local population. Consequently, household waste analyses should also include waste taken to civic amenities (recycling sites such as bottle banks and scrap metal sites) and would also include composting.

Table 3.10 *Comparison of industrial waste generation multipliers*

Industrial sector	Industrial waste generation multiplier (tonnes/employee/year)	
	New York	Wisconsin
Food and similar products	5.26	4.42
Textile mill products	0.24	0.76
Paper and allied products	13.06	13.52
Printing, publishing	0.44	1.03
Chemicals and allied products	0.57	7.45
Petroleum and refining	0.00	26.35
Rubber and plastics	2.36	1.01
Stone, clay, glass and concrete	2.18	10.69
Primary metals industry	21.77	6.09
Fabricated metal products	1.54	3.38
Electrical machinery	1.54	2.43
Transportation equipment	1.18	1.18

Source: Rhyner C.R. and Green B.D., Waste Management and Research, 6, 329–338, 1988. Reproduced with the permission of ISWA – International Solid Waste Association.

3.2.1.3 *Industrial and Commercial Waste Generation Multipliers*

Industrial waste multipliers will vary depending on the type of industry or business sector. For example, the food industry produces much more waste than, for example, the printing and publishing industry (Table 3.10, Rhyner and Green 1988). In addition, the particular waste multiplier used from one city to the next will depend on a number of factors, including variations in manufacturing process, economies of scale of large scale production and regional differences in the mix of industries included in an industrial sector (Rhyner and Green 1988, Table 3.10).

Estimation of industrial and commercial waste multipliers for the UK, the Netherlands and Germany are shown in Table 3.11 (Analysis of Industrial and Commercial Waste 1995). The data for the UK were obtained from two separate sources: raw data from surveys of companies and data obtained from Waste Disposal Plans drawn up by the Waste Regulation Authorities before they were absorbed into the Environment Agency. Table 3.11 shows the differences which can be obtained between different types of estimation. In some sectors the differences in the waste multipliers are enormous, with consequent implications for determining the appropriate waste management operations to treat and dispose of the waste. Generally, the data derived from Waste Disposal Plans tend to be higher than data derived from surveying companies. This may be due in part to the estimation procedure in Waste Disposal Plans leading to over-estimates of the waste generated from industrial and commercial sectors compared with raw data on waste obtained directly from surveys. In some cases the comparison with

Table 3.11 *Industrial and commercial waste multipliers (tonnes/employee/year)*

Industry sector	Holland	Germany	UK survey[1]	UK plan[2]
Mining	–	53.4	–	–
Extraction of mineral oil and gas, mineral oil processing	–	24.5	–	–
Public utilities	–	–	9.1	48.5
Metal manufacturing	12.4	55.3	17.0	83.4
Quarrying	–	46.9	0.25	83.4
Manufacture of non-metallic mineral products	13.4	6.3	16.5	83.4
Chemical industry	7	15.7	9.5	83.4
Manufacture of other metals goods	1.8	8.7	9.3	3.0
Mechanical engineering	0.7	2.8	4.1	3.0
Manufacture of office machinery	–	1.2	0.1	3.0
Electrical and electronic engineering	0.4	1.6	3.5	3.0
Manufacture of motor vehicles and parts and other transport	0.8	3.7	12.5	3.0
Instrument engineering	0.3	1.0	7.8	3.0
Food, drink and tobacco manufacturing industries	18.4	22.3	13.9	4.8
Textile, leather and leather goods industries	1.3	2.4	3.7	4.8
Timber and wooden furniture industries	6	22.3	15.2	4.8
Manufacture of paper and paper products; printing and publishing	5.4	22.7	5.1	4.8
Processing of rubber and plastic	1.7	3.7	5.4	4.8
Other manufacturing industries	0.5	1.4	2.9	4.8
Construction	–	99.2	4.6	25.9

1 Survey data of companies.
2 Data from Waste Regulation Authorities Waste Disposal Plans.

Source: Analysis of Industrial and Commercial Waste going to Landfill in the UK. Department of the Environment, HMSO, London, 1995. Crown copyright is reproduced with the permission of the Controller of Her Majesty's Stationery Office.

Germany and the Netherlands is quite close, for example, the waste generation multipliers for the manufacture of paper, paper products, printing and publishing from the UK survey and from the Netherlands. However, in most cases there are markedly different factors between the industrial and commercial sectors in the three countries. This may in part be due to the different ways in which waste data are produced, but highlights the care which must be exercised when using waste generation multipliers and comparisons of waste arisings data between countries and regions.

Construction and demolition waste generation multipliers have been estimated as 0.63 kg/person/day for cities greater than 10 000 in population size and 0.14 kg/person/day for cities smaller than 10 000 population. Commercial waste multipliers have been estimated and range between 0.42 and 2.04 kg/person/day, with a median of 1.06 kg/person/day (Rhyner and Green 1988).

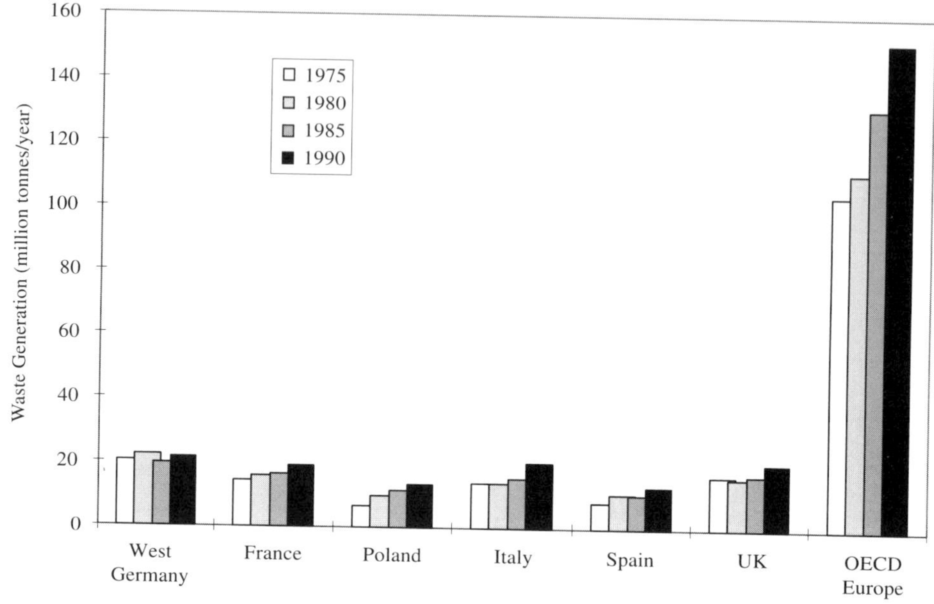

Figure 3.4 *Municipal solid waste generation rates in OECD Europe and selected countries since 1975. Source: Stanners D. and Bourdeau P. Europe's Environment, The Dobris Assessment. European Environment Agency, Copenhagen. Earthscan Publications, London, 1995*

3.2.2 Trends in Waste Generation

Historical trends in waste generation show an increase in the quantities of waste generated for most countries. For example, Figure 3.4, shows historical trends in municipal solid waste generation in OECD (Organisation of Economic Cooperation and Development) Europe and selected countries (Stanners and Bourdeau 1995). These trends in increased waste generation with time are linked to increases in population, but also to increases in the waste generated per head. Other factors which influence the quantity and composition of waste over time are changing consumption patterns, economic growth rates and recessions, and the impact of legislative or economic instruments to increase re-use and recycling for example. Future trends in waste composition will be influenced by the concept of sustainable waste management, with increasing incentives for waste minimisation, re-use and recycling.

There is a strong link between waste generation rates and the economic standing of a country. Figure 3.5 (Stanners and Bourdeau 1995) shows a link between Gross Domestic Product (GDP) and waste production for several countries of the OECD. Increasing economic development represented by GDP is reflected in an increasing rate of waste production.

In addition, waste composition has been shown to vary with seasons of the year. For example, Table 3.12 shows the waste composition of municipal solid waste

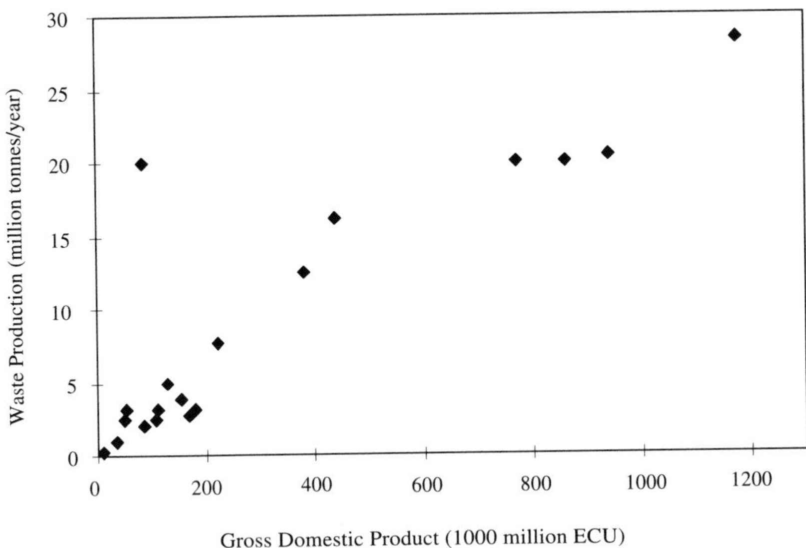

Figure 3.5 *Municipal solid waste production as a function of gross domestic product (GDP). Source: Stanners D. and Bourdeau P., Europe's Environment, The Dobris Assessment, European Environment Agency, Copenhagen. Earthscan Publications, London, 1995*

Table 3.12 *Seasonal variation in US municipal solid waste composition*

Waste component	Autumn	Winter	Spring	Summer	Average
Organics	86.0	86.5	87.7	89.8	87.5
Paper	44.7	45.7	47.5	40.3	44.5
Plastic	6.1	6.5	7.0	6.0	6.4
Yard waste	15.0	15.1	7.4	15.0	13.1
Wood	1.4	0.8	1.0	0.8	1.0
Food	15.2	14.3	18.0	21.5	17.3
Textiles	1.5	1.3	1.3	2.0	1.5
Other organics	2.3	2.9	5.4	4.2	3.7
Inorganics	13.4	12.8	12.2	9.8	12.1
Metals	3.0	3.9	4.0	3.1	3.5
Glass	7.0	7.1	7.1	5.6	6.7
Soil	0.9	0.4	0.0	0.0	0.3
Other inorganics	2.5	1.4	1.1	1.2	1.5
Special wastes	0.5	0.6	0.1	0.3	0.4
Appliances	0.2	0.3	0.0	0.2	0.2
Chemicals	0.1	0.2	0.1	0.1	0.1
Re-useable	0.3	0.1	0.0	0.0	0.1
Total	100	100	100	100	100

Source: Warmer Bulletin, 49, Journal of the World Resource Foundation, High Street, Tonbridge, Kent, UK, May 1996. Reproduced by permission from the World Resource Foundation.

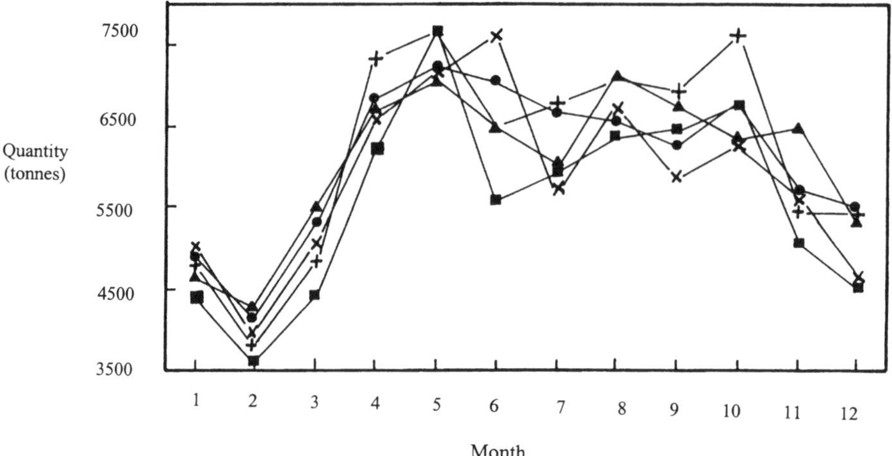

Figure 3.6 Residential waste quantities accepted in landfills at Brown County, Wisconsin, USA, from 1985–1989. (■—■, 1985; +—+, 1986; ●—●, 1987; ▲—▲, 1988; ×—×, 1989). Source: Rhyner C.R., Waste Management and Research, 10, 67–71, 1992. Reproduced by permission of ISWA – The International Solid Waste Association

(local authority waste) in the USA over a one-year period (Warmer Bulletin 49, 1996). Some of the component wastes show significant variation through the year. Short-term seasonal variations of waste are common, for example organic yard waste (garden waste) is known to increase during the growing season, or where there are influxes of visitors into a tourist area then the total quantity of waste will also increase.

Month by month variations in waste arisings can also be detected. Figure 3.6 shows the monthly arisings of residential waste received at landfill sites in Brown County, Wisconsin, in the USA over a five-year period from 1985 to 1989 (Rhyner 1992). The data in Figure 3.6 are for residential waste, but similar fluctuations were seen for commercial, construction and demolition waste. Construction and demolition industries are particularly influenced by seasonal weather conditions. General industrial waste was shown to have little month by month variation in quantities produced. The large swings in the quantities of residential waste being delivered, which represent in the case of the Brown County data increases in May of 33.4% over the average waste delivered to decreases in February of 23.8% of the average, can have significant consequences on the manning levels, equipment use, financing etc. of the waste management facility. Even day to day variations in the generation of waste can be detected.

The UK National Household Waste Analysis Project has shown that some seasonal variation in waste output occurs in the UK, mainly for the better-off households, with higher waste outputs in late summer and spring due mainly to increased garden waste. Also Bank Holidays such as Christmas and Easter tend to produce large increases in waste, due in many cases not only to additional waste being produced, but to the collection system which often collects the equivalent of

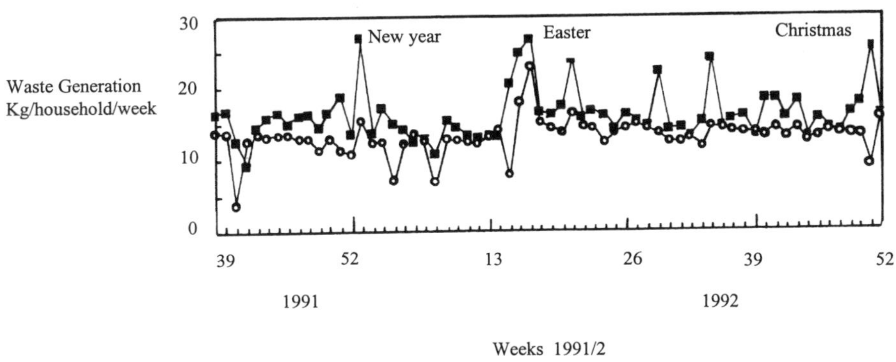

Figure 3.7 *Variation in yearly waste arisings for two waste collection rounds in St. Albans, UK (kg/household/week). Source: The Technical Aspects of Controlled Waste Management. National Household Waste Analysis Project: Phase 2, Vol. 1. Report on Composition and Weight Data. Report No. CWM/082/94, Wastes Technical Division, Department of the Environment, HMSO, London, 1994. Crown copyright is reproduced with the permission of the Controller of Her Majesty's Stationery Office*

two weeks waste in the week following the holiday (Figure 3.7, National Household Waste Analysis Project, 1994).

The significance of changes in composition over short time-spans can be seen in Figure 3.8, which shows the combustible waste fraction of residential solid waste generated in 1989 in the Kita-Ku District of Sapporo, Japan (Matsuto and Tanaka 1993). The collection system in Sapporo is based on separate collections of combustible and non-combustible waste, separated at source, and the disposal system is mainly incineration. Collection in Kita-Ku is twice a week, Monday and Thursday, for the combustible fraction and once a week for the non-combustible fraction. The figure shows the difference in the quantity of the combustible fraction of the waste between two days (Monday and Thursday) in the week. The Monday data obviously contain waste generated at the week-end, which is shown to have a higher combustible fraction. Where heat or electrical output from an incinerator has to be guaranteed, then such variations can significantly influence the economic operation of such a facility.

3.3 Municipal Solid Waste

Municipal solid waste consists mainly of household and commercial waste which is disposed of by or on behalf of a local authority. In many cases data on the breakdown of wastes into household and commercial is not available. Many national and international studies on waste arisings and composition classify municipal solid waste as a specific category of waste. However, comparative data are often difficult to interpret since different types of waste, including in some cases industrial wastes, may be collected by the local authority. Figure 3.9 shows a comparison of municipal solid waste generation per capita per year for various

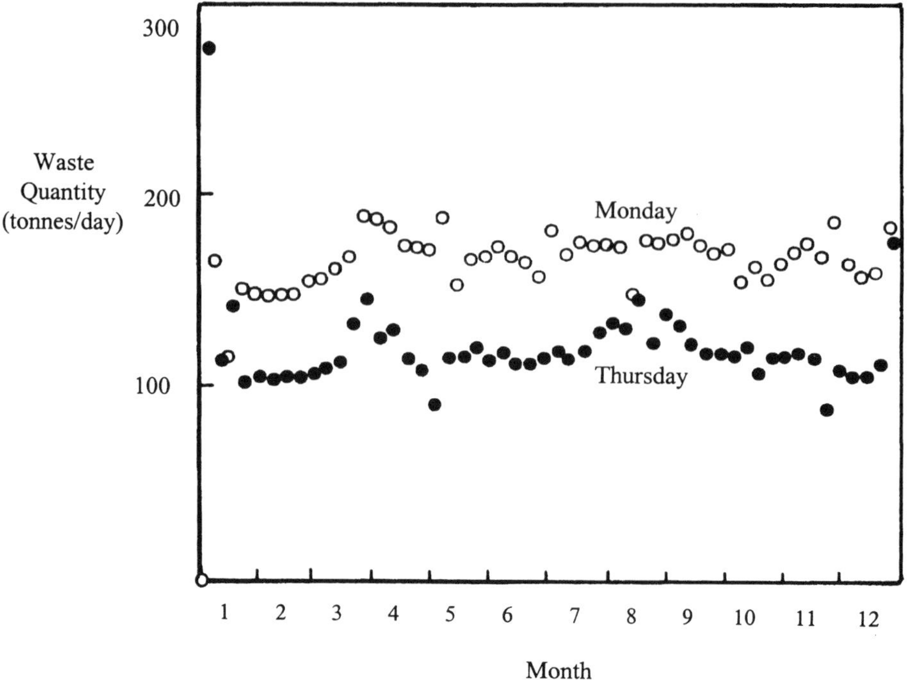

Figure 3.8 *Daily collection of combustible waste in Kita-Ku District, Sapporo, Japan, 1989. Source: Matsuto T. and Tanaka N., Waste Management and Research, 11, 333–343, 1993. Reproduced by permission from ISWA – The International Solid Waste Association*

countries (Environmental Indicators 1991; Stanners and Bourdeau 1995). The USA has more than double the per capita generation of waste than many European countries, including the UK. These data should be compared with those from the developing world, where waste generation weights tend to be low. For example, the annual per capita production of municipal solid waste in Delhi, India, has been estimated at 136 kg/person/year, in Kathmandu, Nepal, at 109 kg/person/year, and in Wuhan, China, at 200 kg/person/year (Rushbrook and Finney 1988).

Figure 3.10 shows the historical change in household or domestic waste arisings in the UK from 1879 to 1990 (The Open University 1993). The dominant component of waste in the late part of the nineteenth and early part of the twentieth centuries was dust and cinder derived largely from the residues of the household coal fire. The dramatic fall in the contribution of dust and cinders arose from the move during the late 1950s and 1960s away from coal fires to the more convenient central heating systems fired by oil and gas. One of the major factors influencing the move way from coal was the introduction of the Clean Air Act of 1956, which sought to reduce the smoke and sulphur dioxide pollution from residential and also industrial sources. This lead directly to the replacement of the coal-fired living room fire by gas and electric fires. Vegetable and putrescible waste has grown from about 8% of domestic waste to about 28% in 1965, after which the percentage contribution has remained fairly constant at between 22 and 25%. The biggest increases in percentage contribution to

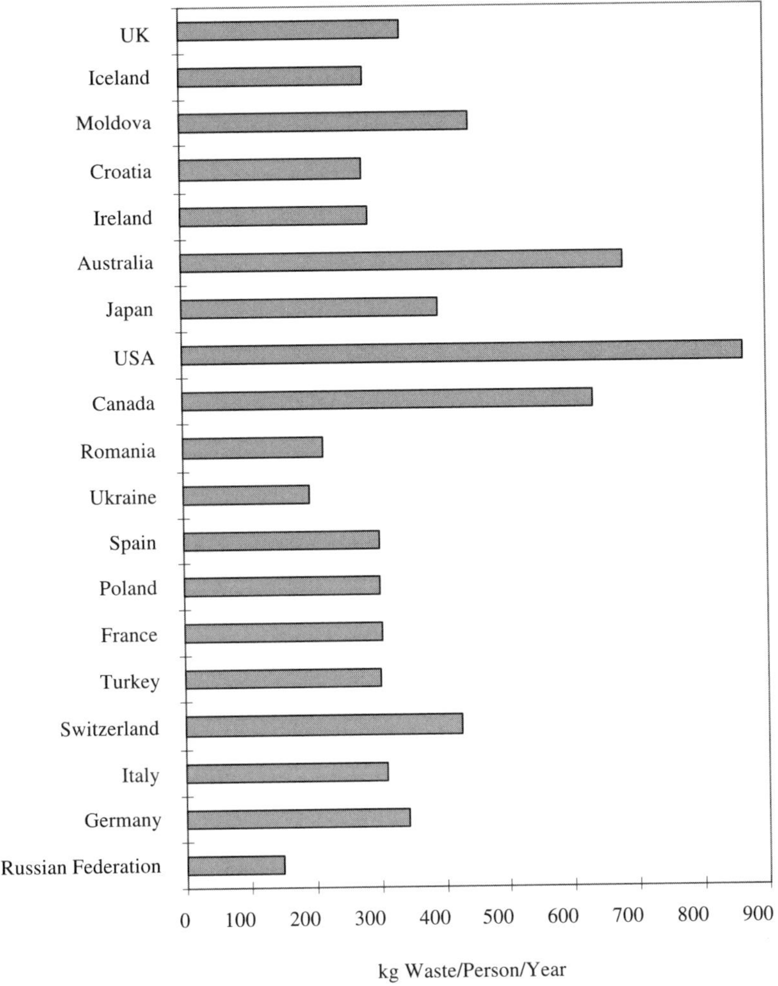

Figure 3.9 *Municipal solid waste generation in relation to per head of population. Sources: [1] Stanners D. and Bourdeau P., Europe's Environment, The Dobris Assessment. European Environment Agency, Copenhagen. Earthscan Publications, London, 1995; [2] Environmental Indicators, Organisation of Economic Cooperation and Development, Paris, 1991*

the domestic dustbin composition have been paper and board, metals, glass and plastics. These wastes are associated with the increases in packaging, newspapers and magazines, plastic bottles, tin cans etc., reflecting our change to a consumer society (The Open University 1993).

Comparison of UK municipal solid waste composition can be made with other countries. Compositions of municipal solid waste in some European countries are compared in Figure 3.11 (Stanners and Bourdeau 1995). In some cases there are large variations in waste compositions. This illustrates the point that the composition of

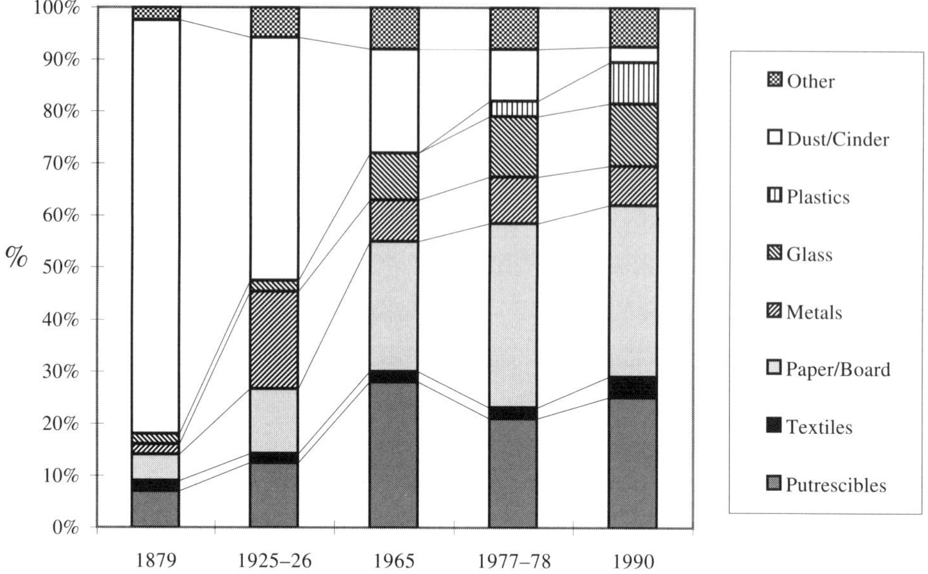

Figure 3.10 *Composition of UK domestic waste 1879–1990. Source: Adapted from Open University Course T237, Environmental Control and Public Health, Municipal Solid Waste Management. The Open University, Milton Keynes, 1993*

municipal solid waste is very dependent on the local conditions. Again, however, the comparison of waste compositions from different countries can be difficult since the methods of reporting and classification and the degree of recycling all influence the reported composition. The increasing trends in waste management are to waste minimisation, and re-use and recycling of waste. This will inevitably influence the quantities and composition of waste. Each town or city produces a different composition of waste, since the inputs depend on socio-economic factors, types of industry and level of industrialisation, geographic location, climate, level of consumption, collection system, population density, the extent of recycling, legislative controls and public attitudes.

The composition of municipal solid waste also has other factors to be considered, particularly where the waste is to be combusted in an incinerator. Incinerator operators are concerned with such aspects as the energy (calorific) value of the waste, the elemental (ultimate) analysis and the proximate analysis of the waste. Proximate analysis is the ash, moisture and combustible fraction of the waste, and in some cases the combustible fraction is subdivided into the volatiles and fixed carbon content. The calorific value may be reported as the as-received figure (gross calorific value), corrected for the moisture content (net or dry calorific value), or corrected for the moisture and ash content (dry, ash-free calorific value). All these factors are important in the design, operation and pollutant emissions of the incinerator. For

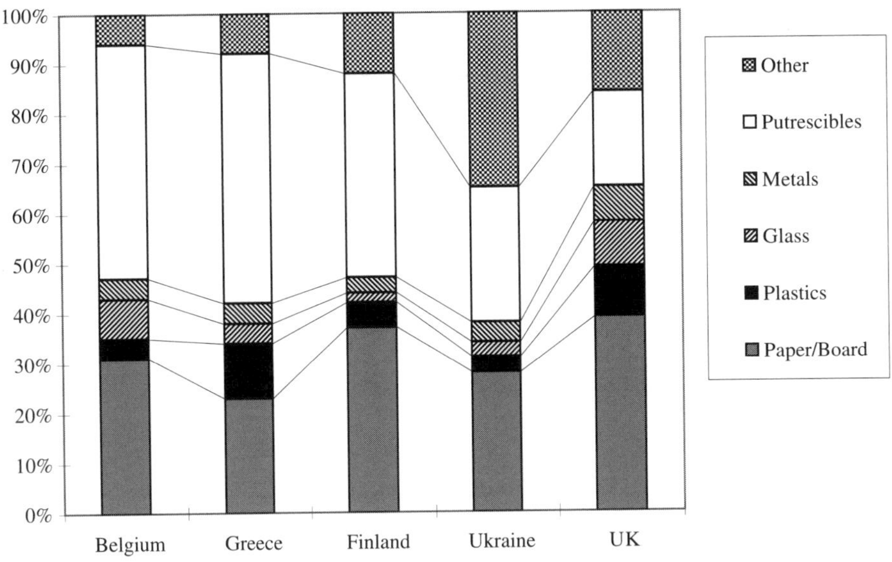

Figure 3.11 *Comparison of municipal solid waste composition in some European countries. Source: Stanners D. and Bourdeau P., Europe's Environment, The Dobris Assessment. European Environment Agency, Copenhagen. Earthscan Publications, London, 1995*

example, the incinerator grate, ceramic lining, furnace chamber and boilers would be designed to cope with a certain calorific value of waste. If the calorific value changed, potentially rapid corrosion of the boiler or ceramic lining could occur. In addition, the heat output from the plant would also be affected. Similarly, the incinerator flue gas clean-up system would be designed to cope with certain pollutants at estimated concentrations. If the nature of the waste changed then the resultant pollutants and their concentration could mean that authorised limits may be exceeded. For example, an increase in chlorinated waste in the incoming stream could lead to increases in hydrogen chloride or dioxins and furans. The typical calorific value of municipal solid waste is approximately 8500 kJ/kg (Royal Commission on Environmental Pollution 1993). This figure compares with about 30 000 kJ/kg for a typical coal and 42 000 kJ/kg for a typical fuel oil. Estimation of calorific value, and ultimate and proximate analyses of municipal solid waste are inherently inaccurate due to the heterogeneous nature of the waste. In addition, the majority of analyses of wastes are standard tests derived from solid fuel analyses such as coal, where the fuel is more homogeneous and sample sizes are typically one gram. Obtaining a representative one gram sample of a dustbin lorry full of waste is very difficult, even if strict sampling procedures and processing of the sample to a representative ground sample are followed. Larger scale (1 kg) instruments to determine the calorific value of waste are now available, specifically designed for waste analysis (Daborn 1988).

Table 3.13 Typical composition and calorific values of UK municipal solid waste

Component	% by weight	Calorific value (kJ/kg) (as received)
Paper/board	33.0	16 900
Plastics	7.0	32 650
Glass	10.0	Nil
Metals	8.0	Nil
Food/garden	20.0	9000
Textiles	4.0	15 580
Other	18.0	10 600

Sources: [1] Waste Management Paper 28, Recycling. Department of the Environment, HMSO, London, 1992; [2] Kaiser E., In: Combustion and Incineration Processes. Niessen, W.R. (Ed.), Dekker, New York, 1978.

Table 3.14 Ultimate and proximate analysis of typical UK municipal solid waste

Ultimate analysis		Proximate analysis	
Carbon	21.5	Combustibles	42.1
Hydrogen	3.0	Moisture	31.0
Oxygen	16.9	Ash	26.9
Nitrogen	0.5		
Sulphur	0.2		

Source: Buekens A. and Patrick P.K., Incineration. In: Solid Waste Management; Selected Topics. Suess M.J. (Ed.) World Health Organisation, Regional Office for Environmental Health Hazards, Copenhagen, Denmark, 1985.

Tables 3.13 and 3.14 show some typical proximate, ultimate and calorific value analyses for the various components of municipal solid waste (Kaiser 1978; Buekens and Patrick 1985; Waste Management Paper 28 1992). Some components have a high calorific value, for example, plastic material at approximately 40 000 kJ/kg, and although present in low concentration in the waste will have a significant influence on the overall calorific value. The volatile, ash and moisture content will vary considerably depending on the source of the waste. For example, commercial waste with a high plastic content will produce a high evolution of combustible volatiles with negligible ash and moisture. In addition, the day to day rainfall influences how wet the waste is and consequently the measured moisture content. Table 3.15 shows a typical breakdown of the components of UK household waste represented by Warrington (National Household Waste Analysis Project 1994). The data are particularly useful for recycling and re-use strategies. For example, the economic potential of recycling of plastic bottles, glass and metal cans can be more easily and accurately assessed. In addition, the potential of the waste in terms of combustibility in an incinerator may also be assessed. Table 3.15 also shows that the

Table 3.15 *Typical composition of UK household waste (Warrington 1992, ACORN B type housing – Modern family housing, higher incomes)*

Category	Sub-category	Composition: Weight percent		Composition: kg/household
Paper/card	Newspapers	27.4	8.6	4.4
	Magazines		4.3	
	Other paper		6.8	
	Liquid cartons		0.6	
	Card packaging		3.3	
	Other card		3.8	
Plastic film	Refuse sacks	3.8	0.3	0.6
	Other plastic film		3.5	
Dense plastic	Clear beverage bottles	6.0	0.6	1.1
	Coloured beverage bottles		0.1	
	Other bottles		1.0	
	Food packaging		1.6	
	Other dense plastic		2.7	
Textiles	Textiles	1.9	1.9	0.3
Miscellaneous Combustibles	Disposable nappies	10.3	7.1	1.7
	Other misc. combustibles		3.2	
Miscellaneous Non-combustibles	Miscellaneous Non-combustibles	4.6	4.6	0.8
Glass	Brown glass bottles	9.7	0.4	1.7
	Green glass bottles		1.5	
	Clear glass bottles		2.2	
	Clear glass jars		1.8	
	Broken glass		3.8	
Putrescible	Garden waste	21.7	3.6	3.6
	Other putrescibles		18.1	
Ferrous metal	Beverage cans	6.1	0.8	1.0
	Food cans		4.2	
	Batteries		–	
	Other cans		0.3	
	Other ferrous		0.8	
Non-ferrous metals	Beverage cans	1.7	0.4	0.3
	Foil		0.5	
	Other non-ferrous metals		0.8	
Fines	10 mm fines	6.8	6.8	1.1
Total	100.0	100.0	100.0	16.6
Collection details:	Number of properties	231	Bulk moisture	31.9
	Total weight collected	3840 kg	Ash	16.15%
	Mean weight/household	16.6	Gross CV	10.93 MJ/kg

Source: The technical aspects of controlled waste management: National household waste analysis project, Phase 2, Vol. 1. Report on composition and weight data. Report No. CWM/082/94. Department of the Environment, Wastes Technical Division, HMSO, London, 1994.

Table 3.16 Proximate analysis of municipal solid waste components (%)

Component	Moisture	Volatiles	Fixed carbon	Ash
Paper/paper products				
Paper – mixed	10.24	75.94	8.44	5.38
Newsprint	5.97	81.12	11.48	1.43
Corrugated boxes	5.20	77.47	12.27	5.06
Plastic coated paper	4.71	84.20	8.45	2.64
Waxed milk cartons	3.45	90.92	4.46	1.17
Junk mail	4.56	73.32	9.03	13.09
Food/garden waste				
Vegetable food waste	78.29	17.10	3.55	1.06
Meat scraps (cooked)	38.74	56.34	1.81	3.11
Fried fats	0.00	97.64	2.36	0.00
Lawn grass	75.24	18.64	4.50	1.62
Leaves	9.97	66.92	19.29	3.82
Green logs	50.00	42.25	7.25	0.50
Evergreen shrubs	69.00	25.18	5.01	0.81
Flowering plants	53.94	35.64	8.08	2.34
Wood and bark	20.00	67.89	11.31	0.80
Household waste				
Leather shoe	7.46	57.12	14.26	21.16
Rubber	1.20	83.98	4.94	9.88
Upholstery	6.90	75.96	14.52	2.62
Polystyrene	0.20	98.67	0.68	0.45
PVC	0.20	86.89	10.85	2.06
Linoleum	2.10	64.50	6.60	26.80
Rags	10.00	84.34	3.46	2.20
Vacuum cleaner dirt	5.47	55.68	8.51	30.34

Source: Kaiser E. In: Combustion and Incineration Processes. Niessen, W.R. (Ed.) Dekker, New York, 1978.

more readily recyclable newspapers and magazines make up the bulk of the paper and card composition of household waste with a figure of 27.4 wt%. The plastic components of waste are almost 10 wt% in total from the Warrington analysis, representing a major and increasing component. Interestingly, disposable nappies make up a significant proportion of the composition of household waste, and also make a significant contribution to the overall calorific value of the waste. Putrescible material and fines also have a certain combustible fraction. The glass content of the waste approaches 10 wt%, and together with the metals component represents a significant proportion of the waste which can be readily recycled.

Detailed analyses of the components of municipal solid waste (Tables 3.16–3.18) show that the calorific value, ultimate and proximate analyses of different sorts of paper, food and garden waste can be very different (Kaiser 1978). Other components that may occur in domestic waste, such as rubber and plastic, can also significantly influence the properties of the waste.

More detailed analyses of household waste have been undertaken for the UK National Household Waste Analysis Project (1995) for the wastes Technical

Table 3.17 *Ultimate analysis (dry) of municipal solid waste components (%)*

Component	Carbon	Hydrogen	Oxygen	Nitrogen	Sulphur
Paper/paper products					
Paper – mixed	43.41	5.82	44.32	0.25	0.20
Newsprint	49.14	6.10	43.03	0.05	0.16
Corrugated boxes	43.73	5.70	44.93	0.09	0.21
Plastic coated paper	45.30	6.17	45.50	0.18	0.08
Waxed milk cartons	59.18	9.25	30.13	0.12	0.10
Junk mail	37.87	5.41	42.74	0.17	0.09
Food/garden waste					
Vegetable food waste	49.06	6.62	37.55	1.68	0.20
Meat scraps (cooked)	59.59	9.47	24.65	1.02	0.19
Fried fats	73.14	11.54	14.82	0.43	0.07
Lawn grass	46.18	5.96	36.43	4.46	0.42
Leaves	52.15	6.11	30.34	6.99	0.16
Green logs	50.12	6.40	42.26	0.14	0.08
Evergreen shrubs	48.51	6.54	40.44	1.71	0.19
Flowering plants	46.65	6.61	40.18	1.21	0.26
Wood and bark	50.46	5.97	42.37	0.15	0.05
Household waste					
Leather shoe	42.01	5.32	22.83	5.98	1.00
Rubber	77.65	10.35	–	–	2.00
Upholstery	47.10	6.10	43.60	0.30	0.10
Polystyrene	87.10	8.45	3.96	0.21	0.02
PVC	45.14	5.61	1.56	0.08	0.14
Linoleum	48.06	5.34	18.70	0.10	0.40
Rags	55.00	6.60	31.20	4.12	0.13
Vacuum cleaner dirt	35.69	4.73	20.08	6.26	1.15

Source: Kaiser E., In: Combustion and Incineration Processes. Niessen, W.R. (Ed.) Dekker, New York, 1978.

Division of the Department of the Environment. Table 3.19 shows a detailed chemical analysis of four waste samples associated with the four ACORN Groups of B, C, F and J (B, modern family housing, higher incomes; C, older terrace housing; F, council estates, Category II; J, affluent suburban housing). This analysis of household waste shows calorific values ranging from 6.94 to 8.95 MJ/kg (6940–8950 kJ/kg). The ash contents are high, and can be compared with coal at typically 10–15 wt% ash content.

In addition to ultimate and proximate analyses, Table 3.19 also shows detailed analyses of halides and heavy metals. The chlorine content of household waste is significant at between 0.21 and 0.97% and has implications for the formation of hydrogen chloride, which condenses to form hydrochloric acid, and for the formation of dioxins and furans if the waste is combusted. Similarly, fluoride may result in the formation of very corrosive hydrogen fluoride. The chloride and fluoride are derived from plastics such as polyvinyl chloride (PVC) and polytetrafluoroethylene (PTFE). However, paper products also have a significant concentration of chloride. Some of the heavy metals found in household waste are in

Table 3.18 Calorific values (CVs) of municipal solid waste components (kJ/kg)

Component	As received	Dry	Moisture/ash free
Paper/paper products			
Paper – mixed	15 750	17 530	18 650
Newsprint	18 550	19 720	20 000
Corrugated boxes	16 380	17 280	18 260
Plastic coated paper	17 070	17 910	18 470
Waxed milk cartons	26 350	27 290	27 660
Junk mail	14 160	14 830	17 210
Food/garden waste			
Vegetable food waste	4170	19 230	20 230
Meat scraps (cooked)	17 730	28 940	30 490
Fried fats	38 300	38 300	38 300
Lawn grass	4760	19 250	20 610
Leaves	18 490	20 540	21 460
Green logs	4870	9740	9840
Evergreen shrubs	6270	20 230	20 750
Flowering plants	8560	18 580	19 590
Wood and bark	19 570	19 940	20 140
Household waste			
Leather shoe	16 770	18 120	23 500
Rubber	25 930	26 230	29 180
Upholstery	16 120	17 320	17 800
Polystyrene	38 020	38 090	38 230
PVC	22 590	22 640	23 160
Linoleum	18 870	19 240	26 510
Rags	15 970	17 720	18 160
Vacuum cleaner dirt	14 790	15 640	23 060

Source: Kaiser E., In: Combustion and Incineration Processes. Niessen, W.R. (Ed.) Dekker, New York, 1978.

significantly high concentrations, for example, iron, copper, zinc, manganese and lead. Of lower concentrations but of higher toxicity are cadmium and mercury, which are mainly derived from batteries.

Table 3.19 shows the variability of chemical analyses of waste from different sources. For example, the ash content shows a greater than 100 wt% variation depending on the type of household sampled. Other elements showing a wide variation in the four samples shown in Table 3.19 are aluminium, calcium, chromium, iron, nickel, arsenic, cadmium and mercury.

3.3.1 Treatment and Disposal Options for Municipal Solid Waste

The options for the treatment and disposal of household and commercial waste in the UK have largely been confined to landfill, which constitutes about 90% of the total for household waste (representing about 18 million tonnes) and 85% of

Table 3.19 *Analysis of bulk waste samples as received from ACORN group classifications in Leeds*

Element	Sample B	C	F	J
Moisture content (%)	37.8	41.4	38.7	37.0
Ash (%)	27.35	14.7	27.28	29.17
Gross calorific value (MJ/kg)	6.94	8.95	7.73	7.76
Fixed carbon (%)	4.59	5.96	4.49	4.1
Volatile matter (%)	30.24	38.32	29.5	29.4
Carbon (%)	20.04	22.76	19.74	20.38
Hydrogen (%)	2.93	3.23	2.73	2.82
Nitrogen (%)	0.40	0.67	0.89	0.72
Sulphur (%)	0.06	0.09	0.05	0.06
Chlorine (%)	0.25	0.31	0.21	0.97
Bromide (%)	0.01	0.01	0.01	0.01
Fluoride (%)	0.01	0.01	0.01	0.01
Silicon (%)	4.13	1.53	4.04	3.74
Mercury (p.p.m.)	0.03	0.12	0.02	0.03
Sodium (p.p.m.)	11783	2220	7977	12022
Magnesium (p.p.m.)	2721	1737	2056	1908
Aluminium (p.p.m.)	12243	14920	8828	8532
Potassium (p.p.m.)	2277	2788	4185	3643
Calcium (p.p.m.)	13116	7821	17251	11723
Chromium (p.p.m.)	48	24	37	61
Manganese (p.p.m.)	383	374	372	572
Iron (p.p.m.)	26609	3217	73066	11127
Nickel (p.p.m.)	24	17	28	105
Copper (p.p.m.)	100	71	44	41
Zinc (p.p.m.)	149	240	201	313
Arsenic (p.p.m.)	2.4	1.8	6.4	10.0
Molybdenum (p.p.m.)	2.8	2.4	1.95	2.75
Silver (p.p.m.)	0.73	0.3	0.62	4.58
Cadmium	0.4	0.5	0.59	1.81
Antimony (p.p.m.)	1.8	0.68	3.4	3.7
Lead (p.p.m.)	84	33	41	247

B – Modern family housing, higher incomes
C – Older terrace housing
F – Council estates – Category II
J – Affluent suburban housing

Source: The technical aspects of controlled waste management. National household waste analysis project, Phase 2, Vol. 3. Chemical analysis data. Report No. CWM/087/94, Department of the Environment, Wastes Technical Division, HMSO, London 1994. Crown copyright is reproduced with the permission of the Controller of Her Majesty's Stationery Office.

commercial waste (representing about 13 million tonnes) (Department of the Environment and Welsh Office 1995). Incineration of household waste is a much under-utilised option, with only 5% of household waste (1 million tonnes) and 7.5% of commercial waste (1.1 million tonnes) being treated in this way. Recycling and re-use of household and commercial waste are similarly under-utilised options, with only 5% (1 million tonnes) and 7.5% (1.1 million tonnes) falling into these

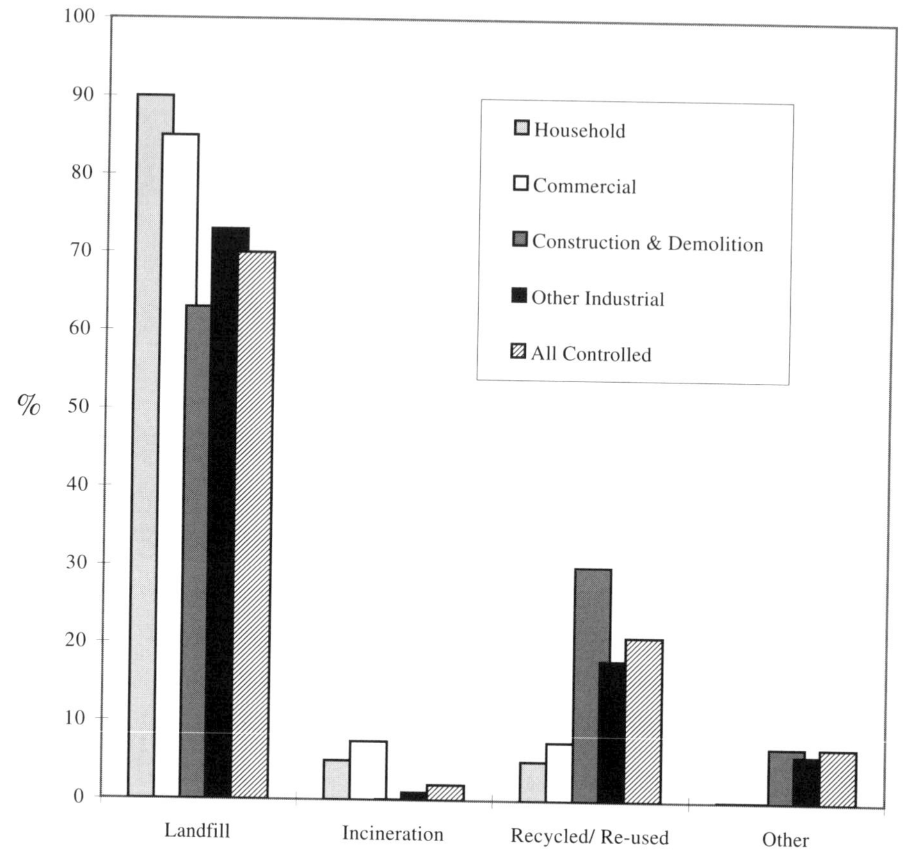

Figure 3.12 *Proportion of UK waste, landfilled, incinerated and recycled by sector. Source: Making Waste Work, Department of the Environment and Welsh Office, HMSO, London, 1995. Crown copyright is reproduced with the permission of the Controller of Her Majesty's Stationery Office*

categories. Figure 3.12 shows a breakdown of the wastes and their treatment and disposal routes (Department of the Environment and Welsh Office 1995).

Landfill is chosen as the most suitable option in most cases because of its low cost, its ready availability and its applicability for a wide range of wastes. Landfill can also be regarded as an environmentally acceptable method of waste disposal, for example where the holes produced from quarries and mineral workings are infilled with waste to produce restored landscapes. In addition, where the landfill gas produced from the normal biodegradation of the organic waste in the site can be utilised for energy recovery, landfill can become the best practicable environmental option. Landfill sites are classified as those accepting inert wastes, biodegradable landfill sites, and co-disposal sites. Inert wastes such as demolition waste will not decompose and produce pollutants. Biodegradable sites accept wastes such as household waste, which will biodegrade within the landfill, and modern sites encourage the biodegradation process by leachate recirculation to produce a 'bioreactor'. Co-disposal is used where industrial and hazardous wastes are co-disposed with biodegradable wastes to encourage the breakdown of all the wastes in the landfill.

The promotion of a sustainable waste management strategy by the UK Government has led to initiation of targets to minimise the production of waste by the use of newer technologies and processes, to minimise the proportion of waste going to landfill and to encourage re-use and recycling of waste. The UK Government has set a target of 40% recovery of municipal solid waste (household, commercial and some industrial waste) by the year 2005 (Department of the Environment and Welsh Office 1995). Recovery in this instance should include household waste composting, and glass, paper and plastic recycling.

One form of recycling encouraged by the UK Government is energy from waste, but the majority of landfill sites in the UK are old and do not have means of using the derived landfill gas for energy. Similarly, only six of the municipal waste incinerators built in the 1960s and 1970s had any form of energy recovery through district heating or electricity generation. The incinerator system was often seen at that time as a disposal route only. However, the majority of landfill sites and modern incinerators designed and engineered today have energy recovery as a essential component of the system, to derive economic benefit from the disposal of the waste.

Incineration utilises the energy content of the waste to produce steam from boilers, which in turn is used for district heating systems or to produce electricity for sale into the national grid. In addition, incineration reduces the waste to about 10% of its original volume and to 30% of its original weight. The residue is then landfilled. Even the residue can be recycled, for example to remove metals. However, the costs of incineration are high. The costs derive from the high plant costs, since the incinerator consists not only of the combustion unit to burn the waste, but also of a very high efficiency, high cost, gas clean-up system. High efficiencies of clean-up are required to meet stringent EC and UK emission limits. A typical modern incinerator, for example, might have electrostatic precipitators to remove particulate material, a wet or dry lime gas scrubbing system to remove acid gases such as hydrogen chloride and sulphur dioxide, and a bag filter system with activated carbon additive to remove fine particulate matter, heavy metals and dioxins.

Municipal solid waste has a number of components which can be readily separated either at source or at the treatment plant. A major component of municipal solid waste is packaging waste, consisting of paper and board, aluminium cans, glass, plastics and steel food cans from household waste, and metal and plastic drums, wooden pallets and board and plastic crates and containers from the commercial and industrial sectors. The EC Directive on Packaging and Packing Waste set a target for recovering 50–65% with a recycling target of 25–45% by the year 2001. To comply with the Directive, the UK Government has set a target of 50% recycling and 8% recovery through energy recovery (Department of the Environment and Welsh Office 1995). The EC Directive also requires the implementation of national programmes for the prevention of packaging waste, and the principle of producer responsibility means that the packaging industry must contribute to the costs of recycling and recovery.

A number of household waste recycling schemes for plastic and glass bottles, paper and metal cans have been established throughout the UK. However, the household waste recycled in the UK is only about 5% and commercial waste recycling is only 7.5% (Department of the Environment and Welsh Office 1995).

Other treatment and disposal options for municipal solid waste include composting and anaerobic digestion, but in the UK these comprise less than 1% of the treatment options whilst it is more common in other countries (Wheatley 1990). Composting is the aerobic biological degradation of biodegradable organic waste such as garden and food waste by micro-organisms, whilst anaerobic digestion is the biological degradation of the organic components of the waste by different groups of micro-organisms which thrive under anaerobic conditions. Composting is a relatively fast biodegradation process, taking typically about 4–6 weeks to reach a stabilised product. Small-scale household composting has been carried out for many years, and large scale composting schemes using organic waste collected from parks and household garden waste collected from civic amenity sites are planned in the UK (Department of the Environment and Welsh Office 1995). The UK Government has set a target of composting 25% of all household waste by the year 2000 (Department of the Environment and Welsh Office 1995). Anaerobic digestion takes place in an enclosed, controlled reactor and produces a product gas rich in methane which can be used to provide a fuel or act as a chemical feedstock. In addition, the solid residue arising from anaerobic digestion can be cured and used as a fertiliser. The biodegradable fraction of the waste requires separation from the other components. Biodegradation takes place in a slurry of the separated organic waste and micro-organisms.

Table 3.20 shows a comparison of the disposal routes for municipal solid waste for other countries (Warmer Bulletin 44 1995; Department of the Environment and Welsh Office 1995). Some countries such as Greece, Canada, Ireland, Finland and Portugal are equally dependent on landfill as the UK. Other countries rely heavily on incineration, such as Japan and Luxembourg. Whilst other countries such as Denmark, the Netherlands and Switzerland have well-developed recycling policies with significant proportions of the municipal solid waste being recycled or composted. In many cases, individual country characteristics dictate the options chosen for municipal waste treatment and disposal. For example, the Netherlands is a densely populated country with a limited landfill capacity, and incineration and waste reduction and recycling are the preferred options. Japan is a highly mountainous and volcanically active country with only approximately 10% of the land being suitable for residential purposes and consequently there is little land available for waste landfill. Waste management policies in Japan are based on waste reduction and recycling to minimise the amount of material that is ultimately destined for landfill, and the main route for waste disposal is incineration either with or without energy recovery. Municipal waste disposal is the responsibility of each municipality in Japan, and each is responsible for the preparation of a waste disposal plan for waste treatment and disposal.

Sweden has a similar policy of municipal responsibility for waste treatment and disposal, and implementation is through a waste disposal plan. Like most countries, the emphasis in Sweden is on waste reduction and recycling; waste incineration incorporates energy recovery, whilst landfilling is in larger sites which increases the possibilities of materials recycling on-site and the development of landfill gas

Table 3.20 Comparison of waste disposal routes (%)

Country	Incineration	Landfilled	Composting	Recycling
Austria	11	65	18	6
Belgium	54	43	0	3
Canada	8	80	2	10
Denmark	48	29	4	19
Finland	2	83	0	15
France	42	45	10	3
Germany	36	46	2	16
Greece	0	100	0	0
Ireland	0	97	0	3
Italy	16	74	7	3
Japan	75	20	5	–[1]
Luxembourg	75	22	1	2
Netherlands	35	45	5	16
Norway	22	67	5	7
Portugal	0	85	15	0
Spain	6	65	17	13
Sweden	47	34	3	16
Switzerland	59	12	7	22
UK	6	88	0	6
USA	16	67	2	15

[1] Japan MSW levels are calculated after removal of recyclable materials.

Source: Warmer Bulletin Information Sheet. The World Resource Foundation, The Warmer Bulletin, 44, February, 1995. Reproduced by permission from the World Resource Foundation. UK data taken from Making Waste Work, Department of the Environment and Welsh Office, Department of the Environment and Welsh Office, HMSO, London, 1995.

utilisation. Canada's waste disposal is primarily managed through municipal and regional government and is dominated by the option of landfill, due in part to the ready availability of suitable cheap land. In addition, there are relatively cheap sources of energy via petroleum, hydroelectricity and nuclear electricity which lowers the incentive for energy from waste schemes such as incineration. The USA has a large proportion of its municipal solid waste going to landfill, with a progression to larger and larger sites. There are strategies in the USA to recover more energy from waste, and it is estimated that incineration of municipal solid waste will increase to 25% by the year 2000. In some countries such as Greece, Ireland and Portugal there is no incineration of municipal solid waste at all (Bonomo and Higginson 1988; Royal Commission on Pollution 1993; Department of the Environment and Welsh Office 1995; Warmer Bulletin 44 1995; Waste Management in Japan 1995).

3.4 Hazardous Waste

Hazardous waste is a term used throughout the world as a term for waste which is dangerous or difficult to keep, treat or dispose of, and which may contain substances which are corrosive, toxic, reactive, carcinogenic, infectious, irritant or harmful to human health, and which also may be toxic to the environment. In the UK hazardous waste has been termed 'special waste' and is subject to the Special Waste Regulations 1996 which replaced the Control of Pollution (Special Wastes) Regulations, 1980. The 1996 Regulations update the 1980 regulations to conform with European legislation and to unify the definition of hazardous waste throughout the European Community (Council Directive 91/689/EEC, 1991). The Regulations list over 200 different types of special waste, reproduced from the list of hazardous wastes in Council Decision 94/904/EC (1994), which establishes the EC list of hazardous wastes in connection with the EC Directive on Hazardous Wastes. In addition, any wastes which have certain hazardous properties, such as being explosive, highly flammable, toxic, carcinogenic, mutagenic etc., are also termed special wastes. A range of inorganic and organic compounds which are hazardous to health or may pose a physical hazard are shown in Table 3.21 (Woodside 1993). The hazardous materials are derived from a variety of industries. Many industries use hazardous materials in their processes, and consequently hazardous waste may be generated as part of the waste stream. Contaminated soils may also be designated as hazardous waste if they contain hazardous materials such as heavy metals or pesticides, or if they are contaminated with tars, oils or other organic materials from old industrial sites, for example, from old gasworks.

The OECD estimates of arisings of hazardous waste in the UK are 4.5 million tonnes (Table 3.22, Yakowitz 1993), of which approximately 2.1 million tonnes (1993/4 data) are designated as special wastes subject to the consignment note system of the Special Waste Regulations (Digest of Environmental Statistics 1996). The difference represents waste treated or disposed of at the point of production. In relation to the total tonnage of waste arising in the UK these tonnages are small, but because of their associated hazard they represent difficult and expensive wastes to treat. The chemical and primary metal industries produce the majority of the wastes, and 80% of the wastes generated are sludges or liquids. Estimates of hazardous waste arisings have shown an approximate 50% rise in quantity in the last 10 years (Digest of Environmental Statistics 1996). Whilst the trends are for increased arisings of special waste, reflecting a general increase in production of such waste, the increase may also represent an increase in the reporting of such wastes because there is now greater awareness by waste producers of their responsibilities and the influence of increased regulation. Estimates of special waste in the UK are made from the associated consignment notes which are required for the handling of special wastes. Large year to year fluctuations in arisings in the UK have occurred because of infrequent disposal of dangerous materials and irregular arisings of contaminated soil from redevelopment. In addition, the influence of economic recessions has slowed the production of industry-related hazardous wastes. The realisation of waste disposal as a cost to industry has also been clearly recognised, and waste minimisation programmes have influenced the upward trend in waste production.

In addition to hazardous waste arisings in the UK, a significant quantity of hazardous waste is imported into the UK for treatment and disposal. The implementation of the EC Directive on Transfrontier Shipment of Waste has tightened up control of the movements of hazardous waste imports via a consignment note

Table 3.21 *Examples of hazardous chemicals and physical hazards*

[A] Health hazard	**Chemicals that create health hazards**
Carcinogen	Aldrin, formaldehyde, ethylene dichloride, methylene, dichloride, dioxin
Toxic	Xylene, phenol, propylene oxide
Highly toxic	Hydrogen cyanide, methyl parathion, acetonitrile, allyl alcohol, sulphur dioxide, pentachlorophenol
Reproductive toxin	Methyl cellosolve, lead
Corrosive	Sulphuric acid, sodium hydroxide, hydrofluoric acid
Irritant	Ammonium solutions, stannic chloride, calcium hypochlorite, magnesium dust
Sensitizer	Epichlorohydrin, fibreglass dusts
Hepatotoxin	Vinyl chloride, malathion, dioxane, acetonitrile, carbon tetrachloride, phenol, ethylenediamine
Neurotoxin	Hydrogen cyanide, endrin, mercury, cresol, methylene chloride, carbon disulphide, xylene
Nephrotoxin	Ethylenediamine, chlorobenzene, dioxane, acetonitrile, hexachlonaphthalene, allyl alcohol, phenol, uranium
Blood damage	Nitrotoluene, benzene, cyanide, carbon monoxide
Lung damage	Asbestos, silica, tars, dusts
Eye/skin damage	Sodium hydroxide, ethylbenzene, perchloroethane, allyl alcohol, nitroethane, ethanolamine, sulphuric acid, liquid oxygen, phenol, propylene oxide, ethyl butyl ketone
[B] Physical hazard	**Chemicals that create the hazard**
Combustible liquids	Fuel oil, crude oil, other heavy oils
Flammable materials	Gasoline, isopropyl alcohol, acetone, butane spray cans
Explosives	Dynamite, nitroglycerine, ammunition
Phyrophoric	Yellow phosphorus, white phosphorus, superheated toluene, silane gas, lithium hydride
Water reactive	Potassium, phosphorus pentasulphide, sodium hydride
Organic peroxides	Methyl ethyl ketone peroxide, dibenzoyl peroxide, dibutyl peroxide
Oxidisers	Sodium nitrate, magnesium nitrate, bromine, sodium permanganate, calcium hypochlorite, chromic acid

Source: Woodside G., Hazardous Materials and Hazardous Waste Management: A Technical Guide. Wiley Interscience, New York, 1993. Copyright © 1993 John Wiley and Sons. Reprinted by permission from John Wiley and Sons Ltd.

system, and a more accurate estimation of the amounts involved has occurred in recent years. Of the order of 68 000 tonnes of hazardous waste were imported into the UK in 1993–4 (Digest of Environmental Statistics 1996). This waste includes inorganic and organic acids, toxic metal compounds, organic and inorganic compounds, PCBs and PCB-contaminated wastes, polymeric materials, fuels, oils and greases, paints and hospital waste. The majority of the waste originates in Germany, Ireland, the Netherlands, Luxembourg and Italy, which together represented approximately 90% of the imports of hazardous waste to the UK in 1993–4. Of the imported waste, approximately 40% is physically or chemically treated, including solidification, and 60% is incinerated (Digest of Environmental Statistics 1996).

A further category of hazardous waste which has significance for local authority waste management is household hazardous waste. Household hazardous waste

Table 3.22 *Inter-country comparisons of hazardous waste arisings within the OECD*

Country	Hazardous waste (thousand tonnes)	Generation population (kg/person)	Generation GDP (kg/1000$)
Austria	200	26.4	1.71
Belgium	915	93.0	6.59
Denmark	112	21.8	1.11
Finland	270	54.7	3.02
France	3000	53.9	3.41
Germany	6000	98.0	5.37
Greece	423	42.3	8.96
Iceland	5	20.3	0.93
Ireland	20	5.6	0.68
Italy	3800	66.3	5.01
Luxembourg	4	10.7	0.67
Netherlands	1500	102.3	7.04
Norway	200	47.7	2.42
Portugal	165	16.1	4.50
Spain	1708	44.0	5.91
Sweden	500	59.5	3.15
Switzerland	400	60.4	2.34
Turkey	300	5.7	4.45
UK	4500	79.0	6.72
OECD Europe	24022	–	–
Canada	3290[1]	128.3	6.86
USA	275000[2]	1127.4	57.23
Australia	300	18.4	1.24
Japan	666	5.5	0.23
New Zealand	60	18.1	1.42

[1] Wet weight.
[2] USA figures include large quantities of dilute wastewaters classified as hazardous waste, whereas in Europe these are managed under water protection regulations.
OECD = Organisation of Economic Cooperation and Development, the members of which include all the above countries.

Source: Reprinted from Yakowitz H., Resources, Conservation and Recycling, 8, 131–178, 1993, with kind permission of Elsevier Science – NL, Sara Burgerhartstraat 25, 1055 KV Amsterdam, The Netherlands.

includes such materials as garden pesticides and herbicides, paints, medicines, oils, batteries, solvents and other materials which put human health or the environment at risk because of their chemical or biological nature. The waste in this category comprises about 5–10 kg/household/year. This is less than 1% of the total domestic waste, but it represents a disproportionately high risk. In addition, household hazardous waste, whilst occurring in small quantities, occurs over a large number of locations, adding to the difficulties of management. Much household hazardous waste is intermixed with other household waste and collected together, and may represent a disproportionate risk to human health and the environment unless correctly handled (Warmer Bulletin 50 1996). In addition, the options for treatment and disposal of such wastes may be limited. For example, high levels of heavy metals may rule out the option of composting the waste since the final product would have heavy metal concentrations outside regulated levels. Similarly, the presence of persistent pesticides could result in groundwater contamination if the mixed waste were landfilled (White et al 1995). Such wastes may be separated by the householder and collected from the household either on demand or via yearly or twice yearly collections or at civic amenity sites. Once sorted, such a concentration of hazardous wastes would be classified as 'special waste' and as such would be subject to the Special Waste Regulations 1996 with its consequent regulation of storage, transport, handling and treatment (White et al 1995; Special Waste Regulations 1996).

The European Commission has proposed that household hazardous waste be separately collected from households and other municipal sources (ENDS 1997). The proposal would also require manufacturers to mark such products with a special logo informing consumers not to dispose of them with ordinary household waste. The proposal lists several categories of such wastes, for example, paint, mineral and synthetic oils and their filters, medicines, aerosols, bleaches, batteries, solvents, adhesives etc. Many European countries, for example, Austria, Denmark, Finland, Germany, Luxembourg and the Netherlands, already have systems for the separate collection and/or delivery of at least some household hazardous waste (ENDS 1997).

Comparison of hazardous waste arisings with other countries throughout the world are shown in Table 3.22 (Yakowitz 1993). OECD Europe generates about 24 million tonnes of hazardous waste per year, or about 55 kg/person/year. As is the case in the UK, landfilling of wastes is by far the preferred option in OECD Europe for the disposal of hazardous waste, representing the disposal route for between 70 and 75% of the total. Approximately 1.5 million tonnes, representing between 5 and 8%, are incinerated, and between 1.5 and 4 million tonnes are subject to some recycling/reclamation or physico-chemical treatment (Yakowitz 1993). However, with increasing regulation, particularly the proposed EC Landfill Directive, waste minimisation and other routes of treatment and disposal such as incineration and physical and chemical treatment are predicted to increase. In addition, more stringent controls of on-site hazardous waste disposal are increasingly being adopted throughout Europe.

3.4.1 Treatment and Disposal Options for Hazardous Waste

Hazardous waste generated in the UK may be treated or disposed of on site at the source of production; the remainder is disposed of mainly through landfill sites by direct disposal and co-disposal with controlled waste and some incineration. Disposal of hazardous waste by sea dumping and also incineration at sea was an option until the early 1990s, after which EC regulation prohibited this route for a variety of wastes. The majority of hazardous waste treatment and disposal in the UK is carried out by the waste producer at source or by private contractors, and only a small quantity is dealt with by the Waste Disposal Authorities (Holmes 1995; Digest of Environmental Statistics 1996). Disposal of hazardous wastes into landfill is subject to additional regulation under the Special Wastes Regulations, in that accurate and permanent records of where in the landfill site the hazardous wastes are disposed of is part of the consignment note regulatory control.

Co-disposal of hazardous waste into landfill is the most common route to hazardous waste disposal in the UK, representing about 50% of the disposal of UK waste (Digest of Environmental Statistics 1996). The principle behind co-disposal is that the landfill site acts as a bio-reactor in which the micro-organisms in the municipal solid waste break down the components of the hazardous waste. There is some evidence that the leachate derived from such co-disposal sites is in fact quite similar to a municipal solid waste site leachate (Waste Management Paper 26F 1996, Ch. 5). The co-disposal of hazardous waste is regarded throughout Europe as a poor disposal option and the proposed EC Landfill Directive seeks to eliminate the practice. The drawback of co-disposal has centred on possible interaction of the wastes to produce a toxic product which may harm the population or environment. Other European countries use highly engineered containment landfills which use natural and synthetic polymer barriers to contain the waste and prevent leachate moving beyond the site boundary. Containment systems such as this require that there is minimal interaction of the barriers with the waste to avoid deterioration of the barrier system. The system has also been criticised in that containment is merely a long-term storage of the waste and does not treat or render the waste harmless (British Medical Association 1991).

Incineration of hazardous waste with energy recovery would be the preferred option for the sustainable disposal of hazardous waste. However, the flue gas emissions from the incinerators require extensive clean-up using a variety of systems such as electrostatic precipitators, scrubbers and bag filters to remove the potentially highly toxic pollutants. The flue gas treatment systems are expensive, and consequently disposal costs via incineration are high and can represent between 10 and 50 times the equivalent cost of landfill depending on the degree of hazard associated with the waste (Ch. 2). However, incineration for certain types of waste such as liquid organic wastes is legally the only option in some countries, including the UK. Incineration of the waste is regarded as a thermal treatment process. Hazardous waste incinerators have used rotary kilns as the preferred design, whilst other designs have included fluidised bed and vortex or other spray combustor type incinerators (Dempsey and Oppelt 1993; La Grega et al 1994). Rotary kiln technology involves a continuously rotating ceramically lined cylinder in which the waste is combusted. The kiln is tilted at an angle and the waste moves down the kiln until burn-out is complete. Where hazardous waste is combusted, a secondary chamber is also involved which combusts the derived gaseous products. Incineration of hazardous waste blended with other fuels in cement kilns is also practised in the UK, utilising high combustion temperatures (>1400 °C) and long residence times in the rotary kiln to destroy the waste. In addition, when chlorinated or fluorinated wastes are combusted, the large mass of alkaline clinker from the process absorbs and neutralises the acidic stack gases. The utilisation of wastes by the cement industry has mutual benefits in that the cement kiln process in very energy-intensive and the organic wastes which are often used have high calorific values (Benestad 1989; Holmes 1995).

The incineration of special or hazardous waste is subject to European Directive control. The Hazardous Waste Incineration Directive (European Directive 94/67/

EEC) applies from 1996 for new plant, and for existing plant from the year 2000. The Directive covers merchant and dedicated chemical waste incineration and hazardous grade clinical waste incineration (Waste Incineration. Environment Agency 1996). The regulations are stringent on the control of emissions to atmosphere, for example the allowable daily average limit for particulate emissions is 10 mg/m^3, volatile organic carbon (as carbon) 10 mg/m^3, hydrogen chloride 10 mg/m^3, mercury 0.05 mg/m^3 and dioxins 0.1 ng/m^3 (TEQ, toxic equivalent) (Waste Incineration. Environment Agency 1996).

Other treatment options for hazardous waste include chemical, biological and physical treatment, which account for approximately 30% of hazardous waste treatment in the UK (Digest of Environmental Statistics 1996).

Biological treatments use microbiological organisms to breakdown the components of the waste into less hazardous products. The ability of the organisms to breakdown the organic waste is dependent on the types of organic compounds present in the waste material. For example, easily biodegradable compounds include alkenes, alcohols and aldehydes, intermediate biodegradable compounds are alkanes, aromatic compounds and nitrogen-containing compounds, and difficult wastes to degrade are those containing halogenated compounds. Some inorganic compounds can also be treated in biological systems, for example, cyanides and metals such as lead and arsenic.

Biological treatment systems are categorised as either aerobic or anaerobic. Aerobic systems are oxygenated and the microbiological organisms convert the organic components of the hazardous waste into carbon dioxide and water. Anaerobic systems are devoid of oxygen and the organisms convert the waste into methane and carbon dioxide. The aerobic activated sludge process is widely used for the treatment of industrial wastewaters containing organic wastes, and also for sewage sludge. The process of waste and microbiological interaction takes place in suspension, and the reactants are continuously mixed in a bioreactor. The solid suspension may separate out as a sludge and will be recycled back to the bioreactor to maintain a suitable concentration. The sludge would have a typical residence time of 20–30 days in the bioreactor, after which it is removed for treatment by dewatering, thickening and final disposal, which is typically via landfill or incineration (Woodside 1993; La Grega et al 1994; Kim and Qi 1995). Other aerobic treatment systems are, for example, the supported sludge processes, where the biological organisms are supported on some sort of solid substrate in a bioreactor and the organic waste material is slowly passed over the solid bed of biological reactant. Because the residence time in the bioreactor is less than that of the activated sludge process, the organic waste may have to be recycled through the reactor several times to achieve the required low concentrations of organic material in the effluent.

Anaerobic processes utilise a different type of microbiological organism since the reacting environment is devoid of oxygen. The bioreactor used is similar to that used in the activated sludge process in sewage sludge treatment works, where the waste and the organisms interact in suspension. The organisms biodegrade the organic components of the waste via a two-step process, firstly converting the waste to organic acids, alcohols, carbon dioxide and water, followed by breakdown of the acids and alcohols to carbon dioxide and methane.

Physical, chemical and physico-chemical treatment processes have all been used for hazardous waste (Woodside 1993; La Grega et al 1994; Kim and Qi 1995). Physical treatment includes carbon adsorption, fractional distillation, solvent extraction and sedimentation. Carbon adsorption uses activated carbon with a very high active surface area, typically 1000–2000 m^2/g, to adsorb physically the organic material from solution or waste water. The waste components interact with the activated carbon either in suspension in a reactor, or as fixed filter beds of activated carbon through which the waste stream flows. Distillation relies on the different boiling points of organic material in the waste. Solvent wastes can be fractionated and separated into different purified organic compounds, or organic material can be separated from waste water streams. Similar in principle is evaporation, where volatile compounds can be separated from non-volatile compounds. Sedimentation is the separation of hazardous suspended fine solid material from solution via settling by gravity.

Chemical treatment includes neutralisation of acidic or alkaline wastes to produce an acidically neutral solution, concentration of the waste from solution by precipitation of the hazardous components, removal of inorganic material such as heavy metals from solution by ion exchange resins which selectively remove such components, and oxidation/reduction reactions to produce less hazardous components or less volatile waste by adding or subtracting electrons. Chemical treatments, which include incineration, have also included thermal processes such as wet-air oxidation, which involves treatment of organic wastes at high temperatures and pressures in the presence of steam, pyrolysis, which is the thermal degradation of organic wastes at high temperature in the absence of oxygen, and vitrification or solidification, where wastes such as contaminated soils and sludges are combined with silica-containing material at high temperature to form a glass.

Solidification and stabilisation are physico/chemical treatment methods. The treatment involves mixing liquid waste with solids to produce a waste material which is more easily handled, can be landfilled and is less susceptible to leaching. The treatment may be physical adsorption or chemical interaction depending on the type of waste and the type of solid material. For example, lime- or cement-based treatment of toxic metal ions has been used (Woodside 1993; La Grega et al 1994).

The treatment and disposal options used in OECD Europe for hazardous waste concentrate on landfill as the main route, with the other options being used to a lesser extent (Stanners and Bourdeau 1995). Landfilling of hazardous waste accounts for 68%, incineration 8%, physico-chemical treatment 10%, recovery 10% and others 4%. Visvanathan (1996) has reviewed the different options available for the treatment and disposal of hazardous waste.

3.5 Sewage Sludge

Sewage sludge is formed as a by-product of the treatment of raw sewage from domestic households, but may also include industrial and commercial effluent. Consequently the sludge composition can vary considerably depending on the main source of the sewage. Where industrial sewage systems contribute to the domestic waste loadings, significantly higher concentrations of heavy metals such as lead, zinc and copper, or high levels of soluble organic matter may result (Dean and Suess 1985; Hall 1992; Hudson 1993; Try and Price 1995; Digest of Environmental Statistics 1996). The heavy metals in particular may be carried through the treatment processes and end up in the final sewage sludge. Where such sludges are produced this limits the disposal options, since application to agricultural land via landspreading may not be allowable within existing UK and EC regulations. The sewage is mostly water, but after treatment, the particulate and colloidal matter is concentrated to form sewage sludge. The treatment of the sewage sludge utilises aerobic digestion in either the activated sludge process or the supported sludge process. The activated sludge process involves the interaction of the sewage with organic micro-organisms in suspension in a bioreactor. The suspension is continuously stirred and aerated by air bubbling through the bioreactor, and the settled sewage sludge is recycled until it has been fully biodegraded. The supported sludge process relies on the biological micro-organisms being supported on a solid substrate contained in a bioreactor through which the sewage waste water is trickled. The supported sludge system has a shorter residence time than the activated sludge process, and the sludge may therefore be recycled through the bioreactor to ensure sufficiently low concentrations of pollutants. The settled solid material is removed for final anaerobic digestion treatment, where the sludge is heated to about 35 °C for approximately 2 weeks until the organic material is broken down to reduce pathogen levels and reduce odours. Mechanical dewatering takes place to reduce the water content of the sludge product. The final material is termed treated sewage sludge, which is either spread on land, landfilled, incinerated or dumped at sea. However, sea dumping is to be phased out by 1998 (Bruce et al 1989; Wheatley 1990; Hall 1992).

Table 3.23 shows the estimated arisings of sewage sludge in the UK from 1985 to 1994 (Digest of Environmental Protection and Water Statistics 1993; Digest of Environmental Statistics 1996). The data are on a dry basis; the water content of raw sewage sludge is estimated as an average 96%, which translates to a total raw sewage sludge production in the UK of about 35 million tonnes per year (see Table 3.4). The production of sewage sludge in Europe is compared in Table 3.24, which represents the sludge as dry solid material (Brunner and Lichtensteiger 1989). The majority of sewage sludge in Europe is disposed of via landspreading or landfill, and in addition incineration disposes of significant concentrations in some countries.

The composition of sewage sludge is shown in Table 3.25 (Yokoyama et al 1987; Frost 1991). The sewage sludge has a high ash content, typically between 35 and 40% of the solid content. This has great significance for incineration processes, the efficient operation of the incinerator and the clean-up of the flue gases, which will contain high concentrations of particulate and associated heavy metals from the sludge. However, the calorific value is moderately high and sewage sludge can be successfully incinerated, even if there is a requirement for supplementary fuel. The nitrogen content of the sludge is high, representing the nutrient value of the sludge when it is used for landspreading. However, high concentrations of nitrogen can lead to high concentrations of polluting nitrogen oxides during incineration processes.

Table 3.23 *Estimated dry sewage sludge arisings: 1985–95*

Year	England and Wales	Scotland	N. Ireland
1985	884	–	–
1986	896	–	–
1987	900	–	–
1988	954	–	–
1989	958	107	27
1990	945	80	27
1991	945	80[1]	33
1992	907	89	32
1993	902	89[2]	33
1994	916	89[2]	33

[1] Estimated from 1990 figures.
[2] Estimated from 1992 figures.

Sources: [1] Digest of Environmental Protection and Water Statistics. Department of the Environment, HMSO, London 1993; [2] Digest of Environmental Statistics 1996, HMSO, London, 1996.

Table 3.24 *Estimated dry sewage sludge production in Europe*

Country	Sewage sludge production (thousand tonnes/year, dry solid)
Belgium	29
Denmark	150
Finland	140
France	600
Greece	15
Ireland	24
Italy	800
Luxembourg	15
Netherlands	250
Norway	82
Spain	210
Sweden	200
Switzerland	250
UK	1100

Sources: [1] Brunner P.H. and Lichtensteiger Th. In: Treatment of Sewage Sludge. Bruce A.M., Colin F. and Newman P.J. (Eds.) Elsevier Applied Science, London, 1989; [2] Making Waste Work, Department of the Environment and Welsh Office, HMSO, London, 1995.

Of more concern to the environmentally acceptable disposal of sewage sludge is the potential concentration of toxic inorganic and organic compounds. Table 3.26 shows the ranges of potentially toxic chemicals which have been shown to occur in raw sewage sludge (Dean and Suess 1985; Hall 1992). Discharges from industrial sources to sewers are subject to their own legislative control, and consequently heavy metal pollutants should be present only in low concentrations. Domestic sewage,

Table 3.25 *Typical composition and properties of sewage sludge*

Property	Typical value
Calorific value	21.3 MJ/kg (dry, ash free)
Ash content	37%
Composition of combustible fraction (dry, ash free)	
Carbon	53.0%
Hydrogen	7.7%
Oxygen	33.5%
Nitrogen	5.0%
Sulphur	0.8%
Organic composition (dry, ash free)	
Crude protein	30.0
Crude fat	13.0
Crude fibre	33.0
Non-fibrous carbohydrate	24.0

Sources: [1] Frost R., University of Leeds Short Course 'Incineration and Energy from Waste'. The University of Leeds, Leeds, UK, 1991; [2] Yokoyama S., Murakami M., Ogi T., Kouchi K. and Nakamura E., Fuel, 66, 1150–1155, 1987.

Table 3.26 *Concentration ranges of potentially toxic contaminants found in sewage sludge (parts per million by weight of dry solid)*

Inorganic contaminant	Concentration	Typical domestic concentration
Cadmium	2–1500	5
Copper	200–8000	380
Nickel	20–5000	30
Zinc	600–20 000	515
Lead	50–3600	120
Mercury	0.2–18	1.5
Chromium	40–14 000	50
Molybdenum	1–40	4
Arsenic	3–30	3
Selenium	1–10	2
Boron	15–1000	50
Fluorine	60–40 000	200
Organic contaminant		**Concentration**
Phthalic acid esters		1–100
Polycyclic aromatic hydrocarbons (PAH)		0.01–50.0
Polychlorinated biphenyls (PCB)		0.16–9.11
Dieldrin (pesticide)		0.018–3.90
Lindane (pesticide)		0.025–0.410
Aldrin (pesticide)		0.02–0.24
Dichlordiphenyltrichloroethane (DDT) (pesticide)		0.02–0.80

Sources: [1] Hall J.E. In: The Treatment and Handling of Wastes. Bradshaw A.D., Southwood R. and Warner F. (Eds.) Chapman and Hall, London, 1992; [2] Dean R.B. and Suess M.J., Waste Management and Research, 3, 251–278, 1985.

whilst containing many metals in significant concentrations, does not approach the levels from industrially derived sewage. Organic compounds occur naturally in sewage sludge, but those listed in Table 3.26 are synthesised compounds which may be resistant to the biodegradation sewage treatment process and may persist in the treated sludge and waste water.

3.5.1 Treatment and Disposal Options for Sewage Sludge

Most sewage sludge in the UK is spread on agricultural land (50%), incinerated (10%), landfilled (10%) or disposed of at sea (30%) (Department of the Environment and Welsh Office 1995). Disposal at sea is regulated by the Ministry of Agriculture, Fisheries and Food under the Food and Environment Protection Act 1985. The Ministry regulates the disposal sites, and the regulations stipulate the volumes of sludge and monitors and controls the levels of certain pollutants including heavy metals, PCBs (polychlorinated biphenyls) and certain pesticides. Following the EC Urban Waste Water Directive (91/271/EEC), the disposal of sewage sludge to sea is to cease by the end of 1998; alternative land-based disposal routes will have to be found. The disposal of sewage sludge to landfill and by incineration are subject to waste regulation as controlled waste and are covered by the 1974 Control of Pollution Act, 1990 Environmental Protection Act and 1995 Environment Act. When spread on agricultural land, the operations are subject to the provisions of Sludge Use in Agriculture Regulations 1989. Sewage treatment works are required to keep records of the quantities, composition and properties of the sludge and details of where it was spread. Where the source of the sewage has been derived from a largely industrial area, the resultant sewage sludge may contain high concentrations of heavy metals or other chemicals which are resistant to the biological treatment processes. Consequently, the use of such sludges for land spreading may not be appropriate and may exceed legislative limits. Whilst land spreading may not at first appear to be the best option for sewage, the sludge in fact contains valuable nutrients such as nitrogen, phosphorus and organic matter in high concentrations, and is a suitable supplement to other fertilisers. Typical application rates of sewage sludge to land are about 50–100 tonnes/hectare (Hall 1992; Try and Price 1995). However, in some cases untreated sewage sludge is used on agricultural land, resulting in high pathogen levels, very unpleasant odours and unwanted solid debris from the sewage system. The European Community Directive 86/278/EEC, implemented in the UK by the Sludge Use in Agriculture Regulations 1989 and 1990 amendment, and accompanying Code of Practice, seeks to promote the use of sewage sludge use on land but with more control on the impact on the environment. There is a requirement to test the sludge and soil to limit the concentration of certain heavy metals, namely, cadmium, mercury, chromium, zinc, nickel, copper and lead. The levels of nitrogen are also subject to limits. The use of untreated sewage sludge on land is prohibited unless the sludge is injected or worked into the soil. The use of sewage sludge in forestry areas has also been undertaken in the UK and has resulted in a Code of Practice to limit the heavy metal and nitrogen levels in forestry soils. The current level of application is

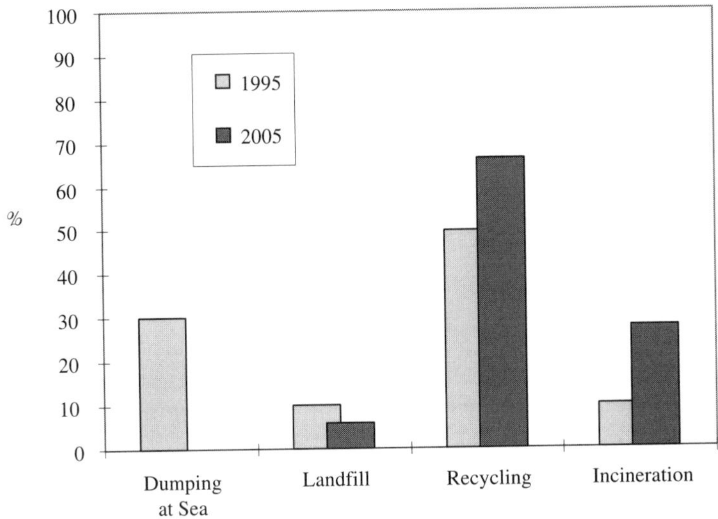

Figure 3.13 *UK waste management of sewage sludge in 1995 and predicted routes by 2005. Source: Making Waste Work, Department of the Environment and Welsh Office, HMSO, London, 1995. Crown copyright is reproduced with the permission of the Controller of Her Majesty's Stationery Office*

that about 1% of the total sewage sludge arising is landspread in forestry areas (Hall 1992; Try and Price 1995).

Landfill is used as a disposal option for sewage sludge, usually where there is no suitable agricultural land nearby for application or where the sludge poses a pollution hazard to the soil because of high concentrations of pathogens or heavy metals. Landfilling of sewage sludge accounts for approximately 40% of the disposal route in Europe (Dean and Suess 1985). The legislation covering landfill sites and the associated authorisations will limit the type and quantities of waste, including sewage sludge, which can be accepted by the landfill. The disposal of sewage sludge in landfill sites by co-disposal is encouraged in some cases since the high organic content of sewage sludge promotes the biological breakdown of other co-disposed wastes.

Incineration of sewage sludge in the UK is the final disposal route for approximately 10% of the total waste arising. This is predicted to rise to about 28% by 2005 (Figure 3.13; Department of the Environment and Welsh Office 1995) as a consequence of the EC Directive which prohibits the disposal of sewage sludge at sea (EC Urban Waste Water Directive, 91/271/EEC). The sludge has a high ash content, which can range between 20% and 50%, and which remains as a residue after incineration and requires disposal. The ash is generally landfilled because it will contain a high concentration of toxic heavy metals. Indeed in some European countries the ash is regarded as toxic waste and may be subject to special regulation (Hall 1992). In addition, the sewage sludge has a very high water content, and incineration of sewage sludge requires that the water content of the sludge be reduced to 70% or less, using mechanical dewatering and thermal drying treatments, resulting in increased processing costs (Frost 1991). Alternatively, support fuel can be added to aid incineration. The dry solid content of the sewage sludge where support fuel is not required is termed the autothermic solids content. In the UK, large scale sewage sludge incinerators are subject to Integrated Pollution Control as a scheduled process and require an authorisation from the Environment Agency. Smaller incinerators also require regulation, but for air pollution control only, by local authorities.

The technologies used for sewage sludge incineration in the UK have been based on multiple hearth designs, but in more recent plants fluidised bed incinerators have been used (Frost 1991). The plants involve a dewatering stage utilising centrifuge, filter belt presses, filter plates or membrane press systems. The solids content of the sludge is increased to between 24 and 95% depending on whether or not supplementary fuel is used, or in the case of fluidised bed systems, on the temperature of the incoming combustion air, which may be up to 600 °C. The advantages of fluidised bed systems are that the sludge can be pumped into the incinerator as a fluid. Combustion of the sludge takes place in the incinerator bed, usually sand, and the temperatures are typically 800–900 °C. The heat generated is usually used to preheat the incoming combustion air or to pre-dry the sewage sludge. Where excess heat is generated this may be used for steam generation for use either as electricity generation or for heating purposes. Since the incinerators are subject to regulation of emissions, clean-up of the flue gases, in particular, requires expensive treatment. Such clean-up may include electrostatic precipitators and scrubbers to remove particulates and heavy metals. Dioxin emissions have also been detected from sewage sludge incineration, although at very low levels, and therefore may also require some form of treatment (Hudson 1993; Review of Dioxin Emissions in the UK 1995). Multiple hearth systems are not popular for new sludge incinerator plants because they have lower combustion efficiencies than fluidised beds, and consequently are likely to have higher emissions and require higher-level treatment to clean up the emissions. In addition, multiple hearth incinerators have higher capital and operating costs (Frost 1991).

With the prohibition of sewage sludge disposal at sea by 1998, alternative methods of disposal are required. The UK Government, in conjunction with the EC, supports the continued and increasing use of sewage sludge spread on land as the best practicable environmental option subject to proper environmental controls. The predictions for the disposal of sewage sludge by the year 2005 are shown in Figure 3.13 (Department of the Environment and Welsh Office 1995). The predicted alternative to sea disposal is via recycling, i.e. spreading of the treated sludge on land, and an increase in incineration. Landfilling of sewage sludge is predicted to decrease as stricter controls of landfill sites and consequent costs rise.

3.6 Clinical Waste

Clinical waste includes wastes from hospitals, doctors and dentists surgeries, health centres, nursing homes and veterinary surgeries, and amounts to approximately 0.3

million tonnes per year in the UK (Wassermann and McCullough 1994). The definition of clinical waste is contained in the Controlled Waste Regulations 1992 (Waste Management Paper 25 1993) and comprises:

(i) any waste which consists wholly or partly of human or animal tissue, blood or other body fluids, excretions, drugs or other pharmaceutical products, swabs or dressings, or syringes, needles or other sharp instruments, being waste which unless rendered safe may prove hazardous to persons coming into contact with it;
(ii) any other waste arising from medical, nursing, dental, veterinary, pharmaceutical or similar practice, investigation, treatment, care, teaching or research, or the collection of blood from transfusion, being waste which may cause infection to any person coming into contact with it.

In general, clinical waste is regarded as 'industrial waste' and as such would be a controlled waste. Some wastes arising from the rest of the health care sector may be separated from clinical wastes and be classified as household or commercial waste depending on where in the health care sector the waste has arisen. For example, newspapers, magazines, flowers or office waste would clearly be non-hazardous and could be classified as commercial waste. Similarly, waste from nurses accommodation and nursing homes would be classified as household. Certain wastes, such as prescription-only medicines, organic solvents, and wastes from natal care or the diagnosis, treatment or prevention of disease in humans, may be classified as Special or hazardous wastes under the 1996 Special Waste Regulations if the wastes exhibit hazardous properties such as being 'infectious' or 'toxic' as defined in the Regulations. Responsibility for defining clinical wastes as hazardous or infectious rests with the clinician involved (Waste Management Paper 25 1993).

Regulation of clinical waste is covered not only by the Environment Agency but also The Health and Safety Commission. The UK Health and Safety Commission identified five different categories of clinical waste in their Guidelines on the safe Disposal of Clinical Waste, 1992 (Waste Management Paper 25 1993).

Group A All human tissue including blood (whether infected or not), animal carcasses and tissue from veterinary centres, hospitals or laboratories, and all related swabs and dressings. Soiled surgical dressings, swabs and other soiled waste from treatment areas. Waste materials, where the assessment indicates a risk to staff handling them, for example, from infectious disease cases.

Group B Discarded syringes, needles, cartridges, broken glass and other contaminated disposable sharp instruments or items.

Group C Microbiological cultures and potentially infected waste from pathology departments (laboratory and post-mortem rooms) and other clinical or research laboratories.

Group D Certain pharmaceutical and chemical wastes.

Table 3.27 *Average composition of hospital waste*

Constituent	Percent present
Paper – clean	9.5
Paper – dirty	6.5
Kitchen waste	22.6
Plastics	14.6
Ward and pathology waste	19.7
Non-combustible waste	27.9

Source: Wallington J.W. and Kensett R.G. In: Engineering for Profit from Waste. Proceedings of the Institution of Mechanical Engineers, Conference, 1988–1. Coventry, March 1988.

Group E Items used to dispose of urine, faeces and other bodily secretions or excretions assessed as not falling within Group A. This includes used disposable bedpans or bed pan liners, incontinence pads, stoma bags and urine containers.

Group E wastes are distinguished from Group A wastes by their low hazard potential, and as such may be separated from the clinical waste stream to be dealt with as commercial waste. This reduces the amount of waste to be handled under the clinical waste definition, which requires more expensive treatment and disposal. However, careful segregation of the different categories of waste are required to prevent hazardous clinical wastes ending up in landfill sites. This may not only be offensive, but may also result in prosecution if the landfill licence has been infringed. Waste Management Paper 25 (1993) outlines the recommended safe storage, labelling, handling and transport of the different categories of clinical waste.

Table 3.27 shows the average composition of hospital waste (Wallington and Kensett 1988). The type and quantity of waste produced will vary according to the institutional type. For example, geriatric and long-stay hospitals produce about 1.0 kg of waste/bed/day, general short-stay hospitals produce about 3.1 kg/bed/day and large teaching hospitals with hostel accommodation and administrative centres can have waste generation rates of over 6.0 kg/bed/day. These figures translate to an average UK generation of clinical waste for the total population of 5.6 kg/person/year (Wassermann and McCullough 1994). The composition of the waste will also change with institution, for example a higher proportion of human tissue will be present in General Hospital waste, whereas there may be none from a geriatric type hospital. Table 3.27 shows a high proportion of non-combustible waste which, since the waste will generally be incinerated, would end up in the ash, to be disposed of via landfill. The plastic content in hospital waste is also high compared with municipal solid waste. Typical hospital waste has a calorific value of about 15 MJ/kg (Wallington and Kensett 1988) compared with municipal solid waste at about 8.5 MJ/kg and a typical coal at 28–32 MJ/kg.

Comparison with the arisings of clinical waste in the UK are compared with other European countries in Figure 3.14. UK clinical waste generation is high in comparison, but definitions of clinical waste in other countries are different from those in the UK and some countries have adopted segregation policies for the waste. As

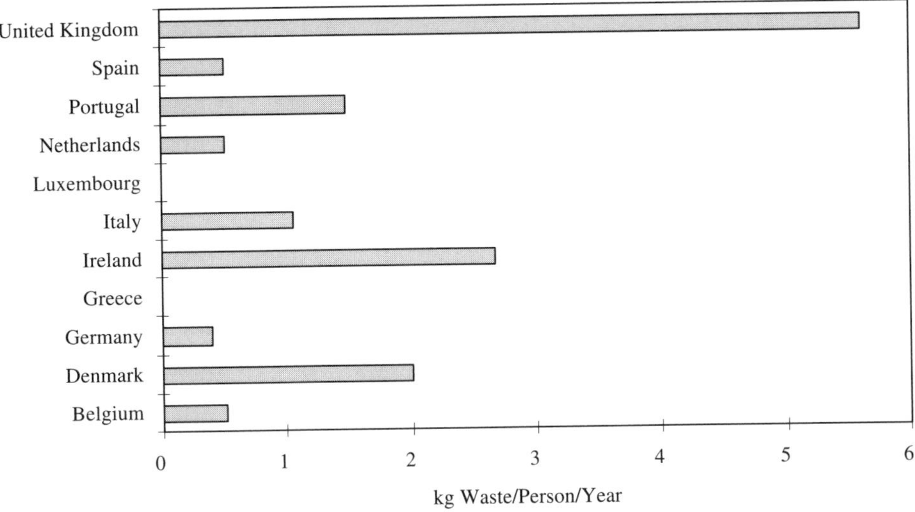

Figure 3.14 *Comparison of UK clinical waste arisings and European 'high risk healthcare' wastes. Source: Wassermann D.R. and McCullough J., Journal of Wastes Management and Resource Recovery, 1, 119–123, 1994*

such, the data represented in Figure 3.14 are 'high risk healthcare' waste. The European Union is currently categorising certain wastes as priority waste streams for special attention. Under the EC Priority Waste Streams Project on Healthcare Wastes, which is in preparation, clinical wastes may in the future be redefined as Healthcare Wastes for those solids and liquids arising from healthcare. A further definition of Healthcare Risk waste would cover additional infectious, chemical, pharmaceutical and toxic wastes, including 'sharps', such as hypodermic needles, and radioactive waste (Wassermann and McCullough 1994; Holmes 1995).

3.6.1 Treatment and Disposal Options for Clinical Waste

Until recently the recommendation by the Health and Safety Commission and the Department of the Environment for the safe disposal of clinical waste was incineration in all cases. However, the Department of the Environment Waste Management Paper 25 on Clinical Wastes (1993) recommends the segregation of the waste and that only Clinical Waste Categories A to D require incineration. Less hazardous Category E waste may then be disposed of via landfill, with consequent lower costs. Obviously this strategy will require the careful segregation of the waste and will involve staff training and strict waste segregation and collection procedures. However, since some of the waste arising in Category E may be offensive, and because of the problems involved in waste segregation, some health authorities are likely to continue with incineration of all clinical wastes. In addition, wastes arising as household hospital wastes, such as that from nurses homes, are similar to other domestic wastes and should be treated via the municipal waste routes.

The design of most clinical waste incinerators is of the two-stage, pyrolytic or starved air or rotary kiln variety, with throughputs typically of between 0.5 and 1 tonnes per hour. Essentially the incinerators use two-stage combustion where the first stage is a combustion zone operated under sub-stoichiometric conditions to produce a reducing atmosphere within the primary chamber. The heat generated breaks down the hydrocarbon in the pyrolysis zone to form combustible gases which pass to the second-stage combustion zone, where temperatures of 1000–1200 °C or higher are found and complete combustion of the volatile material takes place. Chapter 6 discusses the incineration process in more detail.

Sustainable waste management dictates that incineration with heat recovery would be the best practicable environmental option. There is an obligation upon hospital authorities to ensure the safe disposal of clinical waste, and this has largely been via in-house incineration. Until 1991 hospital incinerators were subject to Crown Immunity and were exempt from legislation affecting other types of waste incinerator. Under Crown Immunity, hospitals could not be prosecuted for any violation of environmental law since they were classified as Government property and came under the protection of the Sovereign. Immunity was removed under The National Health Service and Community Care Act of 1990. Consequently, clinical waste incinerators above 1 tonne per hour throughput are now subject to Integrated Pollution Control under the 1990 Environmental Protection Act and require authorisation from the Environment Agency. Smaller incinerators, with less than 1 tonne per hour throughput, are subject to air pollution control only, which is regulated by local authorities. With the loss of Crown Immunity and the legislative requirement to meet more stringent emission controls, particularly from European Directives, with their associated high costs, many incinerators closed down in the years 1995/6. The trend now is to larger clinical waste incinerators serving several institutions. Some authorities utilise private clinical waste contractors who store, collect and incinerate the waste as a full waste management service. It has been estimated that in 1984 in the UK approximately 1000 hospital waste incinerators were in operation. However, following the loss of Crown Immunity and the requirement to comply with emissions legislation that number has dropped to only 40 much larger plants, some operating 24 hours per day (Wasserman and McCullough 1994).

3.7 Agricultural Waste

Agricultural waste consists of organic material such as manure from livestock, slurry, silage effluent and crop residues. Large tonnages are produced in the UK; approximately 80 million tonnes is estimated to come from housed livestock alone (Department of the Environment and Welsh Office 1995). Since all wastes arising from all premises used for agriculture are not deemed to be a controlled waste under the legal definition of wastes, they are not subject to regulation under the Environmental Protection Act, 1990, in terms of authorisations, duty of care, licensing etc.

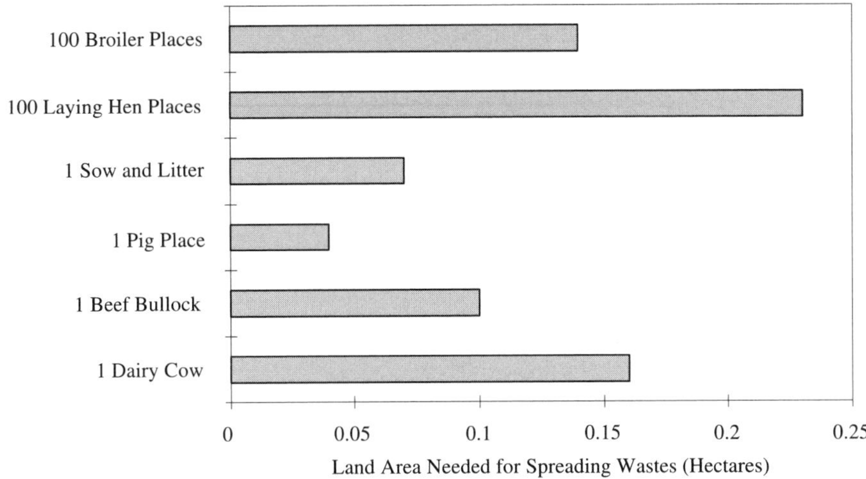

Figure 3.15 *Land area need for spreading animal wastes. Source: Waste Management Planning: Principles and Practice. Department of the Environment, HMSO, London, 1995*

However, the adoption of the EC Framework Directive on Waste (75/442/EEC and amended by 91/156/EEC) by the UK Government will place some agricultural waste such as household and non-organic wastes in the controlled waste category and consequently will bring such wastes within the controlled waste regulations.

Estimates of the production of agricultural waste throughout OECD Europe (Organisation of Economic Cooperation and Development) are approximately 700 million tonnes (Stanners and Bourdeau 1995). Individual countries within Europe show a wide variation in arisings of agricultural waste due to the different extents of agriculture within the economy and differing farming methods. For example, agricultural waste arisings from Spain are estimated at 112 million tonnes/year, France 400 million tonnes/year, Norway 18 million tonnes/year, Sweden 21 million tonnes/year, Finland 23 million tonnes/year and Austria less than 1 million tonnes/year (Eurostat Year Book 1996).

3.7.1 Treatment and Disposal Options for Agricultural Waste

The majority of the agricultural waste is landspread and some is used as animal feed or composting. The organic waste is high in nutrients and provides a substitute for commercial fertilisers; consequently landspreading is regarded as the best practicable environmental option. Landspreading of organic agricultural waste is generally beneficial and has resulted in increased yields and helped to maintain high organic carbon and nitrogen levels. Whilst regulation as a controlled waste does not apply to agricultural waste, the spreading of the waste on land is covered by Codes of Good Practice issued by the Ministry of Agriculture, Fisheries and Food. The codes recommend good practice in the areas of waste storage, management and application of the waste to land. For example, Figure 3.15 shows the typical land area needed for spreading the wastes arising from certain animals (Waste Management Planning 1995).

Poultry litter, consisting of a mixture of bird droppings and wood shavings, is a particular form of agricultural waste that has received interest due to its high generation rates and its high calorific value, and therefore its potential use as a fuel rather than landfilling. In addition, application of broiler litter to land has been shown adversely to affect soil chemistry in that increased levels of phosphorus, nitrogen, potassium, calcium and magnesium have been detected, with their potential for significant environmental impact. Approximately 1.4 million tonnes of poultry litter are generated in the UK each year, with an average calorific value of 13.5 MJ/kg (Dagnall 1993). A number of poultry litter combustion systems exist in the UK which are used to generate either heat for use as space heating within the farm, including for the poultry themselves, or electricity for on-site use or sale into the national grid (Dagnall 1993).

3.8 Industrial and Commercial Waste

Industrial and commercial waste arisings in the UK are shown in Table 3.4 and show that an estimated 70 million tonnes of industrial waste and 15 million tonnes of commercial waste are generated each year (Department of the Environment and Welsh Office 1995). Table 3.28 shows an analysis of the breakdown of the major industrial sectors producing the waste. The estimated data on industrial waste arisings, as was discussed previously, are prone to large errors due to the difficulties in obtaining data from the diverse range of industries and vast number of companies. For example, the data in Table 3.28 were obtained from a survey of analyses of company data on waste. However, estimates of industrial and commercial waste based on Waste Disposal Plans produced by Waste Regulation Authorities produced an estimate of the total arisings of industrial and commercial waste of 121.5 million tonnes per year (Analysis of Industrial and Commercial Waste Going to Landfill in the UK 1995), which is a figure 70% higher than the estimate derived from the company waste survey.

The composition of the various industrial and commercial waste are extremely wide and very variable. Table 3.29 shows the range of industrial and commercial wastes disposed of by landfill in the UK (Analysis of Industrial and Commercial Waste Going to Landfill in the UK 1995). Whilst analyses of industrial and commercial waste compositions include animal and food waste and such chemical groups as acids, alkalis, metal compounds, fuels, oils, greases, chemical waste etc., the largest proportions of wastes composition are in the categories of 'other inorganic compounds' and 'miscellaneous wastes'. These categories represent a large range of wastes which do not naturally fall into any one classification (Analysis of Industrial and Commercial Waste Going to Landfill in the UK 1995).

Table 3.28 Estimated waste arisings of industrial and commercial waste in the UK

Industry group	Waste arising (1000 tonnes/year)
Coal extraction; coke ovens; extraction of oil and gas etc.	1227.3
Production and distribution of electricity, gas and other forms of energy	1628.6
Water supply industry	82.5
Extraction and preparation of metalliferous mineral ores	m
Metal manufacturing	2329.6
Extraction of minerals not elsewhere specified	7.2
Manufacture of non-metallic mineral products	2878.7
Chemical industry	2939.8
Production of man-made fibres	12.3
Manufacture of metal goods not elsewhere specified	2616.2
Mechanical engineering	2775.2
Manufacture of office machinery and data processing equipment	10.5
Electrical and electronic engineering	1758.8
Manufacture of motor vehicles and parts thereof	1896.9
Manufacture of other transport equipment	933.0
Instrument engineering	669.9
Food, drink and tobacco manufacturing industries	7549.4
Textiles, leather and leather goods, footwear and clothing	1635.7
Timber and wooden furniture industries	3262.7
Manufacture of paper and paper products; printing and publishing	2367.6
Processing of rubber and plastics	1116.2
Other manufacturing industries	207.4
Construction	4581.5
Wholesale distribution	4689.8
Dealing in scrap and waste materials	1974.2
Commission agents	m
Retail distribution	11 454.1
Hotels and catering	4114.3
Repair of consumer goods and vehicles	415.5
Transport and communication	3184.6
Banking, finance, insurance, business services, leasing	6370.3
Other services	11 240.5
Total	85 930.3

m = missing.

Source: Analysis of Industrial and Commercial Waste going to Landfill in the UK. Department of the Environment, HMSO, London, 1995.

Estimates of the total production of industrial waste throughout OECD Europe are approximately 330 million tonnes, and Figure 3.16 shows the estimated arisings of industrial waste in certain European countries (Stanners and Bourdeau 1995). The data for the UK represent some disparity between the estimate of UK arisings and those estimated at 70 million tonnes (Department of the Environment and Welsh Office 1995) due to the differences in industrial category definitions and inaccuracies of estimating waste arisings.

Estimated industrial waste arisings in the USA are of the order of 400 million tonnes/year (Bonomo and Higginson 1988), although this figure includes some

Table 3.29 *Distribution of industrial and commercial waste disposed of by landfill in the UK by composition*

Waste category	Range (%)	Median (%)
Inorganic acids	0–2	<1
Organic acids	0	<1
Alkalis	0–2	<1
Toxic metal compounds	0–4	1
Non-toxic metal compounds	0–1	<1
Metals (elemental)	0–19	1
Metal oxides	0	<1
Inorganic compounds	0–1	<1
Other inorganic compounds	17–59	44
Organic compounds	0	<1
Polymeric materials	1–3	1
Fuels, oils and greases	0–4	2
Fine chemicals	0	<1
Miscellaneous chemical waste	0–1	<1
Filter materials etc.	0–27	2
Interceptor wastes	1–4	2
Miscellaneous waste	6–48	38
Animal and food waste	3–48	9
Clinical waste	0–6	<1

Source: Analysis of Industrial and Commercial Waste going to Landfill in the UK. Department of the Environment, HMSO, London, 1995.

wastes such as power generation wastes included in other categories. Over 90% of the industrial wastes arise through just seven industrial sectors: industrial organic chemicals (97 million tonnes/year – 24.8%); primary iron and steel manufacturing and ferrous foundries (60 million tonnes/year – 15.5%); fertiliser and other agricultural chemicals (59 million tonnes – 15.0%); electric power generation (56 million tonnes/year – 14.2%); plastics and resins manufacturing (45 million tonnes/year – 11.5%); industrial organic chemicals (26 million tonnes/year – 6.7%); stone, clay, glass and concrete products (19 million tonnes/year – 4.7%) (Bonomo and Higginson 1988).

3.8.1 Treatment and Disposal Options for Industrial and Commercial Waste

There is very little information available as to the waste treatment and disposal methods used by industry and commerce. However, it is estimated that the majority of such waste in the UK is landfilled. Table 3.30 shows the breakdown of the treatment and disposal routes for industrial and commercial waste in the UK (Department of the Environment and Welsh Office 1995). By far the most popular route is landfill, representing 73% of the total. Recycling and re-use represent a high proportion of the waste arisings, which are mostly recycled back into the industry sector from where the waste arose.

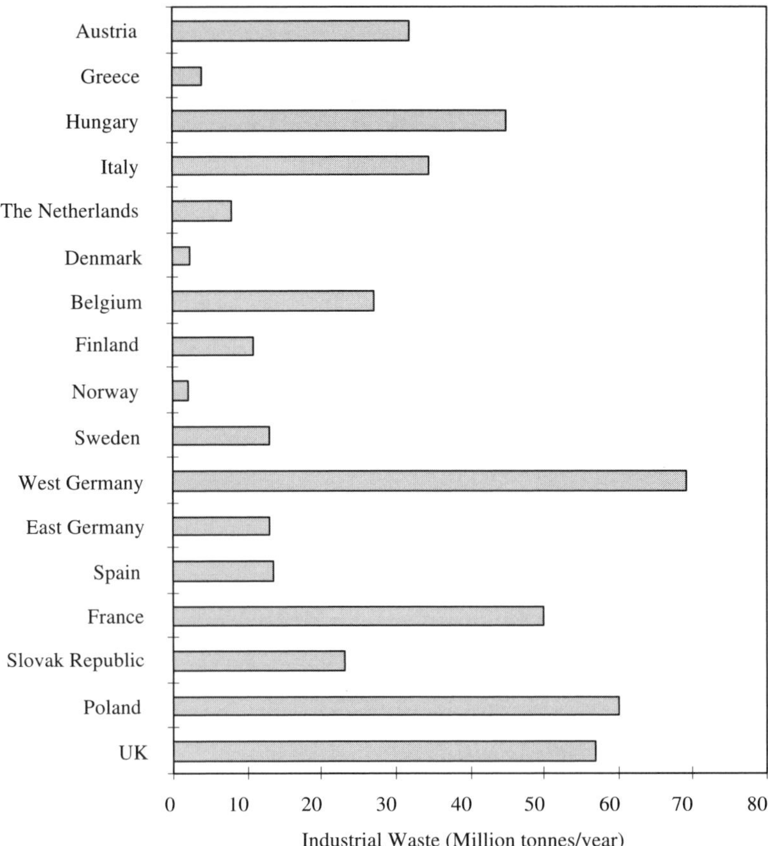

Figure 3.16 *Estimated arisings of industrial waste in certain European countries. Source: Stanners D. and Bourdeau P., Europe's Environment, The Dobris Assessment. European Environment Agency, Copenhagen. Earthscan Publications, London, 1995*

Table 3.30 *Treatment and disposal of industrial and commercial waste in the UK*

	Industrial waste	Commercial waste
Estimated arisings (million tonnes/year)	70	15
Treatment and disposal route		
Landfill (%)	73	85
Incineration (%)	1	7.5
Recycled/re-used (%)	18	7.5
Other (%)	6	0

Source: Making Waste Work. Department of the Environment and Welsh Office, HMSO, London, 1995.

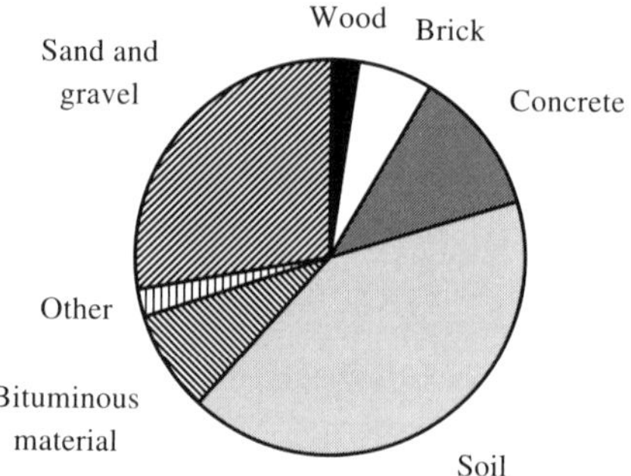

Figure 3.17 Typical composition of UK construction and demolition waste. Source: Warmer Bulletin 47, Warmer Bulletin Information Sheet: Construction and Demolition Waste. Journal of the World Resource Foundation, Tonbridge, 1995. Reproduced by permission from the World Resource Foundation

3.9 Other Wastes

3.9.1 Construction and Demolition Industry Waste

Extremely large tonnages of demolition and construction industry waste arise in the UK each year, of the order of 70 million tonnes/year, representing 16% of the total UK waste arisings (Department of the Environment and Welsh Office 1995). The waste consists of soil, brick, plaster, metal work, concrete, glass, tiles, wood, plastic etc., and is generally bulky and inert. Figure 3.17 shows the typical proportions of the components of construction and demolition waste in the UK (Warmer Bulletin 47 1995). Typically soil, stones and clay form the largest percentage of the composition. The majority of the waste is landfilled, although about half of this is used to provide aggregate for access roads to the landfill and landfill site construction engineering to build the embankment walls of landfill 'cells' and for cover and final site capping. Landfill charges for such inert wastes are low, and where the wastes are used for landfill site construction then waste disposal is usually free. About 30% of the 70 million tonnes of waste arising is used as low-grade bulk fill and construction site engineering within the construction industry (Department of the Environment and Welsh Office 1995). For example, materials may be used for hard core for roadways, landscaping and car parks. Such use involves the crushing, removal of metals and size grading of the waste. Higher-level use of the waste is not possible due to the poor quality and heterogeneous nature of the material compared with primary aggregates used in the industry.

Construction and demolition waste statistics are collated throughout Europe by the Statistical Office of the European Communities as 'demolition waste'. Intercountry comparisons show that Germany has a huge generation of demolition waste of the order of 120 million tonnes/year, Spain generates 22 million tonnes/year, Italy 34 million tonnes/year and Austria 18 million tonnes/year (Eurostat Year Book 1996). Estimates of the production of construction waste throughout OECD Europe are approximately 260 million tonnes (Stanners and Bourdeau 1995). Construction waste generation in Japan is estimated at approximately 30 million tonnes/year (Bonomo and Higginson 1988).

3.9.2 Mines and Quarry Waste

The arisings of mines and quarrying wastes are estimated to amount to 110 million tonnes per year, representing 25% of the total waste generated in the UK (Department of the Environment and Welsh Office 1995). Mines and quarrying waste are not controlled wastes under the 1990 Environmental Protection Act and are also not covered by the EC Framework Directive on Waste (75/442/EEC and amended by 91/156/EEC). The wastes consist mainly of colliery spoil from coal mines, china clay wastes and slate wastes, and are generally inert mineral materials.

The large majority of mines and quarry waste is disposed of as large scale above-ground open tips close to the mine or quarry, which are later landscaped and restored. The size, location and proper restoration of the tips are controlled by the Town and Country Planning Act, 1990, which deals with the strategic and forward planning of land use within district and county council areas. The stability and safety of the waste tips are controlled by the Mines and Quarries (Tips) Act and Regulations of 1969 and 1971.

Estimates of the production of mining waste throughout OECD Europe are approximately 360 million tonnes (Stanners and Bourdeau 1995). Comparison with other countries of Europe show a wide variation in waste arisings from the mining and quarrying industries. For example, Germany produces 68 million tonnes/year, Spain 70 million tonnes/year and France 75 million tonnes/year, whilst Norway produces only 9 million tonnes/year and the Netherlands, Austria and Portugal each produce less than 1 million tonnes/year (Eurostat Year Book 1996). USA generation of mining waste is estimated at 1400 million tonnes/year (Bonomo and Higginson 1988).

3.9.3 Power Station Ash and Iron and Steel Slags

Approximately 12 million tonnes of power station ash from coal-fired electricity power stations and 5.0 million tonnes of slag from the iron and steel industries are generated in the UK each year (Department of the Environment and Welsh Office 1995). Power station ash is generated from the ash in coal, which amounts to about 15% of the original coal. The coal is first pulverised to small grain size before use in the power station. The ash from the furnace chamber is coarse grained and termed coarse slaggy clinker; the fine-grained material captured in the flue gas clean-up system is termed pulverised fuel ash. Iron and steel slags are generated throughout the production of iron and steel from the original iron ore, and the slags are of variable composition and size.

Most of the power station ash is recycled for use in the construction industry as a concrete mix ingredient, in blended cement, for use as structural fill in road building, as a light weight aggregate and for use as a constituent in building blocks. The remainder is landfilled, usually close to the power station. The fine-grained pulverised fuel ash is pumped into settling lagoons in solution or conditioned with water and then landfilled. Iron and steel industry slags are used as aggregate for road construction, as concrete filler and in building blocks. Some slags have a high metal content and can be recycled through the steel-making process.

Intercountry comparisons of power station ash may be made through comparison of waste generated as waste arising through the production of energy reported by the Statistical Office of the European Communities (Eurostat Year Book 1996). These data show that energy production waste arisings in Germany were 19 million tonnes/year, 1 million tonnes in Italy, Austria, the Netherlands and Belgium, and 3 million tonnes in Finland (Eurostat Year Book 1996). Estimates of USA waste arisings from electrical power generation are of the order of 56 million tonnes/year (Bonomo and Higginson 1988).

3.9.4 Tyres

Aproximately 28 million car tyres and 3 million truck tyres are scrapped each year in the UK, representing about 0.4 million tonnes by weight of tyre (Bressi 1995). Total EC scrap tyre arisings are 2 million tonnes per year; this compares with 0.7 million tonnes/year in Japan and 2.5 million tonnes/year in North America (Peat Marwick 1990; Bressi 1995; Williams et al 1995). Typical tyre compositions are shown in Table 3.31 (Ogilvie 1995).

It is estimated that 21% of the scrap tyres are retreaded, 15% incinerated, 10% crumbed, about 50% disposed of in landfill sites, open dumping or stockpiled, and the remainder used in applications such as boat fenders, for fuel in cement kilns etc. (Ogilvie 1995).

Retreading of tyres, where a new rubber tread is bonded to the surface of worn tyres, accounts for about 20% of annual tyre consumption. Car tyres can be retread once only, but truck tyres may be retread up to three times.

Landfilling of tyres is declining as a disposal option, since tyres do not degrade easily in landfills, they are bulky, and can cause instability within the landfill. In addition, they can be a breeding ground for insects and a home for vermin. Many landfill sites refuse to take tyres, and the Department of the Environment has recommended that whole tyres should not be landfilled and that landfilling of shredded tyres should be restricted to 5% of the landfill volume. Consequently the cost of landfilling tyres has risen sharply in recent years. Open dumping and stockpiling have the potential for accidental fires or arson, resulting in high pollution emissions to the atmosphere and water courses.

Incineration of tyres utilises the high calorific value of tyres, about 32 MJ/kg, and currently accounts for about 15% of tyre waste. The main UK incineration plant is

Table 3.31 Approximate proportions of components in tyres

(a) Car tyres

Component	Steel-braced radial	Textile-braced radial	Cross ply
Rubber compound	86	90	76
Steel	10	3	3
Textile	4	7	21

(b) Truck tyres

Component	All-steel radial	Cross ply
Rubber compound	85	88
Steel	15	3
Textile	<0.5	9

(c) Composition of tyre rubber compound

Component	Weight (%)
Rubber hydrocarbon	51
Carbon black	26
Oil	13
Sulphur	1
Zinc oxide	2
Others	7

Source: Ogilvie S.M., Opportunities and Barriers to Scrap Tyre Recycling. AEA Technology, National Environmental Technology Centre, Abingdon, HMSO, 1995.

the Elm Energy plant based in Wolverhampton in the West Midlands, which has a design capacity to recover up to 20% of the total tyre waste arisings in the UK. The plant generates 20 MW of electricity, which is sold into the national grid, and also recovers 20 000 tonnes of steel and 3000 tonnes of zinc. Cement kilns have been used for the disposal of tyres by combustion, but represent only a small proportion of the disposal route for tyre arisings.

Rubber crumb is derived from tyres via shredding and grinding of the rubber part of the tyre to produce a fine-grained product for use in such applications as children's playgrounds, sports surfaces and carpet backing, and other applications such as absorbents for oils and hazardous and chemical wastes.

An alternative technology for the treatment of tyres is pyrolysis, which is the thermal degradation of the tyre in an inert atmosphere. The process produces an oil, char and gas product, all of which have the potential for use. Pyrolysis of tyres has been established for many years, but is currently receiving renewed attention due to the difficulties of landfilling tyres and the limited uses for recycling. Pyrolysis of tyres to produce gases and oils for combustion, the energy from which then is used to produce electricity, may also qualify for a Non-Fossil Fuel Obligation (NFFO) contract with its associated premium prices for the electricity generated. The recovery of oil from pyrolysis of tyres can be as high as 58%; the oil has a high calorific value and can be used as a fuel or chemical feedstock, or can be added to petroleum refinery feedstocks. The char has use as a solid fuel, or can be used as activated carbon or carbon black. The derived gas has a sufficiently high calorific value to provide the energy requirements for the pyrolysis process.

The treatment and disposal option for tyres most commonly used throughout the European Union is landfilling, representing about 46% of total waste tyres. Recycling to recover materials or energy represents about 31% and retreading about 23% (Stanners and Bourdeau 1995). In some European countries, such as Germany, landfilling of tyres is not a significant option and retreading and incineration in cement kilns represent almost 60% disposal of the estimated 0.45 million tonnes/year scrap tyres generated. In France approximately 0.5 million tonnes of scrap tyres are generated each year, with an estimated 60% going to landfill, 10% incinerated, 10% reclaimed and the remainder to other uses (Peat Marwick 1990).

The problem of scrap tyres is even more acute in the USA, where it is estimated that the production of scrap tyres is between 240 and 270 million tonnes per year and that there is a stockpile of 3000 million tonnes awaiting disposal (Peat Marwick 1990). The main method of scrap tyre disposal in the USA is landfilling, followed by illegal stockpiling and incineration. Japan generates approximately 0.7 million tonnes of scrap tyres per year. The majority are recycled through, for example, reclaiming and retreading, representing about 35%. Energy recovery accounts for a further 34% and the remainder is landfilled, stockpiled or exported (Peat Marwick 1990).

3.10 Waste Containers, Collection Systems and Transport

The type of container used to store the waste generated from households, commercial and industrial premises depends on the frequency and efficiency of collection, the amount of waste, the type of housing, the density of the collected wastes, collection vehicle type, vehicle usage and manpower and economics relating to the container and the collection system (Scharff and Vogel 1994). The correct size of waste container is important, since it has been shown that the use of non-standard containers is the greatest cause of litter (Pescod 1991–93).

Household waste containers used in the UK include traditional metal or plastic dustbins, wheeled bins and plastic sacks. The capacity of the household waste storage container depends on how many collections are made per week. With a general increase in the amount of waste generated, there has been a response to use bigger or more containers, which also has the potential to reduce manning costs by less frequent waste collections. Another factor dictating size of container and frequency of collection is climate; in cooler areas such as Northern Europe, where odour from the degradation of the waste occurs more slowly, the frequency of collection may be once or twice per week. Consequently the container must be able to store a full week's volume of waste. In warmer climates such as Southern Europe, collections of waste have to be more frequent, often daily, to minimise nuisance from odours and fly breeding (Pescod 1991–93). Consequently, the waste container can be much smaller than one used for weekly collection. The size of the waste container and the frequency of collection has been shown to influence the quantity of household waste placed in the container. For example, UK experience has shown that a twice-weekly collection produces more waste than a weekly one, and that where wheeled bins are used compared with traditional smaller metal or plastic dustbins, then increases in waste have been up to 30% (Pescod 1991–93). The use of larger wheeled bins has also resulted in large bulkier items being placed in the larger containers. In some UK cities recycling initiatives have led to more than one container. For example, in Leeds two wheeled bins are used in some parts of the city, a green bin to collect recyclable materials, newspaper, metal cans and plastic bottles, and a black bin to collect the remaining waste. The green bin is collected once per month and the black bin weekly. The system is targeted at particular recyclable materials for which there is a proven economic or environmental return. The segregation of waste also reduces the non-recyclable portion of household waste and therefore reduces final waste treatment and disposal costs. In the UK generally, recycling of paper, glass, metal tins and plastic bottles has been by the use of large containers sited throughout the community.

Commercial buildings such as offices and shops, but also larger household areas such as blocks of flats and institutions such as schools, require much larger container capacities than single households. A typical waste container for such buildings would be about 0.85 m^3 capacity. Industrial and trade wastes and large commercial buildings require even larger containers and use waste 'skips' of up to 30 m^3 capacity (Pescod 1991–93).

Waste collection vehicles have compaction rams to reduce the volume of waste. The vehicle has a team of up to six collectors, depending on the type and capacity of vehicles used and the method of collection.

In some areas of the UK, the increasing distance between the city centre generation of the waste and suitable landfill sites has led to the development of transfer stations. Another major advantage is the reduction in waste transport costs by reductions in the number of vehicles travelling to the disposal site and manning levels of personnel travelling with the vehicle. The waste from the collection vehicles is transferred to larger road vehicles with capacities up to 30 tonnes, or to freight containers for movement by either rail or waterway barges (Pescod 1991–93).

A recent survey of the comparison of collection systems used for municipal solid waste in European cities has shown that they are highly efficient, with a combination of containers, vehicles, personnel and logistics individually suited to the local conditions such as population density, residential structure or traffic conditions (Scharff and Vogel 1994). The cities compared in the survey were Berlin, Budapest, Copenhagen, Munich, Paris, Stockholm, Vienna and Zurich. The size of collection container for household waste was calculated in most instances on a figure of 120 litres volume of waste arising per household. Collection frequencies range from Paris, with a daily emptying of containers or the daily collection of waste bags from each collection point, to two to three times per week in other cities, with a once per week collection being the minimum acceptable. Traffic congestion, number of collection vehicles, population density, collection frequency and the treatment and disposal used for the waste greatly influence the daily distance travelled by the collection vehicles. For example, Paris has 500 collection vehicles for a population density four times higher than that of the other cities surveyed, and achieves a daily travelling distance of between 20 and 52 km. Budapest, however, with a similar waste quantity to Paris but a different collection organisation and traffic situation, uses 175 collection vehicles and achieves a daily travelling distance of between 53 and 175 km (Scharff and Vogel 1994). The size of waste containers used in the eight cities surveyed ranged from 30 litres volume to over 1100 litres. The waste collection efficiencies for waste collected per day were found to be greater by up to four times for the larger containers compared with smaller ones.

The USA use a range of collection containers for municipal solid waste collection (Bonomo and Higginson 1988). Residential containers range from plastic or paper bags, to 90 litre volume metal cans, to 350–500 litre volume plastic wheeled carts, which are transferred to the collection point. Larger containers are used for multi-family or apartment blocks, and may have compaction facilities and require automated dedicated vehicles for collection. Commercial waste is collected in 6000 litre volume plastic or metal containers, and up to 30 000 litre containers for commercial waste from, for example, shopping centres.

The collection of household waste in Japan is mostly from collection sites serving between 10 and 40 properties, although in some cases individual household collection may occur (Waste Management in Japan 1995). In Tokyo, for example, there are some 240 000 waste collection points. Waste is segregated by the householder into combustible and non-combustible. The combustible fraction of the waste is the largest and is collected more frequently than the non-combustible waste. In some areas waste is collected up to four times per week as the very narrow streets within the cities and the segregation of waste require frequent visits by very small vehicles. For example, in Tokyo, there are some 3000 waste collection vehicles of either 1.2 tonnes or 2.4 tonnes capacity.

Bibliography

Analysis of Industrial and Commercial Waste going to Landfill in the UK, 1995. Department of the Environment, HMSO, London.

Benestad C. 1989. Incineration of hazardous waste in cement kilns. Waste Management and Research, 7, 351–361.

Bonomo L. and Higginson A.E. 1988. International Overview on Solid Waste Management. Academic Press, London.

Bressi G. 1995. Recovery of Materials and Energy from Waste Tyres: Present situation and future trends. Working Group on Recycling and Waste Minimisation, International Solid Wastes Association, Denmark.

British Medical Association, 1991. Hazardous Waste and Human Health. Oxford University Press, Oxford.

Bruce A.M., Colin F. and Newman P.J. 1989. Treatment of Sewage Sludge. Elsevier Applied Science, London.

Brunner P.H. and Lichtensteiger Th. 1989. In: Bruce A.M., Colin F. and Newman P.J. (Eds.) Treatment of Sewage Sludge. Elsevier Applied Science, London.

Buekens A. and Patrick P.K. 1985. Incineration. In: Suess M.J. (Ed.) Solid Waste Management: Selected topics. World Health Organisation, Regional Office for Environmental Health Hazards, Copenhagen.

Daborn G. 1993. In: Brown A., Evemy P. and Ferrero G.L. (Eds.) Energy Recovery Through Waste Combustion. Elsevier Applied Science, London.
Dagnall, S.P. 1993. Poultry Litter as a Fuel in the UK: A review. Energy Technology Support Unit, Harwell.
Dean R.B. and Suess M.J. 1985. The risk of chemicals in sewage sludge applied to land. Waste Management and Research, 3, 251–278.
Dempsey C.R. and Oppelt E.T. 1993. Incineration of hazardous waste: A critical review update. Air and Waste, 43, 25–73.
Department of the Environment and Welsh Office, 1995. Making Waste Work: A strategy for sustainable waste management in England and Wales. HMSO, London.
Development of a National Waste Classification Scheme, 1995. Stage 2: A system for classifying wastes. Consultation Draft, December 1995, Department of the Environment, Welsh Office, The Scottish Office, Department of the Environment for Northern Ireland, Wastes Technical Division, Department of the Environment, London.
Digest of Environmental Protection and Water Statistics, 1993. Department of the Environment, HMSO, London.
Digest of Environmental Statistics, 1996. Department of the Environment, HMSO, London.
ENDS, 1995. Report 251, Environmental Data Services Ltd., Bowling Green Lane, London, p. 27.
ENDS, 1997. Report 265, Environmental Data Services Ltd., Bowling Green Lane, London, February, pp. 41–42.
Environmental Indicators, 1991. Organisation of Economic Cooperation and Development, Paris.
Eurostat Year Book, 1996. Office for Official Publications of the European Communities, Luxembourg.
Frost R. 1991. Incineration of sewage sludge. University of Leeds Short Course 'Incineration and Energy from Waste', The University of Leeds, Leeds.
Hall J.E. 1992. Treatment and use of sewage sludge. In: Bradshaw A.D., Southwood R. and Warner F. (Eds.) The Treatment and Handling of Wastes. Chapman and Hall, London.
Holmes, J. 1995. The UK Waste Management Industry 1995. Institute of Wastes Management, Northampton.
Hudson J. 1993. Sewage sludge incineration. In: Incineration an Environmental Solution for Waste Disposal. IBC Technical Services Ltd. (London) Conference, February, Manchester.
Kaiser E. 1978. In: Niessen, W.R. (Ed.) Combustion and Incineration Processes. Dekker, New York.
Kim B.J. and Qi S. 1995. Hazardous waste treatment technologies. Water Environment Research, 67, 560–570.
La Grega M.D., Buckingham P.L. and Evans G.J. 1994. Hazardous Waste Management. McGraw-Hill, New York.
Lane P. and Peto M. 1995. Blackstone's Guide to the Environment Act 1995. Blackstone, London.
Matsuto T. and Tanaka N. 1993. Data analysis of daily collection tonnage of residential solid waste in Japan. Waste Management and Research, 11, 333–343.
National Household Waste Analysis Project, 1994. Phase 2, Vol. 1. Report on composition and weight data. Report No. CWM 082/94, Wastes Technical Division, Department of the Environment, HMSO, London.
National Household Waste Analysis Project, 1995. Phase 2, Vol. 3. Chemical analysis data. Report No. CWM/087/94, Wastes Technical Division, Department of the Environment, HMSO, London.
Ogilvie S.M. 1995. Opportunities and Barriers to Scrap Tyre Recycling. AEA Technology, National Environmental Technology Centre, Abingdon. HMSO, London.
Peat Marwick, 1990. The Recycling and Disposal of Tyres. Report PD00808WPC, Peat Marwick Management Consultants, London.
Pescod M.B. (Ed.) 1991–93. Urban Solid Waste Mangement. World Health Organisation, Copenhagen.

Petts J. and Eduljee G. 1994. Environmental Impact Assessment for Waste Treatment and Disposal Facilities. Wiley, Chichester.

Porteous A. 1992. Dictionary of Environmental Science and Technology. Wiley, Chichester.

Review of Dioxin Emissions in the UK, 1995. Environment Agency (Her Majesty's Inspectorate of Pollution), Department of the Environment, HMSO, London.

Rhyner, C.R. 1992. Monthly variations in solid waste generation. Waste Management and Research, 10, 67–71.

Rhyner C.R. and Green B.D. 1988. The predictive accuracy of published solid waste generation factors. Waste Management and Research, 6, 329–338.

Royal Commission on Environmental Pollution, 1993. 17th Report, Incineration of Waste. HMSO, London.

Rushbrook P.E. and Finney E.E. 1988. Planning for future waste management operations in developing countries. Waste Management and Research, 6, 1–21.

Scharff C. and Vogel G. 1994. A comparison of collection systems in European cities. Waste Management and Research, 12, 387–404.

Standard Industrial Classification of Economic Activities, 1992. HMSO, London.

Stanners D. and Bourdeau P. 1995. Europe's Environment. The Dobris Assessment. European Environment Agency, Copenhagen.

The Analysis and Prediction of Household Waste Arisings, 1991. Report No. CWM/037/91, Wastes Technical Division, Department of the Environment, HMSO, London.

The Environment Act 1995, 1995. HMSO, London.

The Open University, 1993. Course T237: Environmental Control and Public Health. Units 8–9, Municipal Solid Wastes Management, The Open University, Milton Keynes.

The Special Waste Regulations 1996, 1996. HMSO, London.

Try P.M. and Price G.J. 1995. Sewage and industrial effluents. In: Hester R.E. and Harrison R.M. (Eds.) Waste Treatment and Disposal. Issues in Environmental Science and Technology, Issue 3, The Royal Society of Chemistry, London.

Visvanathan C. 1996. Hazardous waste disposal. Resources, Conservation and Recycling, 16, 201–212.

Wallington J.W. and Kensett R.G. 1988. Energy recovery from refuse incineration in the National Health Service. In: Engineering for Profit from Waste. Proceedings of the Institution of Mechanical Engineers Conference, 1988, Coventry.

Warmer Bulletin 44, 1995. Information Sheet. The World Resource Foundation, Tonbridge.

Warmer Bulletin 47, 1995. Warmer Bulletin Fact Sheet: Construction and Demolition Waste, Journal of the World Resource Foundation, Tonbridge.

Warmer Bulletin, 49, 1996. Journal of the World Resource Foundation, Tonbridge.

Warmer Bulletin 50, 1996. Household Hazardous Waste. Journal of the World Resource Foundation, Tonbridge.

Wassermann D.R. and McCullough J. 1994. Clinical waste management and disposal. Journal of Wastes Management and Resource Recovery, 1, 119–123.

Waste Incineration. Environment Agency, 1996. Processes Subject to Integrated Pollution Control: Waste incineration. IPC Guidance Note S2 5.01, Environment Agency, Bristol.

Waste Management in Japan, 1995. The Institute of Wastes Management, Department of Trade and Industry, Overseas Science and Technology Expert Mission Visit Report, IWM, Northampton.

Waste Management Paper 28, 1992. Recycling. Department of the Environment, HMSO, London.

Waste Management Paper 25, 1993. Clinical Wastes: A technical memorandum on arisings, treatment and disposal. Consultation Draft, Wastes Technical Division, Department of the Environment, London.

Waste Management Paper 26F, 1996. Landfill Co-disposal (draft). Department of the Environment, HMSO, London.

Waste Management Planning, 1995. Principles and Practice. Department of the Environment, HMSO.

Wheatley A. 1990. Anaerobic Digestion: A waste treatment technology. Elsevier Applied Science, London.

White P.R., Franke M. and Hindle P. 1995. Integrated Solid Waste Management. Blackie Academic and Professional, London.

Williams P.T., Besler S., Taylor D.T. and Bottrill R.P. 1995. Pyrolysis of automotive tyre waste. Journal of the Institute of Energy, 68, 11–21.

Woodside G. 1993. Hazardous Materials and Hazardous Waste Mangement: A technical guide. Wiley Interscience, New York.

Yakowitz H. 1993. Waste management: what now? what next? An overview of policies and practices in the OECD area. Resources, Conservation and Recycling, 8, 131–178.

Yokoyama S., Murakami M., Ogi T., Kouchi K. and Nakamura E. 1987. Liquid fuel production from sewage sludge by catalytic conversion using sodium carbonate. Fuel, 66, 1150–1155.

Yu C-C. and Mclaren V. 1995. A comparison of two waste stream quantification and characterisation methodologies. Waste Management and Research, 13, 343–361.

4

Waste Reduction, Re-Use and Recycling

Summary

Chapter 4 is concerned with waste reduction, re-use and recycling, with the emphasis on recycling. The legislative background to recycling is discussed. Industrial and commercial waste recycling and household waste recycling are discussed in detail with examples. Examples of recycling of particular types of waste, i.e., plastics, glass, paper, metals and tyres, are reviewed. Economic considerations of recycling are discussed. Current examples of recycling schemes for household waste, and particular types of waste are described.

4.1 Introduction

The hierarchy of waste management places waste reduction at the top, followed by re-use, recovery and finally disposal. This chapter deals with waste reduction, re-use and recycling.

4.2 Waste Reduction

Waste reduction is both environmentally and economically beneficial both to society as a whole and to business and the community. Waste reduction is synonymous with waste minimisation and has been defined as any technique, process or activity which either avoids, eliminates or reduces a waste at its source, usually within the confines of the production unit (Crittenden and Kolaczkowski 1995). A large proportion of the costs of waste treatment and disposal are borne by either business or community charge payers. However, recent surveys have shown that many companies do not regard waste management as a key business area, and many do not even measure waste costs. In addition, it has been found that smaller companies have less regard for waste minimisation and recycling as an issue than larger companies (West Yorkshire Waste Management Plan 1996). Waste treatment and disposal by industry has often been regarded as an 'end of pipe' process of disposal rather than waste reduction within the manufacturing process which may also reduce costs and increase process efficiency. Environmental and cost savings arise from less waste being processed with, for example, savings in energy costs, waste storage space, transport costs, administrative costs, and lower emissions to air, water and on to land. Reducing waste in the production process in industry can also reduce the amount of raw material inputs in addition to final disposal costs. Reduction in waste costs need not always equate with reduction in waste amount. Reduction in the toxicity of waste will also reduce costs due to the lower costs of treatment compared with more hazardous wastes (Crittenden and Kolaczkowski 1995).

New incentives to reduce waste at source have been introduced with the increases in disposal costs, for example the landfill tax. Also, the introduction of ever more stringent legislation and regulation of the emissions from industrial processes to air, water and on to land through Integrated Pollution Control has highlighted that waste is a significant cost element for business.

Waste reduction at source involves good practice, input material changes, product changes and technological changes (Figure 4.1) (Crittenden and Kolaczkowski 1995). Good practice can be as simple and effective as good housekeeping, maintaining production procedures, minimising spillage's and proper auditing of input and final destination of raw materials. There are many examples of where a change in input material has resulted in a reduction of waste in the production process. For example, replacement of organic solvents with water-based types, or replacement of hazardous solvents such as benzene and chlorinated organic solvents with less hazardous solvents. Product changes can be used to reduce the production of waste or materials used within the manufacturing process, and also when the consumer eventually disposes of the product. For example, Table 4.1 shows how changes in product design have been used to reduce the weight of packaging material used by industry, with a consequent reduction in the post-consumer waste stream (Making Waste Work 1995). Technological changes include retrofitting and development of cleaner processes. A change in technology can mean a fundamental change in the process, change in process conditions, change to an automated system or re-engineering of the process.

A number of waste reduction projects involving groups of companies have been set up in the UK with Government aid, for example, the Catalyst project based in the Merseyside area, the Aire and Calder project in Yorkshire, and the River Dee project in Wales (Fletcher and Johnston 1993; The Waste Manager 1994a,b; Making Waste Work 1995). The projects highlight waste reduction measures through the identification of the 'best practice' in the manufacturing process which reduces waste and which can be used throughout industry. In most cases the savings from reduction in the waste going for treatment and disposal are small; the largest

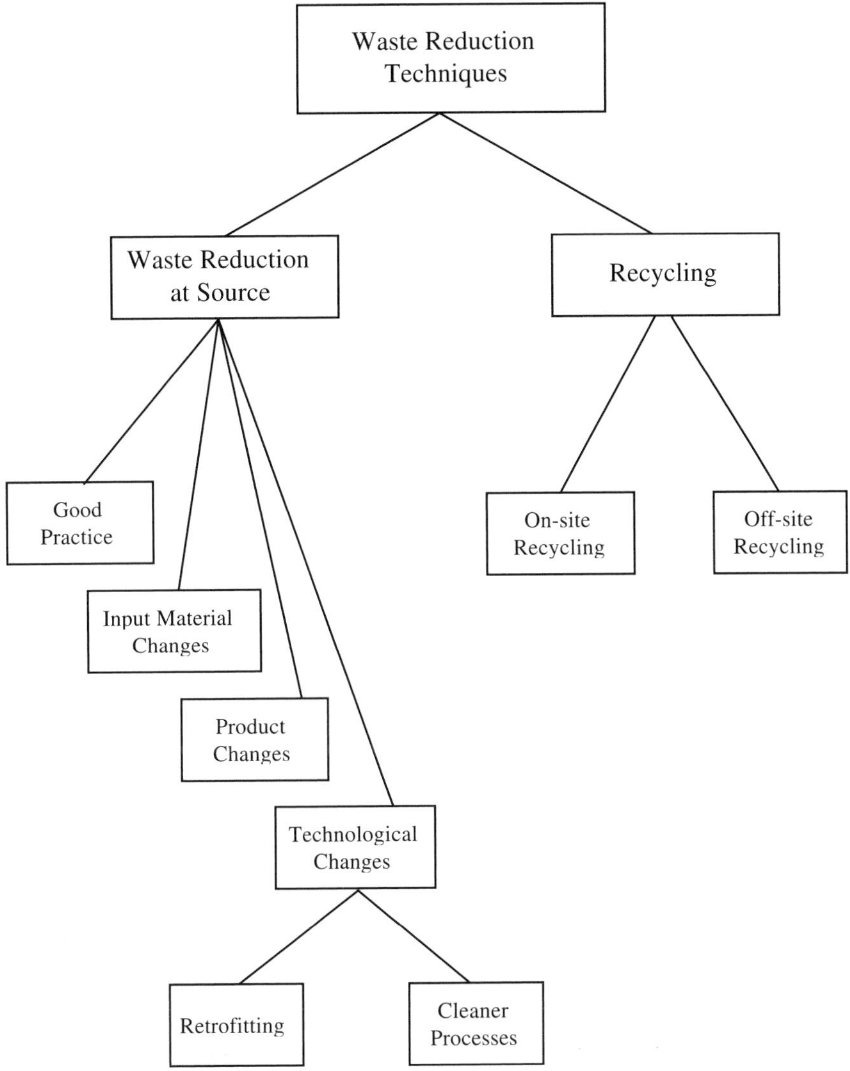

Figure 4.1 *Techniques for waste reduction. Source: Crittenden B. and Kolaczkowski S., Waste Minimisation: A Practical Guide. Institution of Chemical Engineers, Rugby, 1995. Reproduced by permission from the Institution of Chemical Engineers*

savings coming from reduction in the use of raw materials by changes in the process. In addition, targets for reduction of the overall waste produced by industry are to be set by the Government in 1998. In formulating Waste Management Plans and regulating industrial processes under Integrated Pollution Control, the Environment Agency promotes waste reduction as a means of reducing pollution to either air, water or land. Box 4.1 shows examples of waste reduction in industry.

Table 4.1 *Examples of packaging reduction*

Product	Year	Weight/thickness
Cardboard box	1970	559 g
	1990	531 g
Metal food can	1950	90 g
	1980	58 g
Plastic yoghurt pot	1965	12 g
	1990	5 g
Metal drinks can	1960	45 g
	1990	17 g
Plastic carrier bag	1970	47 micron
	1990	25 micron

Source: Making Waste Work. Department of the Environment and the Welsh Office, HMSO, London, 1995.

Box 4.1
Examples of Waste Reduction in Industry

Example 1
The most commonly used salts in industry for chromium plating materials are the hexavalent form, Cr^{6+}. A UK company, Jebron Ltd., decided to introduce the chromium trivalent form, Cr^{3+}, in their new metal finishing plant because of its much lower toxicity than the hexavalent form. Lower costs resulted due to the lower costs of treating the plant waste effluents. Previously, treatment involved a first stage conversion of the Cr^{6+} to the less toxic Cr^{3+} before further treatment prior to the disposal of the effluents. In addition, the company reported better end-results for the plating process, including personnel working in a safer environment, a better coating of chromium and a better colour on the finished product.

Example 2
The US company 3M makes flexible electronic circuit boards from copper sheeting. Cleaning of the sheeting involved the use of ammonium persulphate, phosphoric acid and sulphuric acid, creating a hazardous waste that required expensive special handling and disposal procedures. Replacement of the chemical cleaning process with a physical abrasion process where the sheeting was scrubbed with pumice stone produced a non-hazardous sludge which could be disposed of in a conventional landfill. The costs of disposal were significantly reduced, and the payback period for the installation of the new process plant was less than 3 years.

Example 3
Arjo Wiggins Ltd. operate a paper mill with large throughputs of water, with consequent high charges for the water and also high sewerage charges for the disposal of the effluent water. An audit of the water usage and disposal in the company resulted in the implementation of a programme of replacement of water lines, recycling of cooling water, installation of backwater tanks and changes to the process, control and operation. There resulted a 77% reduction in water consumption and a consequent reduction in waste water effluent, with an overall reduction in costs.

Sources: [1] Crittenden B. and Kolaczkowski S., Waste Minimisation: A Practical Guide. Institution of Chemical Engineers, Rugby, 1995; [2] Pescod M.B. (Ed.), Urban Solid Waste Management, World Health Organisation, Copenhagen, Denmark.

4.3 Waste Re-use

Re-use involves using a product or package more than once or re-using it in another application. Examples include re-using plastic supermarket carrier bags and glass milk bottles, retreading partly worn tyres and re-using car parts via car scrap merchants. Re-using a product or packaging extends the lifetime of the material used and therefore reduces the waste quantity requiring treatment and disposal. In addition, other savings for the environment include the costs of producing the replaced item, which would include energy, materials and transport costs. Re-use of materials can also take the form of new use applications in a different environment than that of the original function, for example, the use of tyres for securing covers on silage mounds and for boat/dock fenders (Making Waste Work 1995).

The re-use of beverage bottles was very common until the 1980s using a 'deposit refund' system, a small charge on the bottle, which was refunded when the bottle was returned. These schemes were widely used and cost-effective, since collecting and washing the returned bottles was more economic than manufacturing new ones. However, with the introduction of new materials and production technology and changes in consumer preferences, the deposit refund system has declined. Some schemes have still survived, such as the traditional re-use of milk bottles, and with a more environmental public attitude, business has responded with the introduction of new schemes. For example, some supermarkets encourage the re-use of plastic carrier bags with the incentive of supporting charities. Other countries have extended the deposit refund scheme, for products such as batteries (Denmark and the Netherlands), disposable cameras (Japan) and even car bodies (Sweden and Norway). Such schemes have mixed success, where the consumer inconvenience of returning the item is not outweighed by financial gain then there is a lower rate of return. Re-use schemes where industry or business have full control of the product or packaging are easier to implement. For example, transit packaging such as polystyrene chips, and cardboard used to transport items between different destinations within the same company can be re-used several times.

Re-use of glass bottles, for example, means that savings are made in raw materials and energy used. However, re-use also has an environmental impact. The re-use of the glass bottle involves the energy used in collecting and transporting, with their associated impacts. In addition, large amounts of detergents, water and chemicals are required to ensure adequate cleaning to meet hygiene regulations. A further factor is that re-useable containers need to be more robust than a single-use container, which means a thicker or heavier container with an increased use of resources.

4.4 Waste Recycling

Recycling is the collection, separation, clean-up and processing of waste materials to produce a marketable material or product. Recycling can take place within the

Table 4.2 *Average US waste composition in relation to waste production source (%)*

Waste component	Multi-family residence	Hotel	Schools	Office building	Shopping centres	Construction site
Mixed paper	10.3	11.3	28.6	43.0	11.8	2.2
Newsprint	33.7	11.5	2.5	3.5	6.7	0.5
Corrugated board	10.0	27.5	26.1	21.5	37.4	17.4
Plastic	8.6	13.7	6.9	3.7	8.0	3.8
Yard waste	5.3	0.5	0.0	0.0	2.8	3.2
Wood	1.7	0.7	0.1	8.5	3.6	27.1
Food waste	13.0	4.9	6.7	2.2	12.4	1.0
Other organics	7.4	7.5	0.1	1.2	1.9	2.3
Total combustibles	90.0	77.6	71.0	83.6	84.6	2.3
Ferrous	2.5	6.9	8.2	0.6	3.6	17.5
Aluminium	1.1	2.7	1.5	0.4	0.7	0.1
Glass	5.9	11.9	6.2	0.8	2.7	1.0
Other inorganics	0.3	0.9	13.1	14.6	8.6	23.9
Total non-combustibles	9.8	22.4	29.0	16.4	15.6	42.5
Total	100	100	100	100	100	100

Source: Warmer Bulletin, 49, Journal of the World Resource Foundation, High Street, Tonbridge, May 1996. Reproduced by permission from the World Resource Foundation.

manufacturing process, such as in the paper industry where surplus pulp fibres, mill off-cuts and damaged paper rolls are recycled back into the pulping process. Alternatively, recycling takes place at the post-consumer stage, where paper can be collected separately or extracted from the waste and then re-enter the paper-making process. The advantages of using recyclable materials means that there is reduced use of virgin materials with consequent environmental benefits in terms of energy savings in the production process, reduced emissions to air and water and on to land. Recycling may not always be the best environmental or economic option for a particular type of waste, and a full analysis of the processes involved in recycling versus treatment and disposal should be made.

Detailed analysis of the waste-producing sectors can highlight the potential areas where recycling of specific materials would be most beneficial. Table 4.2 shows average waste composition from the US in terms of the production source of the waste (Warmer Bulletin 49, 1996a). The opportunities for recycling various components of the waste streams associated with each source are very different. For example, paper recycling can be identified as a potential recycling opportunity from office buildings, family residences and schools, plastic waste and glass from hotels, and wood waste and ferrous metal from construction sites.

Waste recycling in general in the UK has shown a steady increase throughout the 1990s as the environment and environmental issues have achieved a higher prominence in the media and with the public. Table 4.3 shows the proportion of recycling of different wastes in the UK (Making Waste Work 1995). Some wastes, such as construction and demolition waste and sewage sludge, have a relatively high proportion of recycling, whilst in the household and commercial sectors the

Table 4.3 *Proportion of recycling in the UK by waste sector*

Waste sector	Proportion recycled (%)
Household	5
Commercial	7.5
Construction and demolition	30
Other industrial	18
Total controlled wastes	21

Source: Making Waste Work. Department of the Environment and the Welsh Office, HMSO, London, 1995.

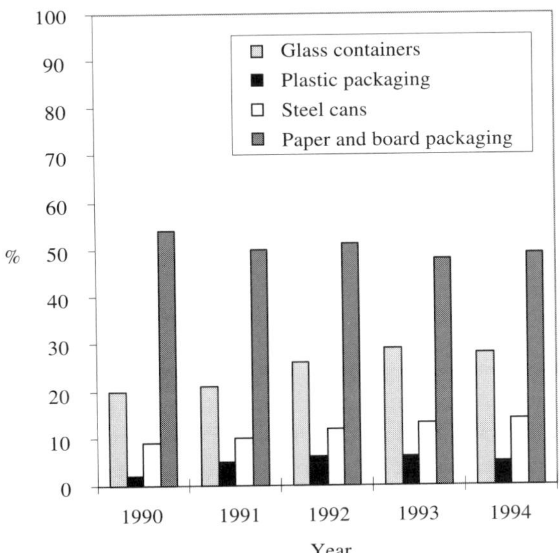

Figure 4.2 *UK recycling rates. Source: Making Waste Work. Department of the Environment and Welsh Office, HMSO, London, 1995. Crown copyright is reproduced with the permission of the Controller of Her Majesty's Stationery Office*

recycling rates are low. However, in some cases individual waste streams have shown high percentage rates of recycling, boosted by increased collection of the streams from household waste sources. This in turn has been encouraged by an increase in the number of locally organised facilities for recycling specific types of waste.

Figure 4.2 shows the recycling rates for glass containers, plastic packaging, steel cans and paper and board packaging through the 1990s (Making Waste Work 1995). By comparison, Figure 4.3 shows the increase in UK recycling facilities (Making Waste Work 1995). The convenience of so many locally situated recycling facilities has clearly influenced the recycling rates of the associated waste materials. The trends in material recycling have also been influenced by the markets for the segregated material. For example, the paper and board packaging recycling rate

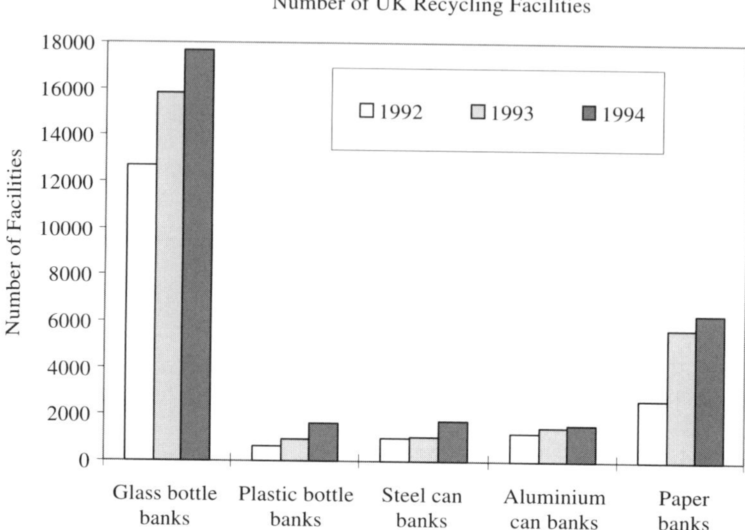

Figure 4.3 *Number of UK recycling facilities. Source: Making Waste Work. Department of the Environment and Welsh Office, HMSO, London, 1995. Crown copyright is reproduced with the permission of the Controller of Her Majesty's Stationery Office*

showed a fall between 1990 and 1994 due to the oversupply of waste paper for the pulp and paper industries. This market has recently recovered, but illustrates that recycling is susceptible to the market price and price stability for the recycled material. In this respect the recycled material is in direct competition with the virgin product. The market is also a two-way process; the industrial user of the recycled material requires a secure supply of material and confidence that the consumer market for the processed recycled material will be stable and grow in the future, since investment costs in the necessary recycling processing plant is high with long pay-back periods. For example, waste paper processing plants in Aylesford in Kent and Gartcosh in Lanarkshire each require investment of over £200 million and have capacities of over 400 000 tonnes per year. Similarly, the recycling of aluminium drinks cans in a plant in Warrington, Lancashire, has investment costs of £28 million and planned throughputs of 50 000 tonnes per year (Waste Management Paper 28, 1992).

Throughout the world, many countries have experienced shortages in waste disposal landfill capacity. The response from these countries has been to implement waste management policies via waste reduction and recycling schemes to reduce the amount of waste entering the waste stream. The recycling rates in the UK may be compared with other countries throughout the world (Table 4.4) (McCarthy 1993). However, the comparisons should be made with some caution as the definitions of recycling will vary from country to country. For example, recycling may or may not include energy recovery as a form of recycling, and some recycling may take place before it is collected in the municipal solid waste stream.

Table 4.4 *Comparison of world-wide recycling rates (%)*

Country	Recycled material		
	Paper and board packaging (1987)	Glass containers (1989)	Aluminium cans (1990)
US	25.6	–	64
Japan	49.6	–	42
Germany	40.5	53	–
Spain	41.4	24	–
Canada	18	–	60
Portugal	44.1	14	–
Netherlands	52.8	57	–
France	36.2	38	–
Italy	23.3	42	10
Sweden	40	34	83
Australia	31.8	–	62
Belgium	33.3	60	–

Source: McCarthy J.E., Resources, Conservation and Recycling, 8, 293–360, 1993.

4.4.1 Legislative Background to Recycling

There are a number of legislative measures in the UK to encourage recycling (Waste Management Paper 28, 1992; Environment Act 1995; Making Waste Work 1995).

The 1990 Environmental Protection Act placed a duty on the Waste Collection Authorities to prepare Recycling Plans for household and commercial waste. The Collection Authorities also have the power to require householders to sort their waste into different recyclable categories. Waste Disposal Authorities also have a duty to provide civic amenity waste disposal sites. The 1990 Act also introduced a system of recycling credits. The system is a mechanism for passing on to the recycler the cost savings arising from the Waste Disposal Authority not having to dispose of so much waste. The recycler may be the Waste Collection Authority or a third party recycler.

The 1995 Environment Act introduced the principle of 'producer responsibility'. The principle currently extends to certain waste sectors, including packaging, newspapers, automotive batteries, consumer batteries, electrical and electronic goods, tyres and motor vehicles. The 1995 Act empowers the Secretary of State for the Environment to impose producer responsibility on these sectors to increase the re-use, recovery or recycling of products which they have produced. The strategy used by the Government to achieve the increases is via targets set in consultation with the industrial or business sectors involved. The initiative seeks to increase the share of responsibility in dealing with the waste which arises from these products. The producer bears some of the costs of waste disposal in that they must introduce measures to re-use and recycle waste to meet the imposed targets. The incentive therefore encourages waste producers to reduce the waste arising to minimise the costs of waste management.

One of the aims of the introduction of the Landfill Tax was to ensure that landfill costs reflect the environmental impact of landfilling waste to business in a cost-effective and non-regulatory manner. The incentive to business to reduce costs is therefore to produce less waste, to re-use and recycle waste, and to dispose of less waste into landfill. As a consequence of the introduction of the landfill tax, some landfill operators are separating mixed wastes to produce hardcore, soils and wood for sale, thereby reducing the amount of waste going to landfill and contributing to waste recycling (ENDS 1997a). The introduction of the proposed EC Landfill Directive with a more stringent monitoring and landfill gas and leachate emission regime will also increase the costs of landfill and move the treatment of waste higher up the waste management hierarchy.

The EC Packaging Waste Directive 1994 requires recovery of between 50 and 65% of all packaging waste, of which between 25% and 45% must be recycled and with a minimum of 15% for each material. In addition, the Directive also requires that Member States of the European Union implement programmes to prevent the production of packaging waste. The UK, in consultation with the packaging industry, through the Producer Responsibility Group has set a target of 58% recovery, of which 50% would be recycled and 8% incinerated, to be achieved by the year 2000. The implementation of the EC Directive within the UK will dramatically influence the recycling of packaging waste, which makes up about 20% of household waste and a major proportion of commercial and industrial waste (Making Waste Work 1995).

Whilst there are a range of regulations and measures in place to encourage recycling, there is still a voluntary sense about recycling and in most sectors of the economy recycling remains relatively low. However, the increasing influence of the European Union through a range of Directives on waste management, for example, the Landfill, Packaging Waste and Waste Framework Directives, and the development of the hierarchy of waste management and introduction of producer responsibility will inevitably lead to a more regulated environment and increase in waste recycling.

4.4.2 Household Waste Recycling

The composition of household waste is shown in Figure 4.4 (Waste Management Paper 28, 1992). The theoretically recyclable components include paper and board, plastics, glass, metals and putrescible materials. However, in some cases it is not possible to recycle some of the wastes due to contamination. Table 4.5 shows an estimation of the *potentially* recyclable components of household waste. Approximately 40% of household waste is potentially recyclable after discounting the contaminated materials. In addition, the putrescible waste such as food and garden waste makes up over 20% of household waste, and it is estimated that about half of this material could be potentially recyclable through the composting process.

Whilst the recyclable materials are present in household waste, they are present in a very heterogeneous matrix, and the segregation of the materials is one of the major factors involved in waste recycling. Two types of system exist to reclaim the

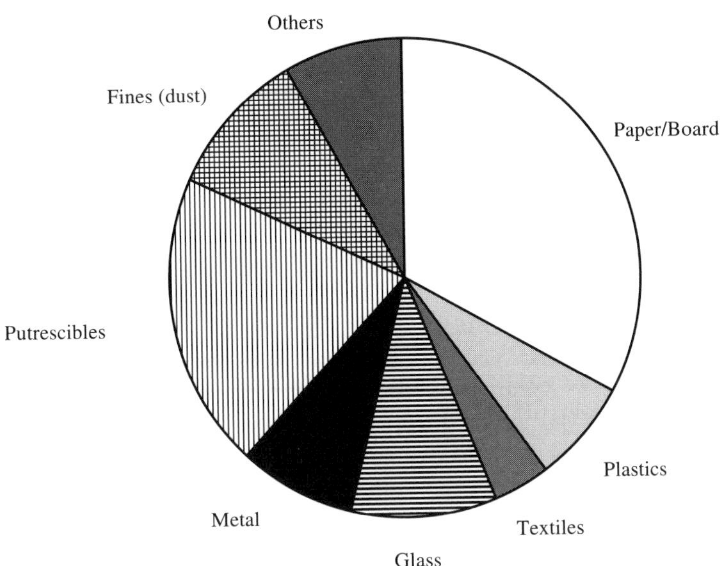

Figure 4.4 *Typical composition of collected household waste in the UK. Source: Waste Management Paper 28, Recycling. Department of the Environment, HMSO, London, 1992. Crown copyright is reproduced with the permission of the Controller of Her Majesty's Stationery Office*

Table 4.5 *Potentially recyclable components of household waste*

Material	Generation rate (kg/household/year)	Maximum recyclable (kg/household/year)	Recyclable percentage (%)
Paper	200	120	60
Plastic film	24	14	60
Plastic articles	18	13	70
Textiles	24	12	50
Glass	60	54	90
Ferrous metal	42	34	80
Aluminium film/cans	4	3	70
Other material[1]	228		
Total	600		

[1] Other material contains putrescible materials, some of which may be compostable.

Source: Waste Management Paper No. 28, Recycling. Department of the Environment, HMSO, London, 1992. Crown copyright is reproduced with the permission of the Controller of Her Majesty's Stationery Office.

materials separately, the 'Bring' and the 'Collect' systems (Waste Management Paper 28, 1992; White et al 1995; ERRA 1994a). The 'Bring' systems involve the segregation of recyclable materials, for example, paper, plastic and glass bottles, metals and textiles from household waste by the public and delivery to a centralised collection site. The sites may be bottle and paper banks situated at the local supermarket, civic amenity sites for the disposal of many types of material, or the local scrap merchant.

This system has the advantage of being low in capital costs, easily accessible and can provide an easy method of segregating clean readily marketable materials. However, the take up of the schemes by the public can be low, with average rates of recovery in the UK being between 15 and 20%. In addition, the sites can become unsightly with litter spillage and can be an attraction for vandalism. The 'Bring' schemes in the UK have involved collaborations between local authorities, waste management companies, voluntary groups and materials manufacturers. The collected material are taken either directly to a materials reprocessor or to larger recycling facilities where further processing takes place to sort, clean and grade the materials before transfer to the reprocessing plant.

The 'Collect' systems involve house to house kerbside collection of designated recyclable materials, source-separated by the householder and placed in separate containers. A number of schemes are operating in the UK, for example, in Sheffield, Milton Keynes, Leeds, Bury, Cardiff, Stocksbridge, Chudleigh (Devon), Adur (West Sussex), Fife, Dundee and Falkirk, and have names such as 'the blue box', the 'green bin' or the 'green bag' scheme. The schemes cover large sections of the city populations, for example, in Leeds, 80 000 households are covered, Milton Keynes, 70 000, Cardiff, 55 000 and Falkirk 42 000. There are a number of variations on the 'Collect' system. For example, the recyclable materials such as paper, plastics and metal cans are all placed in one container, therefore the mixture has to be sorted, either by processing equipment or by hand at the central materials recycling facility. Alternatively, the materials may be sorted at the kerbside by the collector. More sophisticated systems involve the separation of the recyclable materials into several containers or sections of a container by the householder for separate collection. The latter two systems require a more elaborate collection vehicle to collect the separated waste streams. The advantages of the 'Collect' system include convenience for the householder and higher recovery rates of recyclable materials, for example, up to 80% recovery of paper and plastics from the normal household waste stream (Table 4.6). However, collection costs are higher in that separate collections or purpose built vehicles with separate enclosures are required. In addition, costs associated with the sorting of the materials and transport to the reprocessing facility are extra costs. However, these are offset by the income from the sale of the recycled materials.

Table 4.6 shows the UK collection systems for the recycling of household waste (Atkinson and New 1992, 1993; Atkinson et al 1993). The schemes have been successful in increasing the rates of recycling of household waste. Individual materials achieve high rates of reclamation, for example, glass collection rates using the 'Collect' system are up to 71% and paper 67%. Generally 'Bring' systems require more motivation and effort on behalf of the householder and usually achieve lower

Table 4.6 *Comparison of UK 'bring' and 'collect' systems, percentage collection*

	Paper (%)	Glass (%)	Plastic (%)	Cans (%)	Textiles (%)	Diversion rate (%)
Kerbside						
Separate wheeled bins						
Leeds (biweekly)	70–80	n/c	70–80	30–40[1] 50–60[2]	40–50	50
Blue box						
Stocksbridge, Sheffield	28	45	12	17	n/a	6.6
East Sheffield	52	66	28	14	32	15.3
Milton Keynes	57	44	57	24	n/c	18.7
Adur	67	71	60	54	n/c	27
No container (biweekly)						
Chudleigh, Devon	36	55	n/c	21	3	6.9
Plastic sack (biweekly)						
Cardiff	52	52	13	6	21	17.7
Bring system						
Rydale	13	40	n/c	n/c	3	4.1
Richmond, Surrey	18	61	n/c	3	8	8.2

n/c, not collected; n/a, not available.

Sources: [1] Atkinson W., Barton J.R. and New R., Cost assessment of source separation schemes applied to household waste in the UK. Warren Spring Laboratory, Report No. LR 945, National Environmental Technology Centre, Harwell, 1993; [2] Atkinson W. and New R., Kerbside collection of recyclables from household waste in the UK. Warren Spring Laboratory, Report No. LR 946, National Environmental Technology Centre, Harwell, 1993; [3] Atkinson W. and New R., Monitoring the impact of bring systems on household waste in the UK. Warren Spring Laboratory, Report No. LR 944, National Environmental Technology Centre, Harwell, 1992.

levels of reclamation. The exception is glass, where the 'Bring' system is well established in the UK and can achieve recovery rates of over 60%.

Whilst individual materials recovery rates can reach high levels, the overall rate of recycling for household waste in the UK is only approximately 5% (Making Waste Work 1995). However, some local authorities have reached significant levels of recycling and approach the UK Government target of recycling 25% of all household waste by the year 2000. For example, 10 councils in England and Wales have reached recycling rates above 17%, including Adur, Weymouth and Portland, and Castle Morpeth at 21% (ENDS 1997). Comparison of recycling rates is difficult, however, since local authorities use different criteria in calculating rates. For example, some councils include an *estimate* of home composting in their calculations.

Alternative to the bring and collect systems are centralised materials recycling facilities where the household waste is brought to a central plant for reclamation and recycling (ERRA 1994a). The waste can be segregated into recyclable materials, partially segregated or completely unsegregated (ERRA 1994a; Warmer Bulletin 1995a). The number of components in the waste will be influenced by the degree of pre-segregation, which will influence the sorting and separation technology or manpower required for the materials recycling facility. Recycling facilities for unsegregated materials are designed to process household, commercial and industrial wastes. Inevitably, the materials are contaminated, for example, with broken glass, food stuffs etc., and recovery rates of recyclable materials are low, of the order of 10–30% (Warmer Bulletin 1995a). However, where one type of material from a specific source is involved rates can be much higher, for example, general office waste may contain levels of paper of over 80% by weight (Waste Management Paper 28, 1992). Segregated material streams handled by materials recycling facilities process a set of particular materials, typically between three and eight components, which may be separated or mixed. Such facilities handle 'clean' waste and consequently contamination levels are low and recovery rates are high.

The design of a typical unsegregated municipal solid waste recycling facility is shown in Figure 4.5 (Warmer Bulletin 1995a). The example is for a facility which can recover ferrous metals, high density polyethylene (HDPE) and polyethylene terephthalate (PET) plastics, aluminium and several grades of paper, i.e., corrugated, newspaper and high grade paper. The design would include mechanical and manual separation processes. At the input stage, large quantities of pre-segregated materials such as old corrugated cardboard would not require processing and therefore can be fed directly to the baler. The stages of separation include trommel screening, magnetic separation and manual sorting. Manual sorting is necessary to separate different types of plastic and different coloured glass, although the trend is towards an increase in mechanisation of the process. An unsegregated municipal solid waste materials recycling facility would recover approximately 15% of the waste stream as useable materials. The remaining 85% is largely organic and can be used to produce a fuel (refuse derived fuel, RDF), converted to compost or landfilled (Warmer Bulletin 1995a).

Figure 4.6 shows a materials recycling facility to handle a mixture of containers composed of different coloured glass, aluminium and ferrous cans, and plastic containers composed of high density polyethylene (HDPE) and polyethylene terephthalate (PET) (Warmer Bulletin 1995a). Whilst the separation processes are very similar, the pre-segregation of the materials before they arrive at the facility mean that very high recovery rates are possible compared with unsegregated materials recycling facilities. Such systems can achieve recovery rates of 90% of the incoming segregated waste material in the form of useable end products.

In some cases the materials recovery facility may be designed with a compost or a refuse derived fuel (RDF) as the end material, with other recyclable materials as the by-products. For composting from municipal solid waste, the stages of sizing, and metal and glass removal would be essential to provide a compost suitable for marketing (Warmer Bulletin 1995a). More successful schemes have relied on source-separated organic wastes such as garden waste which are processed at centralised composting facilities, for example a local authority scheme operates in the North London area.

There are several recycling schemes throughout Europe and North America (ERRA 1996; Steuteville 1996), and in the UK, at Leeds, Milton Keynes, Hampshire, Cardiff, Sheffield and Adur. Box 4.2 describes the project at Adur. Throughout Europe there are numerous materials recycling schemes, for example, Barcelona in Spain, Prato in Italy, Dunkirk in France, Dublin in Eire and Lemsterland in the

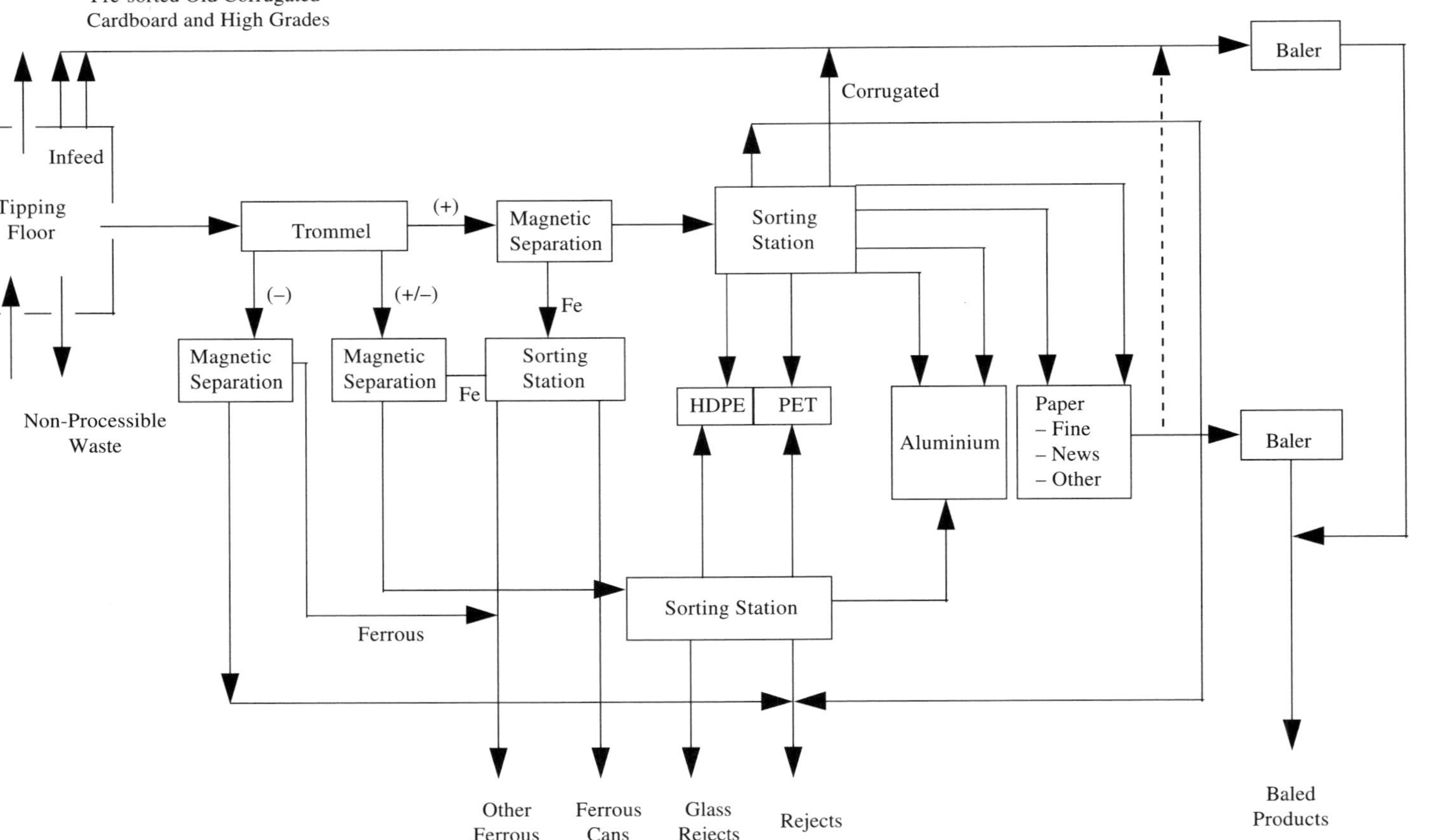

Figure 4.5 *Example of the design of an unsegregated materials recycling facility for municipal solid waste. Source: Warmer Bulletin, Technical Brief. Materials Reclamation Facilities. Warmer Bulletin, Journal of the World Resource Foundation, Tonbridge, 1995. Reproduced by permission from the World Resource Foundation*

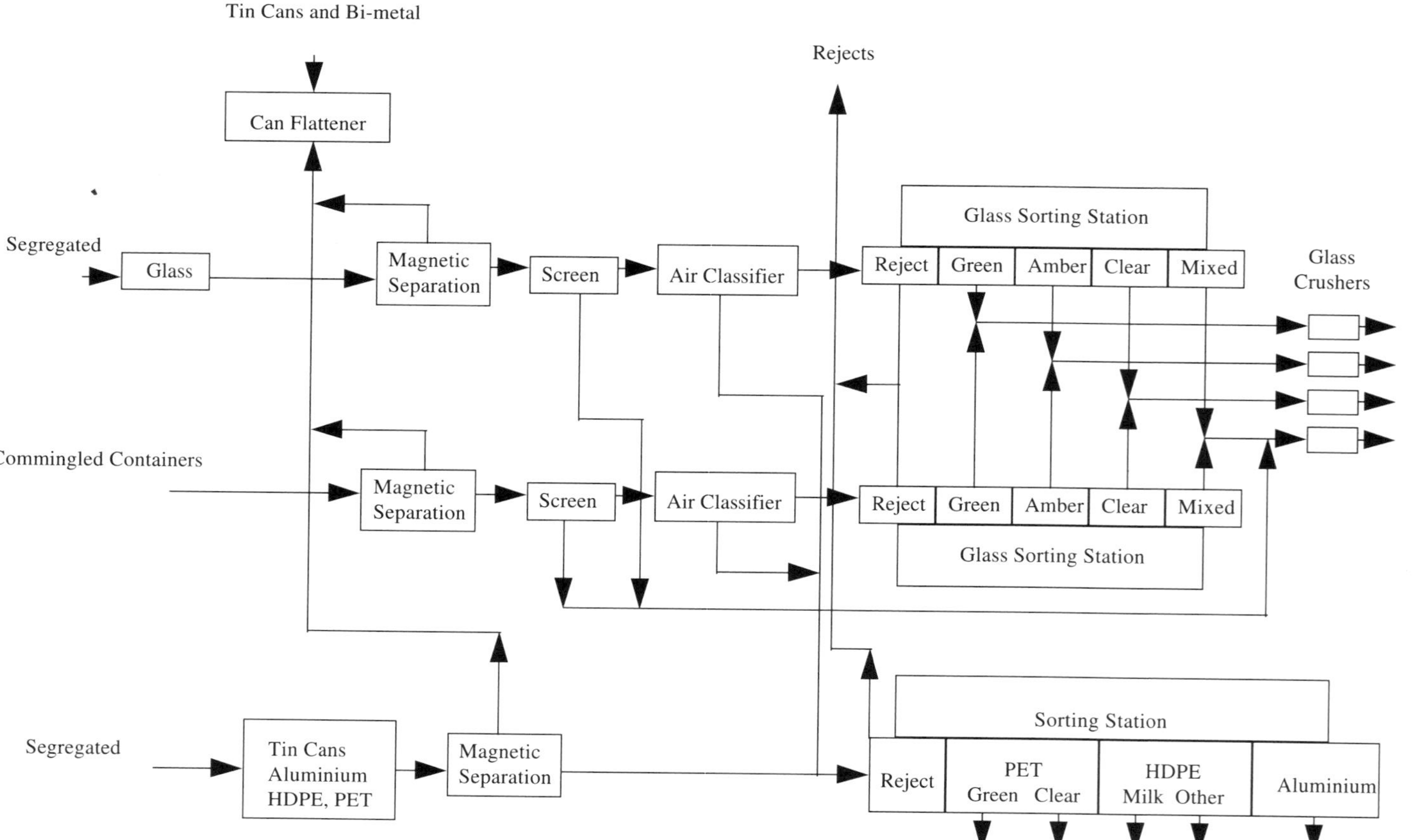

Figure 4.6 *Example of the design of a segregated materials recycling facility for glass, metal cans, high density polyethylene (HDPE) and polyethylene terephthalate (PET) plastic containers. Source: Warmer Bulletin, Technical Brief. Materials Reclamation Facilities. Journal of the World Resource Foundation, Tonbridge, 1995. Reproduced by permission from the World Resource Foundation*

Box 4.2
The Adur District Council, West Sussex, Waste Recycling Plan

The main aim of the Adur District Council Waste Recycling Plan is to reduce the average weekly household waste within the district to no more than 7.0 kg/household/week by the year 2000. The average national weekly waste is 11.5 kg/household/week. The target of 7.0 kg/household/week follows the original Adur Plan of 1991, which attained a reduction in waste to 8.2 kg/household/week in 1994, representing a recycling rate of over 25%. The composition of the waste stream dealt with by the Council, which is typical of UK waste, is shown below. In addition, the ACORN groupings for the area are also shown.

Waste Composition		ACORN Groupings	
Paper and card	33%	B – Modern family housing, higher incomes	13.6%
Garden and kitchen	14%	C – Older housing of intermediate status	25.7%
Glass	9%	E – Council estates, Category I	9.1%
Plastics	9%	F – Council estates, Category II	3.4%
Metals	7%	I – High status non-family areas	0.1%
Textiles	3%	J – Affluent suburban housing	31.9%
Other Waste	25%	K – Better off retirement areas	15.4%
		Other	0.8%

The recycling system is based on a kerbside collection involving four types of recyclables, glass, plastics, cans and newspapers, using a 'blue box' in addition to a 'bank' system and a composting scheme. The segregated waste in the blue box is collected once per week by a specially divided recycling vehicle. The recyclable waste is separated into the four waste streams at the vehicle. The materials are processed at a Materials Recovery Facility operated by West Sussex County Council and a private sector waste management company. The facility further sorts the materials, which are then sent for reprocessing. There is also a system of banks for segregation of the four types of waste for blocks of flats etc. Community recycling schemes operate through community groups and charities. Home composting is promoted by the use of free composting units, and 25% of the households in the district compost their own kitchen and garden waste. Income from the scheme is through sale of recycled materials and from recycling credits. The Adur scheme also receives large grants and sponsorship to offset the costs of the scheme. The economics of the scheme are shown for 1994/5.

	£/household/year
Costs	
Household waste collection cost	16.62
Household waste disposal cost	18.68
Net cost of blue box and mini recycling banks	4.64
Income	
Sale of materials and recycling credits	2.75
Grants and sponsorship (blue box scheme)	8.33
Total paid by Adur council tax payers	4.47

Source: Adur District Council Recycling Plan Summary 1996. Adur District Council, Shoreham-by-Sea, 1996.

Netherlands. The Prato scheme involves a population of approximately 170 000, living mainly in apartment blocks. The recycling scheme involves 11 750 households, with each being provided with a re-useable bag in which to store recyclable materials. The materials chosen for recycling by the local council were paper and board, glass, metal cans (steel and aluminium), beverage cartons, plastic bottles (polyethylene terephthalate (PET), polyvinylchloride (PVC) and polyethylene (PE)). The recycled materials are delivered by the householder to large blue bins located close to the general household waste bins, which are in the street where the waste is normally delivered by the householder. The bins are emptied twice per week and taken to a purpose-built materials recycling facility. The potential diversion rate of recycled materials diverted out of the household waste stream was 29%, but an actual diversion rate of 10% was achieved (ERRA 1994b). Barcelona has a similar household profile of high-rise apartment blocks and involves a similar householder segregation scheme to that of Prato. The recycled materials (paper/board, glass, plastic, beverage cartons, metal cans) are delivered by the householder to large bins located on the street next to the normal household waste bins. Separate containers for glass and paper/board are also used. The materials are collected daily and delivered to a materials recycling facility. The potential diversion rate for Barcelona is 38%, with the actual diversion rate being much lower at 8% (ERRA 1994b).

There are over 7300 kerbside recycling programmes in the USA serving a potential 46% of the population representing 120 million people (Steuteville 1996). In addition there are over 8500 ‘Bring’ recycling sites. Overall recycling rates have been estimated at 27%, which includes composting. It is estimated that in the USA there are approximately 1300 materials recycling facilities processing source-separated materials and 128 processing mixed solid wastes (Steuteville 1996). In over 500 of the residential kerbside recycling programmes it is a requirement by law to recycle selected materials, and in such areas the material recovery rates tend to be significantly higher than in areas operating voluntary schemes (Everett and Peirce 1993). In addition, it is reported that there are 3316 centralised composting facilities treating garden (yard) waste.

There has been a recent trend downwards in the number of landfill sites in the UK, with current estimates of 63% of municipal solid waste going to landfill. In addition, the number of landfill sites has fallen, with a trend to smaller numbers of large landfill sites. Closed landfills are being replaced with waste transfer stations with the opportunity to separate and recover materials. Incineration takes place at 156 municipal waste incinerators, mostly with energy recovery, and the energy recovery aspects are not included as part of the recycling data (Steuteville 1996).

One form of materials recycling facility is the refuse derived fuel plant, where the main end product is the production of a fuel in the form of the combustible fraction of municipal solid waste. The plant would normally process unsegregated municipal solid waste. Consequently, a refuse derived fuel plant seeks to concentrate the combustible fraction of the waste by removal of non-combustible materials such as glass and metals. Further processing to remove low calorific value materials such as putrescibles and very fine material increase the calorific value of the residual product, which consists of paper, plastics, textiles and other combustible material.

Figure 4.7 shows the processing stages in the production of a typical refuse derived fuel which is in a compressed pelletised form of size approximately 1.8 cm diameter by 10 cm long (Barton 1989). The major steps involved are preliminary liberation, where bags of waste are mechanically opened, and size screening, magnetic separation and coarse shredding, a refining separation stage and finally a series of processes to control the physical characteristics of the fuel for ease of combustion. The figure shows the weight and energy content as 100%, and the various stages show how the weight and energy content decrease with processing. The derived pellets decrease in weight to 36% of their original value but contain 70% of their original energy content. The raw waste has a typical calorific value of 9.1 MJ/kg whilst the processed refuse derived fuel pellets have a calorific value of 18.0 MJ/kg. Some plants do not produce pellets but a coarse refuse derived fuel which requires less processing and has a similar calorific value to pelletised fuel but cannot be directly substituted for coal-based combustion equipment. The refuse derived fuel pellets have been successfully fired in different combustion systems as a substitute for coal, although some problems associated with the composition of the fuel have been experienced. Table 4.7 shows a comparison of the characteristics of different refuse derived fuel pellets with a typical UK bituminous coal (Jackson 1988; (Rampling and Hickey 1988; Barton 1989). The refuse derived fuel has a 30% lower calorific value, lower bulk density, higher ash content and higher volatile content. Consequently, combustion equipment which has been designed for coal may experience problems when refuse derived fuels are used, for example, higher throughput requirements to achieve the same thermal output, shorter residence times within the combustor for complete burnout of the pellets, and greater ash and clinker formation. In addition, the formation of deposits on boiler tubes (fouling) due to the high ash with high alkaline metal content, particularly sodium and potassium, has been shown to be severe, resulting in reduced boiler efficiency. Emissions from the combustion of refuse derived fuel have been shown to be higher for heavy metals such as lead, cadmium and zinc, since these metals are in higher concentrations in the raw fuel compared with coal. Hydrogen chloride emissions are also higher due to the higher chlorine content of the fuel from, for example, chlorinated plastics and chloride fillers in paper (Rampling and Hickey 1988).

Several refuse derived fuel schemes were initiated in the UK during the 1980s, for example at Byker in the northeast of England, Doncaster, Eastbourne in Sussex, the Isle of Wight and Govan in Scotland. The schemes were based initially on full-scale municipal solid waste materials recycling facilities, with the planned recovery of several different recyclable materials. For example, the Doncaster plant was designed to recover glass, paper, ferrous and non-ferrous metals, and film plastics and produce refuse derived fuel in pelletised form. However, because of the high costs of processing, the inherent contamination in the derived recovered materials and lack of market demand for the products, those plants still operating have concentrated on the recovery of ferrous metals and the production of refuse derived fuel (Waste Management Paper 28, 1992). Refuse derived fuel plants have been developed in many countries world-wide, for example, the USA, Italy, France, Germany, the Netherlands, Canada, Spain and South Korea (ETSU 1993; Warmer Bulletin 1993). For example, the Herten Plant in Germany can produce up to 15

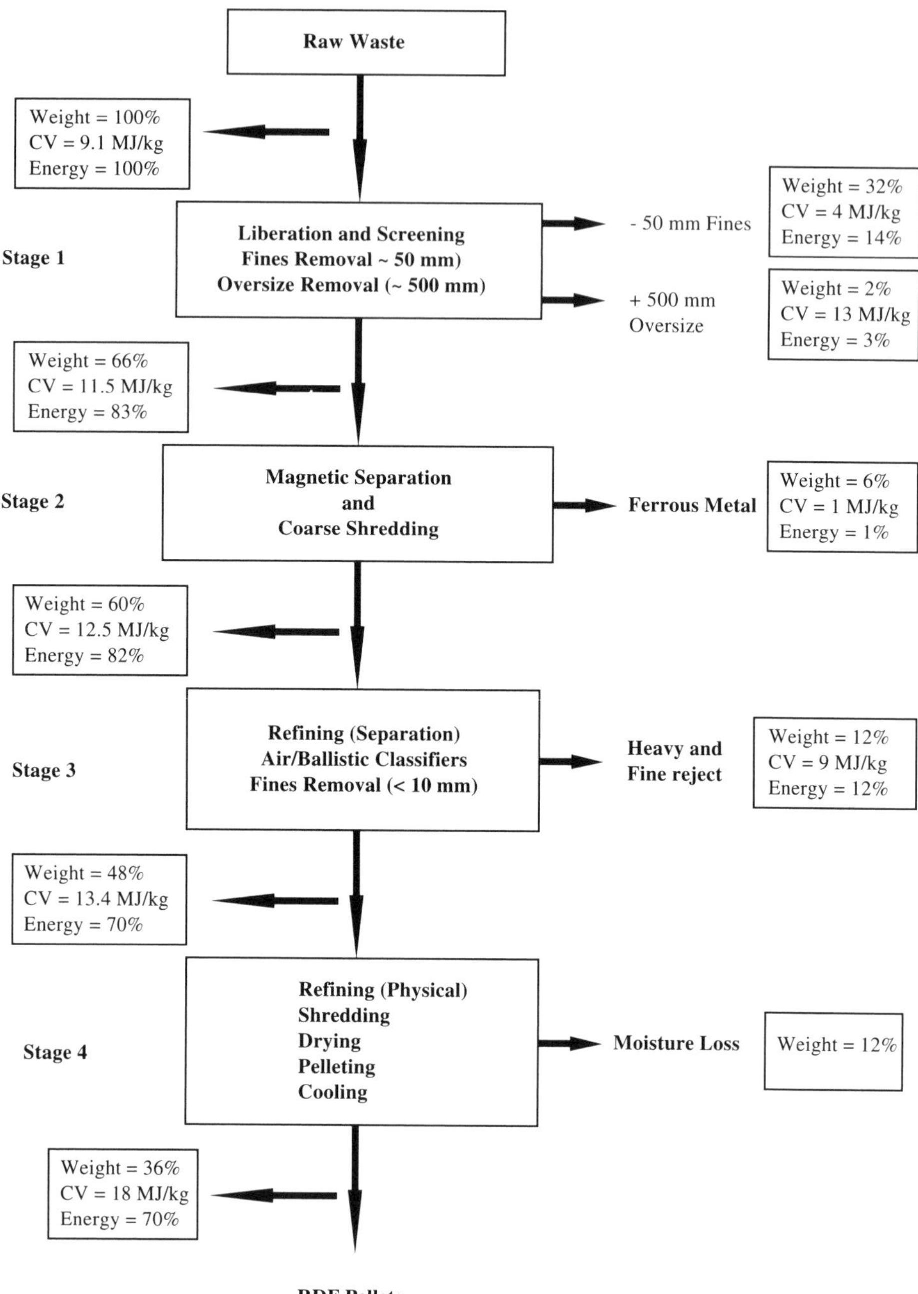

Figure 4.7 *Stages in the production and refining of refuse derived fuel pellets. Source: Barton J.R., In: Pyrolysis and Gasification, Ferrero G.L., Maniatis K., Buekens A. and Bridgwater A.V. (Eds.), Elsevier Applied Science, London, 1989*

Table 4.7 *Analysis of typical refuse derived fuel pellets compared with a typical UK coal*

Analysis	Refuse derived fuel (Byker plant)	UK coal
Calorific value	18.7	27.2
Proximate analysis (wt%)		
Volatile matter	67.5	25.9
Fixed carbon	10.2	55.5
Ash	15.0	10.2
Moisture	7.3	8.4
Ultimate analysis (wt%, dry ash free)		
Carbon	55.0	83.5
Hydrogen	7.3	5.4
Oxygen	35.9	8.4
Nitrogen	0.6	1.7
Chlorine	0.9	0.04
Fluorine	0.01	trace
Sulphur	0.3	1.0
Lead	0.02	0.003
Cadmium	0.0008	0.00001
Mercury	0.0002	0.00007
Bulk density (kg/m^3)	600	900

Sources: [1] Rampling T.W. and Hickey T.J., The laboratory characterisation of refuse derived fuel. Warren Spring Laboratory (National Environmental Technology Centre), Report LR 643 (MR), HMSO, London, 1988; [2] Jackson D.V., A review of the developments in the production and combustion of refuse derived fuel. In: Energy Recovery through Waste Combustion. Brown A., Evemy P. and Ferrero G.L. (Eds.) Elsevier Applied Science, London, 1988; [3] Barton J.R., Processing of urban waste to provide feedstock for fuel/energy recovery. In: Ferrero G.L., Maniatis K., Buekens A. and Bridgwater A.V., Elsevier Applied Science, London, 1989.

tonnes/hour of RDF and 2 tonnes/hour of ferrous scrap, in Udine in Italy a 210 tonne/day plant is in operation, and in the Edin Prairie Minnesota Plant in the USA there is a plant able to process up to 800 tonnes/day of waste (Warmer Bulletin 1993). These schemes have had more success than those in the UK, and combust the derived fuel as either pellets, or in coarse form as a direct substitute for coal, or co-fired with other fuels such as coal and biomass. For example, fluidised bed combustion of coarse refuse derived fuel has been particularly successful in Sweden, where it is co-combusted with peat or wood chips. In the USA, typically coarse refuse derived fuel is co-combusted as a 20% RDF/80% coal mixture in large boilers to produce steam for power generation, although plants combusting 100% coarse refuse derived fuel also exist in the USA (Chappell 1991; Warmer Bulletin 1993).

4.4.3 Industrial and Commercial Waste Recycling

Industrial waste recycling includes direct recycling, where waste material is recycled back into the manufacturing process in-house within the factory. For example,

Table 4.8 *Estimated potential annual recyclable materials from commercial premises*

Waste	Annual potential recyclable arisings (million tonnes)
Paper and board	0.8–0.9
Plastics	0.1–0.2
Glass	0.3–0.4
Aluminium	0.1–0.15

Source: Waste Management Paper No. 28, Recycling. Department of the Environment, HMSO, London, 1992. Crown copyright is reproduced with the permission of the Controller of Her Majesty's Stationery Office.

broken and misshapen glass would routinely be re-melted in the production process. Similarly plastic off-cuts and scrap are also recycled during the manufacturing process. Other industrial sources of waste are routinely recycled within the industry. For example, agricultural waste is mostly landfilled or used as animal feed, and consequently the material does not enter the general waste management process. Similarly construction and demolition waste is often recycled on-site as aggregate or ballast in the construction of new buildings. It is estimated that 30% of construction and demolition waste in the UK is recycled (Making Waste Work 1995). The majority of materials, which would include soil, sub-soils, concrete, bricks, timber, plaster etc., are used for bulk fill such as aggregates for road construction projects. The iron and steel industry also has a significant proportion of in-house recycling of waste metals.

Other industrial wastes are commonly recycled, but indirectly, as post-consumer waste. Commercial and industrial wastes are, by their nature, very variable in composition. Commercial waste would include waste from shops, offices, restaurants and institutions such as schools. Office waste contains a high proportion of waste paper, whilst restaurants will have high proportions of putrescible waste, but also glass, metal cans and plastic packaging. Estimates of the potential amounts of recyclable material from commercial waste are shown in Table 4.8 (Waste Management Paper 28, 1992). The figures amount to about 1.5 million tonnes for the four selected waste streams out of an estimated 15 million tonnes of commercial waste generated each year in the UK. Currently, only approximately 5% of all commercial waste is recycled.

Industrial waste will be heterogeneous in its composition, and depends on the individual business and type of on-site production. Many larger companies have separate waste collection and disposal arrangements, which may include recycling. For example, the scrap metal industry in the UK has a well-defined infrastructure and makes a major contribution to the recycling of metals (Making Waste Work 1995). Table 4.9 shows the recycling of scrap metal in the UK; the data also include some recycled metals from households (Digest of Environmental Statistics 1996).

Table 4.9 *Scrap metal recycling in the UK*

Scrap metal	Post-consumer scrap recycled as a proportion of consumption
Ferrous scrap	42
Aluminium scrap	39
Copper scrap	32
Lead scrap	74
Zinc scrap	20

Source: Digest of Environmental Statistics, HMSO, London, 1996 (1994 data).

4.4.4 Examples of Waste Recycling

4.4.4.1 Plastics

Plastic polymers make up a high proportion of waste and the volume and range used is increasing dramatically. The two main types of plastic are thermoplastics, which soften when heated and harden again when cooled, and thermosetts, which harden by curing and cannot be re-moulded. Thermoplastics are by far the most common types of plastic, comprising over 80% of the plastics used in Europe; they are also the most easily recyclable. Table 4.10 shows typical applications of the main plastic types (Warmer Bulletin 1992). Europe generates about 15 million tonnes of post-consumer plastic waste from agricultural, automotive, construction, distribution and domestic use. Figure 4.8 shows the plastics end-uses in Western Europe (Warmer Bulletin 1992). Most of the post-consumer plastic is landfilled or incinerated, and on average only 7% is recycled to produce low-grade plastic products. About 14% of European plastic waste is incinerated with other wastes, mainly as municipal solid waste, and the 'recovery' of the plastic in such cases is via energy recovery. The EC Packaging and Packaging Waste Directive accepted by the UK Government encourages waste minimisation and waste recycling. The EC Directive sets a target of recovering 50–65% of packaging waste by 2001. Not more than 10% of packaging waste can be landfilled and not more than 30% can be incinerated. In the UK plastic is collected from commercial and industrial sources as separate plastic fractions comprising an estimated 130 000 tonnes, much of which is recycled directly back into the plastic product manufacturing process. Although plastics make up approximately 7–10 wt% of domestic waste, it comprises 20–30% of the volume. The plastics in household waste are also in the form of plastic film and rigid containers. Plastic film comprises about 3–4 wt% of the household waste stream and is almost impossible to recycle. However, plastic containers are more easily collected separately or segregated from the waste stream. There are six main plastics which arise in European municipal solid waste, these are high density polyethylene (HDPE), low density polyethylene (LDPE), polyethylene terephthalate (PET), polypropylene (PP), polystyrene (PS) and polyvinyl chloride (PVC). Figure 4.9 shows the percentage of the main plastics arising in European waste (APME 1993). Separation

Table 4.10 Primary applications of plastics

Plastic type	Typical application
[1] Thermoplastics	
High density polyethylene (HDPE)	Bottles for household chemicals, bottle caps, toys, housewares
Low density polyethylene (LDPE)	Bags, sacks, bin liners, squeezy bottles, cling film, containers
Polyvinyl chloride (PVC)	Blister packs, food trays, bottles, toys, cable insulation, wallpaper, flooring, cling film
Polystyrene (PS)	Egg cartons, yoghurt pots, drinking cups, tape cassettes
Polyethylene terephthalate (PET)	Carbonated drinks bottles, food packaging
Polypropylene (PP)	Margarine tubs, crisp packets, packaging film
[2] Thermosets	
Epoxy resins	Automotive parts, electrical equipment, adhesives
Phenolics	Appliances, adhesives, automotive parts, electrical components
Polyurethane	Coatings, cushions, mattresses, car seats
Polyamide	Packaging film
Polymethylmethacrylate	Transparent all weather electrical insulators
Styrene copolymers	General appliance mouldings

Source: Warmer Bulletin Information Sheet. Plastics Recycling. Warmer Bulletin 32, February 1992, Journal of the World Resource Foundation, Tonbridge.

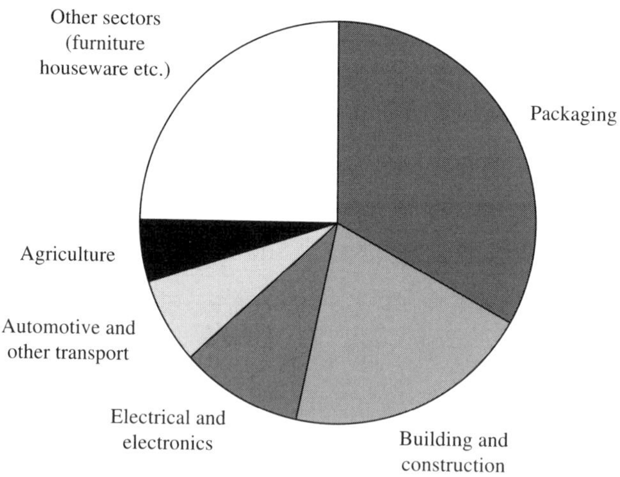

Figure 4.8 *Plastics end-use in Western Europe. Source: Warmer Bulletin Information Sheet, Plastics Recycling. Warmer Bulletin 32, Journal of the World Resource Foundation, Tonbridge, 1992. Reproduced by permission from the World Resource Foundation*

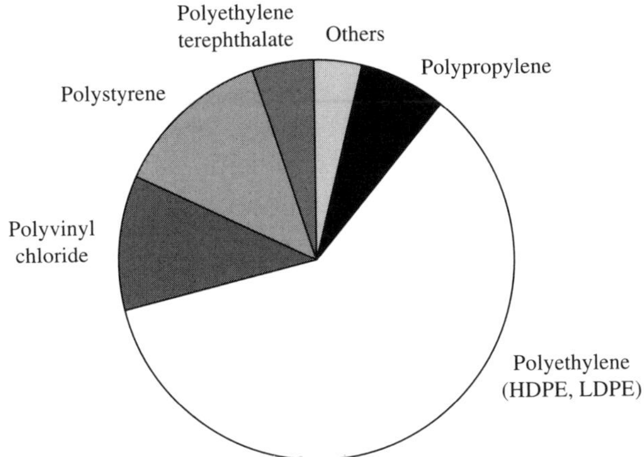

Figure 4.9 *The main plastic types found in municipal solid waste. Source: Plastics Recovery in Perspective. Association of Plastics Manufacturers in Europe, Brussels, 1993. Reproduced by permission from the Association of Plastics Manufacturers in Europe, Brussels*

of the plastics from waste is mainly by hand, either by the householder prior to collection or at a materials recycling facility. New developments attempt to automate the separation of plastics into different types. For example, separation schemes for segregating plastic types using X-ray analysis, flotation and colour sensors have all been researched (Warmer Bulletin 1992; Basta et al 1995).

UK arisings of plastic waste are of the order of 2.2 million tonnes per year. The majority of the plastic waste arises from packaging in municipal solid waste and distribution packaging, and large industry waste plastic. Figure 4.10 shows the distribution of plastic waste from UK sources (APME 1993). The overall recycling rate for post-consumer plastics in the UK, however, is only of the order of 5%. The distribution and industrial sector recycles approximately 30% of the waste, and such a high rate is achieved because the material is predominantly clean and of a single plastic type. Therefore, recycling produces a high quality material which can be readily recycled, often back into the same industrial sector. In-house recycling of scrap generated from the plastics industry as off-cuts and rejected parts are granulated and recycled back into the process.

Recycling of plastics in municipal solid waste is limited in practice to plastic containers, since the remaining plastic is in the form of film which is difficult to separate. Several plastic bottle collection schemes have been initiated in the UK, for example, Leeds, Birmingham, Manchester, Glasgow, Sheffield, Milton Keynes and Dorset. The schemes are based on both 'Bring' and Collect' methods of collection. In 1996, 5400 tonnes of plastic bottles were collected in the UK, which rose to an estimated 8500 tonnes/year by 1997 (ENDS 1997b). This tonnage converted to actual bottles collected is enormous; for example, the number of plastic bottles needed to make one tonne of polyethylene terephthalate is 20 000. The recovery rate for plastic bottles in the UK is about 2.5% of the total 200 000 tonnes of bottles

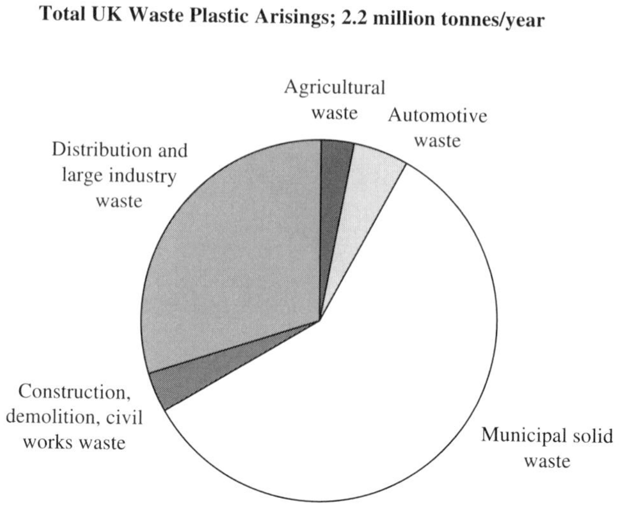

Total UK Waste Plastic Arisings; 2.2 million tonnes/year

Figure 4.10 *UK sources of plastic waste. Source: Plastics Recovery in Perspective. Association of Plastics Manufacturers in Europe, Brussels, 1993. Reproduced by permission from the Association of Plastics Manufacturers in Europe, Brussels*

used each year and represents about 100 million bottles collected. The bottles are transferred to the materials recovery facility where they are separated from any unwanted contaminating material, of which there is from 5 to 20%, and sorted, usually by hand, in to the three main types of bottle, PET, PVC and HDPE. The bottles are compacted and baled to produce bales of approximately one cubic metre and weighing between 150 and 300 kg, and sent for further processing. Processing takes the form of shredding, washing, drying and bagging the plastic, which is then sold on to the end user of the recycled plastic (RECOUP 1995, 1996). Comparison with other countries shows that plastic bottle recycling is widespread throughout Europe and North America. For example, in the USA, many states have a mandatory collection requirement for PET plastic bottles. PET is one of the most expensive plastics to produce, and PET recycling has grown in the USA to over 30% of the PET production. European examples include Munich, Madrid, Nancy and Amsterdam.

In Germany, 420 000 tonnes of plastic waste is separated out from municipal solid waste, representing a 70% recycling rate (Basta et al 1995). Germany's Packaging Waste Regulation requires that 64% of all packaging waste is recycled. In terms of household waste plastics recycling, this constitutes 21%. In the USA, plastics comprise approximately 9.3% of the municipal solid waste stream, representing approximately 20 million tonnes. Plastics recycling in the USA amounts to 4.5%, equivalent to 860 000 tonnes overall, but higher percentage recoveries are made from households, where 19% is reclaimed (Basta et al 1995).

The separated plastic material is processed by the end user by being granulated or pelletised, melted or partially melted and extruded to form the end product. The recycled plastic may be added to virgin plastic during the process. Outlets for single types of recycled plastics include, for example, HDPE for dustbin sacks, pipes and garden furniture, PVC for sewer pipes, shoes, electrical fittings and flooring, and PET for egg cartons, carpets, fibre filling material and audio cassettes (Warmer Bulletin 1992). Applications for plastic mixtures have included plastic fencing, industrial plastic pallets, traffic cones, playground equipment and garden furniture. There is resistance from the customer market for recycling plastics to produce film which may be used for food packaging because of the perceived associated health hazard. There is concern that compounds acquired during use may survive the recycling process to contaminate foods held in containers made from recycled materials (Kuznesof and VanDerveer 1995; Sadler 1995).

The low-grade uses for mixed plastic recycled materials has led to research into alternative processing methods to produce higher value products. One example is via tertiary recycling or feedstock recycling, where the plastic waste materials are processed back to produce basic petrochemicals that can be used as feedstock to make virgin plastic (Lee 1995; Meszaros 1995). The process has the advantage that mixed plastics can be used since all of the feedstock is reduced to petrochemicals. The plastic is identical to virgin plastic and can therefore be used in any application. Tertiary recycling can be via hydrogenation at high temperature and pressure, or via pyrolysis in an inert atmosphere at atmospheric pressure to degrade thermally the plastics. A number of pilot plants and demonstration schemes have been developed world-wide. For example, a European consortium has developed a fluidised bed plastics pyrolysis system at the BP Chemicals site in Grangemouth, Scotland. The system, which can handle mixed plastic waste, thermally decomposes the plastic at between 400 and 600 °C in an inert atmosphere of hydrocarbon gas. The fluidised bed is composed of sand and is fluidised by the hydrocarbon gas which is one of the products of the process. The plastics degrade in the high temperature of the fluidised bed, breaking the bonds of the polymer to produce lower molecular weight oligomers and monomers. The vapours resulting from the process are condensed to produce a waxy hydrocarbon product which has a high degree of purity and can be mixed with naphtha derived from crude petroleum oil at up to 20%. The resulting mixture can be fed directly into a steam cracker at the petroleum refinery to produce a range of petrochemical products, including virgin plastic. Figure 4.11 shows a schematic diagram of the pyrolysis process (Lee 1995).

Tertiary recycling of plastics schemes have been reported in Japan, the USA and Germany (Basta et al 1995).

4.4.4.2 Glass

Glass bottles, jars and other containers comprise about 6–8% of the UK household waste stream and recycling rates for glass containers are almost 30%, which equates to approximately 500 000 tonnes of waste glass or 'cullet' in the UK each year from industrial and household sources (Making Waste Work 1995). The main system for collecting glass in the UK is through bottle banks, of which there are approximately 17 000 in the UK. However, there are several kerbside collection schemes to collect glass from individual households as part of a city recycling scheme, for example Milton Keynes and Adur (Poll 1995). In addition, about 700 000 tonnes of flat glass

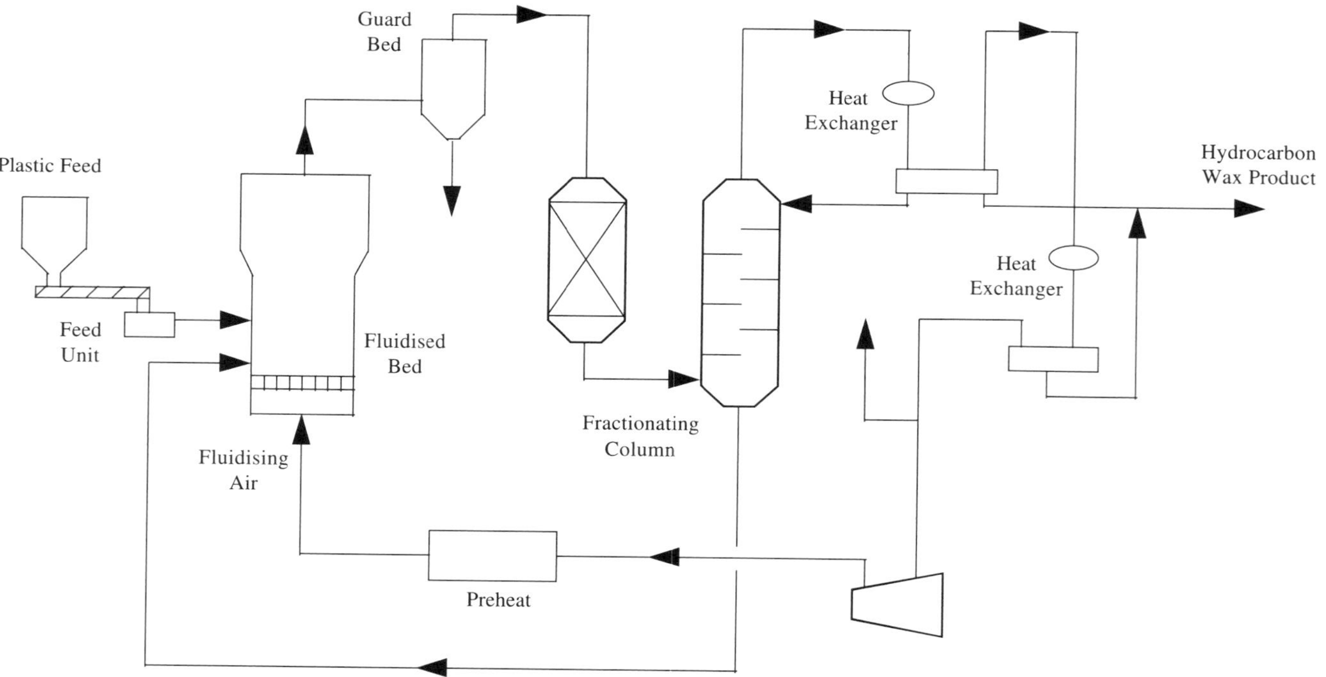

Figure 4.11 *Schematic diagram of the BP Chemicals plastic pyrolysis process. Source: Lee M., Chemistry in Britain, July, 515–517. The Royal Society of Chemistry, London, 1995. Reproduced by permission from The Royal Society of Chemistry, London*

for window and door glazing are produced each year in the UK, of which about 15% is recycled. There will also be some in-house recycling of glass waste within the industrial process itself.

The glass industry has committed itself to an overall recycling target of 50% by the year 2000 (Poll 1995). Recycling targets for glass have to be met through recycling of the material back into the industry, since glass comprises part of the non-combustible fraction of waste and therefore there is no option of energy recovery to meet overall recycling targets. The recycling of glass containers is well developed in the UK via the 'Bring' system, with householders delivering to bottle banks. The banks may be categorised in terms of the colour of the glass. The most common glass used for containers in the UK is clear coloured, although the green and brown bottle is most common in Europe, where it is used for wine and beer. Consequently a high proportion of cullet, over 60%, is made up of green and brown glass for which there is a lower demand in the UK; the new production of clear glass in the UK is 70%. The 'Bring' system has achieved high recovery rates in some areas; for example, Table 4.6 shows recovery rates of 61% for Richmond-upon-Thames (Atkinson and New 1992, 1993; Atkinson et al 1993). Collection schemes of the 'Collect' variety also achieve high rates of recovery, due mostly to the convenience of the system for the householder. Recovery rates of 66% and 71% have been achieved at Sheffield S.E. and Adur, West Sussex, respectively.

To overcome the problem of different coloured glass and the need for sorting, a USA company has developed a process to cover clear glass with coloured organic coatings which, when the glass is being recycled, simply melt away (Warmer Bulletin 49, 1996b). The result is that there would be no need for coloured glass to be manufactured and consequently no limit to the amount of cullet which could be recycled. Other developments include photographic quality ink only labels applied directly to the containers eliminating the need for, labels made of foil, plastic and paper with there consequent need for removal during recycling (Warmer Bulletin 49, 1996b).

Glass is made from relatively cheap raw materials (silica sand, limestone and sodium carbonate), but glass making is energy intensive and glass recycling can reduce the energy used since recycled glass melts at a lower temperature than the raw materials. For example, increasing the amount of glass cullet in the furnace to 50% can result in a 15% saving in energy. In addition, when using virgin raw materials, 15% of the weight of input is lost as waste gases, whilst when using cullet, there are no waste gases and also the water used in processing is reduced by up to 50% (Warmer Bulletin 49, 1996b).

The processing of waste glass consists of a number of stages. The glass from the bottle banks is delivered to the recovery facility, where it is sorted by colour. The glass is stored by colour in separate bunkers until required, when it is fed to conveyor belts, where ferrous materials such as bottle caps are removed by magnetic separation, and hand-sorting is used to remove other unwanted contamination. The glass is then crushed and screened, and light-weight non-ferrous contamination such as aluminium caps, plastics and paper labels are removed by vacuum suction. A final stage of removal of opaque material such as china pieces from the clear glass may be made electronically. The crushed processed glass is then available for

Table 4.11 *Percentage of recycled glass used in glass production in the UK and prices*

	Type of glass		
	Clear	Green	Brown
Amount of cullet (1000 tonnes)	189	242	48
Total production of containers (1000 tonnes)	1168	287	259
Percentage of cullet (%)	16	84	19
Price paid by the glass recycling industry (£/tonne – 1995 prices)	27.75	23.00	25.50

Source: Poll A.J., Opportunities and barriers to glass recycling. National Environmental Technology Centre, AEA Technology, Harwell, Department of Trade and Industry, 1995.

recycling into the glass making process. Usually the glass cullet making process is operated by a separate company from the glass making company. Glass manufacture is fully automated. Three tonne batches of mixed recycled cullet and virgin raw materials are heated rapidly to 1540 °C. The melted materials are blown into glass bottles at the rate of about 150/minute. A hot end-coating process coats the bottles with tin oxide to smoothe the surfaces. The cooled glass is reheated and slowly cooled to anneal them, which reduces stresses in the glass. A coating of polyethylene may be given at the end of the process to reduce surface scratching. Up to 80% of glass cullet can be accepted into the glass making furnace, and there is no limit to the number of times that glass can be recycled (BGRC 1996; Warmer Bulletin 49, 1996b). The proportion of each type of glass cullet used as feedstock along with virgin materials is shown in Table 4.11 (Poll 1995). Also shown are the prices paid by the industry for the recycled glass.

Glass recycling in Europe is well established, with some countries achieving recycling rates of over 75% (Table 4.12) (Warmer Bulletin 49, 1996b). Higher recycling rates achieved in some European countries are due to more effective collection methods, such as higher densities of bottle banks. However, the data may also represent different methods of defining glass recycling in different countries. The influence of the EC Directive on Packaging and Packaging Waste in Europe has meant a strengthening of the glass recycling industry. However, the Directive calls for a minimum of 15% recycling rate for each packaging material, and most countries already achieve this target for glass recycling (Warmer Bulletin 49, 1996b). It is estimated that glass makes up about 22% of household packaging, but commercial packaging contains virtually no glass. Because of the very low value of glass cullet there is virtually no international trade. Recycling of glass containers in the USA is 25%, but represents a much lower proportion (1.5%) of municipal solid waste than in most of Europe due to the growth in alternative containers in competition to glass (Basta et al 1995). Other countries world-wide have similar rates of glass container recycling, for example, Australia has a rate of 26% for non-refillable glass bottles, whilst Japan has a glass bottle recycling rate of 49% (McCarthy 1993).

Alternative uses for glass, instead of its use in containers, have been investigated, and potential applications of finely ground glass cullet in the production of cement,

Table 4.12 *Glass recycling rates in Europe*

Country	Tonnes collected	Recycling rate (%)
Austria	203 000	76
Belgium	235 000	67
Denmark	108 000	67
Finland	28 000	50
France	1 300 000	48
Germany	2 763 000	75
Greece	37 000	29
Ireland	28 000	31
Italy	890 000	54
Netherlands	367 000	77
Norway	36 000	72
Portugal	71 000	32
Spain	371 000	31
Sweden	95 000	56
Switzerland	242 000	84
Turkey	54 000	22
UK	492 000	28
Total	7 320 000	

Source: Warmer Bulletin 49, Glass Re-use and Recycling. Journal of the World Resource Foundation, Tonbridge (1994 data), 1996. Reproduced by permission from the World Resource Foundation.

road surfacing material, building aggregates, bricks, tiles and pipes have been suggested (Poll 1995). However, in many such applications the glass cullet is competing with low-cost products, and hence the requirement for a competitive low-cost material incorporating recycled glass. In addition, in many of the alternative applications the glass has to be ground to a fine particle size, which involves further processing costs.

4.4.4.3 Paper

The UK consumption of paper and board is approximately 10 million tonnes per annum, of which about 50% is imported and 50% produced in the UK (Ogilvie 1995a). Figure 4.12 shows a comparison of the quantities of paper and board use per head of population in various countries (Warmer Bulletin 43, 1994). Of the 50% UK production about 55% is produced from recycled paper and board (Warmer Bulletin 43, 1994; Ogilvie 1995a). There is some variation in paper recycling statistics since some quote recycled paper and board, whilst others quote the amount of recycled fibres used in paper and board production. The latter gives a lower figure due to losses of non-fibrous material during processing (Waste Management Paper 28, 1992). In addition, some estimates include the in-house recycling of paper, resulting in higher rates than post-consumer derived statistics. The majority of recycled paper and board, about 88%, comes from commercial and industrial waste streams and the remaining 12% from the domestic waste stream (Waste Management Paper 28,

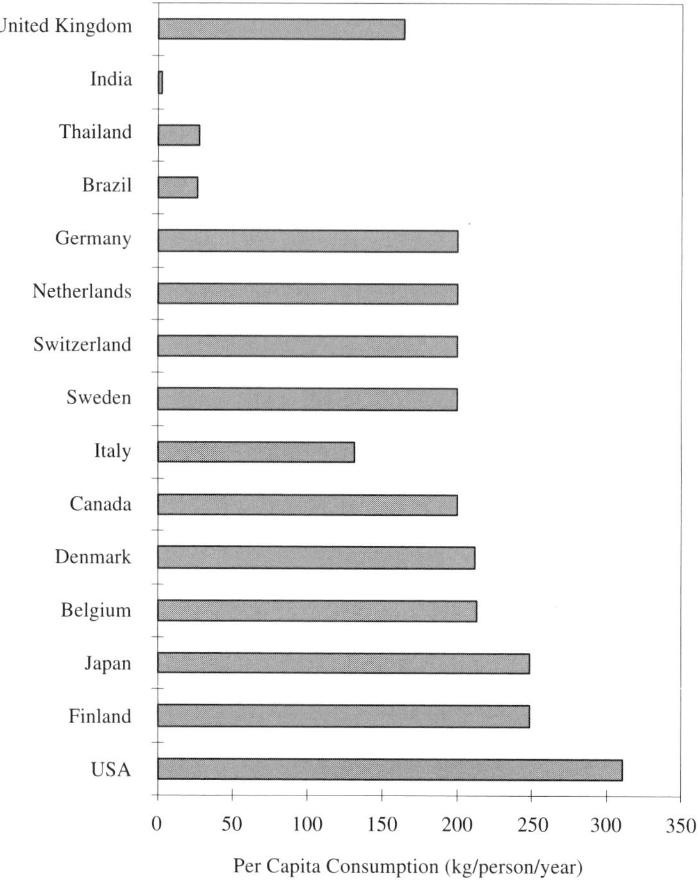

Figure 4.12 *Inter-country comparison of paper and board use (kg/person/year). Source: Warmer Bulletin Information Sheet: Paper Making and Recycling. Warmer Bulletin 43, Journal of the World Resource Foundation, Tonbridge, 1994.*

1992). Waste paper is graded in the UK into 11 different categories based on quality. The degree of reprocessing required depends on the grade of paper collected as waste, and the end use. The higher quality grades, collected, for example, as paper mill production scrap and office waste, require less processing and are used as primary paper pulp substitute for use in applications such as printing paper and tissues. Intermediate grades of waste paper, such as newspapers, require further processing to de-ink the paper and can be recycled back into the newspaper industry for newsprint. Low quality waste paper is used mainly for packaging material. Table 4.13 shows the 11 grades of waste paper used in the UK with examples of the different sub-categories (Waste Management Paper 28, 1992). Some of the sub-groups are paper products from a single source, which are unmixed and have little or no contamination. Post-consumer waste paper derived

Table 4.13 *Standard UK classification of waste paper*

Category	Examples
Group 1: (White wood free unprinted)	White soft tissue, white printers shavings, best white shavings, white envelope cuttings
Group 2: (White wood free printed)	Book quire, white continuous business forms, white carbonless copy paper
Group 3: (White and lightly printed mechanical)	White unprinted news, lightly printed scanboard
Group 4: (Coloured woodfree)	Coloured carbonless copy paper, coloured tissue, coloured shavings
Group 5: (Heavily printed mechanical)	Crushed news, telephone quire, mechanical book quire, newspaper, magazines
Group 6: (Coloured crafts and manilas)	Buff envelope cuttings, buff and coloured tab cards, dark and light coloured manilas
Group 7: (New Kraft lined – KLS)	Double lined kraft
Group 8: (Container waste)	Container waste
Group 9: (Mixed papers)	Mixed papers
Group 10: (Coloured card)	Coloured card
Group 11: (Contaminated grades)	Woodfree paper mixtures, plastic coated paper and board, telephone directories

Source: Waste Management Paper No. 28, Recycling. Department of the Environment, HMSO, London, 1992. Crown copyright is reproduced with the permission of the Controller of Her Majesty's Stationery Office.

from industrial and commercial sources can also be clearly defined and would have a much higher value than, for example, household waste paper, which would consist of several different types. The 11 categories of waste paper can be divided into 'pulp substitute grades' (grades 1–4) for higher quality end-use, and 'bulk' or 'packaging' grades (grades 5–11) for the production of packaging and board products. The large majority of newspaper and magazine collections organised by local community groups and local authorities is used for low grade packaging. The recycled paper used for writing and printing is in fact largely derived from industrial and commercial sources or even recycled paper production waste from the paper mill itself. In other countries different classification schemes exist for different types of paper and board, for example, Germany, USA and Japan all have different systems (McKinney 1995).

The recycling process used depends on the categories of waste paper and the end product. Initially the paper is pulped, followed by various stages of screening to remove contamination, cleaning to remove glues and other contaminants, de-inking

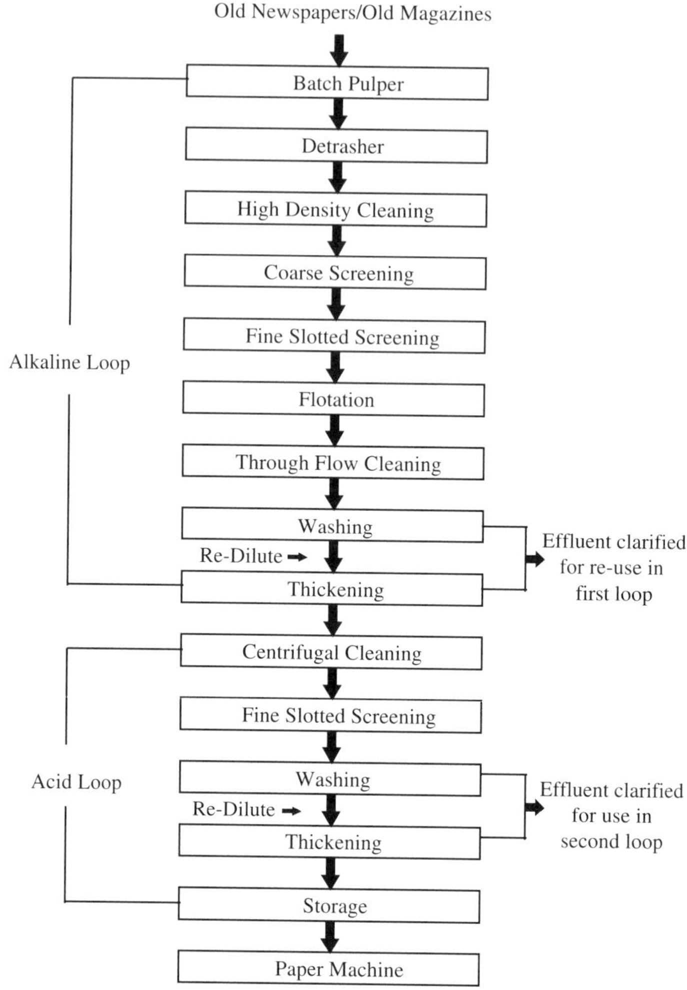

Figure 4.13 *The European de-ink process flow sheet for paper production from recycled newspapers and magazines. Source: Huston J.K., In: Technology of Paper Recycling. McKinney R.W.J. (Ed.) Blackie Academic and Professional (now Chapman and Hall) London, 1995. Reproduced by permission from Chapman and Hall*

and further processing to clean and thicken the pulp. In the case of higher quality papers a final bleaching stage may be included. Figure 4.13 shows the various stages involved in the reprocessing of waste paper to produce recycled fibre for the paper industry (Huston 1995). Waste paper is recycled to recover its cellulosic fibre content, and the pulped fibres are bound together under pressure to form paper. In addition to the cellulose, non-fibrous fillers are used to give added strength to the paper. Virgin paper pulp is made largely from wood derived from softwoods such as spruce or pine or hardwoods such as poplar and eucalyptus, which are farmed as an industrial crop. It is estimated that between 10 and 17 trees are required to produce a tonne of paper. In addition, the production of one tonne of paper requires 130 kg of calcium carbonate, 85 kg sulphur, 40 kg chlorine and 300 000 litres water (Warmer Bulletin 43, 1994). The UK paper industry is one of the largest industrial users of water. The waste water is treated and returned to the water system via sewerage works. The paper manufacturing process also consumes large quantities of energy, with estimates of over 20 GJ per tonne of paper being required.

Recycling waste paper reduces the need for wood pulp from trees, but in some cases the wood is harvested as a commercial farming crop and recycling would clearly influence this market. In addition, recycling can reduce the energy requirements by up to 40% and water consumption by 60%. In addition, emissions to air and water and solid waste can be reduced when recycled paper is used in comparison to virgin paper (Warmer Bulletin 43, 1994). There is a practical limit to the number of times that paper can be recycled because the fibres eventually break down to become too small for the paper making process. Estimates suggest that a maximum number of four recycles would be possible.

Two new waste paper recycling plants have been initiated recently in the UK, Aylesford in Kent and Gartcosh in Lanarkshire, with throughput of over 400 000 tonnes per year. The plants are very capital intensive, with investment costs of over £200 million. The plants are aimed at post-consumer recycled waste paper to produce newsprint and therefore have de-inking stages as part of the recycling process (Waste Management Paper 28, 1992; Making Waste Work 1995).

Figure 4.14 shows a comparison of the proportion of recycled fibre used in paper and board production in different European countries (Warmer Bulletin 43, 1994). Paper and board recycling is well established throughout Europe. The majority of recycled paper comes from industry, where the waste is well categorised and free from contamination. Recycled household waste paper is lower grade, mixed, contains more contamination and is generally used for lower grade packaging applications.

In the USA, many States have set mandatory recycling goals which have resulted in a recent rapid rise in the recycling of paper and board (Warmer Bulletin 50 1996c). In particular, there have been State or Government initiatives to increase the quantities of recycled fibre in newsprint and in Federal Government used paper.

The demand for recycled waste paper is very dependent on market conditions, particularly as waste paper is an internationally traded commodity (Making Waste Work 1995). There have been many fluctuations in the demand for recycled paper and board. For example, over-supply of waste paper in Europe and North America during the early 1990s led to a collapse in the market. Many paper and board recycling companies were charging for collected waste paper and for the use of collection containers. Increases in demand for pulp and paper products since 1995 throughout the world, due to an increase in world-wide economic growth, has led to a large increase in demand for recycled paper. This has also been aided by a decrease in supply of virgin wood pulp by the withdrawal of large areas of forestry, particularly in North America, from the production process. However, 1996 has seen yet another decrease in demand and consequent decline in the price paid for recycled paper and board (Warmer Bulletin 50 1996c). There is some international

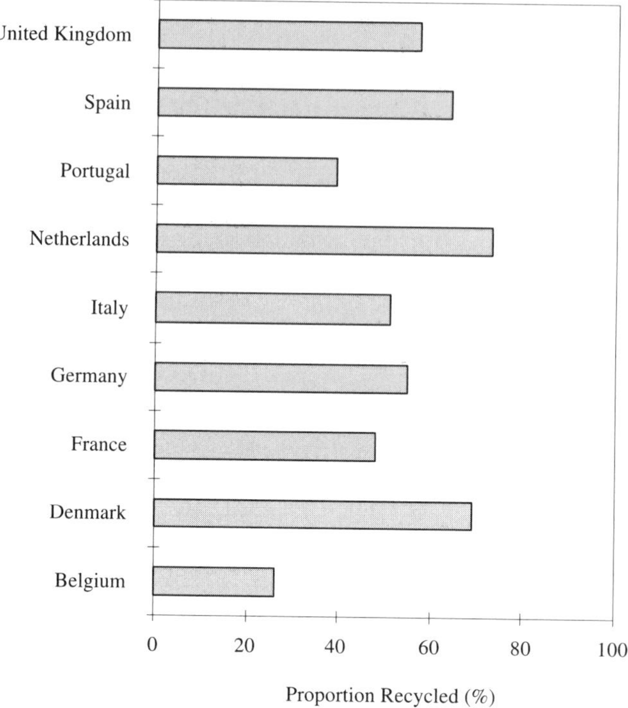

Figure 4.14 *Proportion of recycled paper and board used in production in European countries. Source: Warmer Bulletin Information Sheet: Paper Making and Recycling. Warmer Bulletin 43, Journal of the World Resource Foundation, Tonbridge, 1994. Reproduced by permission from the World Resource Foundation*

trade in recycled paper and board, with the Far East becoming a growing market supplied largely from Europe and the USA.

4.4.4.4 *Metals*

In the UK recycling of metals, particularly scrap iron and steel, is well established. Recycling of iron and steel is approximately 9 million tonnes per year from an annual consumption rate of 15 million tonnes per year, representing a 60% recycling rate (Waste Management Paper 28, 1992; Holt 1995; Making Waste Work 1995). The 9 million tonnes of scrap are either exported as scrap (40%) or incorporated into the UK production of iron and steel (60%). Aluminium consumption in the UK is approximately 0.8 million tonnes per year, and scrap aluminium recycling accounts for approximately 40% (Digest of Environmental Statistics 1996). Aluminium scrap is expensive and produces high per kilogram incomes for recycling schemes. Other valuable metals found in significant quantities in waste streams are copper, with an annual UK consumption estimated at 0.46 million tonnes per year and a recycling rate of 32%, zinc, with an annual consumption of 0.24 million tonnes per year and 20% recycling rate, and lead, with an annual consumption of 0.29 million tonnes per year and 74% recycling rate (Digest of Environmental Statistics 1996).

The ferrous scrap derived from the iron and steel industry comes mainly in the form of bulky waste such as scrapped vehicles and kitchen goods such as washing machines, cookers, fridges etc. About 2 million vehicles and 6 million items of kitchen goods are scrapped each year (Making Waste Work 1995). These items are collected by scrap metal merchants and sorted into various types and grades of scrap metal for transfer up the waste metal infrastructure to the metal refiners. The bulky scrap items are broken down by fragmentisers, which fragment the waste metal in a hammer mill, and sorted by air classifiers to produce a metal-rich stream and a residual stream containing plastics, wood, fabric etc., which is eventually landfilled. The scrap ferrous metal is recycled in steelworks using either the basic oxygen furnace or an electric arc scrap re-melting furnace. Up to 25% scrap may be incorporated into the basic oxygen process and up to 100% in the electric arc process (Waste Management Paper 28, 1992; Holt 1995).

In the UK, steel production has concentrated on higher quality type steels with lower tolerances of contamination, but iron and steel scrap will inherently be contaminated by other metals and materials. Consequently, there is a high export trade of some 40% of scrap iron and steel for lower grade production abroad.

Aluminium is recycled in-house from producers and recycled in secondary smelters using collected scrap aluminium. The aluminium scrap is sorted to remove contaminants and graded before re-melting to produce ingots of aluminium alloys. The major user of aluminium alloys from secondary smelters is the automotive industry.

The metals content of household waste is estimated at between 5 and 10%, and is mainly in the form of tin-plated steel cans used for drinks and tinned foods, and aluminium drinks cans. Recycling of metal cans, both steel and aluminium, represents a significant source of scrap metal. For example, it is estimated that over 1900 million aluminium drinks cans are recycled in the UK each year, representing a recovery rate of 28% and a recycled annual amount of 30 000 tonnes (Warmer Bulletin 50 1996d). Similarly, about 1400 million steel cans are recovered, representing a recovery rate of 13% and an annual amount of 90 000 tonnes (Making Waste Work 1995). The cans are collected by both the 'Collect' and 'Bring' systems. The steel and aluminium can recycling schemes are supported by the main industries involved, namely, British Steel and Alcan, respectively. Steel cans are separated from other metals and materials in the waste stream by magnetic separation. Reprocessing of the cans involves a de-tinning stage using hydrometallurgical processing. Two major detinning plants are located in Hartlepool and Llanelli, producing about 30 000 tonnes/year of steel. A further significant source of ferrous metals is from magnetic separation from mass burn municipal waste incinerator ash. A major aluminium can recycling plant is located in Warrington (Waste Management Paper 28, 1992).

Aluminium can recycling in other countries shows some impressive figures, for example in the USA a recycling rate of 64% has been achieved, representing 55 000

Table 4.14 *Non-ferrous metals in the household waste stream*

Metal	Form	Modes of occurrence
Copper	Metal	Electrical fittings and wire, plumbing fittings, kitchen ware
	Alloy	Constituent of brass, screws, plated products
	Electroplate	Decorative waste
Zinc	Metal	Carbon–zinc batteries
	Alloy	Components in diecastings, door fittings, domestic appliances, toys
	Galvanised	Domestic kitchen and garden ware
Tin	Electroplated	Cans, containers, toys, kitchen ware, electrical contacts
	Alloys	Solder, constituent of bronze
Lead	Metal	Pipes, electric bulb contacts, wine bottle closures
	Alloys	Lead–acid batteries, plumbing products
	Chemical	As oxide and sulphate in lead–acid batteries, lead-based paints
Nickel	Metal	Plating
	Alloys	Cutlery, components of kitchen goods

Source: Waste Management Paper No. 28, Recycling. Department of the Environment, HMSO, London, 1992.

million cans recycled. Similarly Australia has a recycling rate of 62% representing 1500 million cans recycled, and Japan recycles 42% of its aluminium cans representing 3500 million cans recycled (McCarthy 1993).

Other metals in the household waste stream and their mode of occurrence are shown in Table 4.14. The total is of the order of 80 000 tonnes per year occurring in UK household waste. However, because of the diverse nature of the mode of occurrence of these metals, very little is recycled. In-house, industrial and commercial recycling of these metals, however, is relatively high (Digest of Environmental Statistics 1996).

4.4.4.5 Tyres

Between 25 and 30 million scrap tyres are generated in the UK each year, representing about 400 000 tonnes of tyres. Currently about 50% of the scrap tyres are landfilled, stockpiled or dumped, about 21% are retreaded, 15% incinerated and 10% are crumbed (Ogilvie 1995b). Scrap tyres have been designated as a priority waste stream by the European Commission, and as such are subject to recycling targets and recommendations regarding environmentally acceptable treatment and disposal methods. It is estimated that Europe produces about 2 million tonnes of scrap car tyres per year; this compares with estimates of 2.5 million tonnes for North America and 0.5 million tonnes per year for Japan (Bressi 1995; Williams et al 1995).

Landfilling of tyres is declining as a disposal option, since tyres do not degrade easily in landfills, they are bulky, they take up valuable landfill space and they prevent waste compaction. They can also cause instability within the landfill and may 'float' to the surface of the landfill site. In addition, they can be a breeding ground for insects and a home for vermin (Lemieux and Ryan 1993; Bressi 1995). Many landfill sites refuse to take tyres, and the Department of the Environment has recommended that whole tyres should not be landfilled and that landfilling of shredded tyres should be restricted to 5% of the landfill volume. Consequently the cost of landfilling tyres has risen sharply in recent years, with estimates of between £35 and £80 per tonne for disposal for those sites willing to accept tyres (Ogilvie 1995b). Open dumping and stockpiling have the potential for accidental fires or arson, resulting in high pollution emissions to the atmosphere and water courses. Organic compounds identified in the fumes from the combustion of scrap tyres includes a wide range of toxic chemicals such as benzene, chrysene, benzo[b]fluoranthene, benzo[a]pyrene and heavy metals (Lemieux and Ryan 1993; DeMarini et al 1994). The resultant fires are very difficult to extinguish since they burn at very high temperatures producing large volumes of thick smoke. Large tyre dump fires generate pyrolytic or semi-pyrolytic conditions near the base of the fire, generating large quantities of oil which can cause contamination of land and water courses. The fires cause major pollution episodes, for example, the Hagersville tyre dump fire in the USA took 17 days to extinguish and involved 200 fire-fighters. Other tyre fires have been known to smoulder for several years, with the constant risk of re-ignition. Recycling of tyres has consequently increased in prominence in recent years.

Retreading of tyres is well established in the UK, with an annual production of about six million tyres, and involves the removal of the tread surface and replacement with a new tread surface. The process involves grinding the tyre surface to produce a clean surface onto which the new rubber is bonded. The grinding produces a rubber crumb which can be used in other applications. About 80% of retread tyres are for passenger car use and 20% for truck and bus tyres. There is competition for the expansion of retreading tyres from the importation of low-cost tyres from abroad. The tyres are produced from petroleum oil, and it has been estimated that the retreading process achieves savings of 4.5 gallons of oil for a car tyre and 15.0 gallons of oil for a truck tyre compared with the oil used in the production of a new tyre (Ogilvie 1995b).

Crumb is fine-grained or granulated tyre material obtained from the shredding and grinding of tyres. The process also involves the removal of the tyre reinforcing fibre and steel. The UK production of rubber crumb from tyres is about 30 000 tonnes per year (Ogilvie 1995b). The main uses for rubber crumb are as children's playground surfaces, sports surfaces, carpet backing, and other applications such as absorbents for oils and hazardous and chemical wastes.

Rubber reclaim is a method of recycling scrap tyres, but in the UK accounts for less than 5% of scrap tyre treatment. Rubber reclaim consists of devulcanisation of the rubber by temperature and pressure with various reclaiming chemicals and solvent treatments to produce a rubber which can be used in low grade applications. Such applications include cycle tyres, conveyor belts and footwear.

Incineration of tyres utilises the high calorific value of tyres, about 32 MJ/kg, and currently accounts for about 15% of tyre waste. The main UK incineration plant is the Elm Energy plant based in Wolverhampton in the West Midlands, which has a design capacity to recover up to 20% of the total tyre waste arisings in the UK, representing over 90 000 tonnes of tyres (Cooper 1995). The plant obtains income from the fee for the disposal of the tyres. However, the main income is from sales of

Table 4.15 Comparison of emissions to air from the Elm Energy tyre incineration plant, with regulatory limits

Pollutant	Environment agency guidance note limit (mg/m^3)	Elm Energy emissions (mg/m^3)
Particulate matter	30	3.7
Carbon monoxide	50	0.33
Sulphur dioxide	300	138.9
Nitrogen oxides	–	89.2
Hydrogen chloride	50	<1
Volatile organic compounds (as carbon)	20	4.9
Lead + chromium + copper + manganese	5	<0.1
Nickel + arsenic	1	<0.1
Cadmium	0.2	<0.1
Mercury	0.2	<0.1
Dioxin (TEQ)	–	<0.008 ng/m^3

Sources: [1] Cooper J., In: Recovery of Materials and Energy from Waste Tyres. Bressi G. (Ed.) Working Group on Recycling and Waste Minimisation, International Solid Waste Association, Copenhagen, 1995; [2] Chief Inspectors Guidance to Inspectors, Environmental Protection Act 1990. Process Guidance Note IPR 1/6, Combustion Processes: Combustion of fuel manufactured from or comprised of tyres, tyre rubber or similar rubber waste in appliances with a net rated thermal input of 3 MW or more. HMSO, London, 1992; [3] Elm Energy, Scrap Tyres as Fuel. Elm Energy and Recycling Ltd., West Midlands, 1995.

electricity generated by steam turbines powered by steam from boilers heated by the combustion of the tyres. Up to 30 MW of electricity can be generated for export to the national grid. The process consists of five separate process lines, each consisting of a scrap tyre incinerator, steam boiler and flue gas clean-up system. The steam boiler generates steam for the condensing steam turbine to generate the electricity. Some of the electricity is used within the plant and the remainder exported to the national grid to generate income from electricity sales. The scheme benefits from the Non-Fossil Fuel Obligation, which obtains a guaranteed higher price for the electricity compared with fossil fuel generated electricity. In addition, to the electricity, about 15 000–20 000 tonnes of steel wire scrap are produced from the steel reinforcement of the tyres. The tyres also contain significant concentrations of zinc, which can be recovered as a zinc-rich powder from the fly ash. The incineration process is subject to legislation on emission limits like any other large process, and consequently has an advanced flue gas clean-up system consisting of a filter bag house for initial removal of particulate matter, a calcium oxide based semi-scrubber to control sulphur dioxide, and a secondary bag house to remove reaction products from the flue gas prior to discharge. Table 4.15 shows the emissions of the Elm Energy tyre incineration plant compared with the Environment Agency, Chief Inspectors Guidance Notes limits for the emission to air from tyre combustion (Process Guidance Note 1992; Cooper 1995; Elm Energy 1995).

Comparison of the recycling options for scrap tyres used by other countries are shown in Figure 4.15 (Peat Marwick 1990). Difficulties arise in comparing different countries due to the fact that in many cases a large number of scrap tyres remain

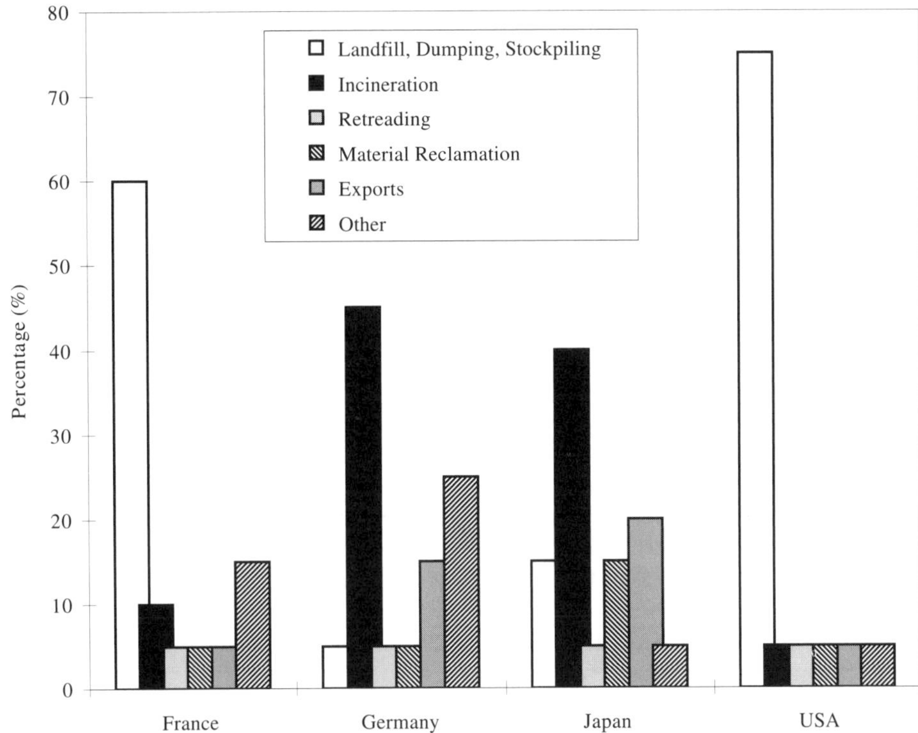

Figure 4.15 *Treatment and disposal options for scrap tyres in several countries. Source: Peat Marwick, The Recycling and Disposal of Tyres. Peat Marwick Management Consultants, London, 1990*

unaccounted for. In the USA and France most tyres are dumped, stockpiled or disposed of by landfill; by comparison, the largest route for disposal in Germany and Japan is incineration. Recycling through retreading and material reclamation is relatively low, but in all four countries retreading of truck and commercial vehicles is much higher, achieving between 30 and 50% of the replacement market.

An alternative technology for the treatment of tyres is pyrolysis, which is the thermal degradation of the tyre in an inert atmosphere. Pyrolysis of tyres has been established for many years, but is currently receiving renewed attention due to the difficulties of landfilling. The process produces an oil, char and gas product, all of which have the potential for use. The recovery of oil can be as high as 58%, it has a high calorific value (of the order of 42 MJ/kg) and can be used as a fuel or chemical feedstock, or added to petroleum refinery feedstocks (Williams et al 1995). The char has use as a solid fuel, or can be used as activated carbon or carbon black. The derived gas has a sufficiently high calorific value and can provide the energy requirements for the pyrolysis process. A number of small-scale commercial units have been developed. For example, in the UK the Beven Recycling plant, based in Harwell, Oxfordshire, operates a 1–2 tonne per day batch tyre pyrolysis unit (Williams et al 1995). The unit consists of a 4.3 m^3 capacity gas-tight kiln into which partly shredded and whole tyres are loaded. The system is purged with nitrogen to eliminate any oxygen and to ensure pyrolysis conditions are maintained. The kiln is heated initially by oil burners, but as the pyrolysis of the tyres proceeds the oil is replaced by the pyrolysis gas derived from the scrap tyre pyrolysis. The evolved pyrolysis oil vapours and gases pass through a heat exchanger to cool the vapours and into an oil condensing tank, the non-condensed gases then pass to a scrubbing unit to clean them, before passing to the burner, or to storage. The oil from the condensing tank is pumped to storage, or can also be used to provide auxiliary fuel for the burners. The pyrolysis conditions can be controlled to produce predominantly a char, oil or gas product. The Beven Recycling plant is aimed at producing a char product which has applications for use as carbon black or activated carbon. The batch pyrolysis system alleviates many of the problems encountered with continuously fed systems, such as non-maintenance of pyrolysis conditions, feed problems and in-bed fouling by the metal core of the tyres. Figure 4.16 shows a schematic diagram of the Beven Recycling unit for processing tyres (Williams et al 1995). In addition, other commercial or demonstration units for tyre pyrolysis are available world-wide which are based on different designs. For example, vacuum pyrolysis of tyres has been used in Canada for the production of high value chemicals such as limonene in the derived pyrolysis oil (Pakdel et al 1991). German designs include fluidised bed pyrolysis of tyres, where a whole tyre is pyrolysed in a matter of minutes in a bed of fluidised sand at temperatures of between 600 and 800 °C (Kaminsky and Sinn 1980).

4.4.6 Economic Considerations

Waste recycling relies on several inter-related requirements, all of which must be in place for an economically successful scheme to be realised. These requirements are (Waste Management Paper 28, 1992):

1. a secure and stable supply of waste materials;
2. a suitable collection system and transportation to the materials recovery facility;
3. a reliable materials separation and clean-up process to produce the end recycled materials and products;
4. secure and stable markets for the raw materials and products.

Secure and stable supplies of waste are required for the market to invest in the long-term development of recycling process facilities. Over-supply of waste or loss of markets for the end recycled products means that disposal costs for the treatment of the un-recycled waste become a factor in the assessment of the economic appraisal of the project. The generation of waste is a continuous process, and treatment and disposal options in the case of fall-off in the demand for recycled products should be insured against by ready alternatives being available. The collection and transportation of the waste to the recycling facility should also be stable and able to undertake preliminary sorting of the waste. The 'Bring' and 'Collect' schemes have a

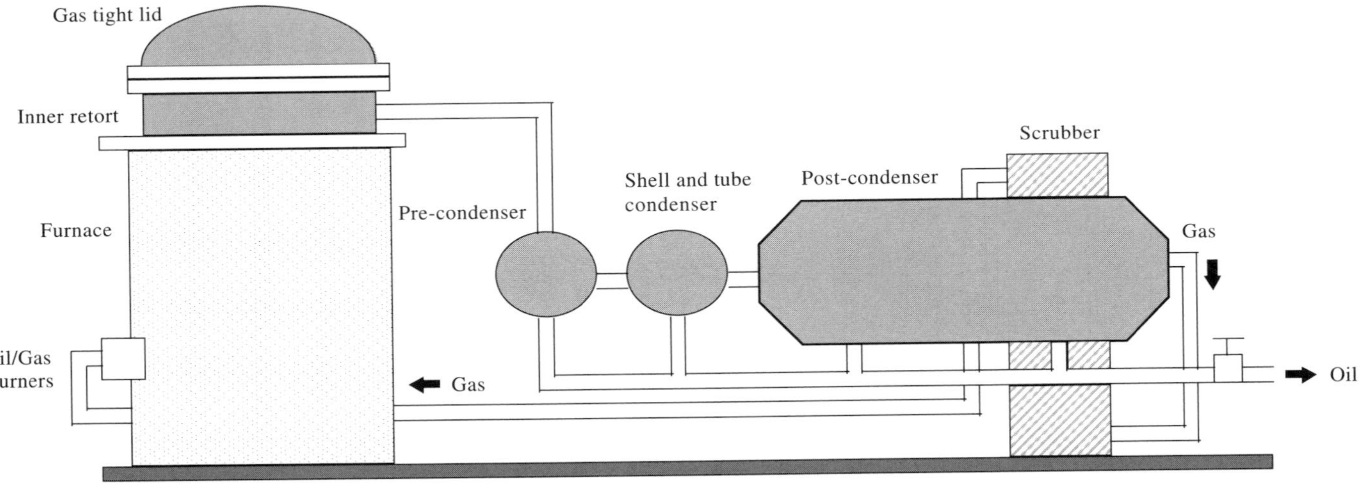

Figure 4.16 *Schematic diagram of the Beven recycling tyre pyrolysis unit. Source: Williams P.T., Besler S., Taylor D.T. and Bottrill R.P., Journal of the Institute of Energy, 68, 11–21, 1995. Reproduced by permission from the Institute of Energy, London*

number of advantages but also disadvantages. For example, 'Bring' systems are cheaper, but rely on the public to supply the recycled materials and the recovery rates for such schemes tend to be lower than 'Collect' systems and are very dependent on public attitudes. 'Collect' systems generally produce higher recovery rates than 'Bring' systems but are more expensive. The materials recycling facility should be able to handle the collected recyclable material and produce a marketable end-product from the process.

Contamination of the materials is also a factor in determining the economic viability of a recycling scheme. Placing non-recyclable waste into the recycling collection container can mean at best a significant increase in the time required for sorting and a consequent increase in costs, and at worst the scrapping of the whole container load. The level of contamination by dirt, grease, food waste etc. on the recyclable materials means an increase in the level of clean-up of the materials and a further increase in costs. The level of tolerable contamination is interlinked with the collection system, the processing facility design and the end market recycled product. The price of recycled end products have generally to be competitive with products made from virgin materials, since they are usually of lower quality and therefore are often sold at lower prices. The presence of contaminants means that recycled materials are impure compared with virgin materials and therefore tend to be aimed at lower specification markets which are also lower value. However, for some products such as recycled paper, which for some grades may be in short supply, higher prices can be obtained than for virgin paper. For example, the demand by consumers for quality recycled writing paper increased the cost of recycled paper above that of the equivalent paper made from virgin wood pulp. Recycled materials, like any other commodity traded in the market place, are subject to supply and demand with the additional proviso that there will be competition from virgin materials. In addition, some recycled materials are traded internationally and therefore subject to competition from recycling schemes in other countries which may be subsidised or which produce recycled materials of higher quality. However, international trade is a two-way process and excess supply of recycled materials can be exported. Examples include scrap ferrous metal, aluminium and waste paper. The interlinking of all the stages of the recycling process means that at times of low demand collection systems, particularly where private companies are involved, can sometimes be curtailed.

Costs of recycling are difficult to assess and compare due to differences in the cost factors included in the assessments. For example, some recycling costs are not separated from the general costs of waste management, some schemes include the income from the sale of the recycled materials with the costs of collection, others do not, and some report the costs of collection only (Warmer Bulletin 45, 1995).

Difficulties in the comparison of the costs of recycling has led to the introduction of the terms 'diversion rate' and 'cost difference' by the European Recovery and Recycling Association (ERRA) so that costs can be compared adequately (Warmer Bulletin 48, 1996e). Costs, whether of recycling or waste management, are very dependent on local conditions such as the price of land, labour costs, equipment costs, local taxes and subsidies, treatment and disposal costs etc. The use of the 'diversion rate' and the 'cost difference' involves the use of ratios to compare directly

$$\textbf{Diversion Rate (\%)} = \frac{\textbf{A}}{\textbf{B}} \times \textbf{100}$$

$$\textbf{Difference in Cost (\%)} = \frac{\textbf{C} - \textbf{D}}{\textbf{D}} \times \textbf{100}$$

A = Amount of material recovered as recycled materials
B = Total amount of waste generated
C = Cost of waste management with recycling
D = Cost of waste management without recycling

Figure 4.17 *Calculation of 'diversion rate' and 'cost difference' for comparison of recycling schemes. Sources: Warmer Bulletin 45, White P.R., Hummel J. and Wilmore J., Journal of the World Resource Foundation, Tonbridge, 1995; [2] Warmer Bulletin 48, White P.R., Hummel J. and Wilmore J., Journal of the World Resource Foundation, Tonbridge, 1996. Reproduced by permission from the World Resource Foundation*

the costs before and after implementation of a recycling scheme. The 'diversion rate' uses the ratio of the amount of materials recovered as recyclable material to the total amount of waste generated. The 'cost difference' is the cost of waste management with recycling minus the cost of waste management without recycling ratioed to the cost of waste management without recycling (Figure 4.17) (Warmer Bulletin 48, 1996e). Therefore the cost of recycling becomes a percentage increase or decrease in overall waste management costs. Similarly the effectiveness of the recycling system gives a percentage recovery of diversion of the recyclable materials out of the waste stream instead of going to final disposal. Thereby comparison of recycling schemes can be made without the distortion of local variations.

Figure 4.18 shows the use of the 'cost difference' and 'diversion rate' to compare different 'Bring' and 'Collect' collection systems (Warmer Bulletin 48, 1996e). The results show that 'Bring' schemes which are close to home achieve low diversion rates of the recyclable materials but are relatively low in the cost difference compared with waste management schemes without recycling. Higher diversion rates are found for kerbside collection, but significant differences can be seen depending on the container for the recyclable materials, the kerbside bag in particular achieving low rates of recovery or diversion but resulting in high comparative cost.

A number of factors can influence the diversion rate and cost difference, and these can be identified as internal and external factors. Internal factors include changes in the operation of the collection system or the re-processing facility or an expansion of the recycling scheme. External factors include the market prices paid for the recycled materials and the costs of the final disposal of the waste. Figure 4.19 shows the analysis of the impact of these internal and external factors on the diversion rate and cost difference for a number of recycling schemes throughout

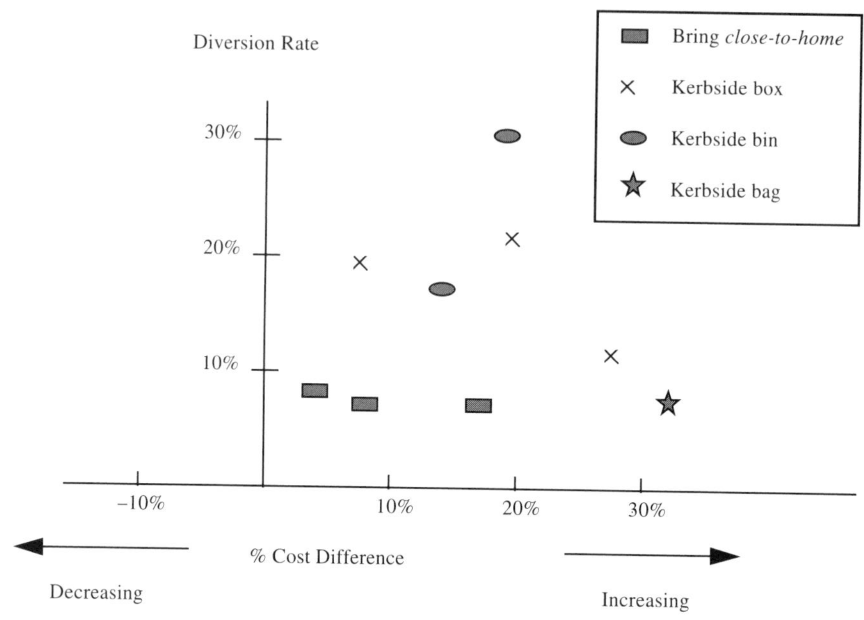

Figure 4.18 *'Cost difference' versus 'diversion rate' for different recycling collection systems. Source: Warmer Bulletin 48, White P.R., Hummel J. and Wilmore J., Journal of the World Resource Foundation, Tonbridge, 1996. Reproduced by permission from the World Resource Foundation*

Europe (Warmer Bulletin 48, 1996e). For a particular programme two scenarios can be identified. Scenario A is for a city with low waste management costs and where there is a small market for the recyclable end-product. It would be difficult for a recycling programme to produce very much recycled material without drastically increasing the costs of waste management (line A in Figure 4.19). Scenario B is for a city with high costs of waste management and where there is a strong market for the recycled end-products. Here higher rates of recycling (diversion) can be obtained for relatively small increases in cost difference until a point is reached where increasing the rate of diversion by a small percentage results in a large increase in the cost difference (line B in Figure 4.19). Area C in Figure 4.19 represents the scenario where implementation of a recycling programme results in a *decrease* in the cost difference of waste management for the city. Area K represents the area of potential improvement in cost difference or diversion rate for a city recycling programme (x), which might be via decreasing the cost difference whilst maintaining the diversion rate (k1), or increasing the diversion rate whilst maintaining the cost difference (k2).

Figure 4.20 shows the results of the implementation of internal and external factors in a set of four European recycling programmes to influence the cost difference whilst maintaining the level of material diversion (Warmer Bulletin 48, 1996e).

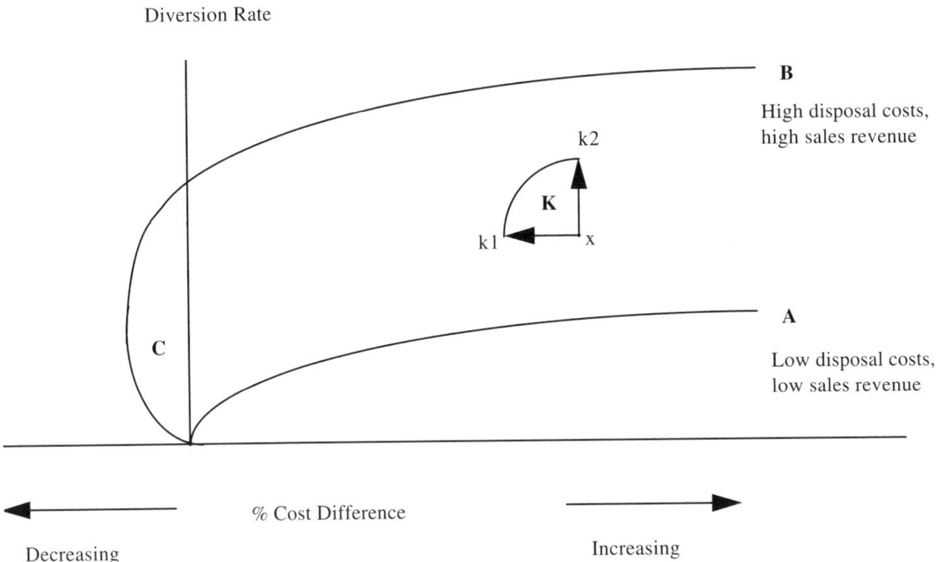

Figure 4.19 *The influence of internal and external factors on the 'cost difference' versus 'diversion rate' for recycling. Source: Warmer Bulletin 48, White P.R., Hummel J. and Wilmore J., Journal of the World Resource Foundation, Tonbridge, 1996. Reproduced by permission from the World Resource Foundation*

The initial capital investment required for a recycling scheme is high, with the added uncertainty as to the constancy of the income. The costs of materials recovery facilities are very high, and significant costs are incurred in setting up the household recyclable sorting bins and collection vehicles. In addition, prices obtained for materials recovered from waste have shown fluctuations, influencing the overall economics of recycling. Price fluctuations result from fluctuations in the market price of the recovered material and also with the degree of contamination. To minimise such price fluctuations, the recycling industry is seeking to obtain longer contracts with agreed specifications of the recycled materials. The required specification may be either on the recycled materials sold on to the end user, or indeed placed on the waste collector as specifications for the quality of the incoming waste to the materials recycling facility.

A recent study on recycling costs and revenues for household waste for five cities in the UK – Birmingham, Leeds, Bristol, Cardiff and Dundee – has reported estimated costs for 'Bring' and 'Kerbside' collection schemes (Atkinson et al 1996). The recycling was based on recycling of four categories of household waste – newspapers and magazines, glass bottles and jars, metal cans and plastic bottles. The report was based on certain assumptions, such as collection rate and recovery rate for each recycled material and costs associated with the type of collection system used – 'bring' or 'kerbside'. The costs associated with the 'bring' system included the capital and operating costs for the bank system of collecting recyclable materials. The 'kerbside' system estimates included estimates of the capital and operating costs

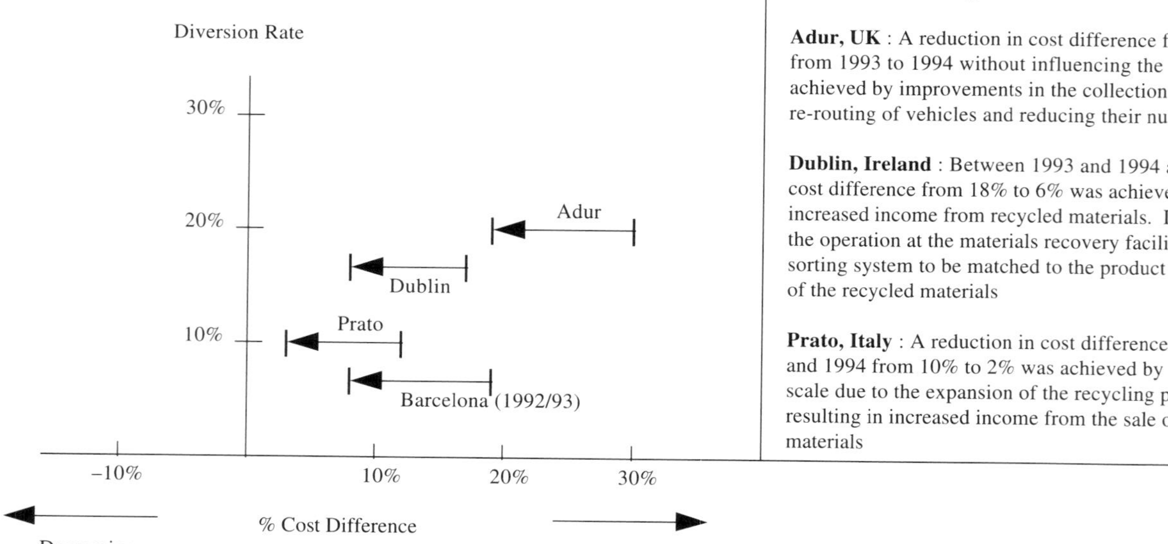

Figure 4.20 *Change in the 'cost difference' and 'diversion rate' for four European recycling programmes. Source: Warmer Bulletin 48, White P.R., Hummel J. and Wilmore J., Journal of the World Resource Foundation, Tonbridge, 1996. Reproduced by permission from the World Resource Foundation*

Table 4.16 *Estimated costs for 'bring' and 'kerbside' recycling systems in the UK*

	Birmingham	Leeds	Bristol	Cardiff	Dundee
Population (1000s)	1001	686	396	292	172
No. of households (1000s)	388	266	396	116	75
Bring system (low density)					
Recovery rate (%)	4.7	4.9	5.3	5.3	4.4
Cost/tonne (£)	44.21	43.01	39.89	37.88	48.61
Cost/household/year (£)	1.31	1.32	1.30	1.32	1.37
Bring system (high density)					
Recovery rate (%)	7.8	8.1	8.8	8.8	7.3
Cost/tonne (£)	64.47	69.95	74.68	82.58	133.41
Cost/household/year (£)	3.18	3.58	4.05	4.77	6.26
Kerbside system (separate collection)					
Recovery rate (%)	10.0	11.0	13.0	13.0	9.0
Cost/tonne (£)	153.38	144.48	139.84	141.12	231.44
Cost/household/year (£)	9.97	10.12	10.94	11.93	13.80
Kerbside system (integrated collection)					
Recovery rate (%)	10.0	11.0	13.0	13.0	9.0
Cost/tonne (£)	82.92	79.09	81.31	86.95	154.64
Cost/household/year (£)	5.39	5.54	6.36	7.35	9.22

Source: Atkinson W., New R., Papworth R., Pearson J., Poll J. and Scott D., The impact of recycling household waste on downstream energy recovery systems. ETSU Report B/RI/00286/REP, Energy Technology Support Unit, Department of Trade and Industry, London, 1996.

of either a separate or integrated collection system together with a materials recovery facility. Estimates of revenue from the sale of recycled materials was also reported. Table 4.16 shows the results of the estimate in terms of a comparison of the 'bring' and 'kerbside' systems of recycling the four categories of household waste. The 'bring' system was based on a high density bank system of one site of recycling banks per 500 households, and a low density bank system of one site per 3500 households. The 'kerbside' system was based on either a separate collection system with additional vehicles to collect recyclable materials, or an integrated collection system where the recyclable and other waste fractions are collected simultaneously by one vehicle. The separate collection system had higher cost implications due to the necessity to replicate waste collection.

The recovery rates for the 'bring' and 'kerbside' collection systems show that higher recovery rates of recyclable materials are found for the 'kerbside' system, reflecting the convenience for the householder. Recovery rates are also higher where more recycling sites are available to the public at the high density rate of one site per 500 households rather than one site per 3500 households. Recovery rates for the two 'kerbside' systems are identical, since the only difference is the method of vehicle collection system. Higher costs are associated with separate 'kerbside' collection involving two collections per household, one for recyclable materials and the other for other waste. The differences in cost/household/year for the five different cities reflects in many respects economies of scale due to the population differences and

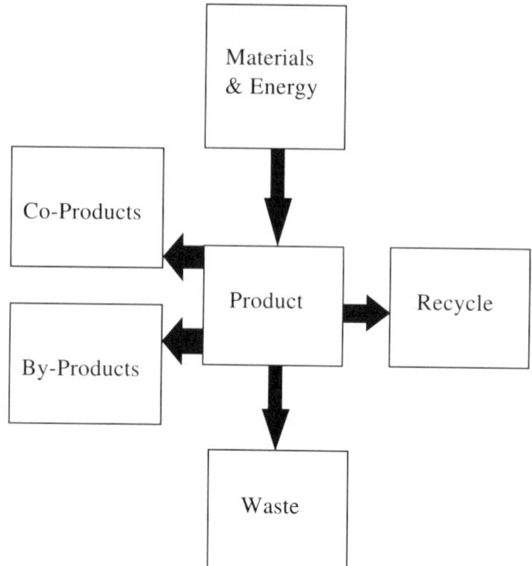

Figure 4.21 *Life cycle analysis. Source: Warmer Bulletin Information Sheet: Life Cycle Analysis and Assessment. Warmer Bulletin 46, Journal of the World Resource Foundation, Tonbridge, 1995. Reproduced by permission from the World Resource Foundation*

number of households in each city. Whilst cost/household/year is lower for 'bring' systems, their recovery rates are also lower.

4.4.5.1 *Life Cycle Analysis of Materials Recycling*

Life cycle analysis is the analysis of a product throughout its lifetime to assess its impact on the environment (Figure 4.21) (Warmer Bulletin 46, 1995c). The concept of life cycle analysis is a useful one in waste management and aids in the determination of whether waste reduction, re-use, recovery or disposal is the best practicable environmental option. Life cycle analysis has been comprehensively applied by White et al (1995) in the area of solid waste management. The analysis quantifies how much energy and raw materials are used and how much solid, liquid and gaseous waste is generated at each process stage of the product's life (Figure 4.22) (Warmer Bulletin 46, 1995c). It can be a particularly useful tool in the assessment of the full environmental impact of the production of a product via materials recycling versus product production from virgin materials.

The life cycle analysis of a product involves making detailed measurements during the manufacture of the product from the mining of the raw materials, including the energy inputs used in its production and distribution, through to its use, possible re-use or recycling, and its final disposal. Defining the boundaries of the life cycle analysis and the methodologies involved can vary from analysis to analysis. For example, some analyses have included the environmental impacts in terms of emissions to air, water and on to land when the final waste is disposed of in landfill

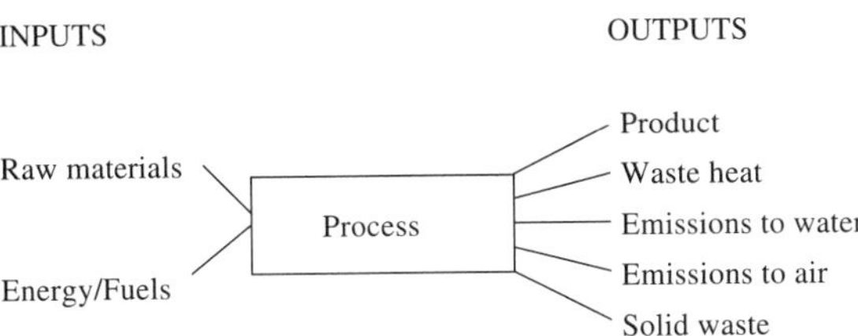

Figure 4.22 *The processes examined in conducting a life cycle analysis of a product. Source: Warmer Bulletin Information Sheet: Life Cycle Analysis and Assessment. Warmer Bulletin 46, Journal of the World Resource Foundation, Tonbridge, 1995. Reproduced by permission from the World Resource Foundation*

compared with incineration. Others may include the life cycle analysis of the machinery used in the manufacture of the product.

The first stage of the analysis involves the collation of the data relevant to the processes involved in the manufacture of the product; the second stage of the life cycle analysis is the interpretation of the data. For example, the production of an aluminium drinks can involves production of aluminium oxide from the raw material bauxite, which involves various stages of reaction with inputs of calcium oxide, sodium hydroxide, high temperatures and pressure. The aluminium oxide is then electrolysed in a solution containing cryolite (calcium aluminium fluoride) and fluorspar (calcium fluoride), and aluminium is produced at the carbon cathode with carbon dioxide given off at the anode. In addition, fluorine is also given off during the electrolysis process. The aluminium metal is then used in the manufacture of the aluminium can, which may involve rolling, pressing and fusing of the metal. Therefore, a full life cycle analysis of the aluminium drinks can should include the mining of the bauxite, the mining of rock salt and limestone used to produce the sodium hydroxide and calcium oxide, respectively, and the energy used in their production. The production of aluminium oxide would involve temperature, pressure and emissions, such as the sludge by-product from the process. The electrolysis stage includes the production of the carbon anodes and cathodes from petroleum, the energy used in the electrolysis process, and emissions of carbon dioxide, fluorine and other gases. Finally the production process from the aluminium involves the production of aluminium sheeting, and rolling and pressing, with the associated energy inputs and waste and emissions outputs. In addition, not included in this survey would be the environmental impact of transport at each stage. Recycling of the product can also influence the interpretation of the life cycle of the product. For example, the aluminium drinks can may be recycled, perhaps several times.

Once collected, the data can also be liable to different interpretations, for example, comparison of the environmental impact of high energy demand in one process with high water demand in another. Similarly, it is difficult to compare the environmental impact of, for example, sulphur dioxide, nitrogen dioxide or hydrogen fluoride. To

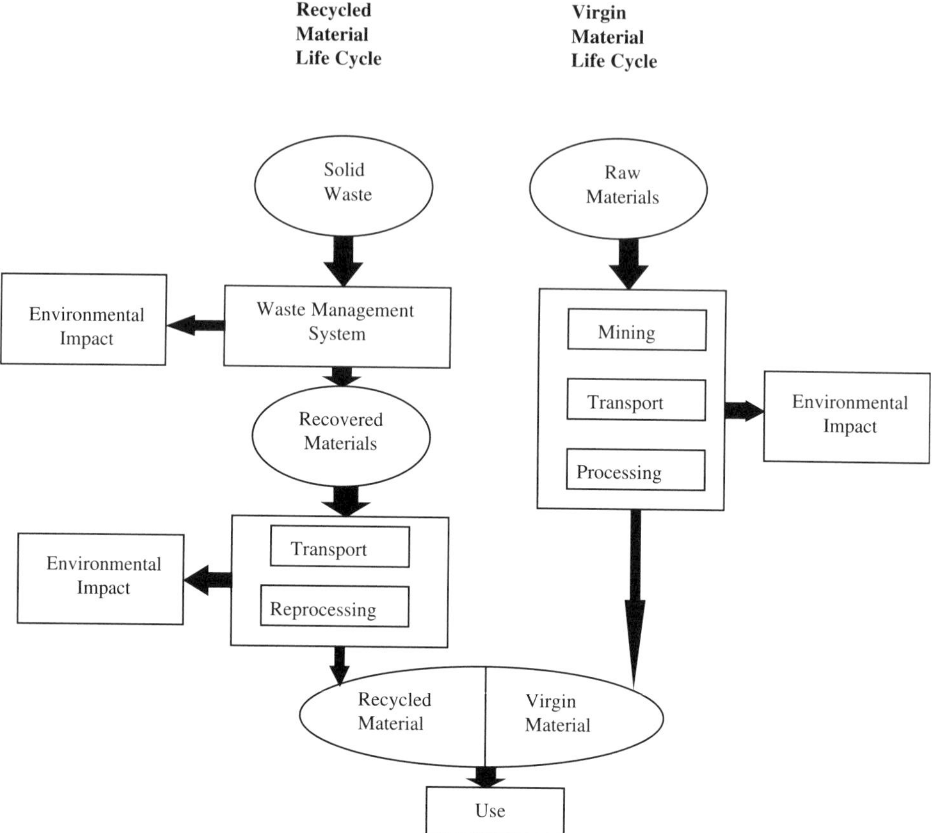

Figure 4.23 *Life cycle analysis for recycled and virgin materials. Source: White P., Franke M. and Hindle P., Integrated Solid Waste Management; A life cycle inventory. Blackie Academic and Professional (now Chapman and Hall), London, 1995. Reproduced by permission from Chapman and Hall*

overcome these problems some analyses aggregate the various environmental impacts into categories such as the impact on the ozone layer, or the contribution to acid rain.

Life cycle analyses comparing recycling with manufacturing of the product from virgin materials have been used to highlight the benefits of recycling. Figure 4.23 shows the comparative life cycle analyses for products derived from recycled material and from virgin materials (White et al 1995). The life cycle for recycled products includes the assessment of the environmental impacts in terms of energy used and emissions at each of the processes involved in recycling. These would include the separation of the recyclable materials from the solid waste at the waste treatment plant, transportation of the recovered materials to the processing plant, and the various processes involved in reprocessing the recovered materials into

Table 4.17 *Energy consumption and emissions: comparison of recycled paper with virgin paper production*

Source	Recycled paper (per tonne produced)	Virgin paper (per tonne produced)	Savings (per tonne produced)
1. Energy consumption (GJ)			
	14.4	22.7	8.3
2. Air emissions (g)			
Particulate	357	4346	8.3
CO	383	3165	2782
NOx	2295	5114	2819
N_2O	280	345	65
SOx	6054	10868	4814
HCl	0	4	4
HF	0.004	0.01	0.006
H_2S	0	15	15
HC	4195	6258	2063
NH_3	2.9	3.4	0.5
Hg	0	0.004	0.004
3. Water emissions (g)			
BOD	1	2921	2920
COD	3	25423	25420
Suspended solids	1	1	0
TOC	25	30	5
AOX	0	3	3
Ammonium	0.331	0.876	0.545
Chloride	9	22	13
Fluoride	0.714	1.89	1.18
Sulphide	0	7	7
4. Solid waste (kg)			
Total solid waste	70.6	150.2	79.6

Source: White P., Franke, M. and Hindle P., Integrated Solid Waste Management: A lifecycle inventory. Blackie Academic and Professional, London, 1995. Now Chapman and Hall. Reproduced by permission from Chapman and Hall.

useable materials. The recycling of a product can be compared with the production based on virgin materials using life cycle analysis to determine which process has the minimum environmental impact. Table 4.17 shows a comparison of a life cycle assessment for recycled paper versus virgin paper using data from Swedish and Swiss sources. The table covers the energy and materials inputs and the emissions outputs, but the energy used and emissions from the collecting and sorting of the material and the transport to the reprocessing plant are not included. The estimates also depend on the types of process plant used and the grade of input paper and required grade of output paper. Consequently energy and emissions data can vary from analysis to analysis. Table 4.18 shows a comparison of the energy and emissions for recycled versus virgin production of a variety of materials (White et al 1995).

Table 4.18 *Energy and emissions savings of recycling versus virgin production*

Material	Process energy saved (GJ/tonne)	Air emissions for recycling (GJ/tonne)	Water emissions for recycling (GJ/tonne)	Solid waste for recycling (kg/tonne)
Paper[1]	8.3	Generally lower	Generally lower	80 – reduction
Glass[2]	3.8	Generally lower	Generally lower	25 – increase
Metal – Fe[3]	13.5	Generally lower	Generally lower	278 – reduction
Metal – Al	156	Generally lower (except HCl)	Generally lower	639 – reduction
Plastic (LDPE)[4]	15.4	Generally lower (except for CO_2)	Few data	93 – increase
Plastic (HDPE)[5]	25.6	Generally lower	Poor data, may be higher for recycled	184 – increase
Textiles[6]	52–59	No data	No data	No data

[1] Pulp and paper making included.
[2] Process to finish container included. Data for 100% virgin extrapolated as all glass making uses some used material.
[3] Data for tinplate recycling up to production of new tinplate.
[4] Incomplete data for reprocessing of low density polyethylene.
[5] Incomplete data for reprocessing of high density polyethylene.
[6] Energy range for woven and knitted wool only.

Source: White P., Franke M. and Hindle P., Integrated Solid Waste Management: A lifecycle inventory. Blackie Academic and Professional, London, 1995. Now Chapman and Hall. Reproduced by permission from Chapman and Hall.

Bibliography

APME, 1993. Plastics Recovery in Perspective. Association of Plastic Manufacturers in Europe (APME), Brussels.

Atkinson W. and New R. 1992. Monitoring the impact of bring systems on household waste in the UK. Warren Spring Laboratory, Report No. LR 944, National Environmental Technology Centre, Harwell.

Atkinson W. and New R. 1993. Kerbside collection of recyclables from household waste in the UK. Warren Spring Laboratory, Report No. LR 946, National Environmental Technology Centre, Harwell.

Atkinson W., Barton J.R. and New R. 1993. Cost assessment of source separation schemes applied to household waste in the UK. Warren Spring Laboratory, Report No. LR 945, National Environmental Technology Centre, Harwell.

Atkinson W., New R., Papworth R., Pearson J., Poll J. and Scott D. 1996. The impact of recycling household waste on downstream energy recovery systems. ETSU Report B/RI/00286/REP, Energy Technology Support Unit, Department of Trade and Industry, London.

Barton J.R. 1989. Processing of urban waste to provide feedstock for fuel/energy recovery. In: Ferrero G.L., Maniatis K., Buekens A. and Bridgwater A.V. Pyrolysis and Gasification. Elsevier Applied Science, London.

Basta N., Fouhy K. and Moore S. 1995. Prime time for post-consumer recycling. Chemical Engineering, February, 31–35.

BGRC, 1996. Fact File. British Glass Recycling Company, Sheffield.

Bressi G. (Ed.) 1995. Recovery of Materials and Energy from Waste Tyres. Working Group on Recycling and Waste Minimisation, International Solid Waste Association, Copenhagen.

Chappell P. 1991. A review of municipal waste combustion technology. In: Energy from Waste. Clean, Green and Profitable. Institute of Energy Seminar, The Institute of Energy, London.

Cooper J. 1995. The safe disposal of tyres in the UK. In: Bressi G. (Ed.) Recovery of Materials and Energy from Waste Tyres. Working Group on Recycling and Waste Minimisation, International Solid Waste Association, Copenhagen.

Crittenden B. and Kolaczkowski S. 1995. Waste Minimisation: A practical guide. Institution of Chemical Engineers, Rugby.

DeMarini D.M., Lemieux P.M., Ryan J.V., Brooks L.R. and Williams R.W. 1994. Mutagenicity and chemical analysis of emissions from the open burning of scrap rubber tyres. Environmental Science and Technology, 28, 136–141.

Digest of Environmental Statistics, 1996. Department of the Environment, HMSO, London.

Elm Energy, 1995. Scrap Tyres as Fuel. Elm Energy and Recycling, West Midlands.

ENDS, 1997a. Report 265. Environmental Data Services Ltd., Bowling Green Lane, London.

ENDS, 1997b. Report 266. Environmental Data Services Ltd., Bowling Green Lane, London.

ERRA, 1994a. Technology Manual. European Recovery and Recycling Association (ERRA), Reference Multi-Material Recovery, Brussels.

ERRA, 1994b. Critical Factors 1994. European Recovery and Recycling Association (ERRA), Reference Multi-Material Recovery. Brussels.

ERRA, 1996. Programme Evaluation 1996. European Recovery and Recycling Association (ERRA), Reference Multi-Material Recovery, Brussels.

ETSU, 1993. Assessment of d-RDF Processing Costs. Energy Technology Support Unit (ETSU), Department of Trade and Industry, HMSO, London.

Everett J.W. and Peirce J.J. 1993. Curbside recycling in the USA: Convenience and mandatory participation. Waste Management and Research, 11, 49–61.

Fletcher G.A. and Johnston N. 1993. The Aire and Calder Project: A regional waste minimisation club. In: Effluent Treatment and Waste Minimisation. Institution of Chemical Engineers Symposium Series 132, Institution of Chemical Engineers, Rugby.

Holt G. 1995. Opportunities and Barriers to Metals Recycling. Recycling Advisory Unit, National Environmental Technology Centre, AEA Technology, Harwell.

Huston J.K. 1995. Manufacture of newsprint using recycled fibres. In: McKinney R.W.J. (Ed.) Technology of Paper Recycling. Blackie Academic and Professional, London. 1995

Jackson D.V. 1988. A review of the developments in the production and combustion of refuse derived fuel. In: Brown A., Evemy P. and Ferrero G.L. (Eds.) Energy Recovery through Waste Combustion. Elsevier Applied Science, London.

Kaminsky W. and Sinn H. 1980. In: Jones J.L. and Radding S.B. (Eds.) American Chemical Society Symposium Series 130, American Chemical Society, Washington, DC.

Kuznesof P.M. and VanDerveer M.C. 1995. In: Rader C.P., Baldwin S.D., Cornell D.D., Sadler G.D. and Stockel R.F. (Eds.) Plastics, Rubber and Paper Recycling. American Chemical Society, Symposium Series 609, American Chemical Society, Washington, DC.

Lee M. 1995. Feedstock recycling; new plastic for old. Chemistry in Britain, July, 515–516.

Lemieux P.M. and Ryan J.V. 1993. Characterisation of air pollutants emitted from a simulated scrap tyre fire. Journal of the Air and Waste Management Association, 43, 1106–1115.

Making Waste Work, 1995. Department of the Environment and the Welsh Office, HMSO, London.

McCarthy J.E. 1993. Recycling and reducing packaging waste: How the United States compares to other countries. Resources, Conservation and Recycling, 8, 293–360.

McKinney R.W.J. 1995. Waste paper recovery and collection. In: McKinney R.W.J. (Ed.) Technology of Paper Recycling. Blackie Academic and Professional, London.

Meszaros M.W. 1995. Advances in plastics recycling. In: Rader C.P., Baldwin S.D., Cornell D.D., Sadler G.D. and Stockel R.F. (Eds.) Plastics, Rubber and Paper Recycling.

American Chemical Society, Symposium Series 609, American Chemical Society, Washington, DC.

Ogilvie S.M. 1995a. Opportunities and Barriers to Wastepaper Recycling. Recycling Advisory Unit, National Environmental Technology Centre, AEA Technology, Harwell.

Ogilvie S.M. 1995b. Opportunities and Barriers to Scrap Tyre Recycling. Recycling Advisory Unit, National Environmental Technology Centre, AEA Technology, Harwell.

Pakdel H., Roy C., Aubin H., Jean G. and Coulombe S. 1991. Environmental Science and Technology, 25, 1646–1649.

Peat Marwick, 1990. The Recycling and Disposal of Tyres. Peat Marwick Management Consultants, London.

Pescod M.B. (Ed.) 1991–93. Urban Solid Waste Management. World Health Organisation, Copenhagen.

Poll A.J. 1995. Opportunities and Barriers to Glass Recycling. Recycling Advisory Unit, National Environmental Technology Centre, AEA Technology, Harwell.

Process Guidance Note, 1992. IPR 1/6, Chief Inspectors Guidance to Inspectors. Environmental Protection Act 1990, Combustion Processes, Combustion of Fuel Manufactured from or Comprised of Tyres, Tyre Rubber or Similar Rubber Waste in Appliances with a Net Rated Thermal Input of 3 MW or More. HMSO, London.

Rampling T.W. and Hickey T.J. 1988. The Laboratory Characterisation of Refuse Derived Fuel. Warren Spring Laboratory (National Environmental Technology Centre), Report LR 643 (MR), HMSO, London.

RECOUP, 1995. News Sheet. Recycling of Used Plastics (Containers) Ltd. (RECOUP), Peterborough.

RECOUP, 1996. News Sheet. Recycling of Used Plastics (Containers) Ltd. (RECOUP), Peterborough.

Sadler G.D. 1995. Recycling of polymers for food use: A current perspective. In: Rader C.P., Baldwin S.D., Cornell D.D., Sadler G.D. and Stockel R.F. (Eds.) Plastics, Rubber and Paper Recycling. American Chemical Society, Symposium Series 609, American Chemical Society, Washington, DC.

Steuteville R. 1996. The state of garbage in America. Biocycle, April, 54–61.

The Environment Act, 1995. HMSO, London.

The Waste Manager, 1994a. Keeping the lid on waste. The Waste Manager, July/August, 18–20.

The Waste Manager, 1994b. Less waste the key to lower bills. The Waste Manager, April, 22–23.

Warmer Bulletin, 1992. Information Sheet. Plastics Recycling. Bulletin 32, February, Journal of the World Resource Foundation, Tonbridge.

Warmer Bulletin, 1993. Fact Sheet. Refuse Derived Fuel. Bulletin 39, Journal of the World Resource Foundation, Tonbridge.

Warmer Bulletin 43, 1994. Warmer Information Sheet; Paper Making and Recycling. Journal of the World Resource Foundation, Tonbridge.

Warmer Bulletin, 1995a. Technical Brief. Materials Reclamation Facilities. The World Resource Foundation, Tonbridge.

Warmer Bulletin 45, 1995b. The cost of recycling: How to ask the right questions. White P., Hummel J., Willmore J., Journal of the World Recycling Foundation, Tonbridge.

Warmer Bulletin 46, 1995c. Information Sheet. Life Cycle Analysis and Assessment. Journal of the World Resource Foundation, Tonbridge.

Warmer Bulletin, 49, 1996a. Journal of the World Resource Foundation, Tonbridge, May.

Warmer Bulletin 49, 1996b. Glass Re-use and Recycling. Journal of the World Resource Foundation, Tonbridge.

Warmer Bulletin 50, 1996c. Markets for Waste Paper. Cooper I., Journal of the World Resource Foundation, Tonbridge.

Warmer Bulletin 50, 1996d. Journal of the World Resource Foundation, Tonbridge, p. 12.

Warmer Bulletin 48, 1996e. Affordable recycling – the critical factors. Hummel J., White P.R. and Willmore J., Journal of the World Resource Foundation, Tonbridge.

Waste Management Paper 28, 1992. Recycling. Department of the Environment, HMSO, London.

West Yorkshire Waste Management Plan, 1996. The Environment Agency (West Yorkshire Waste Regulation Authority), Wakefield.

White P.R., Franke M. and Hindle P. 1995. Integrated Solid Waste Management: A lifecycle inventory. Blackie Academic and Professional, London.

Williams P.T., Besler S., Taylor D.T. and Bottrill R.P. 1995. Pyrolysis of automotive tyre waste. Journal of the Institute of Energy, 68, 11–21.

5

Landfill

Summary

Introduction. Site selection and assessment. Landfill design and engineering, considerations for landfills, operational practice. Types of waste landfilled, inert wastes, bioreactive wastes. Landfill design types, attenuate and disperse landfills, containment landfills, landfill liner materials, landfill liner systems, co-disposal landfills, entombment landfills, sustainable landfills (the controlled flushing bioreactor landfill). Landfill gas, landfill gas migration, management and monitoring of landfill gas. Landfill leachate, leachate management and treatment. Landfill capping. Landfill site completion and restoration. Energy recovery. Old landfill sites. Economics. Proposed European legislation on waste landfills.

5.1 Introduction

Landfill is by far the largest route for the disposal of waste in the UK. Figure 5.1 shows the proportion of different wastes landfilled in the UK (Making Waste Work 1995). About 120 million tonnes of controlled waste per year are landfilled in the UK, which includes 90% of household waste, 85% of commercial waste, 63% of construction and demolition waste and 73% of other industrial waste. Sewage sludge also has a proportion disposed of via landfill, representing about 10% of the total arisings of wet sludge (4% solids content), equivalent to 3.5 million tonnes per year. In addition, mines and quarries produce about 110 million tonnes of waste each year, which is mostly disposed of in stable, above-ground tips at the production site (Making Waste Work 1995). The development of a strong emphasis on landfilling of wastes has been largely due to the inter-relationship with the large mineral extractive industry in the UK. The geological make-up of the UK is unique in providing such a wide diversity of rock types in such a small area, many of which are economically valuable for the minerals industry. Extraction of rock materials has produced, and

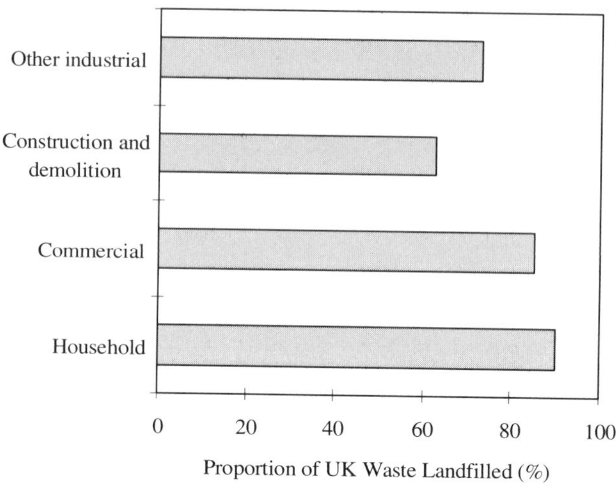

Figure 5.1 *Proportion of different controlled wastes disposal by landfill. Source: Making Waste Work. Department of the Environment and Welsh Office, HMSO, London, 1995. Crown copyright is reproduced with the permission of the Controller of Her Majesty's Stationery Office*

continues to produce, large holes in the ground which at some stage require infilling. Planning permission for minerals extraction usually has conditions which require the restoration of the site after completion of the extraction. Infilling of the mineral workings by waste is therefore an advantage economically for the site developer. The UK is favoured by the stable geology and the type of hydrogeological conditions, which provide a natural, leachate-impenetrable site for the safe deposit of wastes. Consequently, the UK has a long and established tradition of landfilling of wastes, which represents a low-cost disposal option. Currently there are approximately 4000 licensed landfill sites in the UK (Digest of Environmental Protection and Water Statistics, 1994).

Throughout Europe and in the USA there is also a strong emphasis on landfilling of wastes. Figure 5.2 shows the proportion of landfilling of municipal solid waste for various countries (Warmer Bulletin 44, 1995). The geological and hydrogeological conditions found in the UK, which allowed the development of a cheap disposal route for wastes, are not found to the same extent in some other countries such as Japan and the Netherlands. Japan places more reliance on incineration as the main waste treatment and disposal option, due in part to the scarcity of suitable land. Most of the sites are small, with about 60% being less than 1 hectare and 30% between 1 and 5 hectares (Ikeguchi 1994). In addition, because of the scarcity of land, there has been a development of off-shore landfills which are much larger, for example the Tokyo Bay site covers approximately 340 hectares and receives approximately 3 million tonnes of waste per year (Waste Management in Japan 1995). The USA is highly dependent on landfilling of waste, with an estimated 5300 municipal solid waste landfills which dispose of approximately 67% of the waste

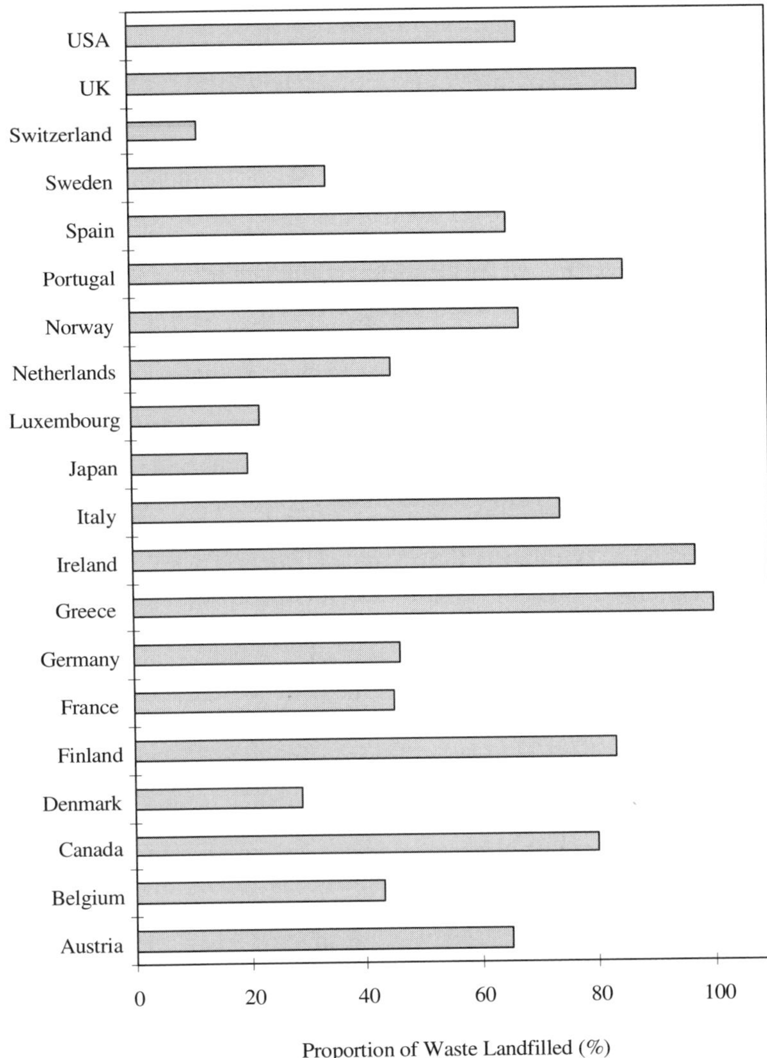

Figure 5.2 *Proportion of municipal solid waste disposal by landfill for different countries. Source: Warmer Bulletin 44, Information Sheet. Journal of the World Resource Foundation, Tonbridge, 1995*

generated (Hickman 1996). There is a trend in the USA towards a smaller number of larger landfills. For example, the Freshkill landfill, which handles the majority of New York's waste, covers an area in excess of 1200 hectares on Staten Island and has a daily waste intake of more than 20 000 tonnes (The Waste Manager 1994a). This site opened in 1948, and currently operates for 24 hours per day, 6 days per week and 52 weeks per year, with waste delivered by road and water-borne barge. The design concept for waste landfill in the USA is one of developing an envelope or barrier system around the waste, thereby isolating it from the local environment.

The objective of such design is to minimise leachate and gas migration into the surrounding soil, and gas migration into the atmosphere, and to minimise the formation of leachate by limiting water access (Ham 1993).

Landfill management in the European community is very variable in terms of siting policies, lining requirements, leachate control requirements and approaches to waste input (Hjelmar et al 1995). Uncontrolled waste landfills are very common in some countries. However, in most Member States of the Community new landfills for municipal waste and hazardous waste are required to have a bottom liner and a leachate collection system, and to remove leachate for disposal. Some countries require complete encapsulation (dry tomb) of hazardous waste.

The major advantages associated with landfilling of wastes are the low costs of landfill compared to other disposal options and the fact that a wide variety of wastes are suitable for landfill. It should also be remembered that ultimately, many other waste treatment and disposal options require that the final disposal route for the residues requires landfill. For example, with incineration, bottom and fly ashes are disposed of in landfill sites. Another important point is that in many cases there is a strong interaction between the minerals extraction process and the infilling of the void space with waste. Within the UK there is a plentiful supply of suitable sites for landfill, although this may not be the case for all locations. The collection and utilisation of landfill gas as a low-polluting fuel for energy generation is also an advantage. Increasingly, there is an emphasis on regarding the modern landfill as a fully designed and engineered process with high standards of management.

There are, however, some disadvantages with landfill. Older sites, which in some cases are still in use or have long been disused, were constructed before the environmental impacts of leachate and landfill gas were realised. Many of these sites are now sources of pollution with uncontrolled leakages. Landfill gas, in particular, can be hazardous, since the largest component, methane, can reach explosive concentrations. This problem is emphasised when it is realised that many of the older sites were constructed close to areas of housing, or in some cases housing sites have been built on disused landfill sites. Therefore, all landfill sites are required to be monitored for landfill gas, and the gas from operational sites must be controlled via proper venting. Landfill methane gas is also a 'greenhouse gas', leading to the problems of global warming but with about 30 times the effect of carbon dioxide. The contribution to total methane emissions in the UK from landfill gas has been estimated by one study the be as high as 40%, representing 2000 kt/year and by far the largest anthropogenic source (Williams 1994). This contrasts with an earlier estimate of only 23% (Digest of Environmental Protection and Water Statistics 1991) and an EU estimate of 32% (Commission of the European Communities, COM(97) 105 Final, 1997).

Whilst there is a plentiful supply of landfill sites in some parts of the UK, other locations, particularly the southeast of England, have existing or predicted future shortages of suitable landfill sites close to the source of waste generation. The Landfill Tax, which came into operation in October 1996 at the level of £7/tonne for 'bioreactive' wastes and £2/tonne for inert wastes, also adds a disincentive to landfill. In addition, landfill achieves a lower conversion of the wastes into energy, with about one-third less energy recovery per tonne from landfill gas than incineration. This is mainly due to the conversion of the organic materials in the waste into non-combustible gases and leachate, and general losses from the system. The proposed EC Landfill Directive (Commission of the European Communities, COM(97) 105 Final, 1997) requires a high level of environmental protection, and includes reduction in the amount of biodegradable waste going to landfill, pre-treatment of the waste, and sampling and analysis of leachate and landfill gas throughout site operation and throughout a significant aftercare period. In addition, the proposal seeks to introduce permit applications and technical requirements for the design, operation, monitoring, closure and post-closure care for landfills, and defines different categories of landfill sites and the types of wastes that can or cannot be accepted in the various categories. Such measures will undoubtedly influence the economic choice of landfill versus other waste treatment and disposal options. The European Waste Landfill Directive is discussed in more detail in Section 5.13.

5.2 Site Selection and Assessment

The selection of a site for a waste landfill depends on a wide range of criteria, including the proximity of the site to the source of waste generation, suitability of access roads, the impact on the local environment of site operations, and the geological and hydrogeological stability of the site. Site assessment is linked to the information requirements of various bodies involved in the planning, development and operation of the site.

The main aim of the landfill site assessment investigation is the identification of the possible pathways and receptors of landfill gas and leachate in the surrounding environment and the environmental impact of site operations (Tchobanoglous and O'Leary 1994; Waste Management Paper 26B 1995). Site assessment involves appraisal of geological and hydrogeological conditions at the site. This may include the use of existing surveys, aerial photography, boreholes, geophysical investigations, geological mapping and sampling etc. The information allows an assessment of soil and bedrock grain sizes, mineralogy and permeabilities, and ground water levels. In addition, the previous use of the site, meteorological data, transport infra-structure and planning use designations, and the planning strategy of the area would also be assessed. A topographical survey is undertaken to calculate the available void space and therefore the waste capacity of the site. Since daily, intermediate and final covering materials will be required extensively in the operation of the site, the availability of these materials in natural form should be assessed. At an early stage background levels of water and air quality may be taken to assess the impact of the site in the long term.

For large landfill sites an environmental assessment is also required to determine the impact on the environment. An environmental assessment is required under the Town and Country Planning (Assessment of Environmental Effect) Regulations 1988, which implements the EC Directive (85/337/EEC) requiring environmental assessments to be made for certain prescribed processes, which include landfills receiving over 75 000 tonnes/year of waste and sites in environmentally sensitive areas. Environmental assessment involves a description and assessment of the direct and indirect effects of the project on human beings, fauna, flora, soil, water, air, climate and landscape, material assets and the cultural heritage (Petts and Eduljee 1994; Energy from Waste: Best Practice Guide 1996).

5.3 Landfill Design and Engineering

Landfill disposal is seen in many respects as the bottom rung of the hierarchy of waste disposal options when considering the concept of sustainable waste management. However, the modern landfill site has developed from a site used merely for dumping waste with little or no thought, to a site which is an advanced treatment and disposal option designed and managed as an engineering project. The development of the engineered landfill site came about through a series of legislative measures throughout the 1970s and 1980s. The 1974 Control of Pollution Act introduced daily and final covering of the landfill to prevent the problems of litter, accidental fires and arson, flies, vermin, scavenging birds etc. In addition, there was a trend to move to larger well-managed landfill sites rather than local sites as a result of the changes in local government organisation which moved management responsibility of waste from local to regional control. A series of incidents involving uncontrolled escapes of gas from landfill sites, most notoriously the Loscoe explosion in 1986 where a bungalow was devastated by an explosion from the leakage of gas from a nearby landfill site (see Box 5.5), led directly to the introduction of guidance from the Department of the Environment on the control of landfill gas (Waste Management Paper 27, 1994). Further legislation via the 1990 Environmental Protection Act produced tighter controls on leachate and landfill gas as part of the site and waste management licensing scheme.

5.3.1 Considerations for Landfills

A waste landfill is a major design and engineering project and there are a number of points to be considered as part of the process.

Final landform profile The profile of the final landform is a key factor in design in that it dictates the after-use of the site, the waste capacity of the site, and settlement of the site after completion and landscaping. Final landform gradients after emplacement of the capping material would normally be between 1 in 4 and 1 in 40, depending on the final use for the site, to ensure adequate safety of the steep slopes and a minimum gradient for suitable drainage (Waste Management Paper 26E, 1996). However, steep slopes greater than 1 in 10 may require control to offset erosion of the site.

Site capacity The capacity of the site is clearly a key factor in site design, and the determination of how much waste can be accommodated in the site depends on waste density, the amount of intermediate and daily cover, the amount of settlement of the waste during the operation of the site, and the thickness of the capping system (Waste Management Paper 26B, 1995).

Settlement Settlement of the waste in the landfill occurs initially due to physical rearrangement of the waste soon after emplacement. As the biological, physical and chemical degradation processes proceed further settlement occurs from overburden pressure due to compaction by its own weight. Typical long-term settlement values for municipal solid waste are 15–20% reduction, although values of up to 40% have been reported where there is a high organic content in the waste (Tchobanoglous and O'Leary 1994; Waste Management Paper 26B 1995). Settlement can take place over periods of time up to 50 years, but the major settlement period (up to 90%) occurs within the first 5 years of the final emplacement of the waste. Inert wastes, which do not biodegrade significantly and tend to be more dense than municipal solid waste, have low settlement values (Sharma and Lewis 1994; Waste Management Paper 26B 1995).

Waste density The density of the waste within the landfill will vary depending on the degree of pre-compaction of the waste before emplacement, the variation in the components within the waste, the progression of biodegradation, the amount of daily and intermediate cover, and the mass of overlying waste. The degree of pre-compaction of the waste influences the amount of waste that can be accepted into the landfill, and also influences to a marked degree the amount of settlement of the landfill (Tchobanoglous and O'Leary 1994). Typical waste densities range from 0.65 to 0.85 tonnes/m^3, although different types of waste may reach densities as low as 0.4 tonnes/m^3 or up to 1.23 tonnes/m^3 depending on the amount of biodegradable and inert waste present. Inert wastes have higher densities, typically about 1.5 tonnes/m^3 (Waste Management Paper 26B 1995).

Materials requirements The containment landfill requires various materials for site development, operation and restoration. Included in these requirements are the natural fill materials such as clay, sand, gravel and soil, which are used in various applications such as sand for lining the site to protect the liner materials, clay to provide an additional low permeability layer to the site, gravel for drainage for leachate collection, clay for capping material and restoration soils. The availability of such materials on site increases the ease of operation and also reduces costs (Waste Management Paper 26B 1995).

Drainage Drainage of the rainwater falling on the site is required to ensure that excessive water does not infiltrate the waste directly or from run-off from surrounding areas. Cut-off drains both around and inside the site will keep the waste from becoming too wet and increasing the production of leachate (Tchobanoglous and O'Leary 1994; Waste Management Paper 26B 1995)

5.3.2 Operational Practice

The typical modern landfill site consists of a secure, fenced, landscaped site with access routes for waste transport vehicles. The sequence of operations for an incoming waste vehicle may include the weighing of the lorry on a weighbridge, document inspection and waste inspection. Once cleared, the lorry would move to the waste disposal area where the waste is tipped, the wheels of the lorry are cleaned, and the lorry is weighed out of the site to determine the weight of waste deposited. The driver collects any necessary documentation. The tipped waste is compacted using specially designed compacting vehicles, and a daily cover of soil or alternative material is added at the end of each working day (Pescod 1991–93).

The landfill site is normally developed, operated and restored in a series of phases to allow the most beneficial use of the site area, a method used throughout the UK, Europe and North America (Tchobanoglous and O'Leary 1994; Sarsby 1995; Waste Management Paper 26B 1995). Each phase may last from 12 to 18 months (Waste Management Paper 26B 1995). This serves to concentrate the waste disposal operation to specific areas and spread the costs of leachate and landfill gas control systems throughout the lifetime of the operation. Therefore, within a typical site, part of the site may be being prepared with liner material and leachate and landfill gas collection systems to accept waste, part may be being filled with waste, part may be being capped, and part may be fully capped and restored. The phases are separated by the use of separation bunds. Each phase would be a sub-area of the site, whereas cells are subdivisions of the phase. Each phase is generally completed from preparation through to capping and restoration but with a temporary unrestored open face (Waste Management Paper 26B 1995).

Within each phase are cells, which are subdivisions of the phase. The size of each cell depends on the volume of waste being deposited and is typically between 3 and 5 m, whilst other influencing factors include rainfall and the absorptive capacity of the waste, and thereby the minimisation of leachate production, and additionally the number of vehicles bringing waste to the site. The cell delineates the tipping area for a week or month. The waste is deposited, usually on shallow working faces, and compacted using a variety of specially designed compacting vehicles which break up and compact the waste. The daily cover is added at the end of the working day and is used to prevent windblown litter and odours, prevent fires, deter scavengers, birds and vermin, and improve the visual impact of the site. The cover material usually consists of about 15 cm of soil; alternative cover materials such as shredded green waste and re-useable plastic sheeting and geotextiles have been used. As the number of cells in a phase increases, intermediate cover material is deposited. This is about 1 m thick and is used to minimise the ingress of rainwater and thereby increase the production of leachate.

5.4 Types of Waste Landfilled

All landfill sites in the UK require a licence to operate, and these are issued by the Environment Agency. The licensing of waste disposal on land was introduced in the Environmental Protection Act 1990 and the licensing system set out in the Waste Management and Licensing Regulations 1994 (Waste Management Paper 4, 1996).

For large facilities an Environmental Assessment may be required to obtain planning permission (Energy from Waste: Best Practice Guide, 1996). The waste management licence includes details of the types and quantities of waste which are permitted to be handled at the site. Three categories of waste have been defined, and each landfill site licence applies only for the categories listed thereon.

1. Inert wastes: i.e., wastes which will not chemically react, decompose by biodegradation or leach pollutants into the environment, and which therefore do not pose an environmental risk either now or in the future. Inert wastes, as their name suggests, are wastes of no or low reactivity, i.e., they do not undergo significant chemical, biological or physical degradation to yield polluting materials. For example, certain construction materials and incinerator bottom ashes are classified as inert. Consequently, only passive control systems are required.

   Inert wastes are disposed of in attenuate and disperse type of sites provided that they are totally inert and unreactive. They may also be disposed of in the higher level type of landfill, the containment landfill.
2. Bioreactive wastes: i.e., wastes which undergo biodegradation within the landfill environment. Municipal solid waste is an example of a bioreactive waste. Biodegradation involves detoxification in order to stabilise the landfill. Stabilisation refers to the degradation of organic matter in the waste to stable products and the settlement of the material in the site to its final rest level.

   Bioreactive wastes are disposed of in containment landfills where a barrier system of liner materials contains the leachate and landfill gas generated.
3. Hazardous/industrial or special waste: i.e., wastes which may be acceptable at co-disposal sites where co-disposal is with biodegradable wastes such as municipal solid waste. The biodegradable waste aids the decomposition of the hazardous waste. The hazardous, industrial or special waste may be solid, sludge or liquid material.

   Hazardous/industrial or special wastes are co-disposed in containment-type landfills with a high specification liner system to contain the derived leachate and landfill gas.

Other landfill classifications based on the type and range of wastes have been used (Waste Management Paper 4, 1996). Approximately 18 million tonnes of household waste are disposed of in landfill sites in the UK each year (Making Waste Work 1995). Approximately 67% of household waste is biodegradable in the form of putrescible matter such as food and garden waste (20%), paper and board (33%) and textile material (4%); in addition, much of the fine material (10%) is biodegradable (Waste Management Paper 28, 1992). Household waste is disposed of in biodegradable landfill sites.

The tonnages of commercial and industrial waste disposed of to landfill are difficult to estimate, since in each sector there are very large numbers of companies generating waste and using a variety of disposal options. A recent analysis of industrial and commercial wastes going to landfill in the UK has been undertaken

Table 5.1 *Industrial and commercial waste going to landfill in the UK, estimated by company surveys and waste disposal plan (WDP) data (tonnes/year)*

Industry group	Tonnes of waste landfilled (1000 tonnes/year)	
	Survey data	WDP data
Coal extraction; coke ovens; extraction of oil and gas etc.	896	2600
Production and distribution of electricity, gas and other forms of energy	1189	3832
Water supply industry	60	949
Extraction and preparation of metalliferous mineral ores	m	m
Metal manufacturing	1701	8334
Extraction of minerals not elsewhere specified	5	1720
Manufacture of non-metallic mineral products	2101	10611
Chemical industry	2145	18771
Production of man-made fibres	9	363
Manufacture of metal goods not elsewhere specified	1910	622
Mechanical engineering	2026	1517
Manufacture of office machinery and data processing equipment	8	160
Electrical and electronic engineering	1284	1118
Manufacture of motor vehicles and parts thereof	1385	504
Manufacture of other transport equipment	681	501
Instrument engineering	489	189
Food, drink and tobacco manufacturing industries	5511	1903
Textiles, leather and leather goods, footwear and clothing	1194	1535
Timber and wooden furniture industries	2382	751
Manufacture of paper and paper products; printing and publishing	1728	1629
Processing of rubber and plastics	815	727
Other manufacturing industries	151	247
Construction	11 872	11 872
Wholesale distribution	3752	1188
Dealing in scrap and waste materials	500	500
Commission agents	m	44
Retail distribution	9163	3133
Hotels and catering	3291	1614
Repair of consumer goods and vehicles	332	259
Transport and communication	2548	1255
Banking, finance, insurance, business services, leasing	5096	887
Other services	8992	3930
Total	73 216	83 265

m, missing.

Source: DOE, Analysis of Industrial and Commercial Waste Going to Landfill in the UK. Department of the Environment, HMSO, London, 1995. Crown copyright is reproduced with the permission of the Controller of Her Majesty's Stationery Office.

and the results are presented in Table 5.1 (DOE 1995a). The analysis was based on published data from regional waste disposal plans (WDP) and from surveys of individual industrial and commercial premises. Table 5.1 shows that there are discrepancies between the two methods of estimation which result in a total of 10 million tonnes difference between the two estimates. This illustrates the difficulties in estimating not only waste arisings, but also the treatment and disposal options used.

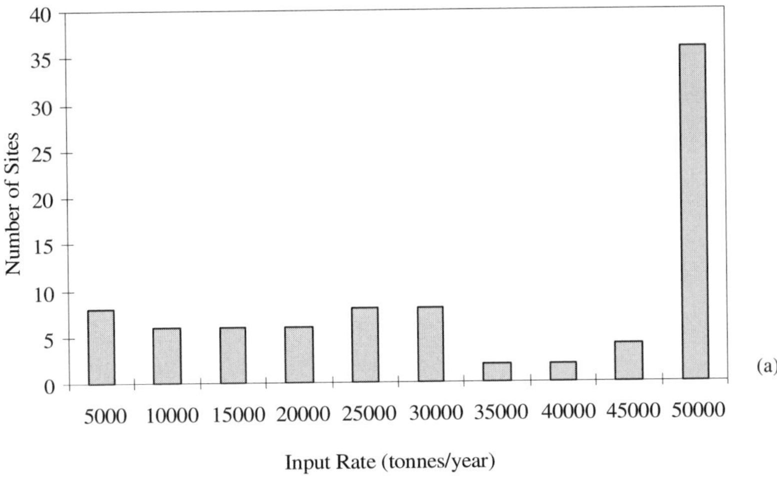

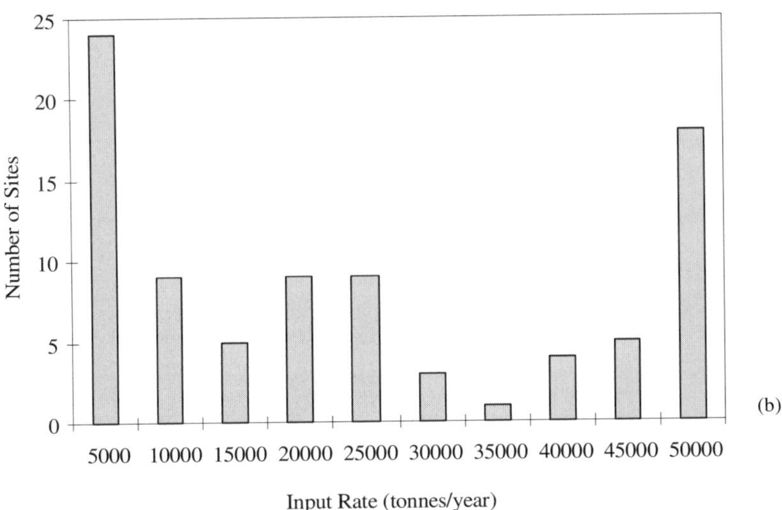

Figure 5.3 *Types of waste landfilled in the survey of 100 UK landfills. (a) Household waste; (b) commercial/industrial waste. Source: DOE, Characterisation of 100 UK Landfill Sites. Report CWM 015/90, Wastes Technical Division, Department of the Environment, London, 1995. Crown copyright is reproduced with the permission of the Controller of Her Majesty's Stationery Office*

Industrial and commercial wastes are disposed of in biodegradable or co-disposal landfill sites depending on the composition of the waste being landfilled.

Figures 5.3–5.5 show the results of a survey of the types of waste landfilled in the UK (DOE 1995b). The survey, conducted by the Department of the Environment, investigated 100 typical landfill sites in the UK, particularly those receiving domestic and commercial wastes, but also including a number of those accepting co-disposal

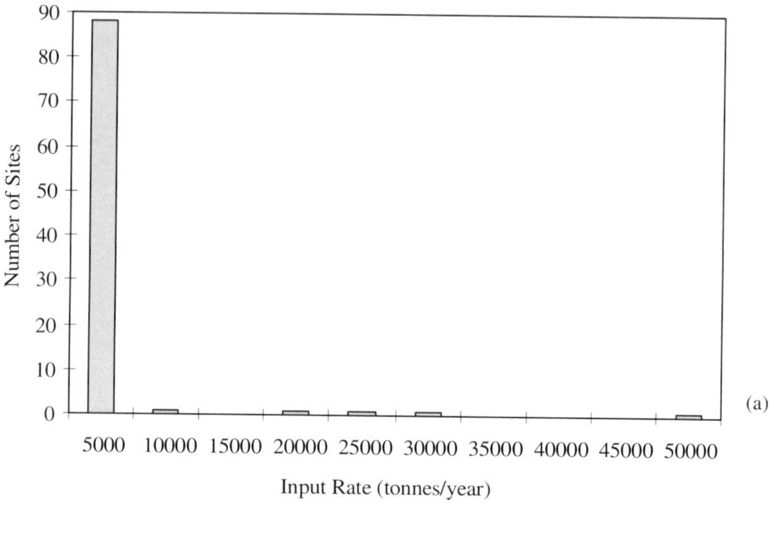

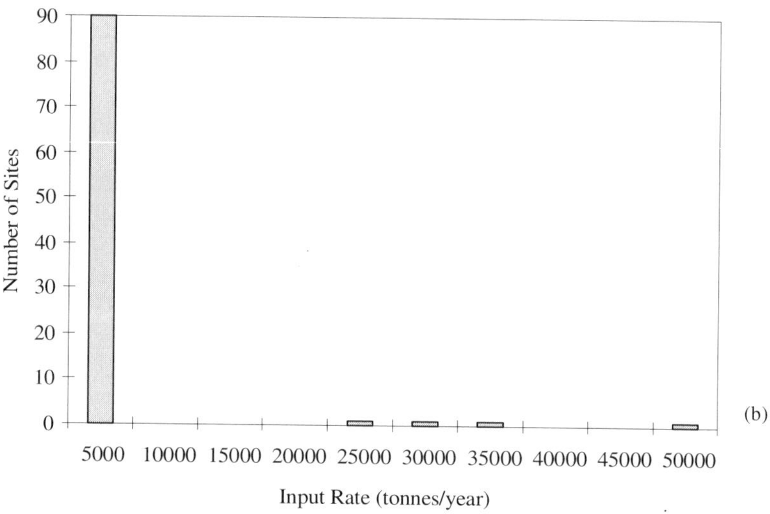

Figure 5.4 *Types of waste landfilled in the survey of 100 UK landfills. (a) Solid hazardous waste; (b) liquid hazardous waste. Source: DOE, Characterisation of 100 UK Landfill Sites. Report CWM 015/90, Wastes Technical Division, Department of the Environment, London, 1995. Crown copyright is reproduced with the permission of the Controller of Her Majesty's Stationery Office*

of industrial wastes. The sites investigated were mainly the newer and larger UK sites, but also included smaller and older sites and some restored sites (Figure 5.6). Figures 5.3–5.5 show that the majority of waste input is domestic and commercial/industrial waste, i.e., municipal solid waste. However, a significant proportion of solid and liquid hazardous waste is landfilled.

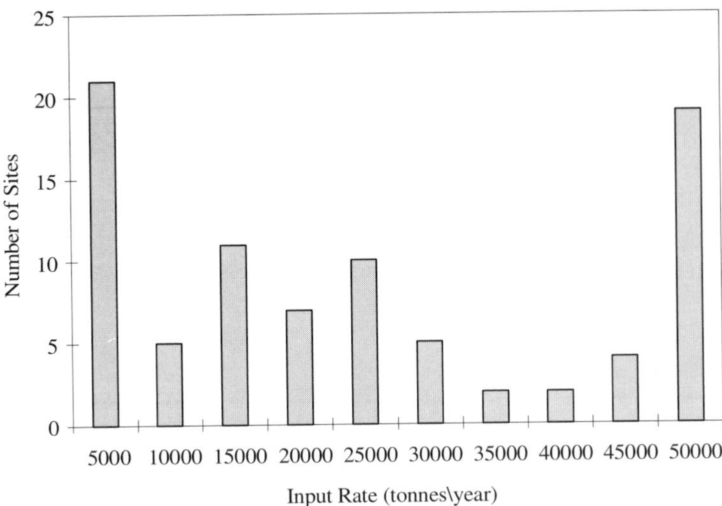

Figure 5.5 *Types of waste landfilled in the survey of 100 UK landfills: inert waste. Source: DOE, Characterisation of 100 UK Landfill Sites. Report CWM 015/90, Wastes Technical Division, Department of the Environment, London, 1995. Crown copyright is reproduced with the permission of the Controller of Her Majesty's Stationery Office*

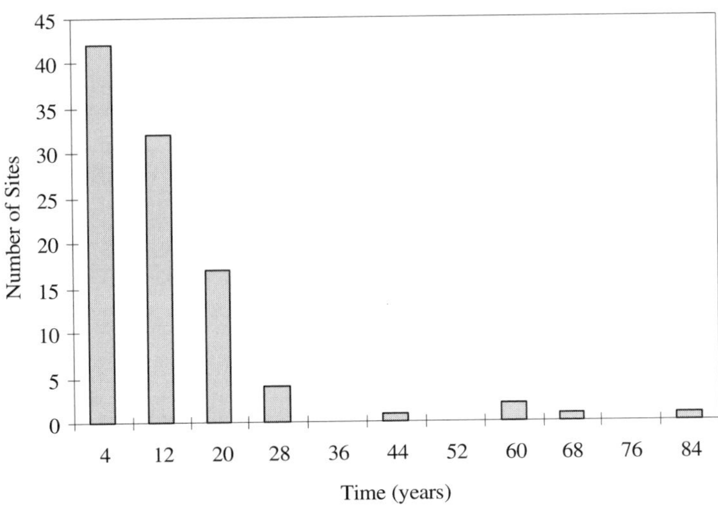

Figure 5.6 *Age of sites examined in the survey of 100 UK landfills. Source: DOE, Characterisation of 100 UK Landfill Sites. Report CWM 015/90, Wastes Technical Division, Department of the Environment, London, 1995. Crown copyright is reproduced with the permission of the Controller of Her Majesty's Stationery Office*

Table 5.2 *Landfill gas composition at six inert waste landfill sites (%)*

Gas	Inert landfill site					
	1	2	3	4	5	6
Methane	0	0	0	2	4	55
Carbon dioxide	3	1	0	10	5	35

Source: DOE, Environmental Impacts from Landfills Accepting Non-Domestic Wastes. Report CWM 036/91, Wastes Technical Division, Department of the Environment, London, 1991, published 1995.

5.4.1 Inert Wastes

Inert wastes are defined as those wastes which do not undergo any significant physical, chemical or biological transformations when deposited in a waste landfill. Typical inert wastes are materials such as brickwork, concrete, sand, cement, clay, plaster, soil, hard core, ash, clinker etc. The major source of inert wastes are from the construction and demolition industries, but other sources include, for example, ash from waste incineration. Approximately 1400 sites in England and Wales, irrespective of size and throughput, are inert sites (Waste Management Paper 26F (Draft) 1994). In the majority of cases insignificant quantities of landfill gas are generated, but in one particular case high and potentially hazardous levels of gas were generated (Table 5.2) (DOE 1995c). In many cases a robust definition of what constitutes inert wastes and different interpretations of the term allow a wide variety of wastes to be included in the definition.

Table 5.3 shows the variation in leachate composition for inert landfill sites in the UK (DOE 1995c). The data shown represent a different set of landfill sites from that in Table 5.2. Whilst the concentrations in the leachate are generally low, significant contaminants are found, indicating that biodegradation is occurring to some extent in the so-called 'inert' wastes.

5.4.2 Bioreactive Wastes

5.4.2.1 Decomposition Processes of Bioreactive Wastes in Landfills

The processes of degradation of bioreactive waste in landfills involves not only biological processes, but also interrelated physical and chemical processes (Waste Management Paper 26, 1986; Pescod 1991–93; Gendebien et al 1992; Tchobanoglous and O'Leary 1994; McBean et al 1995; Waste Management Paper 26B, 1995; Westlake 1995; Christensen et al 1996a). The stages involved in the degradation of bioreactive solid wastes can take many decades to complete. Biodegradable materials found in domestic wastes amount to over 65% dry weight, and include not only the putrescible food and garden wastes, but also paper, cardboard and even to some extent wood and textiles (Waste Management Paper 28, 1992). Industrial and

Table 5.3 Leachate composition from inert waste landfills (mg/l)

Parameter	Inert landfill site					
	7	8	9a	9b	9c	10
pH	8.81	7.83	7.7	8.5	7.92	7.82
COD	600	85	100	95	300	–
Total organic carbon	290	31	43	32	20	140
Acetic acid	0	0	< 20	< 20	< 20	0
Butyric acid	0	0	0	0	0	0
Propionic acid	0	0	0	0	0	0
Valeric acid	0	0	0	0	0	0
Iso-valeric acid	0	0	0	0	0	0
Phosphate	13	0.01	0.2	10.3	0.3	0.7
Chloride	1700	130	94	32	99	180
Sulphate	220	51	330	250	300	120
Nitrate	0.3	52	< 0.1	0.5	< 0.1	1.9
Ammonia	95	26	5.2	3.6	0.4	39
Calcium	110	150	460	340	380	570
Copper	< 0.1	< 0.1	0.5	< 0.1	< 0.1	< 0.1
Iron	1.2	1.5	380	1.8	5.4	30
Potassium	180	38	25	16	12	26
Magnesium	110	38	45	20	20	47
Manganese	0.3	0.3	3	1.2	2.6	2.1
Sodium	3000	150	65	45	60	200
Lead	< 0.1	< 0.2	0.4	< 0.2	< 0.2	< 0.2
Zinc	0.3	0.2	2.8	< 0.1	< 0.1	0.3

Source: DOE, *Environmental Impacts from Landfills Accepting Non-domestic Wastes.* Report CWM 036/91, Wastes Technical Division, Department of the Environment, London, 1991, published 1995. Crown copyright is reproduced with the permission of the Controller of Her Majesty's Stationery Office.

commercial wastes contain up to 62% and 66%, respectively, of dry weight biodegradable organic material, and consequently would be acceptable in a bioreactive waste landfill site (Waste Management Paper 26B, 1995). All these organic materials can be degraded by micro-organisms in the landfill. The organic materials occurring in waste can be classified into broad biological groups, represented by proteins, carbohydrates and lipids or fats. Carbohydrates are by far the major component of biodegradable wastes, and include cellulose, starch and sugars. Proteins are large complex organic materials composed of hundreds or thousands of amino acids groups. Lipids or fats are materials containing fatty acids. Figure 5.7 shows the decomposition pathways of the major organic and inorganic components of biodegradable wastes, and Figure 5.8 shows the process in more detail (Waste Management Paper 26B, 1995). Throughout the process of degradation, because of the heterogeneous nature of waste all the different stages may be progressing simultaneously until all the waste has reached Stage V and stabilisation of the landfill has been reached.

Stage I. Hydrolysis/aerobic degradation The hydrolysis/aerobic degradation stage occurs under aerobic (in the presence of oxygen) conditions. This occurs during the

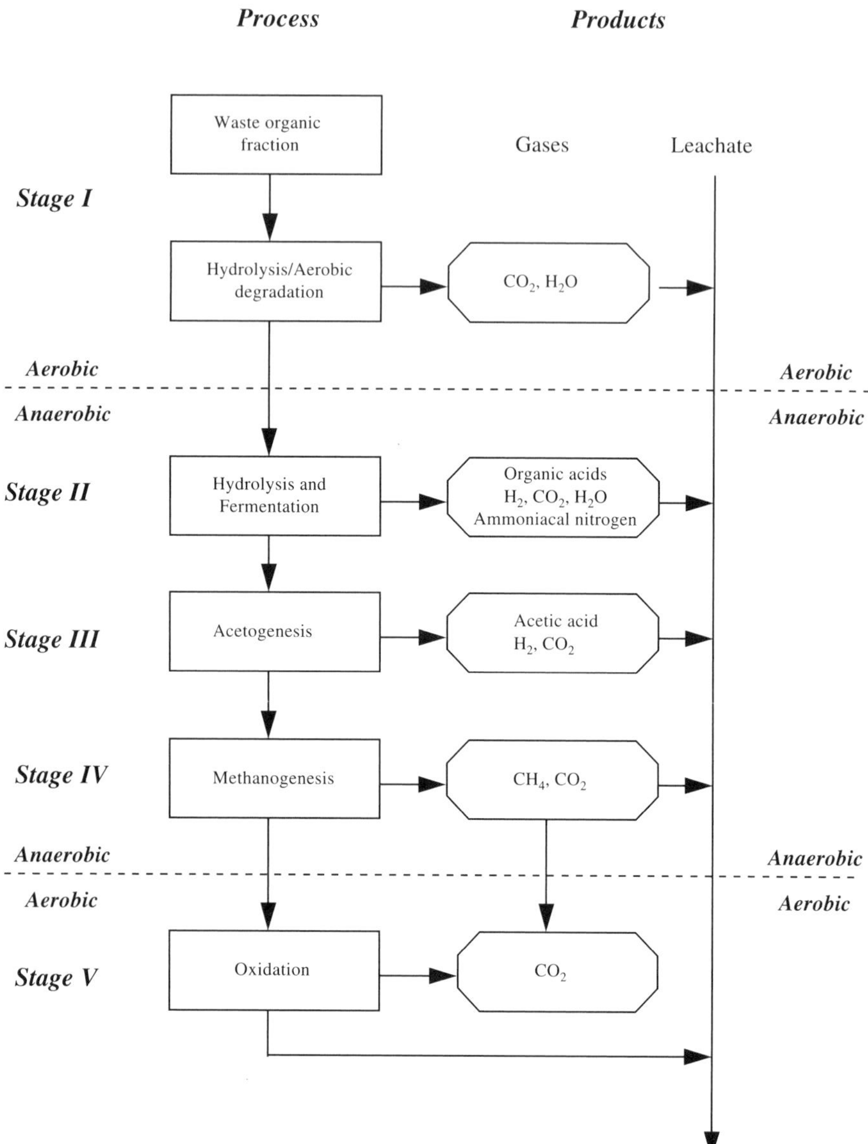

Figure 5.7 *Major stages of waste degradation in landfills. Source: Waste Management Paper 26B, Landfill Design, Construction and Operational Practice. Department of the Environment, HMSO, London, 1995. Crown copyright is reproduced with the permission of the Controller of Her Majesty's Stationery Office*

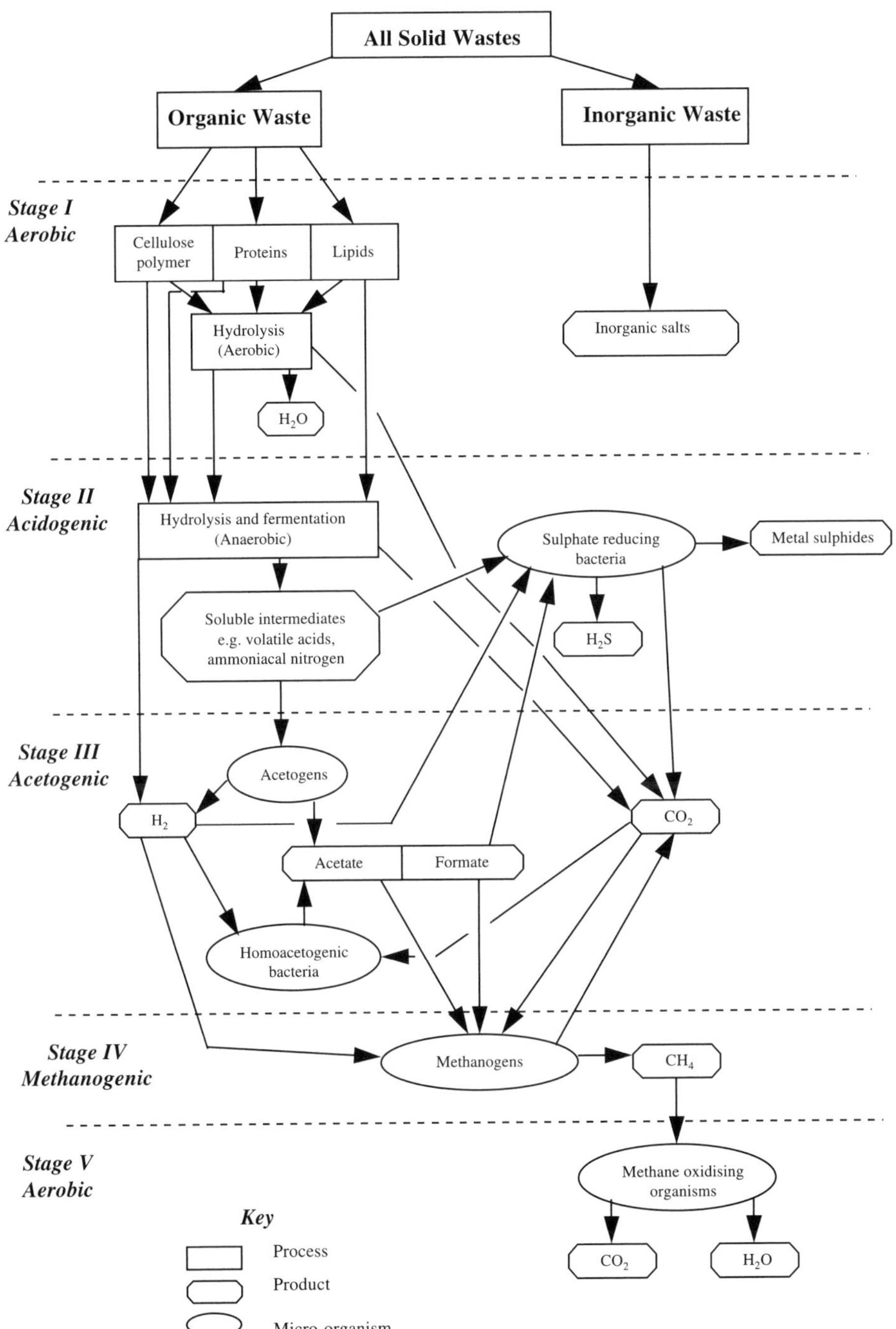

Figure 5.8 *Details of the stages involved in the degradation of bioreactive solid wastes in landfills. Source: Waste Management Paper 26B, Landfill Design, Construction and Operational Practice. Department of the Environment, HMSO, London, 1995. Crown copyright is reproduced with the permission of the Controller of Her Majesty's Stationery Office*

emplacement of the waste and for a period thereafter which depends on the availability of oxygen in the trapped air within the waste. The micro-organisms are of the aerobic type, that is, they require oxygen and they metabolise the available oxygen and a proportion of the organic fraction of the waste to produce simpler hydrocarbons, carbon dioxide, water and heat. The heat generated from the exothermic degradation reaction can raise the temperature of the waste to up to 70–90 °C (McBean et al 1995; Waste Management Paper 26B, 1995). However, compacted waste achieves lower temperatures due to the lower availability of oxygen. Water and carbon dioxide are the main products, with carbon dioxide released as gas or absorbed into water to form carbonic acid, which gives acidity to the leachate.

The aerobic stage lasts for only a matter of days or weeks depending on the availability of oxygen for the process, which in turn depends on the amount of air trapped in the waste, the degree of waste compaction and how quickly the waste is covered.

Stage II. Hydrolysis and fermentation Stage I processes result in a depletion of oxygen in the mass of waste and a change to anaerobic (absence of oxygen) conditions. Different micro-organisms, the facultative anaerobes, which can tolerate reduced oxygen conditions become dominant. Carbohydrates, proteins and lipids are hydrolysed to sugars, which are then further decomposed to carbon dioxide, hydrogen, ammonia and organic acids. Proteins decompose via deaminisation to form ammonia and also carboxylic acids and carbon dioxide. The derived leachate contains ammoniacal nitrogen in high concentration. The organic acids are mainly acetic acid, but also propionic, butyric, lactic and formic acids and acid derivative products, and their formation depends on the composition of the initial waste material. The temperatures in the landfill drop to between 30 and 50 °C during this stage. Gas concentrations in the waste undergoing Stage II decomposition may rise to levels of up to 80% carbon dioxide and 20% hydrogen (Waste Management Paper 26, 1986; Waste Management Paper 26B, 1995).

Stage III. Acetogenesis The organic acids formed in Stage II are converted by acetogen micro-organisms to acetic acid, acetic acid derivatives, carbon dioxide and hydrogen under anaerobic conditions. Other organisms convert carbohydrates directly to acetic acid in the presence of carbon dioxide and hydrogen. Hydrogen and carbon dioxide levels begin to decrease throughout Stage III. Low hydrogen levels promote the methane-generating micro-organisms, the methanogens, which generate methane and carbon dioxide from the organic acids and their derivatives generated in the earlier stages. The acidic conditions of the acetogenic stage increase the solubility of metal ions and thus increase their concentration in the leachate. In addition, organic acids, chloride ions, ammonium ions and phosphate ions, all in high concentration in the leachate, readily form complexes with metal ions, causing further increases in solubilisation of metal ions. Hydrogen sulphide may also be produced throughout the anaerobic stages as the sulphate compounds in the waste are reduced to hydrogen sulphide by sulphate-reducing micro-organisms (Christensen et al 1996a). Metal sulphides may be a reaction product of the hydrogen sulphide and metal ions in solution. The presence of the organic acids generate a very acidic solution which can have a pH level of 4 or even less (Moss 1997).

Stage IV. Methanogenesis The methanogenesis stage is the main landfill gas generation stage, with the gas composition of typical landfill gas approximately 60% methane and 40% carbon dioxide. The reactions are relatively slow and take many years for completion. The conditions maintain the anaerobic, oxygen-depleted environment of Stages II and III. Low levels of hydrogen are required to promote organisms, the methanogens, which generate carbon dioxide and methane from the organic acids and their derivatives such as acetates and formates generated in the earlier stages. Methane may also form from the direct micro-organism conversion of hydrogen and carbon dioxide to form methane and water. Hydrogen concentrations produced during Stages II and III therefore fall to low levels during this fourth stage. There are two classes of micro-organisms which are active in the methanogenic stage, the mesophilic bacteria, which are active in the temperature range 30–35 °C and the thermophilic bacteria, active in the range 45–65 °C. Therefore landfill gas can be generated during the methanogenic stage over a temperature range of 30–65 °C, with an optimum temperature range of gas generation between 30 and 45 °C. In fact, most landfill sites fall within this temperature range, with an average range for UK landfill sites of between 30 and 35 °C. Where temperatures in the mass of waste drop significantly, for example, to below 15 °C in cold weather in shallow sites, then the rate of biological degradation falls off. The organic acids formed during Stages II and III are degraded by the methanogenic micro-organisms, and as the acid concentration becomes depleted the pH rises to about pH 7–8 during the methanogenesis stage. Ideal conditions for the methanogenic micro-organisms are a pH range from 6.8 to 7.5, but there is some activity between pH 5 and pH 9. Stage IV is the longest stage of waste degradation, but may not commence until 6 months to several years after the waste is placed in the landfill, depending on the level of water content and water circulation. Significant concentrations of methane are generated after between 3 and 12 months depending on the development of the anaerobic micro-organisms and waste degradation products. Landfill gas will continue to be generated for periods of between 15 years and 30 years after final deposition of the waste, depending on waste and site characteristics (DTI 1996). However, low levels of landfill gas may be generated up to 100 years after waste emplacement.

Stage V. Oxidation The final stage of waste degradation results from the end of the degradation reactions, as the acids are used up in the production of the landfill gas methane and carbon dioxide. New aerobic micro-organisms slowly replace the anaerobic forms and re-establish aerobic conditions. Aerobic micro-organisms which convert residual methane to carbon dioxide and water may become established.

Figure 5.9 shows the changes in composition of landfill gas and leachate as the five stages of waste degradation progress with time (Waste Management Paper 26A, 1995, 26B, 1995). Initial formation of hydrogen and carbon dioxide in the hydrolysis/aerobic degradation, hydrolysis and fermentation and acetogenesis stages is followed by the main landfill gas generation stage, the methanogenesis stage. The

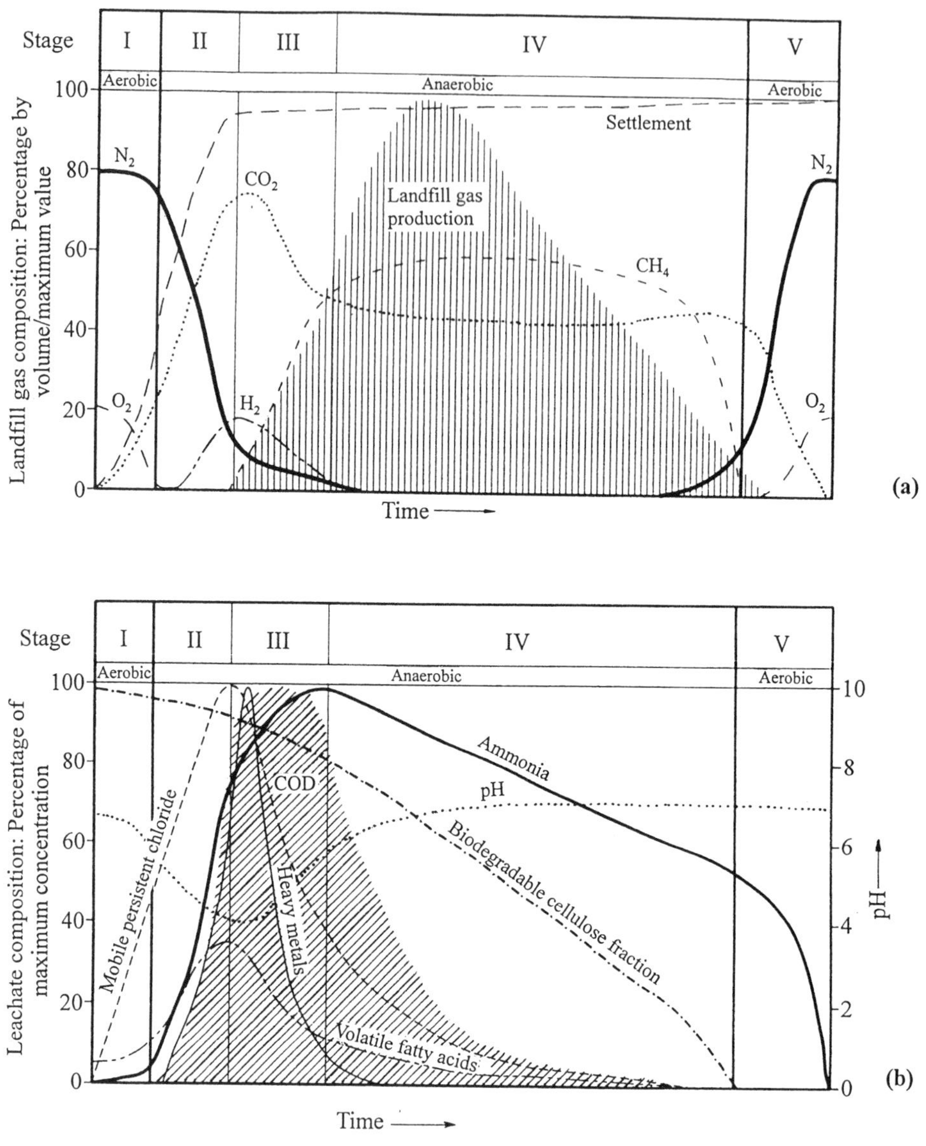

Figure 5.9 *(a) Landfill gas composition and (b) leachate composition in relation to the degradation of bioreactive solid wastes. Source: Waste Management Paper 26B, Landfill Design, Construction and Operational Practice. Department of the Environment, HMSO, London, 1995. Crown copyright is reproduced with the permission of the Controller of Her Majesty's Stationery Office*

Table 5.4 *Comparison of early- and late-stage leachate composition*

Stage II/III leachate	Stage IV leachate
High content of fatty acids	Low content of fatty acids
Acidic (low pH)	Neutral/alkaline (pH 7–8)
High BOD/COD ratio	Low BOD/COD ratio
High ammoniacal nitrogen	Lower levels of ammoniacal nitrogen
High organic nitrogen	
Heavy metals, e.g., Cr, Fe and Mn in solution	

Source: Waste Management Paper 26B, Landfill Design, Construction and Operation Practice. Department of the Environment, HMSO, London, 1995. Crown copyright is reproduced with the permission of the Controller of Her Majesty's Stationery Office.

characteristic landfill gas composition is methane and carbon dioxide with other minor components and water vapour. The final stages mark the end of the reaction and a return to aerobic conditions. Hydrogen sulphide gas may also form, derived from sulphate-reducing micro-organisms in wastes with a high concentration of sulphate.

Changes in leachate composition throughout the five-stage degradation period are also shown in Figure 5.9. Throughout the five stages cellulose becomes depleted by reaction with the various micro-organisms. Fatty acids in the leachate, initially formed from the action of anaerobic micro-organisms in Stages II and III, become depleted in the methanogenic stage as they are converted to methane and carbon dioxide. The leachate becomes acidic in Stage II due to the derived organic acids. The presence of the organic acids results in solubilisation of heavy metals such as chromium, iron and manganese into the acidic leachate. The organic acids become depleted in the methanogenic phase as the micro-organisms convert the acids to methane and carbon dioxide. As the pH begins to rise again, the heavy metals come out of solution as sulphide, hydroxide and carbonate precipitates. The chemical oxygen demand (COD) increases through the second and third stages but decreases in the methanogenic Stage IV (Waste Management Paper 26, 1986; McBean et al 1995; Waste Management Paper 26B, 1995; Westlake 1995). Table 5.4 compares leachate composition in the early and later stages of waste degradation. COD and BOD are standard parameters which measure the quality of water and are discussed in Box 5.1 (Tchobanoglous and Burton 1991; Fifield and Haines 1995).

5.4.2.2 *Factors Influencing Waste Degradation in Landfills*

There are numerous factors influencing the degradation of the waste, and these have been reviewed by Westlake (1995), Christensen et al (1996a), McBean et al (1995), Waste Management Paper 26 (1986), Waste Management Paper 26B (1995), Waste Management Paper 27 (1994) and Gendebien et al (1992).

Site characteristics Landfill sites with waste depths exceeding 5 m tend to develop anaerobic conditions and greater quantities of landfill gas. Shallower sites allow air interchange and lower anaerobic activity, and consequently lower landfill gas

Box 5.1
BOD and COD

Biochemical oxygen demand (BOD) is a standard test for the presence of organic matter in water. High levels of organic matter in water cause pollution problems since micro-organisms in the water biodegrade the organic material and thereby use up the dissolved oxygen in the water, leaving insufficient for fish and other aquatic life. The requirement for oxygen is the biochemical oxygen demand. The standard BOD test involves a sample of water in a completely filled sealed bottle which is left in the dark for 5 days at 20 °C. The dissolved oxygen at the start and end of the test is measured using either a dissolved oxygen electrode or a titrimetric method, the Winkler method. The BOD is expressed in g/m^3, representing the amount of oxygen used up by the micro-organisms in biodegrading the organic materials in the water.

Chemical oxygen demand (COD), as its name suggests, is a chemical method of determining the total organic material in a water sample which can be oxidised chemically rather than biologically. It is a more rapid test, taking only a few hours rather than the 5 days of the BOD test. A strong oxidising agent such as potassium dichromate is used to react with the oxygen in the organic material, and titrimetric analysis indicates the amount of oxidation which has taken place. COD is generally higher than BOD since more compounds can be chemically oxidised than can be biologically oxidised.

Sources: [1] Tchobanoglous G. and Burton F.L., Waste Water Engineering, Treatment, Disposal and Re-Use. McGraw-Hill, (Metcalf and Eddy Inc.), New York, 1991; [2] Fifield F.W. and Haines P.J., Environmental Analytical Chemistry. Blackie Academic and Professional, London, 1995.

production. However, if the site is well capped, anaerobic conditions will be created. Similarly, rapid covering of the waste will reduce the aerobic phase, and since this is the increasing temperature phase this will tend to keep waste temperatures down. Also, rapid covering of the waste will reduce the chance of rainfall increasing the moisture content of the waste, which in turn reduces the initial rate of biodegradation.

Waste characteristics The major components of municipal solid waste include the biodegradable fraction, that is, the paper and board, food and garden waste and textiles, and non-biodegradable components plastics, glass and textiles. The amount of biodegradation will vary depending on the proportion of biodegradable components in the waste. This fraction has been shown to vary depending on a number of factors, for example, higher concentrations of garden waste are produced in spring and autumn, and more industrially developed countries produce more paper. Also, older landfill sites have been shown to contain lower proportions of biodegradable waste than modern sites due to the changing nature of waste over the last few decades. In addition, the composition of the organic components, that is, the proportion of cellulose, proteins and lipids, will similarly influence the degradation pathway.

Shredding or pulverisation of the waste prior to landfilling results in increased available surface area and consequent increased homogeneity and increased rates of biological degradation. The density or degree of compaction of the waste in the landfill will increase the amount of biodegradable material available for degradation, and therefore increase the production of landfill gas per unit volume of void space in the landfill. Too high a degree of compaction, however, may limit the percolation of water through the site, which is necessary for the free flow of nutrients for the micro-organisms.

Moisture content of the waste The waste biodegradation process requires moisture, and this is in fact a major factor in determining the production of landfill gas and leachate. Even in certain types of containment landfill where dry conditions are a design requirement, the degradation processes will continue albeit at much slower rates of reaction. Increased rates of gas production are found with high moisture content landfill sites. The moisture content within the site will depend on the inherent moisture content of the waste, the level of rainfall in the area, percolation of surface and groundwaters into the site and, since water is a degradation product, the rate of biodegradation of the waste. The range of moisture content for typical municipal solid waste ranges from 15 to 40%, with a typical average value of 30% (Moss 1997). However, it is not only the moisture content of the waste that is important, but also the movement of the moisture to distribute the micro-organisms and nutrients and flush away the degradation products. Artificially induced continuous flushing of the site with leachate or water increases the rate of degradation by flushing out degradation products and replenishing nutrients for the micro-organisms. Whilst this flushing may increase the rate of gas production, leachate production will also increase.

Temperature The temperature range indicates the types of micro-organisms which are active. Initially, aerobic bacteria may increase the temperature up to levels of 80 °C if the waste is left well aerated, as the micro-organisms break down the waste to produce methane and carbon dioxide. However, compacted waste achieves lower temperatures due to the lower availability of oxygen. As aerobic conditions are replaced by anaerobic conditions throughout Stages II–IV, and the aerobic micro-organisms are replaced by anaerobic micro-organisms, the temperature drops to between 30 and 50 °C (Waste Management Paper 26B, 1995). The active temperature phase for the methanogenic micro-organisms fall into two ranges, 30–35 °C for the mesophilic-type bacteria and 45–65 °C for the thermophilic bacteria. The majority of landfill sites have temperatures between 30 and 35 °C during the main landfill gas generation phase. If the site is cold, then significantly less gas is produced than at higher ambient temperatures. Chaiampo et al (1996) have monitored the temperature changes with depth throughout a 20-m-deep municipal solid waste landfill in Italy. They showed that the first 1–2 m were in the temperature range 10–15 °C, but the temperature increased to 35–40 °C at 3–5 m depth and to 45–65 °C in the 5–20 m depth region. They equated the temperature regions with the mesophilic bacteria in the 1–5 m range and thermophilic bacteria in the deeper layers.

Acidity The acidity of the landfill site influences the activity of the various micro-organisms and therefore determines the rate of biodegradation. The pH of a typical landfill site would initially be neutral, followed by acidic phases, Stages II and III, where organic acids are produced from waste degradation by the acetogenic micro-organisms and the pH falls to as low as 4. The resultant organic acids provide the nutrients for the methanogenic bacteria, and as the acids are consumed, the pH rises. The methanogenic bacteria are most active in the pH range 6.8–7.5, and if the pH rises or falls outside this optimum range, then gas production is significantly reduced. The formation of organic acids and drop in pH is an essential step in the waste biodegradation process, in that the organic acids provide the nutrients for the main gas-generation stage IV micro-organisms, the methanogens.

5.5 Landfill Design Types

The main types of landfill site are the attenuate and disperse type and the containment type (Westlake 1995; McBean et al 1995; Waste Management Paper 26, 1986; Pescod 1991–93; Bagchi 1994).

Attenuate and disperse landfills Older designs where the site is usually unlined and there is uncontrolled release of leachate and landfill gas to the environment. The movement of leachate is allowed away from the site into the surrounding environment over a long period of time so that the leachate is diluted, reduced in toxicity and dispersed. This type of landfill is decreasing in use throughout the UK, USA and Europe.

Containment landfills The majority of UK landfills are now designed as containment landfills. A liner material contains the leachate within the landfill site boundary. The degradation processes take place within the landfill mass until stabilisation is complete. The leachate and landfill gas are therefore collected and treated. Three other types of containment landfills are described below:

- Co-disposal landfills. Co-disposal is the disposal of industrial and commercial wastes, which may be in solid or liquid form and may be hazardous, with bioreactive wastes such as municipal solid waste. The biodegradation of the municipal solid waste by micro-organisms with the associated biological, physical and chemical processes will also treat the co-disposed wastes and therefore reduce the hazard.
- Entombment landfills. The entombment or dry landfill type of landfill site aims to contain the waste in a relatively dry form for long periods of time by preventing biodegradation and the formation of leachate and landfill gas. The waste is therefore effectively stored rather than treated. The practice is common in the USA and France (Moss 1997).
- Sustainable landfills. The design and development of the flushing bioreactor landfill site in the UK is driven by the strategy of sustainable waste management. The strategy requires that waste should be dealt with by the present generation and not be left for future generations to deal with. Therefore, the new design of landfill site should incorporate this strategy and seek to implement landfill

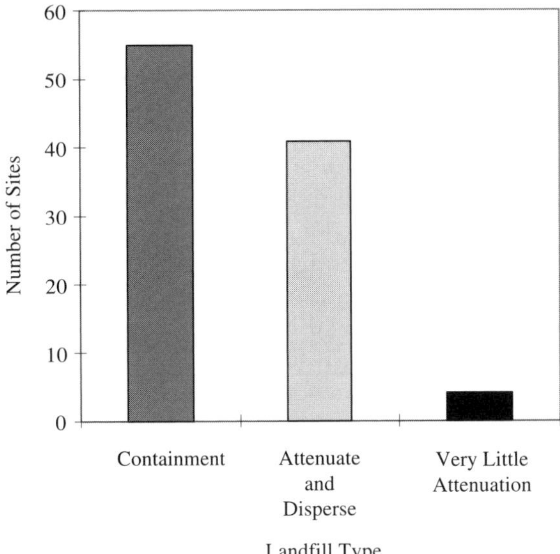

Figure 5.10 *Types of landfill sites in the UK: survey of 100 UK landfills. Source: DOE, Characterisation of 100 UK Landfill Sites. Report CWM 015/90, Wastes Technical Division, Department of the Environment, London, 1995. Crown copyright is reproduced with the permission of the Controller of Her Majesty's Stationery Office*

designs, construction and operational practices which will stabilise waste within a period of 30–50 years (Waste Management Paper 26B, 1995). This treats the landfill site as a controlled bioreactor rather than an uncontrolled biodegradation process. Stabilisation of waste within a generation (Waste Management Paper 26B, 1995)requires a mixed strategy of

- Inert wastes: landfilling low-reactivity inert wastes which produce negligible or low levels of polluting leachate and landfill gas.
- Pre-treatment of wastes: for example by incineration to reduce the volume and mass of waste and produce non-biodegradable, more homogeneous waste.
- The flushing bioreactor: to maintain a high rate of degradation, the landfill site is flushed with water or leachate to remove decomposition products and re-distribute nutrients and micro-organisms.

Figure 5.10 shows the distribution of types of landfills in a survey of 100 typical landfill sites in the UK (DOE 1995b). The majority are containment-type landfills, which include biodegradable waste types and co-disposal types, with the attenuate and disperse type the second most common, but decreasing in popularity. The entombment type is not common in the UK. The sustainable-type landfill is a new development, but is becoming the recommended type for sustainable waste management and stabilisation of the site within 30–50 years.

5.5.1 Attenuate and Disperse Landfills

Attenuate and disperse landfills rely on the very slow 'uncontrolled' release of leachate into the surrounding geological and hydrogeological environment. The leachate is firstly diluted by the groundwater in the surrounding environment. Secondly, the action of the biological, physical and chemical processes on the leachate as it migrates through the surrounding geological strata reduce its polluting potential (Waste Management Paper 26, 1986; Bagchi 1994; McBean 1995; Westlake 1995). This process is known as attenuation. Theoretically the leachate levels become so low as to pose no significant hazard to the environment. The attenuate and disperse site was common throughout history when the environmental implications of landfill gas generation, and in particular leachate, were not realised. Following the rise in environmental concern throughout the 1970s and 1980s, fuelled by a series of highly publicised pollution incidents associated with waste disposal, the attenuate and disperse type of landfill design came under increased criticism as an environmentally acceptable form of waste disposal. Even so the attenuate and disperse type of site is still found throughout the world, although it is becoming less acceptable, particularly throughout Europe and North America. There are a number of uncertainties associated with the attenuate and disperse landfill, for example, the level of leachate pollutants which can be accepted into the surrounding environment before unacceptably high migration occurs. In addition, there is great uncertainty as to the degree to which the leachate pollutants are reduced by interaction with the surrounding soils and geological strata. A further factor impacting on the use of attenuate and disperse landfill designs was the introduction by the EC of the Groundwater Directive in 1980 (EC Directive 80/68/EEC 1980). The Directive prohibited the direct or indirect discharge of a whole range of pollutants into groundwater (Waste Management Licencing Regulations 1994; Waste Management Paper 4, 1996). The range of pollutants were classified in 'List I' or 'List II' depending on their polluting potential. The pollutants specified include ammonia and nitrates, organohalogens and organophosphates, mercury, cadmium, zinc, copper and lead. Many of these compounds are found in landfill leachates and consequently further restricted the use of the attenuate and disperse type of landfill. Research into the effects of leachate on water courses during the 1980s showed that the degree to which leachate was attenuated was very variable, and was dependent on the variations in the surrounding geological and hydrogeological environment. This led to the development of containment sites and the increasing sophistication of the design and engineering of the sites. Consequently, impermeable liner materials and collection systems for leachate and landfill gas were introduced, and the containment type of landfill became the accepted design for waste disposal by landfill.

The attenuate and disperse type of site has been used in the past throughout the UK, Europe and the USA, but it would not be acceptable under current legislation and Government Guidance, and containment-type landfills are the norm. One exception is where the waste has been proven to be totally inert and unreactive; then the attenuate and disperse type of landfill may be used. Since the attenuate and disperse landfill was the most common landfill type up to the late 1980s, the consequences of this design are still impacting on the environment. In addition, it is

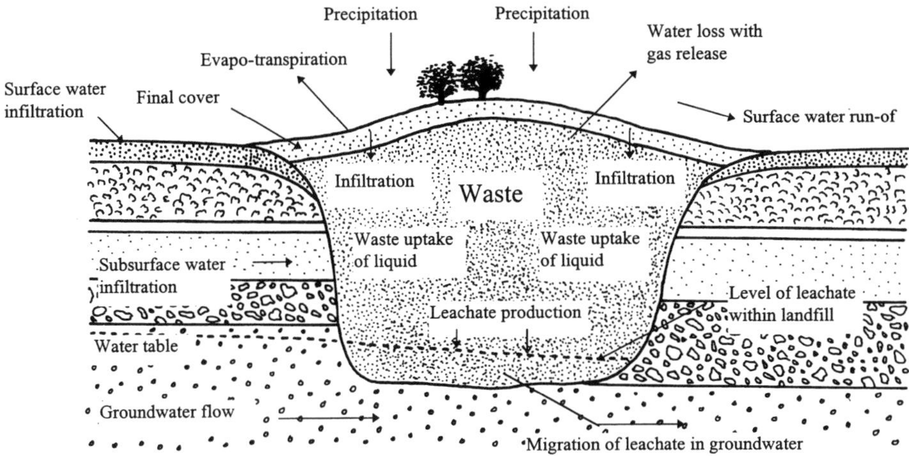

Figure 5.11 *Schematic diagram of the water balance for an attenuate and disperse landfill site. Source: Waste Management Paper 26, Landfilling Wastes. Department of the Environment, HMSO, London, 1986. Crown copyright is reproduced with the permission of the Controller of Her Majesty's Stationery Office*

important to understand the processes involved in leachate interaction with the surrounding geological and hydrogeological environment to assess the impact of any liner failure in containment-type landfills.

Figure 5.11 shows a schematic diagram of the water balance for an attenuate and disperse landfill site (Waste Management Paper 26, 1986). Water input from rainfall (precipitation), surface water and sub-surface water infiltrate into the landfill waste and, together with the water produced during the degradation process, produce the leachate. The leachate migrates through the mass of waste and into the groundwater, where the attenuation and dispersion process is transferred from within the waste mass to the surrounding geological and hydrogeological environment.

The biological, physical and chemical attenuation process acting on the leachate within the landfill mass and outside in the surrounding environment depend on the nature and quantity of waste and the surrounding geological, geochemical and hydrogeological conditions. Table 5.5 shows the range of biological, physical and chemical attenuation processes (Waste Management Paper 26, 1986).

Physical attenuation processes As the leachate migrates from the landfill site it is diluted by the surrounding groundwater. A plume of leachate spreading from the landfill site will disperse in the direction of groundwater movement and also laterally. Dispersion takes place on both a macroscale and a microscale. For example, differential dispersion will occur in different types of rock; it will also be influenced by rock grain size, pore size distribution, concentration of clay minerals in the rock etc. Leachate may be absorbed or adsorbed with the waste itself and the surrounding geological environment. The degree of absorption or adsorption increases with higher contents of clay minerals or increased concentrations of

Table 5.5 *Attenuation processes operating with an attenuate and disperse type of landfill*

Biological	Aerobic and anaerobic micro-organism degradation
Physical	Absorption, adsorption, filtration, dilution, dispersion
Chemical	Acid–base interaction, oxidation, reduction, precipitation, co-precipitation, ion-exchange, complex ion formation

Source: Waste Management Paper 26, Landfilling Wastes. HMSO, London, 1986.

organic carbon in soils. Such processes are easily reversible and allow easy removal of the pollutants at later stages as conditions alter (Waste Management Paper 26, 1986; Bagchi 1994; McBean et al 1995; Westlake 1995).

Chemical attenuation processes Chemical attenuation processes rely on an interaction between the leachate and the surrounding geochemical environment to chemically alter or fix the leachate. The interaction of cations and anions in the leachate with those in the soil or rock may occur via ion-exchange. For example, heavy metals may ion-exchange with cations found naturally in soils, clays or different consolidated rock types. Metals may also be removed by precipitation reactions; for example, many metal carbonates, hydroxides and sulphides are insoluble. A chemical reaction between metals in the leachate and such anions in solution results in precipitation. Co-precipitation may also adsorb or occlude trace metals from the leachate within the primary precipitate. Acidic conditions tend to solubilise metals, whereas more alkaline conditions induce precipitation. The formation of large ion complexes which include metal cations in the structure will effectively remove the metals from the environment by fixation in a large complex molecule. Oxidation–reduction reactions may also occur between inorganic species in the leachate and the surrounding geological and soil environment. Some elements and compounds can exist in more than one oxidation state, and changes in the redox potential may influence their mobility into and out of solution. For example, iron is readily oxidised to the ferric, Fe^{3+}, state in alkaline and mildly acidic conditions, and these conditions may then cause precipitation of ferric hydroxide from solution (Waste Management Paper 26, 1986; Bagchi 1994; McBean et al 1995; Westlake 1995).

Biological attenuation processes As the leachate passes into the surrounding geological and hydrogeological environment, the aerobic and anaerobic biodegradation of the organic materials in the leachate will continue. The circulation of groundwater serves to disperse the micro-organisms and nutrient organic material and remove degradation products (Waste Management Paper 26, 1986; Bagchi 1994; McBean et al 1995; Westlake 1995).

Rugge et al (1995) have reported on the leachate plume emanating from an old landfill site in Denmark which accepted municipal solid waste from the 1930s until its closure in the 1970s. They measured the 'plume' of leachate which penetrated the subsurface to a downgradient distance of between 200 and 250 m. The extent of the plume was detected by increased concentrations of inorganic compounds such as chloride. More than 15 organic compounds of potential harm to the environment, including benzene, toluene, xylenes, chlorinated hydrocarbons and polycyclic aromatic hydrocarbons, were identified at the subsurface border of the waste and sub-rock. However, 60 m downgradient of the landfill, the plume of hydrocarbons contained negligible concentrations of these hydrocarbons due to the natural dispersion and attenuation processes operating in the subsurface environment.

An example of an attenuate and disperse type of landfill site is described in Box 5.2 (DOE 1995d).

5.5.2 Containment Landfills

The containment landfill represents the current type of landfill design predominantly used in the UK, Europe, North America and Japan (Bagchi 1994; Hjelmar et al 1995; Sarsby 1995; Waste Management in Japan 1995; Waste Management Paper 26B, 1995; Westlake 1995; Hickman 1996). The modern landfill is often described as an engineering project involving construction engineering and process engineering. The project involves design details and drawings, construction of the base, installation of the lining, operation of the site throughout its lifetime, involving emplacement of the waste and reacting to biological, chemical and physical changes of the waste, monitoring and control of gas and leachate, and finally aftercare.

Containment landfills aim to contain the waste within the boundaries of the site by managing the leachate and landfill gas. In most cases the containment landfill for municipal solid waste contains moisture from the inherent moisture content of the waste and through rainwater. Therefore, biodegradation of the waste occurs, and the leachate is collected and treated until the landfill becomes stabilised. This process may take many decades to achieve. A number of other different types of landfill are included within the category of containment landfills.

The co-disposal landfill With the increasing complexity of landfill liner systems, the possibility of co-disposal of industrial/hazardous waste together with municipal waste is realised. The biodegradation chemical and physical processes operating within the mass of biodegradable waste also act on the industrial/hazardous waste, thereby reducing the hazard.

The entombment (dry) landfill In some countries, for example, the USA and France (Moss 1997), the approach is to have a sophisticated liner and capping system which ensures that the waste is kept dry and that leachate generation is minimal: the entombment or dry type of landfill. In theory, the waste cannot generate leachate and therefore will not pose an environmental problem. However, inevitably some moisture may be present inherent in the waste, and therefore the waste may take hundreds of years to become stabilised. However, this creates problems for future generations if the landfill liner system fails and is therefore inconsistent with sustainable waste management. In the UK some liner and cover systems are also creating landfills which prevent stabilisation, and consequently will still be generating leachate and landfill gas for many decades into the future (Joseph and Mather 1993).

Box 5.2

The Attenuate and Disperse Landfill Site at Gorsethorpe, Nottinghamshire

The Gorsethorpe landfill is an attenuate and disperse type landfill located in a disused sandstone quarry. It is unlined, and the bed rock is the Sherwood Sandstone. Landfilling commenced in 1969 with the disposal of domestic waste with incinerator ash and non-hazardous industrial materials. The site area, approximately 100 000 m^2, was landfilled in three phases and completely filled by 1983. It was then covered with 1 m of colliery shale. The area has been grassed

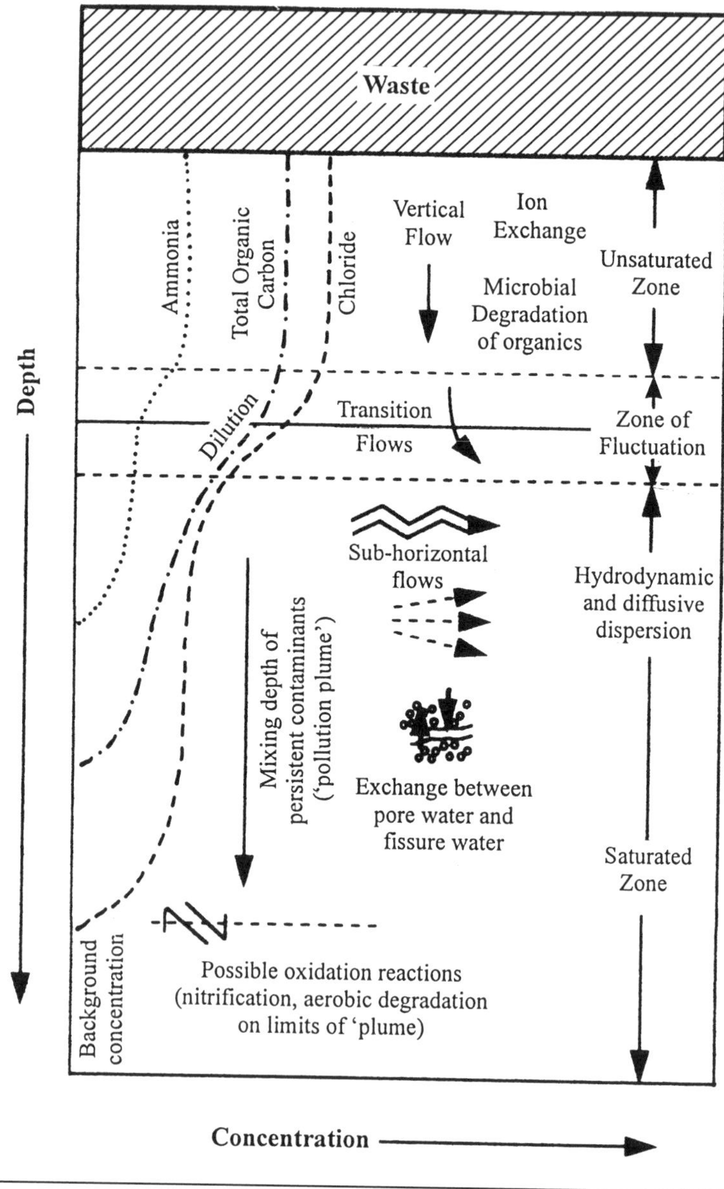

Box 5.2
(continued)

and trees planted. The water table ranges between 8 and 20 m below the base of the landfill. The main source of water for the leachate is rainwater at an estimated 243 mm per year average, rather than percolation of ground water. Calculations of the production of leachate indicate an annual production rate of 10 000–20 000 m^3.

The site was monitored for leachate and landfill gas composition over the period from 1986 to 1992. The results show that leachate penetrates the unsaturated zone (i.e., the zone above the water table) below the landfill mass, and when the saturated zone is reached attenuation and dispersion of the leachate components occurs. The passage of the leachate through the unsaturated zone below the landfill results in some attenuation of the contaminants. High concentrations of methane and carbon dioxide are found within the landfill mass, representing the methanogenic biodegradation of the waste. Below the waste, in the unsaturated zone, significant concentrations of methane and carbon dioxide are found forced through the pores by the advance of the liquid leachate front.

The figure shows the processes of attenuation and degradation of the leachate components in the unsaturated zone immediately below the landfill, which suggest that microbiological degradation and ion exchange are the major decomposition mechanisms. When the leachate reaches the saturated zone (i.e., the water table, where the pores are saturated with water) the contaminants are dispersed by dilution in the ground water, described as a leachate 'plume', which disperses within the saturated zone. Within the plume further biological degradation and other chemical and physical degradation mechanisms will further attenuate the leachate components. It is estimated that mixing of the leachate extends the plume to a depth of 10 m below the water table and also extends it laterally in the direction of groundwater flow.

Source: DOE, Landfill Monitoring Investigations at Gorsethorpe Landfill, Sherwood Sandstone, Notts, 1978–1992. Report CWM/034/94, Wastes Technical Division, Department of the Environment, London, 1994, published 1995. Crown copyright (figure) is reproduced with the permission of the Controller of Her Majesty's Stationery Office

The sustainable (flushing bioreactor) landfill Finally, the containment type of landfill includes the sustainable landfill, or controlled flushing bioreactor. The sustainable landfill is designed and operated to achieve stabilisation within a 30–50-year time span. The landfill is designed as a controlled bioreactor, to accelerate the biodegradation process by continuously recirculating water and/or leachate through the waste.

5.5.2.1 Landfill Liner Materials

There are a large variety of natural and synthetic mineral materials and synthetic polymeric materials which are used in the construction of landfill containment sites (Bagchi 1994; Sharma and Lewis 1994; McBean et al 1995; Christensen et al 1996b). The choice of liner barrier and liner system will depend on the type of waste, the geological and hydrogeological conditions in the surrounding environment, the prediction of the properties of the derived leachate, and the resistance of the liner to the leachate. A critical factor in the type of liner material used for containment types of landfill is the permeability, which is measured as 'hydraulic conductivity'. Hydraulic conductivity is discussed in Box 5.3.

Natural clay Clays are unconsolidated rocks composed of clay minerals formed as breakdown products from the weathering of pre-existing rocks. The clay minerals include, for example, montmorillanite, illite and kaolinite, and are of extremely fine grain size. The very fine grain size means that porosity is also extremely low and consequently permeability is very low. Clay minerals are formed from sheets of alumino-silicates stacked in layers bonded by cations. Water molecules are absorbed between the layers, which cause swelling and hence the observed low permeability. In addition, in situ clay may be utilised as the underlying material of the landfill if the local geological environment lends itself to this choice of site selection, the clay acting as a further low permeability barrier beneath the liner system. However, when used as a design barrier, the properties of the clay should be carefully evaluated due to potential inhomogeneities in the clay strata. In such cases, the in situ clay would be excavated and re-laid. Normal practice for natural clay liners in landfill sites is to use local or imported clay, which is compacted into layers of between 0.6 and 1.0 m thick to form a homogeneous low-permeability layer (Sharma and Lewis 1994; McBean et al 1995).

The clay liner consists of a mixture of clay minerals and fine silt particles which blend to form a clay or clay soil with suitable low permeabilities. The factors which can affect the suitability and performance of a particular clay or clay soil in its use in a landfill liner system include porosity and permeability, which in turn depend on clay mineralogy, particle size distribution, plasticity, strength, moisture content and compaction. Moisture is an important factor in determining permeability, and adjustment of the moisture content of the clay material to obtain suitable permeability may be required before use. Optimum moisture content is determined by a standard test which produces a maximum dry density when a clay soil is compacted in a standard mould. In many cases, lower permeabilities may be obtained by clay liners which are slightly wetter than the optimum moisture content, in which case care must be taken to stop this water from migrating away. Clay mineralogy also determines the permeability of a clay, for example, clays with a higher proportion of montmorillanite have lower permeabilities than clays with a high proportion of illite, which in turn have lower permeabilities than clays with a high proportion of kaolinite (Bagchi 1994).

The clay material is excavated from the source site and blended to form a homogeneous material. It may also be necessary to sieve and remove large rocks. The clay material is prepared by adjusting the moisture content to achieve the lowest permeability. The clay is then transported and spread by bulldozers or scrapers at the site. The clay liner is then compacted on site by large roller vehicles to form a more homogeneous layer by breaking up large pieces of clay and thereby greatly reducing the voidspace between the pieces and also serving to increase the density.

Bentonite-enhanced soils Bentonite is a mixture of clay minerals, principally of the montmorillanite type. Sodium bentonites and calcium bentonites exist with the sodium form having lower permeabilities. Where the naturally occurring clay soil

Box 5.3
Hydraulic Conductivity (Permeability)

Darcy's Law is an empirical law describing the flow of a fluid through a porous material. The law relates the flow rate of the fluid to a cross-sectional area of the porous material and the hydraulic gradient by way of a constant, the coefficient of permeability.

$$Q = kiA$$

Q = flow rate
k = coefficient of permeability, permeability or hydraulic conductivity
i = hydraulic gradient (the pressure difference between the top and bottom of the layer of material)
A = cross-sectional area

Hydraulic conductivity or permeability therefore represents the ease with which a fluid such as leachate will flow through the liner material. The units of measurement are typically cm/s or m/s. Typical hydraulic conductivities of natural and synthetic or processed materials and waste are given below.

Material	Hydraulic conductivity (permeability) (m/s)
Natural materials	
Well-graded, clean gravels, gravel–sand mixture	2×10^{4}
Poorly graded, clean sands, gravely sands	5×10^{-4}
Silty sands, poorly graded, sand–silt mixture	5×10^{-5}
Inorganic silts and clayey silts	5×10^{-8}
Mixture of inorganic silt and clay	2×10^{-9}
Inorganic clays of high plasticity	5×10^{-10}
Synthetic or processed materials	
Compacted clay liner	1×10^{-8}–10^{-10}
Bentonite-enhanced soil	5×10^{-10}
Geosynthetic clay liner	1×10^{-10}–10^{-12}
Flexible membranes	1×10^{-13}
Geotextile	1×10^{-4}–10^{-5}
Geonet	2×10^{-1}
Waste	
Municipal solid waste as placed	1×10^{-5}
Shredded municipal solid waste	1×10^{-4}–10^{-6}
Baled municipal solid waste	7×10^{-6}

Sources: [1] Sharma H.D. and Lewis S.P., Waste Containment Systems, Waste Stabilisation, and Landfills: Design and Evaluation. John Wiley and Sons Ltd., New York, 1994; [2] McBean E.A., Rovers F.A. and Farquar G.J., Solid Waste Landfill Engineering and Design. Prentice Hall, Englewood Cliffs, NJ, 1995.

does not have a high enough level of clay minerals to produce a suitably low permeability, bentonite clay is added to form a bentonite-enhanced soil. The bentonite-type clay minerals have a high swelling characteristic on absorbing moisture, which forces the hydrated bentonite clay around the soil particles to form a synthetic clay. In addition, the bentonite-enhanced soil further swells under pressure and consequently lower permeabilities are obtained as the mass of waste builds up in the overlying landfill. Sodium bentonite is added at between 5 and 15% depending on the type of original soil. Where the calcium form of bentonite is used, larger quantities are required due to its lower swelling properties and therefore higher permeability. To ensure the formation of a homogeneous low permeability clay, the bentonite–soil mixture should be thoroughly mixed and the moisture content adjusted to produce the lowest permeability-enhanced soil. Mixing is usually carried out before application to the site location. Suitable host soils for bentonite application include sands and silty sands, whereas more cohesive materials may be difficult to mix, producing an inhomogeneous variable permeability mixture (McBean et al 1995; Waste Management Paper 26B, 1995).

Geosynthetic clay liners Geosynthetic clay liners are a mixture of bentonite clay mechanically or chemically adhered to a geotextile fabric. Alternatively, the bentonite layer may be sandwiched between two layers of the geotextile fabric where the layers are joined by adhesives, needlepunching or stitching. Typically the geosynthetic clay liners are approximately 1 cm thick and are available in 5 m × 30 m rolls (Sharma and Lewis 1994). The rolls are laid out on natural clay soils or polymeric flexible membranes and joined by bentonite cement. Geosynthetic clay liners are often used as alternatives to natural compacted clay liners, but since they are a relatively new material in landfill applications, their long-term stability in landfills has not been tested over very long periods. Table 5.6 shows a comparison of the natural and geosynthetic types of clay liner (Sharma and Lewis 1994).

Flexible membrane liners Flexible membrane liners are synthetic, polymeric plastic materials with extremely low permeabilities. Whilst permeabilities are extremely low, there is some diffusion of leachate through the membrane. Also, if the liner became punctured with *small* holes, during either manufacture or emplacement, leakage of leachate would also occur. There are a variety of membrane liners available and the major types used in landfill applications are shown in Table 5.7 (Sharma and Lewis 1994). The commonest types used are high-density polyethylene and polyvinyl chloride. The membranes come in sheets or rolls ranging from 5 to 15 m wide and up to 500 m in length, and range in thickness typically from 0.75 mm to 3.00 mm. There are a range of properties which define the suitability for use of the various membranes in landfill applications, including density, tensile strength, puncture resistance, tear resistance, resistance to ultraviolet light and ozone and chemical resistance (Sharma and Lewis 1994; Tchobanoglous and O'Leary 1994; Waste Management Paper 26B, 1995). Membrane chemical resistance is very important since the leachate may contain a range of organic and inorganic acids and alkalis and organic hydrocarbons. There are a range of standard tests available to

Table 5.6 *Comparison of geosynthetic clay liners and natural clay liners*

Characteristic	Geosynthetic clay liner	Natural clay liner (compacted)
Materials	Bentonite clay, adhesives, geotextiles and geomembranes	Native soils or blend of soil and bentonite
Construction	Manufactured and then installed in the field	Construction in the field
Ambient	Installation at low temperature permissible	Installation at or below freezing temperature not permissible
Thickness	Approximately 10 mm	Approximately 0.5–1 m
Clay hydraulic conductivity	10^{-8}–10^{-10} cm/s typical	10^{-7}–10^{-8} cm/s typical
Speed and ease of construction	Rapid, simple installation	Slow complicated installation
Water content	Essentially dry, cannot desiccate	Nearly saturated
Settlement	Adjusts to differential settlement	Performance is poor in cases of differential settlement
Leachate	Cannot be used in direct contact with most leachate	Can be used in direct contact with most leachate

Sources: [1] Sharma H.D. and Lewis S.P., Waste Containment Systems, Waste Stabilisation, and Landfills: Design and Evaluation. John Wiley and Sons Ltd., New York, 1994; [2] Bagchi A., Design Construction and Monitoring of Landfills (2nd edn), John Wiley and Sons Ltd., Chichester, 1994.

Table 5.7 *Types of synthetic flexible membrane liners*

Thermoplastic polymers[1]	Thermoset polymers[2]	Combinations
Polyvinyl chloride (PVC)	Butyl or isoprene-isobutylene (IIR)	PVC–nitrile rubber
Polyethylene, e.g., LDPE[3], HDPE[4]		
Chlorinated polyethylene (CPE)	Ethylene propylene diene monomer (EPDM)	PVC–ethyl vinyl acetate
Elasticized polyolefin (3110)	Polychloroprene (neoprene)	Cross-linked CPE
Ethylene interpolymer alloy (EIA or XR-5)	Ethylene polypropylene terpolymer (EPT)	Chlorosulphonated polyethylene
Polyamide	Ethylene vinyl acetate	(CSPE or Hypalon)

1 Thermoplastic polymers are those that can be repeatedly re-worked by heating and cooling.
2 Thermoset polymers can be processed only once; reheating results in polymer degradation.
3 LDPE, low density polyethylene.
4 HDPE, high density polyethylene.

Source: Sharma H.D. and Lewis S.P., Waste Containment Systems, Waste Stabilisation, and Landfills: Design and Evaluation. John Wiley and Sons Ltd., New York, 1994.

Table 5.8 *Advantages and disadvantages of commonly used synthetic flexible membranes*

Synthetic flexible membrane	Advantages/disadvantages
Butyl rubber	Good resistance to ultraviolet (UV), ozone and weathering elements Good performance at high and low temperatures Low swelling in water Low strength characteristics Low resistance to hydrocarbons Difficult to seam
Polychlorinated polyethylene (CPE)	Good resistance to UV, ozone and weather elements Good performance at low temperatures Good strength characteristics Easy to seam Poor resistance to chemicals, acids and oils Poor seam quality
Chlorosulphonated polyethylene	Good resistance to UV, ozone and weather elements Good performance at low temperatures Good resistance to chemicals, acids and oils Good resistance to bacteria Low strength characteristics Problem during seaming
Ethylene propylene rubber (EPDM)	Good resistance to UV, ozone and weather elements High strength characteristics Good performance at low temperatures Low water absorbance Poor resistance to oils, hydrocarbons and solvents Poor seam quality
LDPE and HDPE	Good resistance to most chemicals Good strength and seam characteristics Good performance at low temperatures Poor puncture resistance
Polyvinyl chloride (PVC)	Good workability High strength characteristics Easy to seam Poor resistance to UV, ozone, sulphide and weather elements Poor performance at high and low temperatures

Source: Bagchi A., Design, Construction and Monitoring of Landfill. John Wiley and Sons, New York, 1994. Copyright © 1994 John Wiley and Sons Inc., Reprinted by permission from John Wiley and Sons Inc.

determine the properties of the membranes. The membrane sheets or rolls are seamed together using heat-sealing or liquid solvents, usually carried out on site. Fast rates of seaming can be achieved, ranging from 30 m/h to 100 m/h depending on the system of seaming (Sharma and Lewis 1994). Table 5.8 compares the advantages and disadvantages of commonly used synthetic flexible membrane liners (Bagchi 1994).

Geotextiles Geotextiles are fabric materials used as protection for polymeric plastic membranes and filtration material to filter out fine-grained particles from the leachate so that drainage layers do not become blocked. Geotextiles are composed of polypropylene or polyester fibres which are manufactured to form a fabric-type material. Manufacture consists of either weaving, using traditional weaving techniques, or needlepunching through the fibre web to produce an interlocked fabric. Geotextiles are not used for containment and therefore have relatively high permeabilities. As with polymeric plastic membranes, there are a wide variety of tests which may be carried out on the geotextile material to determine, for example, tensile strength, tear resistance, burst strength, chemical resistance etc. Seaming of the geotextile may be by overlapping or simple stitching of the fabric (Sharma and Lewis 1994; McBean 1995; Christensen et al 1996b).

Geonets Geonets are porous sheets of plastic netting used as drainage layers to carry leachate or landfill gas. The nets are typically about 5 mm thick and are usually composed of polyethylene (Sharma and Lewis 1994). To prevent the net from clogging due to particles in the leachate there is usually a geotextile fabric material bonded to the geonet. The main role of geonets is drainage, and they are used as alternatives to naturally well-drained materials such as coarse sands or gravels, but require less thickness to achieve the same effectiveness. For example, a 4.5 mm thick geonet layer would have a similar drainage function as 300 mm of sand.

5.5.2.2 Landfill Liner Systems

There are a variety of landfill liners available, and they may be designed in different ways to produce a large number of different liner systems (Sharma and Lewis 1994; Tchobanoglous and O'Leary 1994; McBean 1995; Waste Management Paper 26B, 1995). Increasing complexity of liner design and the use of multilayers of materials increases costs, and a balance has to be achieved between cost and the protection of the environment. To this end, risk assessment is used as the criteria for selection of the most appropriate landfill liner system. The assessment of risk is on a site-specific basis in relation to the surrounding geology and hydrogeology and the type, composition and amount of waste. Risk assessment is used in the determination of a suitable landfill system in defining the acceptable seepage rate of leachate into the surrounding environment so that the pollutants do not cause an unacceptable level of contamination. This would take into account the processes of leachate formation and the dilution, dispersion and attenuation in the surrounding environment. The pathways for leakage of landfill gas and impacts on the environment should also be assessed. Therefore the geological environment, geochemistry and hydrogeology should be part of the assessment.

All liner materials allow a certain low level of seepage, the extent of which is determined by their permeability. Similarly, the system should not allow groundwater to seep into the landfill and increase leachate levels. The liner system also minimises the release of landfill gas. The design of a liner system should also be resistant to the variety of chemical properties of the leachate throughout the lifetime of the site, which may be over a 50-year period. Selection of the liner system will also be influenced by local geological conditions, for example, if clay is the local environment, then clays will be used as a component of the system.

The liner systems rely on combinations of liner materials and gas and liquid collection layers to contain and collect the leachate and landfill gas. There are several different types of liner systems, for example (Waste Management Paper 26B, 1995): single liner system; composite liner system; double liner system; multiple liner system.

Single liner system In most cases it is not sufficient to use only one primary liner since failure of the liner then results in leachate escape. However, the single-liner system may be appropriate for certain low-risk wastes in sites where escape of leachate poses negligible risk of contamination. Figure 5.12(a) shows a example of the single-liner system (Waste Management Paper 26B, 1995). The single liner of the system is composed of a primary barrier consisting of a layer of clay, bentonite-enhanced soil or hydraulic asphalt. Above and below the primary liner or barrier would be a separation/protection layer of geotextile material, for example, non-woven, needlepunch fabrics composed of polyester or polypropylene fibre. The material acts as a protective layer and filter for fine suspended solids. Between the waste and the separation/protection layer would be a leachate collection system consisting of a series of drainpipes or a drainage layer. Similarly, beneath the liner, if necessary, may be a groundwater collection system such as a drainage layer of gravel or a synthetic, polymeric net material (geonet).

Composite liner system Figure 5.12(b) shows a schematic diagram of a composite liner system (Waste Management Paper 26B, 1995). The use of two different types of liner material, a clay-based mineral layer and a polymeric membrane layer, provides the composite liner system with a more secure containment than the single liner system and more protection for the environment. The primary liner or barrier in the case of the composite liner system is a synthetic polymeric flexible membrane liner with a separation/protective layer of thin polymeric or geotextile material. The polymeric membrane liner consists of, for example, high-density polyethylene (HDPE), low-density polyethylene (LDPE), polyvinyl chloride (PVC) etc. The secondary barrier would be a layer of clay, bentonite-enhanced soil or geosynthetic clay liner. Below the secondary barrier would be a separation/protection layer of geotextile material followed by a groundwater collection system. Between the waste mass and the liner system is the leachate collection system.

Double liner system Figure 5.12(c) shows a schematic diagram of a double liner system (Waste Management Paper 26B, 1995). This system incorporates an intermediate high-permeability drainage layer between the primary and secondary liner barriers. The intermediate drainage system is in addition to the leachate drainage collection system, and the groundwater drainage collection system, and is used to monitor and remove leachate and landfill gas from between the barriers. Where a multi-barrier system is used there is the possibility of build-up of leachate or landfill gas between the low permeability layers which may cause problems of lateral

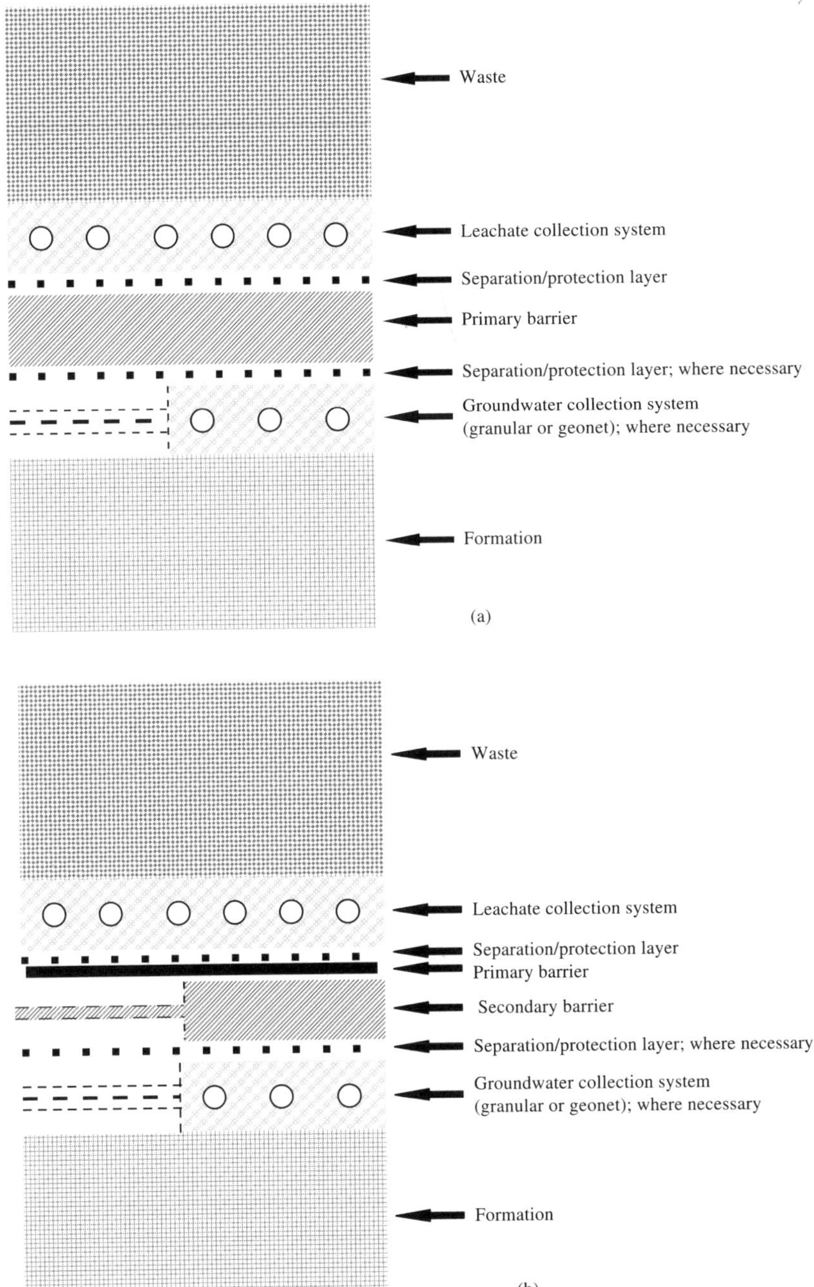

Figure 5.12 *Schematic diagrams of typical liner systems. (a) Single liner system; (b) composite liner system; (c) double liner system; (d) multiple liner system. Source: Waste Management Paper 26B, Landfill design, construction and Operational Practice. Department of the Environment, HMSO, London, 1995. Crown copyright is reproduced with the permission of the Controller of Her Majesty's Stationery Office*

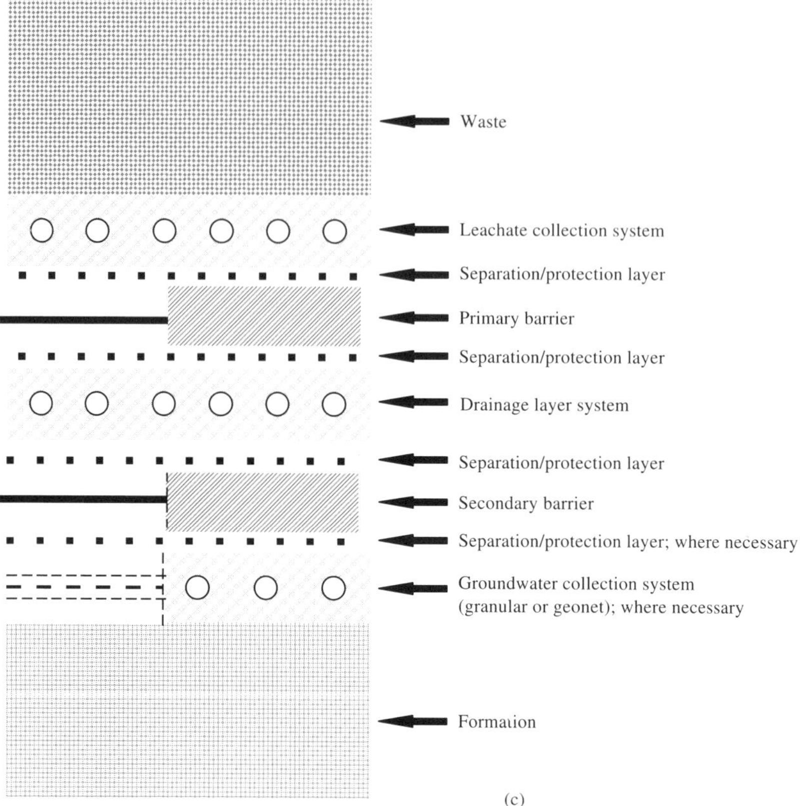

Figure 5.12 *(continued)*

migration of leachate or gas, and hence the need for an intermediate drainage layer. The primary and secondary barriers may be composed of mineral materials such as clay, bentonite-enhanced soil or a geosynthetic clay liner such as bentonite matting, or alternatively a synthetic, polymeric membrane liner is used. As before, each layer is separated by a separation/protection layer of geotextile fabric filter material.

Multiple liner system Figure 5.12(d) shows a schematic diagram of the multiple liner system (Waste Management Paper 26B, 1995). The multiple liner system combines some of the attributes of the composite and double liner systems. Primary, secondary and tertiary liner barriers are incorporated, with intermediate drainage to remove leachate and landfill gas from between the barriers. The primary barrier is usually composed of a synthetic polymeric membrane in intimate contact with the secondary liner barrier composed of mineral materials such as clay, bentonite-enhanced soils or bentonite matting. Intimate contact between primary and secondary barriers is required to prevent lateral movement of leachate or gas. As before, separation/protection layers of geotextile fabric filters and leachate drainage collection and groundwater drainage collection layers are also incorporated.

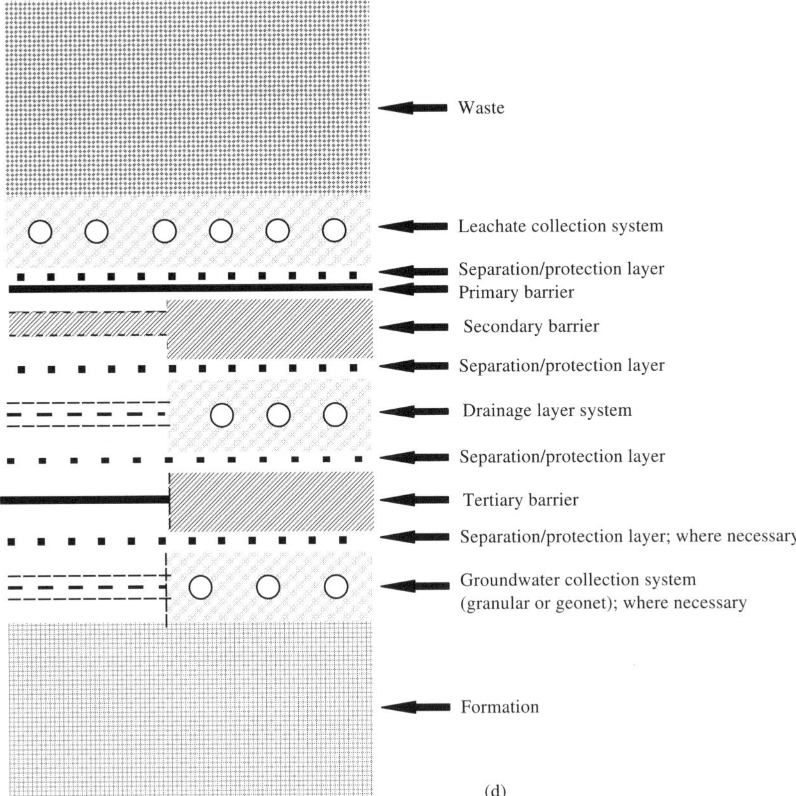

Figure 5.12 *(continued)*

Table 5.9 summarises the typical liner materials used for the various liner barriers in the four systems discussed (Waste Management Paper 26B, 1995). Several other liner systems have been described for different countries and a range of different waste types (see, for example, Ham 1993; Bagchi 1994; LaGrega et al 1994; Sharma and Lewis 1994; Tchobanoglous and O'Leary 1994; Bishop and Carter 1995; McBean et al 1995).

5.5.2.3 Co-disposal Landfills

Co-disposal of a restricted range of industrial/hazardous (special) wastes with municipal wastes in specifically licensed sites is a practice acceptable in the UK, but is subject to increasing opposition by the European Commission (Waste Landfill Directive 1997, see Section 5.13). The proposed EC Waste Landfill Directive states that there should be no joint disposal of hazardous and non-hazardous waste in landfills which would prevent the co-disposal of hazardous/industrial with municipal waste. There are currently about 360 co-disposal landfill sites in the UK which are licensed to accept industrial waste, mostly in solid form (Waste Management Paper

Table 5.9 *Typical liner materials used in landfill liner systems*

Liner type	Primary barrier	Secondary barrier	Tertiary barrier
Single liner system (Figure 5.12a)	Clay Bentonite-enhanced soil or hydraulic asphalt	Not applicable	Not applicable
Composite liner system (Figure 5.12b)	Flexible membrane liner	Clay Bentonite-enhanced soil or geosynthetic clay liner	Not applicable
Double liner system (Figure 5.12c)	Clay Bentonite-enhanced soil Geosynthetic clay liner or flexible membrane liner	Clay Bentonite-enhanced soil Geosynthetic clay liner or flexible membrane liner	Not applicable
Multiple liner system (Figure 5.12d)	Flexible membrane liner	Clay Bentonite-enhanced soil or geosynthetic clay liner	Clay Bentonite-enhanced soil, geosynthetic clay liner or flexible membrane liner

Source: Waste Management Paper 26B, Landfill Design, Construction and Operation Practice. Department of the Environment, HMSO, London, 1995. Crown copyright is reproduced with the permission of the Controller of Her Majesty's Stationery Office.

26F, 1994). With increasing sophistication in landfill liner system design, many sites designed to accept municipal solid waste will also be able to accept industrial waste/hazardous waste because of the high specifications of the liner system. The action of co-disposal is to utilise the biological, physical and chemical degradation processes acting on and within the municipal solid waste to act also on the industrial waste, and thereby minimise the potentially hazardous and polluting components to environmentally acceptable levels. In about 50 UK co-disposal sites liquid wastes may also be accepted, although again there is European Union opposition to the practice (Waste Management Paper 26F, 1994). Not all industrial/hazardous wastes are suitable for co-disposal; Table 5.10 shows examples of waste types that may be suitable for co-disposal and Table 5.11 shows those which are not suitable (Waste Management Paper 26F, 1994). In particular, for certain industrial/hazardous wastes, co-disposal is recommended in the UK as the best practicable environmental option for wastes containing components of acids, heavy metals, phenols, degradable organic compounds, small amounts of grease, oil and hydrocarbons, and dilute concentrations of cyanide. About 2.7 million tonnes of industrial/hazardous solid wastes are generated annually in the UK, and the large majority are co-disposed with biodegradable municipal solid waste at input rates of co-disposal of between 5 and 20% (Waste Management Paper 26F, 1994). In addition, about 0.5 million tonnes per year of liquid industrial/hazardous waste is co-disposed at input rates of between 1 and 5%.

Table 5.10 *Examples of waste types that may be suitable for co-disposal*

Bottom ash from waste incineration
Flyash from waste incineration
Air pollution control residues from waste incineration
Products from waste solidification processes
Contaminated soils
Industrial effluent treatment sludges and filter cakes
Biological treatment sludges and filter cakes, including sewage sludge
Animal and food industry wastes
Tannery and fell-mongering wastes
Brewery wastes
Adhesive wastes
Detergents, fats and greases in water
Paint spray booth effluents
Interceptor wastes and tank sludges
Alkaline degreasants
Cutting oils/cooling oils
Metal finishing wastes
Aqueous effluents containing degradable organics from chemical manufacture
Acids, alkalis

Source: Waste Management Paper 26F, Landfill Co-Disposal (draft). Department of the Environment, HMSO, London, 1994. Crown copyright is reproduced with the permission of the Controller of Her Majesty's Stationery Office.

Table 5.11 *Examples of waste types that are not suitable for co-disposal*

Organic liquids	
Aqueous wastes with:	>100 mg/l strongly complexed heavy metals >100 mg/l total cyanide as CN significant organic content ($>5\%$ total)
Solid waste which:	is hydrophobic is in massive non-leachable form fails long-term leaching criteria contains >100 mg/kg total cyanide as CN contains asbestos
Any waste containing:	PCBs, PCTs, PBBs, PCNs, PAH organometallic compounds non-degradable List I organic compounds
Any waste which is:	explosive oxidising reactive flammable has a flash point lower than 55 °C

Source: Waste Management Paper 26F, Landfill Co-Disposal (draft). Department of the Environment, HMSO, London, 1994. Crown copyright is reproduced with the permission of the Controller of Her Majesty's Stationery Office.

Table 5.12 Comparison of leachate composition from municipal solid waste landfills and co-disposal landfills (mg/l)

Parameter	Municipal landfill		Co-disposal landfill	
	Mean	SD	Mean	SD
pH	7.22	0.47	7.38	0.31
Conductivity (μS/cm)	6686	4125	11 518	3953
Chloride	1039	745	2131	745
Sulphate	144	207	113	100
Ammoniacal nitrogen	438	393	665	320
COD	3275	7304	2366	3002
TOC	733	1345	661	634
BOD	689	1366	674	1290
Sodium	725	548	1513	826
Potassium	444	327	603	386
Calcium	293	360	156	57
Magnesium	139	97	176	100
Iron	71	147	16	16
Manganese	2.6	5.5	0.79	0.88
Zinc	0.39	0.54	0.82	1.67
Copper	0.03	0.012	0.06	0.039
Nickel	0.09	0.081	0.15	0.068
Chromium	0.07	0.041	0.12	0.138
Cadmium	< 0.01	–	< 0.02	–
Lead	0.10	0.053	0.11	0.060
Tin	6.7	12.5	2.4	2.98
Boron	2.7	2.1	17.2	35.1
Vanadium	0.54	0.60	1.18	0.91
Arsenic (μg/l)	6.5	5.7	13.8	13.5
Mercury (μg/l)	< 0.1	–	< 0.1	–
Total cyanide (as CN)	< 0.05	–	0.09	0.11
Monohydric phenols	0.43	0.99	0.45	0.63
Organotin (μg/l)	0.34	0.34	0.24	0.34
AOX (μg/l)	232	231	340	251

Source: Waste Management Paper 26F, Landfill Co-Disposal (draft). Department of the Environment, HMSO, London, 1994. Crown copyright is reproduced with the permission of the Controller of Her Majesty's Stationery Office.

Whilst there is European opposition to co-disposal, experience and research by the UK Department of the Environment suggests that the attenuation processes operating within the waste produces leachate which is similar in composition to that from a municipal waste site (Table 5.12) (Waste Management Paper 26F, 1994). Because of the different degradation processes occurring in the biodegradable waste, depending on, amongst other parameters, the age and hence the stages of degradation reached by the mass of waste, the co-disposed industrial/hazardous waste will be attenuated to different degrees. It has been recommended that industrial/hazardous wastes are better attenuated in mature wastes of between 1 and 5 years of age than in newly emplaced wastes. The recent Landfill Co-disposal Waste Management Paper (26F, 1994) suggests that for suitable co-disposal the sustainable type of flushing bioreactor is the most effective in treating co-disposed industrial/hazardous wastes.

Table 5.13 *Recommended maximum loading rates for industrial/hazardous waste emplacement in co-disposal landfill sites*

Waste	Loading rate
Acid wastes	100 eq/tonne (0.1 m^3 acid[1]/tonne of MSW)
Heavy metal wastes	100 g soluble chromium, copper, lead, arsenic, nickel, or zinc/tonne of MSW
	10 g soluble cadmium/tonne of MSW
	2 g soluble mercury/tonne of MSW
Phenolic wastes	2 kg total phenols/tonne of MSW
Cyanide wastes	100 g/tonne of MSW
Total organic carbon	5 kg/tonne of MSW
Oil, grease and hydrocarbon wastes	2.5 kg waste/tonne of MSW

[1] Acid represented by 1 Molar H^+ (e.g., 3.5% hydrochloric acid waste).

Sources: [1] Waste Management Paper 26, Landfilling Wastes. Department of the Environment, HMSO, London, 1986; [2] Waste Management Paper 26F, Landfill Co-Disposal (draft). Department of the Environment, HMSO, London, 1994.

Consequently new recommendations for co-disposal sites are based on accelerated stabilisation, with the design and operation of the site being based on the principles of the landfill as a flushing bioreactor with high rates of leachate recirculation.

Co-disposal of industrial/hazardous wastes requires that the input rate of the waste does not disrupt the biodegradation process, which degrades not only the municipal solid waste but also the industrial/hazardous waste. For example, industrial/hazardous wastes may have high or low pH which may disrupt biological, physical or chemical processes which are dependent on the acidity level within the waste. In addition, the interaction of the wastes should not produce reactions which may generate heat, toxic gases, flammable gases or, in extreme cases, cause fires or explosions. Industrial wastes may contain high concentrations of certain chemicals which may react with the liner material. For example, organic chemical waste reacts with synthetic polymeric plastic liners and decreases the containing nature of the liner (Waste Management Paper 26F, 1994).

Landfill sites which are licensed as co-disposal sites may accept a wide variety of industrial/hazardous wastes, and the amount recommended for acceptance will depend on the ability of the mass of waste to take and attenuate that load of waste without producing unacceptable effects on the environment. This is known as the loading rate. The loading rates for various industrial/hazardous wastes are shown in Table 5.13 (Waste Management Paper 26, 1986; Waste Management Paper 26F, 1994).

The biological, chemical and physical degradation reactions which occur in biodegradable waste are many and very complex. Where the wastes are made more complex with an input of industrial/hazardous waste, the attenuation and degradation processes within the waste become difficult to predict and render the co-disposal site a potentially more environmentally risky waste treatment option. However, it is recognised that the degrading mass of biodegradable waste is a very active medium offering a wide range of processes for the degradation of the input industrial/hazardous waste. The degradation and attenuation processes operating within the

Table 5.14 *Characteristics of landfills which define the acceptability of a site for co-disposal*

Hydraulic characteristics
Mass of municipal solid waste in place (volume of reactor)
Degree of saturation
Degree of containment
Retention times
Mixing characteristics
Operational methods (e.g., for liquid waste disposal)
Leachate recirculation
Bioreactive characteristics
Status of municipal solid waste degradation
Temperature and rates of biological activity
pH and redox conditions
Buffering capacity of the municipal solid waste
Adsorption potential

Source: DOE, A Review of the Technical Aspects of Co-Disposal. Department of the Environment, Wastes Technical Division, Report CWM/007/89, 1989.

landfilled municipal solid waste and which would attenuate or degrade the industrial/hazardous waste include biodegradation, filtration, redox reactions, complexation, ion exchange, adsorption, precipitation, neutralisation etc. The site would normally be lined with a more complex liner system than for municipal waste only, and the leachate management system would be optimised to produce attenuation of the industrial/hazardous waste.

Table 5.14 highlights the characteristics of landfills which define the acceptability of a site for co-disposal (DOE 1989). Site selection, engineering and operation for a landfill site become even more important where co-disposal is to be practised. Major considerations are the geology and hydrogeology, and engineering and operational considerations such as the liner materials and system, leachate and landfill gas management system, on-site waste verification and documentation, waste emplacement techniques, and intermediate and final capping procedures. Liner materials recommended for co-disposal landfills include, for example, natural clay, bentonite-enhanced soils or polymeric flexible membranes or liner systems incorporating composites of these. Low permeability of the resultant liner is essential, and the recommended hydraulic conductivity is not greater than 10^{-9} m/s. An important parameter in co-disposal landfills is the hydraulic retention time, which represents the time the leachate remains in the landfill before removal for treatment or disposal. Therefore it represents the reaction time of the leachate within the landfill. Hydraulic retention times of between 1 and 5 years are recommended, and consequently to achieve the minimum retention, leachate recirculation may be practised (Waste Management Paper 26F, 1994).

Operational procedures include the assessment of all wastes entering the co-disposal site, but particular attention is paid to the input of industrial/hazardous waste by pre-assessment of the waste, on-site documentation and record keeping, and verification of the waste on site. On-site waste verification may include visual inspection, and chemical and physical testing. An inventory of the different types of waste disposed in the site is also required. The industrial/hazardous waste is deposited in specific areas in the waste landfill to ensure adequate mixing with the biodegradable waste. Therefore in most cases the wastes are emplaced at a joint tipping face with the biodegradable wastes, and in the case of solid wastes they are intimately mixed. In the case of liquid wastes, trenches, approximately 1–2 m wide and 3 m deep, are excavated into previously deposited waste and the liquid pumped into the trench. The trench may be open or covered depending on the type of industrial/hazardous waste. The system of waste emplacement in phases and cells is used, but care is taken to ensure that free leachate movement throughout the landfill is not restricted by barriers.

The physical, chemical and biological degradation processes operating in the co-disposal landfill are identical to those operating in a biodegradable municipal solid waste landfill site. Consequently, provided that the loading rates for the industrial/hazardous wastes remain within the recommended limits, the leachate should be similar to that obtained with no industrial/hazardous waste emplacement. Therefore no special leachate management system should be required. However, it is likely that the number of leachate monitoring points and the range of components analysed would be higher. Analyses of co-disposal landfill leachate and municipal solid waste landfill leachate show that the concentrations of various components are indeed very similar, showing that degradation of the wastes produces similar reaction end-products (see Table 5.12). Whilst some heavy metal results are higher from co-disposal sites, they are within the statistical limits for municipal solid waste sites. In addition, there will be high concentrations of contaminants in the leachate within the landfill close to where the industrial/hazardous waste has been emplaced. However, as the various degradation processes act on the co-disposed waste, the concentrations become greatly reduced as the leachate moves through the municipal solid biodegradable waste. With the increasing emphasis on sustainable waste management, the use of leachate recirculation to achieve accelerated stabilisation is receiving attention. Recommended practice is that co-disposal sites be operated as flushing bioreactors. The advantages and disadvantages of leachate recirculation are shown in Table 5.15 (Waste Management Paper 26F, 1994).

In the USA, Pavelka et al (1993) have examined the leachate from a number of hazardous-waste-only landfill sites compared with co-disposal sites landfilling both municipal solid waste or non-hazardous industrial waste and hazardous waste. They showed that significantly higher concentrations of a number of analysed species were found in the leachate from the hazardous-waste-only landfills compared with the landfill sites where co-disposal was practised (Table 5.16, Pavelka et al 1993). Clearly co-disposal has a beneficial influence on the degradation processes of the hazardous waste resulting in minimisation of the environmental impact.

Box 5.4 discusses an example of a co-disposal landfill site (DOE 1989).

5.5.2.4 *Entombment (Dry) Landfills*

The entombment type of landfill is a containment landfill that is designed and operated on the principle that the landfill is contained indefinitely and the waste is

Table 5.15 *Advantages and potential problems and concerns of leachate recirculation*

Advantages

- Encourages early establishment and maintenance of methanogenesis. A high moisture content and the movement of moisture have both been shown to promote methanogenesis
- Develops a more uniform quality of leachate, so that the design and operation of treatment and disposal facilities is easier
- Minimises dry zones in the wastes, which could otherwise remain largely undegraded for many years
- Takes up the absorptive capacity of the biodegradable waste and reduces fluctuations in leachate flow rate
- Promotes enhanced evaporative losses by surface spraying
- Provides temporary storage of short-lived peak flow rates, allowing treatment facilities to be designed for flows closer to average values

Potential problems and concerns

- Surface flooding may be caused either by irrigation rates being locally too high or by the formation of inorganic solid layers
- Spray drift from leachate recirculation may result in health concerns and increased smells, particularly during the acetogenic phase
- Break-outs of leachate accumulated as perched water from the side slopes of landfills may occur, increased by the presence of compacted or low-permeability layers within the waste
- Clogging of sub-surface recirculation systems may occur
- Extremely high concentrations of dissolved salts may occur in sites accepting predominantly inorganic waste

Source: Waste Management Paper 26F, Landfill Co-Disposal (draft). Department of the Environment, HMSO, London, 1996.

Table 5.16 *Comparison of leachate from co-disposal landfill sites with hazardous waste only landfill sites*

Constituent	Co-disposal leachate (mean concentration μg/l)	Hazardous waste-only leachate (mean concentration μg/l)
Methyl ethyl ketone	1 550	19 800
Methyl isobutylketone	1 840	19 700
Acetone	790	17 400
Phthalic acid	10 600	19 300
Phenol	3 620	21 700
Arsenic	37	17 000
Nickel	264	2 160
Zinc	2 120	950

Source: Pavelka C., Loehr R.C. and Haikola B., Hazardous waste landfill leachate characteristics. Waste Management, 13, 573–580, 1993.

kept dry. The entombment type of landfill is common in the USA and has been the preferred design for new landfills since 1991, following a recommendation from the US Environmental Protection Agency on the disposal of solid waste (Lee and Jones-Lee 1993). The typical features of an entombment type landfill are shown in Table 5.17 (Lee and Jones-Lee 1993). The dry waste approach prevents the infiltration of rain water, ground water and surface water, and consequently the waste will not

Box 5.4
The Stewartby, Bedfordshire, Co-Disposal Landfill

The Stewartby co-disposal landfill is operated by Shanks and McEwan (Southern) Ltd., and is located in a former clay quarry, the natural clay barrier giving containment of the waste. The site is approximately 8 million m^3 in void space availability. Domestic waste disposal began in 1976 and co-disposal of aqueous liquid wastes in 1978. An inventory of typical industrial/hazardous waste input is shown in the table below. Approximately 12 000–26 000 m^3/year of liquid waste and 8000 tonnes/year solid industrial/hazardous waste were deposited in the site. The municipal solid waste into which the industrial/hazardous waste is co-disposed is mostly in the methanogenic stage, although early co-disposal was into waste in the acetogenic stage. The municipal waste averages 25 m in thickness. Some of the leachate is removed for treatment and disposal and some is recirculated. Leachate analysis from the Stewartby site is also shown below.

Typical liquid waste input to the Stewartby 'L' field (1978–80)

Waste	Percentage of liquid waste (%)
Tannery and fellmongers wastes	23
Oil/water mixtures	17
Adhesive wastes	17
Miscellaneous chemical waste	11
Miscellaneous wastes	9
Tank sludge/interceptor wastes	7
Effluent treatment plant sludge	5
Paint waste	5
Other wastes	6
Total	100

Leachate analysis of the Stewartby co-disposal landfill 1989

Parameter	Concentration (mg/l)	Parameter	Concentration (mg/l)
pH	7.7	Alkalinity	8418
Conductivity	17 600	Hardness	1174
Chloride	3187	Sodium	2561
Ammoniacal nitrogen	1390	Potassium	922
Oxidised nitrogen	0.7	Zinc	0.65
Sulphate	182	Copper	0.08
TOC	1076	Nickel	0.27
COD	2676	Chromium	0.20
BOD	269	Lead	0.11
Phenols	0.32	Cadmium	<0.02
Total cyanide	0.25	Iron	7.9
		Manganese	0.15

Source: DOE, A Review of Technical Aspects of Co-Disposal. Report CWM/007/89, Wastes Technical Division, Department of the Environment, London, 1989.

Table 5.17 *Typical features of the entombment-type (dry) landfill*

Liner – The liner is typically a composite liner composed of a compacted clay underlayer with an overlying flexible plastic membrane sheeting liner. The liner is designed to prevent escape of leachate from the landfill to the surrounding environment, and also serves as a foundation for the leachate collection and removal system.

Leachate collection and removal system – The leachate collection and removal system is a drainage system placed between the mass of waste and the liner system, and is designed to collect and transport the leachate to where it can be removed by pumping or gravity flow.

Cover – A low-permeability covering is placed over the landfill once it has been filled, and is designed to keep rainwater and run-on water away from the landfill.

Groundwater monitoring wells – A groundwater monitoring program is relied upon to signal the failure of the liner system to control the containment of the leachate.

Other measures – Other systems may be incorporated in a 'dry tomb' landfill design to enhance the ability of the system to repel moisture or to manage leachate.

Source: Lee F.G. and Jones-Lee A., Landfill and groundwater pollution issues: 'Dry tomb' vs. wet-cell landfills. Christensen T.H., Cossu R. and Stegmann (Eds.) Proceedings Sardinia 93, Fourth International Landfill Symposium, Cagliari, Sardinia, 11–15 October 1993, pp. 1777–1796.

decompose and produce leachate or landfill gas. Consequently the entombment waste landfill acts as a long-term storage site. Whilst preventing the infiltration of water is the aim of the design, in practice water will inevitably be present with the waste, and some degradation of the waste will occur, with the consequent production of leachate and landfill gas, therefore monitoring, collection and control of leachate and landfill gas is still required. To this end the liner materials and liner system are designed to ensure that no percolation of water into or out of the site boundary occurs. Any leachate that forms is quickly removed and treated to prevent increased biodegradation and further formation of leachate and landfill gas (Westlake 1995).

Waste storage has the advantages that the environmental hazard is 'contained', and that at some time in the future new technologies may be available to treat the waste. In addition, because the dry tomb approach produces low levels of leachate and landfill gas, the leachate and gas collection, control and treatment requirements are much reduced and consequently cheaper. The principle of the entombment landfill that formation of leachate and landfill gas is minimised by limiting the amount of water penetrating the site emphasises the importance of the liner system. Developments in entombment design have therefore concentrated on increasingly sophisticated liner and capping systems. In most cases thick polymeric plastic liner materials are the basis of the liner system and are used to line the base, sides and cap of the site. In addition, the liner system uses layers of soil, clay, sand, gravel and geotextiles to protect the liner material and aid the containment of the waste and prevent ingress of water. In addition, the siting of the landfill is increasingly towards local climates and topographies which are naturally drier. The conflict of landfilling biodegradable wastes but trying to prevent or limit the biodegradation of the wastes has been recognised, particularly as the integrity of the landfill liner system must be maintained for decades or even centuries into the future (Joseph and Mather 1993). The stabilisation and completion of the landfill may therefore extend far into the future compared with the sustainable landfill, where accelerated biodegradation is encouraged and where lifetimes of between 30 and 50 years are envisaged. The obligations and liabilities for the monitoring, collection and treatment of leachate and landfill gas will therefore also extend far into the future, leading to environmental and economic implications for the operator.

Whilst the increasing sophistication of the landfill liner system gives theoretical containment lifetimes of leachate of the order of thousands of years, in practice the liner will eventually fail at some time in the future. Therefore, the entombment-type landfill liner systems will not be expected to contain the leachate forever. Some ways for the liner system eventually to fail and emit water may be, for example, through burrowing animals breaking through the capping material, the plastic liner may contain flaws or become damaged during emplacement resulting in small holes in the liner, and freeze–thaw of water and long-term corrosion from the chemical and physical properties of the leachate will cause damage to the liner material. Whilst the time period may be many decades or even centuries, the liner material will fail and release the leachate to the environment. In addition, the long-term ability of the leachate control and removal system and the monitoring system to continue operation for very long periods of time are questioned, particularly as such systems are prone to clogging, and maintenance of such systems would be difficult or impossible to carry out due to accessibility problems. Further problems associated with the long-term containment of waste are that the conditions within the landfill may change, resulting in environmental damage. Consequently, the entombment type of landfill is increasingly being criticised as not an environmentally sustainable route to waste management (Lee and Jones-Lee 1993).

It can be seen from Table 5.17 that the features of a US entombment-type landfill are quite similar to a UK design of landfill. In the UK, the liner systems recommended to contain waste leachate and landfill gas are increasing in complexity so that the hydraulic conductivities become lower and lower. Such systems limit the level of moisture in the landfill and therefore increase the time required for stabilisation of the landfill. Rates of waste degradation will therefore be lowered, and leachate and landfill gas will be generated long after the last emplacement of waste. Time scales of well over one hundred years are predicted, and the management of such long-term operations and allocation of responsibilities over such time scales are extremely difficult (Joseph and Mather 1993). Therefore, the UK has attempted to introduce the flushing bioreactor type of landfill, partly to increase the stabilisation time to a more reasonable time scale of between 30 and 50 years.

5.5.2.5 *Sustainable Landfills (the Controlled Flushing Bioreactor Landfill)*

Sustainable waste management, a concept adopted by many countries throughout the world, has led to the development of the sustainable waste landfill. The basis of the strategy is that the present generation should deal with the waste it produces and not leave it to future generations. In this context, a generation is regarded as between 30 and 50 years (Waste Management Paper 26B, 1995). Consequently, a

Table 5.18 *Hydrogeological requirements for sustainable landfill design*

Parameter	Design requirement
Infiltration rate of water/leachate	3000–10 000 mm/year (dependent on landfill depth and flushing techniques)
Leachate control system	To allow the handling of 4000–11 000 l/tonne leachate
Bulk waste hydraulic conductivity	Between 10^{-5} and 10^{-6} m/s
Homogeneity of waste	Pulverisation or shredding of the waste to increase homogeneity and reactive surface area
Depth of waste	Limited to 20–30 m to maintain hydraulic conductivity between 10^{-5} and 10^{-6} m/s
Water/leachate distribution	Even distribution to wet all parts of the waste spaced approximately 10 m apart using injection pipes, trenches or wells
Daily cover	Restricted to high permeability or biodegradable materials, or removal of previous day's cover before emplacement of waste
Waste density	Low density, e.g., < 0.8 tonnes/m^3 (dry density) to maintain hydraulic conductivity between 10^{-5} and 10^{-6} m/s

Source: Beaven R.P., A hydrogeological approach to sustainable landfill design. In: Harwell Waste Management Symposium: Progress, Practice and Landfill Research. Harwell, May 1996.

sustainable type of landfill is designed and operated to produce stabilisation of each phase of the site 30–50 years after completion of landfill operations. This aim may be readily achievable for inert wastes, but biodegradable wastes may take many more than 50 years to achieve stabilisation. Therefore, new design and operational techniques are required for better control of the biodegradation process. The sustainable landfill is therefore designed and operated as a controlled bioreactor, to accelerate the biodegradation process by continuously recirculating water and/or leachate through the waste. The sustainable type of landfill is therefore sometimes referred to as the 'controlled flushing bioreactor'. The controlled flushing bioreactor is still in the early stages of development as the preferred type of landfill, but it has become the recommended practice for new landfills in the UK (Waste Management Paper 26B, 1995; Westlake 1995), and has also received attention in, for example, the USA (Lee and Jones-Lee 1993; Maier et al 1995; Reinhart 1996; Townsend et al 1996), Denmark (Hjelmar et al 1995) and Australia (Van den Brook et al 1995; Yuen et al 1995)

Continuous flushing of the site with leachate increases the rate of degradation by flushing out degradation products and replenishing nutrients for the micro-organisms, and thereby achieve faster stabilisation of the waste. Increased biodegradation results in an increase in the rate of gas generation, but inevitably also an increase in the generation of leachate. A leachate collection and recirculation system is therefore required which evenly distributes leachate throughout the mass of waste. Table 5.18 shows the hydrogeological requirements for the sustainable type of landfill (Beaven 1996). Depending on the size of landfill, infiltration rates of 3000–10 000 mm/year of leachate are required. This compares with infiltration rates from non-flushing landfills of between 50 and 250 mm/year. To obtain such high levels of liquid throughput, the hydraulic conductivity of the waste should be greater than 10^{-7} m/s and ideally in the range 10^{-5}–10^{-6} m/s. Such high hydraulic conductivities are not normally found in waste landfills. Therefore the implication is that waste processing and operational parameters may have to be changed to achieve stabilisation within a generation. Such waste processing techniques may include pulverising or shredding the waste to produce a homogeneous, high-surface-area material which is more readily biodegraded. In addition, compaction of the waste to densities of greater than 0.8 tonnes/m^3 may reduce the bulk hydraulic conductivity to less than the minimum 10^{-7} m/s required. Similarly, pulverisation of waste causes more dense compaction at depth, and the requirement for a minimum hydraulic conductivity may limit the depth of a sustainable landfill site to between 20 and 30 m. Much higher levels of leachate are generated from the flushing bioreactor, and generation rates of leachate for the lifetime of the waste are between 4000 and 11 000 litres per tonne of waste. This compares with the leachate from the containment type of landfill without flushing of about 276 litres per tonne of waste for the lifetime of the site. Consequently, the leachate management system for the flushing bioreactor is substantially greater than for the conventional systems (Beaven 1996).

Accelerated stabilisation using the flushing bioreactor implies accelerated biodegradation of the waste and a consequent increase in the rate of landfill gas generation. Consequently, the gas collection and control system is required to have a higher capacity since the potential volume of gas is evolved in a shorter period of time.

Biodegradation to produce landfill gas is largely due to methanogenic degradation over a temperature range of 30–65 °C, with an optimum temperature range of gas generation between 30 and 45 °C. Most landfill sites in the UK have an average temperature range of between 30 and 35 °C, but the mass of waste may drop significantly in temperature at the surface and at the base of the landfill. Accelerated biodegradation can therefore be enhanced by maintaining temperature control throughout the mass in the range 30–45 °C by the use of insulating capping and base liner systems.

The system for the even distribution of leachate throughout the waste requires a horizontal system of closely spaced injection pipework, trenches or wells. Even and uniform liquid distribution is dependent on the spacing of the injection system and the permeability of the waste. The recirculation system may be surface or sub-surface. Surface systems use rain-guns or perforated pipes to spray the water/leachate onto the surface of the waste. Sub-surface systems use perforated pipes, and have the advantage that they may be placed below the final cap.

The concept of the sustainable landfill via the flushing bioreactor technique is recommended for new landfill sites. However, for existing sites leachate recirculation systems would require to be retro-fitted. For deep sites the leachate extraction system would require excavation of a well and installation of a leachate injection system. Since the existing site would not have been designed as a flushing bioreactor, homogeneity of the waste and low compaction of the waste would not have been factors in their design and operation. Consequently, the hydraulic conductivities recommended as being the minimum for accelerated biodegradation via flushing may not be achievable without extensive excavation of the site or unless the site has a shallow depth of waste (Beaven 1996).

A number of case studies have been reported involving leachate recirculation at landfill sites to achieve increased waste stabilisation, for example, in the UK at the Brogborough landfill site in Bedfordshire and the Lower Spen Valley landfill, West Yorkshire (Blakey et al 1995), in the USA at the Alachua County Southwest landfill in Florida (Townsend et al 1996), at eight other sites throughout the USA, including the landfills at Central Facility, Maryland, Winfield, Florida, Pecan Row, Georgia, Lemons, Missouri, Mill Seat, New York State etc. (Reinhart 1996; Reinhart and Al-Yousfi 1996), and in Melbourne, Australia (Yuen et al 1995). In summary, most case studies report that accelerated stabilisation of waste and increased gas production occurs at large scale landfill sites where leachate recirculation is employed. Further points arising from the case studies suggest that, whilst operating parameters were variable from case to case, leachate quality may not be significantly affected, and that permeability of the waste is of major importance to allow efficient recirculation of leachate. On-site storage of leachate is also recommended to reduce off-site charges for water and sewerage.

Whilst leachate recirculation techniques and the controlled flushing bioreactor have been adopted as a technique for increased rates of waste stabilisation in landfills in many countries, the process may be threatened in the European Community. The proposed Waste Landfill Directive (Commission of the European Communities, COM(97) 105 Final, 1997) has proposed that the amount of biodegradable waste going to landfill be reduced significantly by the year 2010 (see Section 5.13). This would have consequences for the controlled flushing bioreactor, whose design is to stabilise biodegradable waste using leachate recirculation. Increased costs of the process through leachate management systems and water and sewerage charges have been estimated to involve high capital expenditure and would markedly increase the costs of landfill (ENDS 1997a). Such investment in the technology would be uncertain if the possibility of the amount of bioreactive wastes being permitted in landfill sites are to be reduced in the future.

5.6 Landfill Gas

Gases arising from biodegradation landfills consist of hydrogen and carbon dioxide in the early stages, followed mainly by methane and carbon dioxide in the later stages. The formation of landfill gas has been discussed previously in Section 5.4. What is known as 'landfill gas' is a product mainly of the methanogenic stage of degradation of biodegradable wastes. Landfill gas is produced from household and commercial wastes which contain a significant proportion of biodegradable materials. However, certain industrial and commercial wastes, which have been estimated to contain 62% and 66% biodegradable components, respectively (Waste Management Paper 26B, 1995), may be acceptable at containment landfills or at co-disposal landfill sites and may also generate landfill gas. The main gases are methane and carbon dioxide, but a wide range of other gases can potentially be formed. In addition, the gas is usually saturated with moisture. Table 5.19 shows the composition of the major constituents

Table 5.19 Typical landfill gas composition

Component	Typical value (% by volume)	Observed maximum (% by volume)
Methane	63.8	88.0
Carbon dioxide	33.6	89.3
Oxygen	0.16	20.9
Nitrogen	2.4	87.0
Hydrogen	0.05	21.1
Carbon monoxide	0.001	0.09
Ethane	0.005	0.0139
Ethene	0.018	–
Acetaldehyde	0.005	–
Propane	0.002	0.0171
Butanes	0.003	0.023
Helium	0.00005	–
Higher alkanes	<0.05	0.07
Unsaturated hydrocarbons	0.009	0.048
Halogenated compounds	0.00002	0.032
Hydrogen sulphide	0.00002	35.0
Organosulphur compounds	0.00001	0.028
Alcohols	0.00001	0.127
Others	0.00005	0.023

Source: Waste Management Paper 27, Landfill Gas. Department of the Environment, HMSO, London, 1994. Crown copyright is reproduced with the permission of the Controller of Her Majesty's Stationery Office.

of landfill gas and Table 5.20 shows the range of trace components (Waste Management Paper 26, 1986; Waste Management Paper 27, 1994). The major constituents of landfill gas, methane and carbon dioxide are odourless, and it is the minor components such as hydrogen sulphide, organic esters and the organosulphur compounds which give landfill gas a malodorous smell. Landfill gas contains components which are flammable and when mixed with air can reach explosive concentrations in confined spaces. There have been problems associated with uncontrolled leakages of landfill gas into houses, shafts, culverts, pipework etc., with potentially devastating effects. Box 5.5 describes the notorious Loscoe landfill gas explosion (Williams and Aitkenhead 1991). The lower flammable limit, where ignition of the gas mixture can occur, is 4% for hydrogen and 5% for methane. In addition, the gas can cause asphyxiation where levels accumulate in such areas as manholes and culverts (Waste Management Paper 26, 1986; Waste Management Paper 27, 1994). This is particularly a problem where certain mixtures of landfill gas components result in the gas having a higher or lower density than air and thus stratification of the air and gas. An asphyxiation hazard can occur in a confined space where the oxygen level has fallen from 21% to 18%. Some of the trace components of landfill gas have a toxic effect and may be hazardous if high enough concentrations are reached, for example, hydrogen sulphide (Waste Management Paper 27, 1994; Rettenberger and Stegmann 1996). Aromatic hydrocarbons are in low concentration but may potentially have an adverse effect on the workforce of the landfill site.

Table 5.20 *Trace components found in landfill gas*

Component	Concentration range (mg/m^3)	Component	Concentration range (mg/m^3)
Alkanes		*Alkenes*	
Propane	<0.1–1.0	Butadiene	<0.1–20
Butanes	<0.1–90	Butenes	<0.1–90
Pentanes	1.8–105	Pentadienes	<0.1–0.4
Hexanes	1.3–628	Pentenes	<0.5–2
Heptanes	4–1054	Hexenes	<0.5–136
Octanes	8.5–675	Heptadienes	<0.1–1.9
Nonanes	31–226	Heptenes	0.3–103
Decanes	81–335	Octenes	<1–144
Undecanes	12–164	Nonadienes	<0.1–9
		Nonenes	5.2–7.5
		Decenes	13–188
		Undecenes	<2–54
Cycloalkanes		*Cycloalkenes*	
Cyclopentane	<0.2–6.7	Limonene	2.1–240
Cyclohexane	<0.5–103	Other terpenes	14.3–311
Methylcyclopentane	<0.1–79	Methene	<0.1–29
Dimethylcyclopentanes	0.1–330		
Ethylcyclopentane	<0.1–<2	*Aromatic hydrocarbons*	
Methylcyclohexane	1.5–290	Benzene	0.4–114
Trimethylcyclopentanes	<0.1–58	Toluene	8–>460
Dimethylcyclohexanes	<2–54	Styrene	<0.1–7
Trimethylcyclohexanes	<0.1–27	Xylenes	34–470
Propylcyclohexanes	<0.5–8	Ethylbenzene	17–330
Butylcyclohexanes	<0.1–4	Methylstyrene	<0.1–15
		Propylbenzenes	36–292
		Butylbenzenes	5.8–138
		Pentylbenzenes	0.4–17.5
Halogenated compounds		*Organosulphur compounds*	
Chloromethane	<0.1–1	Carbonyl sulphide	<0.1–1
Chlorofluoromethane	<0.1–10	Carbon disulphide	<0.1–2
Dichloromethane	<0.1–190	Methanethiol	<0.1–87
Chlorodifluoromethane	<0.1–16	Ethanethiol	<0.1–<2
Dichlorofluoromethane	<0.1–93	Dimethyl sulphide	<0.2–60
Chloroform	<0.1–0.8	Dimethyl disulphide	0.1–40
Dichlorodifluoromethane	<0.1–48	Diethyl disulphide	<0.1–0.6
Trichlorofluoromethane	<0.1–20	Butanethiols	<0.1–2.4
Chloroethane	<0.1–46	Pentanethiols	<0.1–1.2
1,1-Dichloroethane	<0.1–130		
1,2-Dichloroethane	<0.1–8	*Alcohols*	
Vinylchloride	<0.1–32	Methanol	<0.1–210
1,1,1-Trichloroethane	<0.1–177	Ethanol	<0.1–>810
1,2-Dichloroethylenes	<0.1–302	Propan-1-ol	<0.1–110
Trichloroethylene	<0.1–170	Propan-2-ol	<0.1–>46
Tetrachloroethylene	<0.1–350	Butan-1-ol	<0.1–>19

continues

Table 5.20 *(continued)*

Component	Concentration range (mg/m^3)	Component	Concentration range (mg/m^3)
1,1-Dichlorotetrafluoroethane	<0.1–1	Iso-butan-1-ol	<0.1–>5.3
1,2-Dichlorotetrafluoroethane	<0.1–10	Butan-2-ol	<0.1–210
1,1,1-Trichlorotrifluoroethane	<0.1–70		
Bromoethane	<0.1–<2	*Ethers*	
Chloropropanes	<0.1–<2	Dimethylether	0.02–<2
Dichlorobutanes	<0.1–<2	Methylethylether	<0.1–<2
Chlorobenzene	<0.1–2.1	Diethylether	<0.1–12
Dichlorobenzenes	<2–16	Dipropylethers	<0.1–220
Esters		*Other oxygenated compounds*	
Ethyl acetate	<0.1–64	Acetone	<0.1–3.4
Methyl butanoate	<0.1–15	1,3-Dioxolane	<0.1–5
Ethyl propionate	<0.1–136	Butan-2-one	0.4–38
Propyl acetate	<0.1–50	Tetrahydrofuran	<0.1–<2
Isopropyl acetate	<0.1–6	Pentan-2-one	<0.1–4.2
Methyl pentanoate	<0.1–22	Methyl furans	<0.1–0.8
Ethyl butanoate	<0.1–350	Dimethyl furans	<0.1–12
Propyl propionate	<0.1–200	Camphor/fenchone	<0.1–13
Butyl acetate	<0.1–60	Carboxylic acids	<0.1–<2
Ethyl pentanoate	<0.1–27		
Propyl butanoate	<0.1–100		

Source: Waste Management Paper 26, Landfilling Wastes. Department of the Environment, HMSO, London, 1986. Crown copyright is reproduced with the permission of the Controller of Her Majesty's Stationery Office.

Chlorinated hydrocarbons are important because of their potential harm to the environment, and when landfill gas is used as fuel in landfill gas utilisation schemes there is the potential to form hydrogen chloride (Rettenberger and Stegmann 1996). Young and Blakey (1996) also report a wide range of trace components in landfill gas, including trace organosulphur compounds.

The major components of landfill gas, methane and carbon dioxide, are 'greenhouse gases'. The greenhouse effect is produced by certain gases in the atmosphere which allow transmission of short wave radiation from the sun but are opaque to long wave radiation reflected from the earth's surface, thereby causing warming of the earth's atmosphere. A molecule of methane has approximately 30 times the greenhouse effect of a molecule of carbon dioxide (Porteous 1992).

The quantities of gas produced from waste depend on the biodegradable fraction of the waste, the presence of micro-organisms, and suitable aerobic and anaerobic conditions and moisture. The results of theoretical work on the production of carbon dioxide and methane from the major fractions of biodegradable wastes, i.e., degradation of representative carbohydrates (cellulose), proteins and fats in the methanogenic stage of biodegradation of wastes in landfills, are shown in Table 5.21 (McBean et al 1995). This table shows that the composition of the waste influences the composition and production of landfill gas. Estimates of the theoretical production of

Box 5.5
The Loscoe Landfill Gas Explosion, 1986

On the 24 March 1986, at 6.30 a.m., the bungalow at 51 Clarke Avenue, Loscoe, in Derbyshire, was completely destroyed by an explosion of methane landfill gas and the three occupants of the bungalow were injured. The bungalow was situated only 70 m from the Loscoe landfill site. In fact, the site was surrounded by housing. The Loscoe landfill was an old quarry which had been worked for clay, stone and coal since before 1879. Infilling of the quarry with waste materials commenced in 1973 and by 1979, 100 tonnes/day of domestic waste was being deposited in the quarry. Disposal of waste ceased in 1982 and the site was covered with a light covering of low permeable material in 1984, followed by a more extensive covering in 1986.

The identification of landfill gas as the cause of the explosion was from the gas composition evidence of 60% methane and 40% carbon dioxide, which is characteristic of landfill gas. In addition, prior to the explosion there had been evidence of localised damage to vegetation which was later ascribed to landfill gas. Examination of the geological characteristics of the rocks underlying the Loscoe site showed that they consisted of permeable sandstones and coal seams, allowing gas migration. In addition, blasting during quarrying operations and excavated wells may also have formed migration pathways for landfill gas. The figure shows a geological cross-section through the Loscoe site. Landfill gas from the landfill site migrated through the permeable sandstone beds, resulting in a build-up of gas to form an explosive mixture with air.

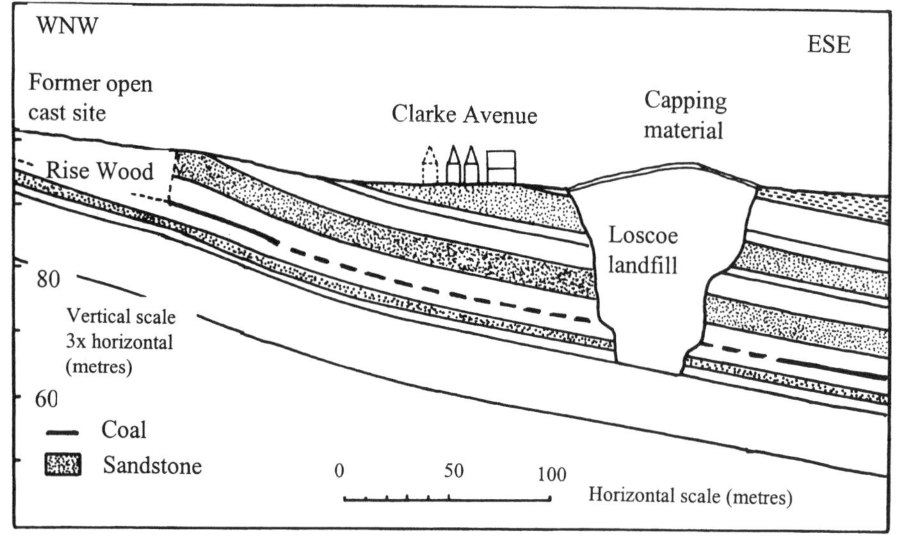

Source: Williams G.M. and Aitkenhead N., Lessons from Loscoe: The uncontrolled migration of landfill gas. Quarterly Journal of Engineering Geology, 24, 191–207, 1991. Figure reproduced by permission of the Director, British Geological Survey.

landfill gas from municipal solid waste with a typical biodegradable composition indicate that between 300 and 500 m^3/tonne would be generated throughout the lifetime of the site. Actual measurements of landfill gas generation rates, however, are highly variable (between 39 and 390 m^3/tonne) due to the range of waste compositions, the fact that all the waste may not decompose, and that gas may not always be the end-product of degradation (McBean et al 1995).

Table 5.21 *Theoretical production of carbon dioxide and methane from the major representative components of waste*

Component	Total Carbon dioxide + methane (m^3/tonne)	Gas composition	
		Carbon dioxide (%)	Methane (%)
Cellulose (carbohydrate)	829	50.0	50.0
Protein	988	48.5	51.5
Fat	1430	28.6	71.4
Typical waste	300–500 (estimated)		
Typical waste	39–390 (measured)		

Source: McBean, E.A., Rovers F.A. and Farquar G.J., Solid Waste Landfill Engineering and Design. Prentice Hall, Englewood Cliffs, NJ, 1995.

5.6.1 Landfill Gas Migration

Gases generated in the landfill will move throughout the mass of waste in addition to movement or migration out of the site. The mechanism of gas movement is via gaseous diffusion and advection or pressure gradient. That is, the gas moves from high to low gas concentration regions or from high to low gas pressure regions. Movement of gas within the mass of waste is governed by the permeability of the waste, overlying daily or intermittent cover, and the degree of compaction of the waste. Lateral movement of the gases is caused by overlying low permeability layers such as the daily cover and surface and sub-surface accumulations of water. Vertical movement of gas may occur through natural settlement of the waste, between bales of waste if a baling system is used to compact and bale the waste, or through layers of low permeability inert wastes such as construction waste rubble. Where landfill gas extraction is practised to recover the gas for energy use, the gas is collected in gas wells, and piped to the surface (Waste Management Paper 27, 1994).

Gas migration out of the mass of waste into the surrounding environment may occur from older sites where containment was not practised, from attenuate and disperse sites, or through containment sites where significant leakage has occurred. Migration of gas outside the site requires migration pathways such as high-permeability geological strata, through caves, cavities, cracks in the overlying capping layer, through man-made shafts such as mine shafts and service ducts etc. In the attenuate and disperse type of site, landfill gas may move into the unsaturated zone beneath the mass of waste and move laterally. In addition, leachate movement out of such sites may cause later degradation to landfill gas. Figure 5.13 shows the possible gas migration pathways for a restored site (Waste Management Paper 27, 1994). Gas may migrate considerable distances from the boundaries of the site through these possible pathways.

It has been reported that changes in the major and trace components of landfill gas occur during sub-surface migration (Ward et al 1996). They showed that reduction in methane concentration occurs due to oxidation, and some alteration of trace landfill gases due to adsorption onto soil particles, oxidation, degradation, condensation and dissolution.

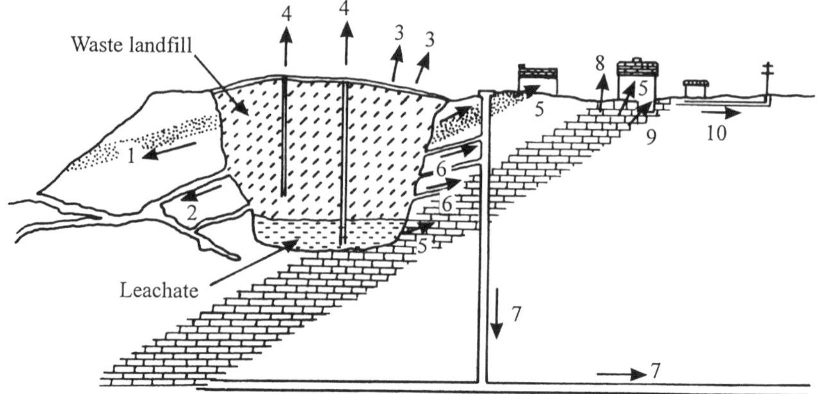

Gas pathways to atmosphere

1. Through high permeability strata down the bedding plane
2. Through caves/cavities
3. Through dessication cracks, tree roots, etc.
4. Through site features such as gas and leachate walls
5. Through high permeability strata up the bedding plane
6. Through fissures caused by explosives, etc.
7. Along man-made shafts, etc.
8. Through highly fissured strata into the atmosphere or buildings
9. Into underground rooms
10. Along underground services

Figure 5.13 *Possible landfill gas migration pathways for a closed site. Source: Waste Management Paper 27, Landfill Gas. Department of the Environment, HMSO, London, 1994. Crown copyright is reproduced with the permission of the Controller of Her Majesty's Stationery Office*

5.6.2 Management and Monitoring of Landfill Gas

With the recognition of the formation of landfill gas and its associated hazards, and the potential to utilise the energy content of the gas, the modern landfill site is designed to trap the gases for flaring or use in energy recovery systems. However, the priority is for control of the gases to protect the environment and prevent unacceptable risk to human health rather than utilisation, and therefore where energy recovery is practised, there would also be a control system alongside. In addition, control mechanisms are required to minimise the risk of migration of the gases out of the site. Control measures to prevent or minimise the leakage of landfill gas from the site are deemed to be effective if measurements around the site using sub-surface probes or boreholes produce levels which never exceed 1% landfill gas by volume of methane or hydrogen, or 1.5% by volume carbon dioxide (Waste Management Paper 27, 1994; Waste Management Paper 26B, 1995).

There are three types of system used to control landfill gas migration (Pescod 1991–93; Tchobanoglous and O'Leary 1994; Waste Management Paper 27, 1994;

McBean 1995; Waste Management Paper 26B 1995): passive venting; physical barriers; pumping extraction systems.

Passive venting Passive venting systems are only recommended for old sites in the late stages of gas generation where gas generation rates are low, or where inert wastes are landfilled and similarly low, or where negligible rates of gas generation are found. The passive venting pit consists of a highly permeable vent of gravel material encased in a geotextile fabric to prevent ingress of fine material and reduction of permeability. The gases flow up the highly permeable layer and vent passively into the atmosphere through a permeable capping layer of sand and granular soil or crushed stone. The vent may also be constructed of granular material, but with a central perforated plastic pipe, the pipe venting directly to the atmosphere. Construction of the passive venting system may be as emplacement of the waste proceeds or afterwards by drilling or excavation into the mass of waste. Typically the vents are placed at intervals of between 20 and 50 m. Other designs of passive gas venting systems include trenches which are excavated into or at the boundary of the waste. The trench is lined at the outer edge with a low-permeability barrier, and the trench filled with a high-permeability gravel, or perforated pipes are used to vent the migrating gas to the surface (Waste Management Paper 27, 1994; Waste Management Paper 26B, 1995).

Physical barriers Physical barriers use low-permeability barriers of, for example, flexible polymeric geomembranes, bentonite cement or clay, to contain and restrict the gas migration. Whilst these barriers might form part of a leachate containment system, they are less effective in containing gas. Coefficients of permeability for gas containment are required to be lower than 10^{-9} m/s. Efficiencies of barriers are improved if they are combined with a means of removing the gas by either passive venting or pumped extraction.

Pumping extraction systems Pumping extraction systems pump the gas out of the landfill. The gas migrates to gas pits or wells within the waste which consist of highly permeable gravel, stones or rubble with a central perforated plastic pipe. The gases pass through the high-permeability vent to a plain unperforated pipe which draws the gases through to the pump. Leachate vapour may also be pumped out with the gas, and because this vapour has a high moisture content a leachate condensation trap is required. Figure 5.14 shows a typical pumping extraction well. The gas pumped to the surface is either flared by self-sustaining combustion or the use of a support fuel, utilised in an energy recovery system, or if the gas concentrations are sufficiently low it is discharged to the atmosphere. Where flaring is used to dispose of the gas minimum flame temperatures of between 850 and 1100 °C are recommended to destroy any hazardous trace components.

Monitoring of landfill gas A monitoring programme for landfill gas at waste landfill sites is recommended in Waste Management Paper 27 (1994). In addition, the 1990 Environmental Protection Act places an obligation on the Environment Agency to arrange regular inspection of sites to determine whether landfill gas is

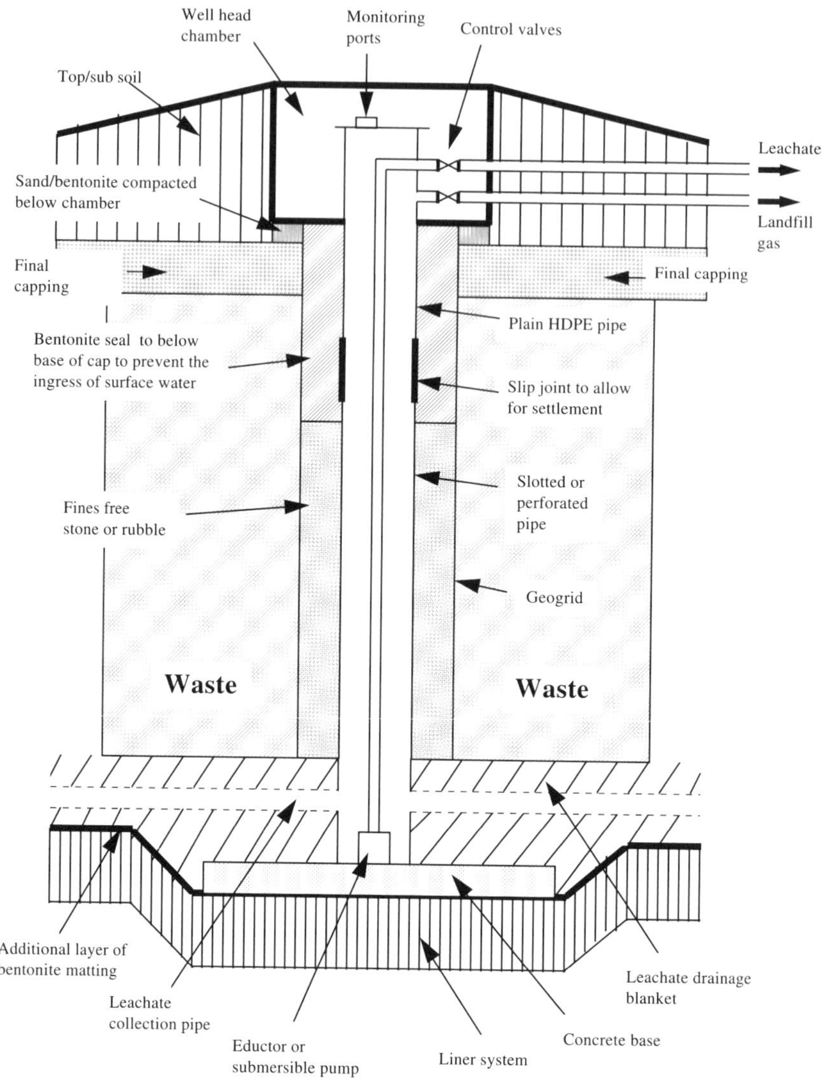

Figure 5.14 *Typical combined leachate and landfill gas collection well. Source: Waste Management Paper 26B, Landfill Design, Construction and Operational Practice. Department of the Environment, HMSO, London, 1995. Crown copyright is reproduced with the permission of the Controller of Her Majesty's Stationery Office*

causing a hazard to human health or the environment. However, responsibility for monitoring rests with the landowner for closed sites and the licence holder for operating sites. The monitoring programmes are used to assess the effectiveness of gas control measures to minimise risk to human health and the environment. Monitoring takes place throughout the operation of the plant and for many years during the post-closure period, until emission levels are below 1.5% by volume of carbon dioxide and less than 1% by volume concentration of flammable gases.

Monitoring takes place within the landfill and outside the site boundary. The monitoring programme, including the frequency of monitoring, will be dependent on the age of the site, the type and mix of waste, the control measures installed, the development of buildings surrounding the site and local geology (Waste Management Paper 27, 1994). Frequency of measurements can be weekly, monthly or even quarterly depending on site-specific characteristics. More frequent monitoring may be required where migration of gas is suspected.

Monitoring techniques include, for example, surface monitoring, sub-surface probes, gas monitoring wells and boreholes. Surface monitoring with portable instruments is mainly used to detect the presence of gas leaks throughout the site. Sub-surface monitoring using gas probes is used to monitor gas production and migration at depths of between 1 and 10 m in the mass of waste and in the surrounding environment. The probes may be left for long periods of time to monitor and map the production of gas from the site throughout site operation and post-closure. The probes are constructed of steel and plastic pipe, and consist of a porous lower section and a gas transfer pipe which transfers the gas to the surface where the gas sample is taken for analysis. Gas monitoring wells and boreholes consist of a porous plastic casing in direct contact with the waste or geological strata. Probes or tubes may be permanently installed. They are installed within the mass of waste and in the surrounding environment.

Gas sample analysis may take the form of portable instruments for gas analysis or laboratory-based analysis. Portable analysers may be simple devices such as gas indicator tubes which produce a colour change to indicate a concentration of a particular gas in a sample. The gas is drawn through the tube on-site and an immediate indication of gas concentration is obtained. However, this method is subject to error. More sophisticated instruments are available such as infrared gas analysers and flame ionisation detectors, which would normally be housed in a portable laboratory or at the analytical laboratory. The gas sample is piped directly to the analyser, or else a sample of the gas is taken in a suitable sealable container, such as 'Teflon' bags or glass sample tubes, and the sample transferred to the instrument for analysis. The most accurate and reliable technique for gas analysis is gas chromatography. A sample of the gas is taken in a suitable container and taken to the laboratory for analysis. Gas chromatography can separate out individual gas components and provide an accurate analysis, even at trace concentrations.

5.7 Landfill Leachate

Leachate represents the water which passes through the waste and water generated within the landfill site, resulting in a liquid containing suspended solids, soluble components of the waste and products from the degradation of the waste by various micro-organisms. The composition of the leachate will depend on the heterogeneity and composition of the waste and whether there is any industrial/hazardous waste co-disposal, the stage of biodegradation reached by the waste, moisture content and operational procedures. The decomposition rate of the waste also depends on aspects such as pH, temperature, aerobic or anaerobic conditions, and the associated types of micro-organisms. Associated with leachate is a malodorous smell, due mainly to the presence of organic acids.

The characteristics of the leachate are influenced by the waste material deposited in the site. For example, inert wastes will produce a leachate with low concentrations of components, whereas a hazardous waste leachate tends to have a wide range of components with highly variable concentrations.

The most common types of landfill site in the UK are municipal solid waste sites and co-disposal sites of municipal solid waste and industrial/hazardous waste.

Leachate from municipal solid waste sites The nature of the leachate changes with time as the waste degrades through the various stages of biodegradation discussed earlier. Table 5.22 compares the typical leachate of the acetogenic Stage III with the methanogenic Stage IV (DOE 1995e). The table shows that the pH of the early formed leachate is acidic/neutral with a pH range between 5.12 and 7.8, equating with the formation of acetic acid and other organic acids by the acetogenic micro-organisms under anaerobic conditions. The organic material of Stage III is very high, in the range 1010–29 000 mg/l for total organic carbon. Ammoniacal nitrogen levels tend to be higher in Stage III due to the biodegradation of the amino acids of proteins and other nitrogenous compounds in the waste. The presence of organic acids of the acetogenic stage increase the solubility of metal ions into the leachate. BOD and COD levels are high, with high BOD:COD ratios, indicating that a high proportion of the organic materials in solution are readily biodegradable. Methanogenic leachate has a neutral/alkaline pH reflecting the degradation of the organic acids of Stage III to methane and carbon dioxide by the methanogenic micro-organisms. As a consequence, the total organic carbon in the leachate decreases compared with the acetogenic stage. Metal ions continue to be leached from the waste, but as the pH of the leachate increases the metal ions become less soluble and decrease in concentration in the leachate. The concentration of ammoniacal nitrogen in the leachate decreases slightly, but remains high. BOD and COD levels decrease compared with acetogenic leachates.

In addition to the components listed in Table 5.22, a wide range of minor components have been detected in leachate from municipal solid waste. Table 5.23 shows the concentrations of trace organic compounds found in leachate from a municipal solid waste landfill site (Rugge et al 1995; White et al 1995).

Leachate from co-disposal (municipal solid waste/industrial waste) landfill sites Co-disposal of industrial wastes with municipal solid wastes produces a leachate which differs from that produced from municipal solid waste depending on the quantity and composition of the industrial waste input. Table 5.12 showed the composition of leachate comparing municipal solid waste leachate with leachate from co-disposal sites with various industrial waste inputs (Waste Management Paper 26F, 1994). Clearly the composition of the leachate will be influenced by the composition of the industrial waste input to the biodegradable landfill site. The potential for large variations in concentration of individual components is more pronounced at co-disposal sites. However, Table 5.12 shows that leachate from co-disposal sites is in

Table 5.22 Composition of acetogenic and methanogenic leachate from large landfill sites with high waste input rate and relatively dry environments (mg/l)

Parameter	Acetogenic			Methanogenic		
	Minimum	Maximum	Mean	Minimum	Maximum	Mean
pH value	5.12	7.8	6.73	6.8	8.2	7.52
COD	2740	152 000	36 817	622	8000	2307
$BOD_{5\ day}$	2000	68 000	18 632	97	1770	374
Ammoniacal-N	194	3610	922	283	2040	889
Chloride	659	4670	1805	570	4710	2074
$BOD_{20\ day}$	2000	125 000	25 108	110	1900	544
Total organic carbon	1010	29 000	12 217	184	2270	733
Fatty acids (as C)	963	22 414	8197	<5	146	18
Alkalinity (as $CaCO_3$)	2720	15 870	7251	3000	9130	5376
Conductivity (μS/cm)	5800	52 000	16 921	5990	19 300	11 502
Nitrate-N	<0.2	18.0	1.80	0.2	2.1	0.86
Nitrite-N	0.01	1.4	0.20	<0.01	1.3	0.17
Sulphate (as SO_4)	<5	1560	676	<5	322	67
Phosphate (as P)	0.6	22.6	5.0	0.3	18.4	4.3
Sodium	474	2400	1371	474	3650	1480
Magnesium	25	820	384	40	1580	250
Potassium	350	3100	1143	100	1580	854
Calcium	270	6240	2241	23	501	151
Chromium	0.03	0.3	0.13	<0.03	0.56	0.09
Manganese	1.40	164.0	32.94	0.04	3.59	0.46
Iron	48.3	2300	653.8	1.6	160	27.4
Nickel	<0.03	1.87	0.42	<0.03	0.6	0.17
Copper	0.020	1.10	0.130	<0.02	0.62	0.17
Zinc	0.09	140.0	17.37	0.03	6.7	1.14
Cadmium	<0.01	0.10	0.02	<0.01	0.08	0.015
Lead	<0.04	0.65	0.28	<0.04	1.9	0.20
Arsenic	<0.001	0.148	0.024	<0.001	0.485	0.034
Mercury	<0.0001	0.0015	0.0004	<0.0001	0.0008	0.0002

N.B. Between 13 and 35 samples of acetogenic leachate and between 16 and 29 samples of methanogenic leachate were analysed to obtain minimum, maximum and mean results. Units mg/l except for pH and conductivity.

Source: DOE, A review of the composition of leachates from domestic wastes in landfill sites. Department of the Environment Research Report No. CWM 072/94. In: Waste Management Paper 26B, Landfill Design, Construction and Operational Practice. Department of the Environment, HMSO, London, 1995.

fact similar to that from sites operating with municipal solid waste alone. The attenuation processes operating within the mass of waste reduce the potentially hazardous pollutants from the industrial waste.

5.7.1 Leachate Management and Treatment

Landfill leachate has been shown to contain a wide variety of toxic and polluting components. A leachate management and treatment system would be required to

Table 5.23 Trace organic components found in municipal solid waste leachate

Component	Concentration (mg/l)
1,1,1-Trichloroethane	0.086
1,2-Dichloroethane	0.01
2,4-Dichloroethane	0.13
Benzo[a]pyrene	0.00025
Benzene	0.037
Chlorobenzene	0.007
Chloroform	0.029
Chlorophenol	0.00051
Dichloromethane	0.44
Endrin	0.00025
Ethylbenzene	0.058
2-Ethyltoluene	0.005
Hexachlorobenzene	0.0018
Isophorone	0.076
Naphthalene	0.006
Polychlorinatedbiphenyls (PCBs)	0.00073
Pentachlorophenol	0.045
Phenol	0.38
1-Propenylbenzene	0.003
Tetrachloromethane	0.2
Toluene	0.41
Toxaphene	0.001
Trichloroethane	0.043
Vinylchloride	0.04
Xylenes	0.107
Dioxins/furans, toxic equivalent (TEQ)	0.32 ng

Sources: [1] Rugge K., Bjerg P.L. and Christensen T.H., Environmental Science and Technology, 29, 1395–1400, 1995; [2] White P.R., Franke M. and Hindle P., Integrated Solid Waste Management: A Lifecycle Inventory. Blackie Academic and Professional, London, 1995.

collect the leachate emanating from the mass of waste and treat it before discharge to a sewer. The generation rate of leachate is estimated based on such factors as the rainfall, the amount of the rainfall infiltrating to the waste through the cover, the absorptive capacity of the waste, the input of co-disposed liquid waste, the weight of absorptive waste and any removal of the leakage via seepage or discharge. Figure 5.15 shows the calculation of the amount of leachate estimated to be generated from a landfill (Waste Management Paper 26B, 1995).

Estimates of leachate generation for a typical municipal solid waste landfill site using the estimation formula in Figure 5.15 are shown in Figure 5.16 (Waste Management Paper 26B, 1995). Figure 5.16 shows that the average production of leachate throughout the 30-year design life of the landfill is 276 litres per tonne of landfilled waste. Because of the uncertainties involved in the leachate generation process from real sites, the estimated leachate generation rate would include

$$\textbf{Lo} = [\textbf{ER} + \textbf{LIW} + \textbf{IRA}] - [\textbf{LTP} + \textbf{aW} + \textbf{DL}]$$

Lo = Free leachate retained at the site, (leachate production minus leachate leaving the site.
ER = Effective rainfall (or actual on an active surface area); this may need to be modified to account for run-off, especially after capping.
LIW = Liquid industrial waste, including any surplus water from sludges with a high moisture content).
IRA = Infiltration through restored and capped area.
LTP = Discharge of leachate off-site.
a = Unit absorptive capacity of wastes
W = Weight of absorptive waste
DL = Designed seepage (if appropriate)

Figure 5.15 *Equation to estimate leachate generation from landfills. Source: Waste Management Paper 26B, Landfill Design, Construction and Operational Practice. Department of the Environment, HMSO, London, 1995*

varied inputs to provide a worst-case scenario for sizing the leachate output and getting discharge consent to allow the leachate into the sewer. For example, the leachate produced from a landfill in terms of volume is subject to large seasonal variations.

The leachate management system consists of leachate drainage, collection and treatment systems. The drainage of leachate is via gravity flow through drainage gradient paths, which consist of a permeable granular system containing perforated pipes, to collection sumps at low points in the waste mass. The leachate collected in the sumps is then removed by either pumping, gravity drains or side slope risers at the site perimeter.

The leachate generated from a landfill site will vary in volume and composition depending on the age of the site and stages of biodegradation reached. Because of the changes in leachate composition with time, the leachate control systems should adapt to these changes. Leachate treatment is required to remove any contaminating components of the leachate and bring it to a standard whereby it can be released to a sewer, a water course, land or tidal water. Before release, a discharge consent or agreement is required from the local water company or the Environment Agency. The consent or agreement may cover a range of potentially polluting components, for example, pH, concentration of organic material, ammonium and nitrate, suspended solids and metal content. Treatment processes for leachate are shown in Table 5.24 (Waste Management Paper 26B, 1995)

Table 5.25 shows the properties of different leachate types formed from different waste types, and Table 5.26 gives a summary of the probable leachate treatment requirements for different categories of leachate (The World Resource Foundation 1995a).

Hjelmar et al (1995) have reviewed the composition and management of leachate from waste landfill sites from within the European Community. They report that the most common form of active leachate management is abstraction and discharge to a

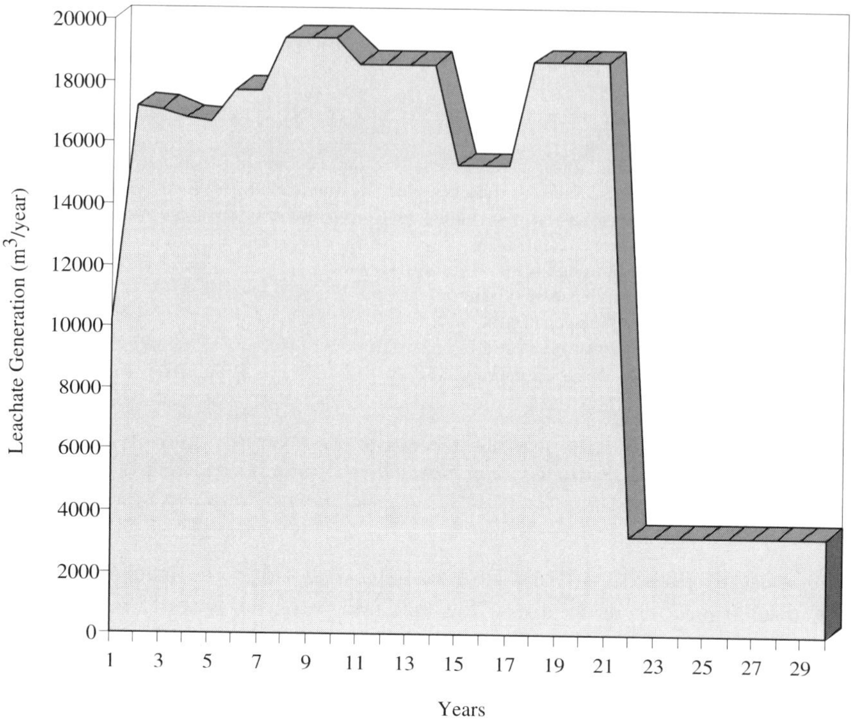

Figure 5.16 Example of the estimation of leachate generation from landfills. Source: Waste Management Paper 26B, Landfill Design, Construction and Operational Practice. Department of the Environment, HMSO, London, 1995

sewer, usually without pre-treatment, and this is particularly the case for municipal solid waste leachate. On-site treatment of leachate is not common at landfill sites throughout the Community but is an increasing practice, mostly at municipal solid waste landfills and at some smaller hazardous waste sites.

The discussion of the management and treatment of leachate is concerned with containment-type landfill sites where there is provision for the collection and treatment of the leachate produced. In older sites, however, where no lining system was employed, the leachate passes directly into the local sub-surface environment. In such cases, the biological, physical and chemical degradation processes discussed in Section 5.5.1 become the treatment option.

5.8 Landfill Capping

Final cover or capping of the landfill site is required after the final waste has been deposited. The purpose of the cap is to contain and protect the waste, prevent rainwater and surface water from percolating into the site and influencing the

Table 5.24 *Treatment processes for leachate*

Process	Examples
Physico-chemical	
Air stripping of ammonia	Leachate pH adjusted to 11 followed by aeration and release of ammonia gas to a scrubbing unit or the atmosphere
Activated carbon adsorption	Highly porous activated carbon adsorbs organic components, used for final stages of treatment
Reverse osmosis	Ultrafiltration membranes concentrate pollutants into a concentrated solution for disposal, used for suspended material, ammoniacal nitrogen, heavy metals
Evaporation	Concentration of contaminants by evaporation or distillation for disposal
Oxidation	Addition of oxidising agents such as hydrogen peroxide or sodium hypochlorite solution. Used for sulphides, sulphite, formaldehyde, cyanide and phenolics
Wet air oxidation	Used for high organic content leachates, based on combustion with air at temperatures up to 310 °C and 200 bar.
Coagulation, flocculation and settling	Addition of reagents followed by mixing and settlement
Attached growth processes	
Trickling filters	Trickling or percolating filters allow the leachate to pass over a substrate containing aerobic micro-organisms which biodegrade the organic components of the leachate
Rotating biological contactors	Rows of rotating discs with attached aerobic micro-organisms alternately exposed to air and leachate as they rotate
Non-attached growth processes	
Aeration in lagoons or tanks	Aerobic micro-organisms in suspension biodegrade the organic constituents of the leachate in aerated lagoons or tanks
Anaerobic treatment	
Anaerobic biodegradation	Utilises the anaerobic methanogenic-type micro-organisms to biodegrade the constituents of the leachate; not effective for ammoniacal nitrogen
Anaerobic/aerobic treatment	
Reed bed biodegradation	Reed bed plant systems stimulate the growth of aerobic micro-organisms at the root system and anaerobic micro-organisms in soil areas away from the roots. The range of micro-organisms biodegrade the leachate components, other contaminants may be immobilised or absorbed by the plants

continues overleaf

Table 5.24 *(continued)*

Process	Examples
Land treatment	
Spraying leachate onto land	Spray irrigation of leachate to grassland and woodland; used for low-contaminant leachate. Treatment processes include micro-organism biodegradation in the soil, plant uptake of contaminants, oxidation, absorption, transpiration, evaporation, precipitation, nitrification and denitrification
Leachate recirculation	
Recirculation of leachate	Recirculation of the leachate through the mass of waste, utilising the landfill as an uncontrolled reactor for further biodegradation

Source: Waste Management Paper 26B, Landfill Design, Construction and Operational Practice. Department of the Environment, HMSO, London, 1995.

generation of leachate, control the release of landfill gas, and prevent ingress of air and disruption of the anaerobic biodegradation process. In addition, the final cover is landscaped and provides a soil for the establishment of the restored site plant materials. The design of the cover system of lining materials used to cap the site depends on the nature of the waste, for example, whether they are inert or biodegradable. Figure 5.17 shows various components which may be used in a capping system (Waste Management Paper 26B, 1995). Overlying the main body of waste may be the gas collection layer, depending on the nature of the waste. The gas collection layer is a porous material such as geotextile, geonet or coarse sand through which the gas can easily permeate to the gas collection and control system. A barrier layer is a low permeability layer such as a plastic polymer geomembrane, a geosynthetic clay liner of bentonite/geotextile fabric, or compacted natural clay. The barrier layer serves a two-fold purpose: to prevent ingress of water and the egress of landfill gas. The barrier layer may have a protective geotextile layer above and below. The drainage layer/pipework zone may be required to minimise the amount of percolating water reaching the barrier layer. The water is drained off through porous pipes set in a porous layer of coarse sand or gravel, geotextile, geonet etc. The protection layer protects the underlying liner system from plant root systems, burrowing animals and man-made intrusions. The protection layer consists of soils and may be an extension of the restoration layer. The restoration layer is the top soil, which may be landscaped, contoured for ease of surface water run-off and used for growing plants depending on the end use of the restored site.

5.9 Landfill Site Completion and Restoration

At the end of the life of a landfill, the landfill operator must demonstrate that the site has physically, chemically and biologically stabilised and no longer poses a risk

Table 5.25 *Categories of leachate*

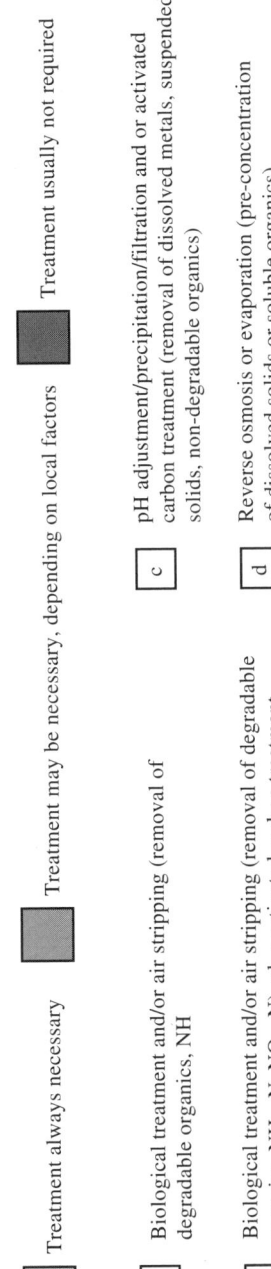

	Disposal to sewer	Disposal to fresh water	Disposal to salt water
Hazardous waste leachate	b + c &/or d	b + c &/or d	b + c &/or d
MSW/co-disposal leachate	a or b or a + c	b + c &/or d	a or b or a + c
Non-hazardous low organic waste leachate	c	a or b + c &/or d	a or a + c
Inorganic waste leachate	c or d	c &/or d	
Inert waste leachate			

Shading key: Treatment always necessary; Treatment may be necessary, depending on local factors; Treatment usually not required

a — Biological treatment and/or air stripping (removal of degradable organics, NH

b — Biological treatment and/or air stripping (removal of degradable organics, NH_3 -N, NO_3 -N), plus activated carbon treatment (removal of non-biodegradable organics)

c — pH adjustment/precipitation/filtration and or activated carbon treatment (removal of dissolved metals, suspended solids, non-degradable organics)

d — Reverse osmosis or evaporation (pre-concentration of dissolved solids or soluble organics)

Source: The World Resource Foundation, Technical Brief. Landfill Techniques. The World Resource Foundation, Tonbridge, 1995. Reproduced by the permission from the World Resource Foundation.

Table 5.26 *Summary of probable treatment requirements for different categories of leachate*

Category	Description
Hazardous waste leachate	Leachate with highly variable concentrations of a wide range of components. Extremely high concentration of substances such as salts, halogenated organics, and trace elements can occur.
MSW/co-disposal leachate	Leachate with high initial concentrations of organic matter (COD >20 000 mg/l and a BOD/COD ratio >0.5) falling to low concentrations (COD in the range of 2000 mg/l and a BOD/COD ratio <0.25) within a period of 2–10 years. High concentrations of nitrogen (>1000 mg/l) of which more than 90% is ammonia-N. This type of leachate is relatively consistent for landfills receiving MSW, mixed non-hazardous industrial and commercial waste and for many uncontrolled dumps.
Non-hazardous low organic waste leachate	Leachate with a relatively low content of organic matter (COD does not exceed 4000 mg/l and it has a typical BOD/COD ratio of 0.2) and a low content of nitrogen (typically total nitrogen is in the range of 200 mg N/l, but can be as high as 500 mg N/l. Relatively low trace element concentrations are observed. This type of leachate comes from landfills receiving only non-hazardous waste exclusive of MSW.
Inorganic waste leachate	Leachate with relatively high initial concentrations of salts (chlorides + sulphates in the range of 15 000 mg/l and a low content of organic matter (typically COD <1000 mg/l) and low content of nitrogen (total N <100 mg/l). Trace element concentrations are often negligible. This type of leachate is typical of landfills for MSW incineration ash.
Inert waste leachate	Leachate with low strength of any component. This type of leachate is representative for inert waste landfills.

Source: The World Resource Foundation, Technical Brief, Landfill Techniques. The World Resource Foundation, Tonbridge, 1995. Reproduced by permission from the World Resource Foundation.

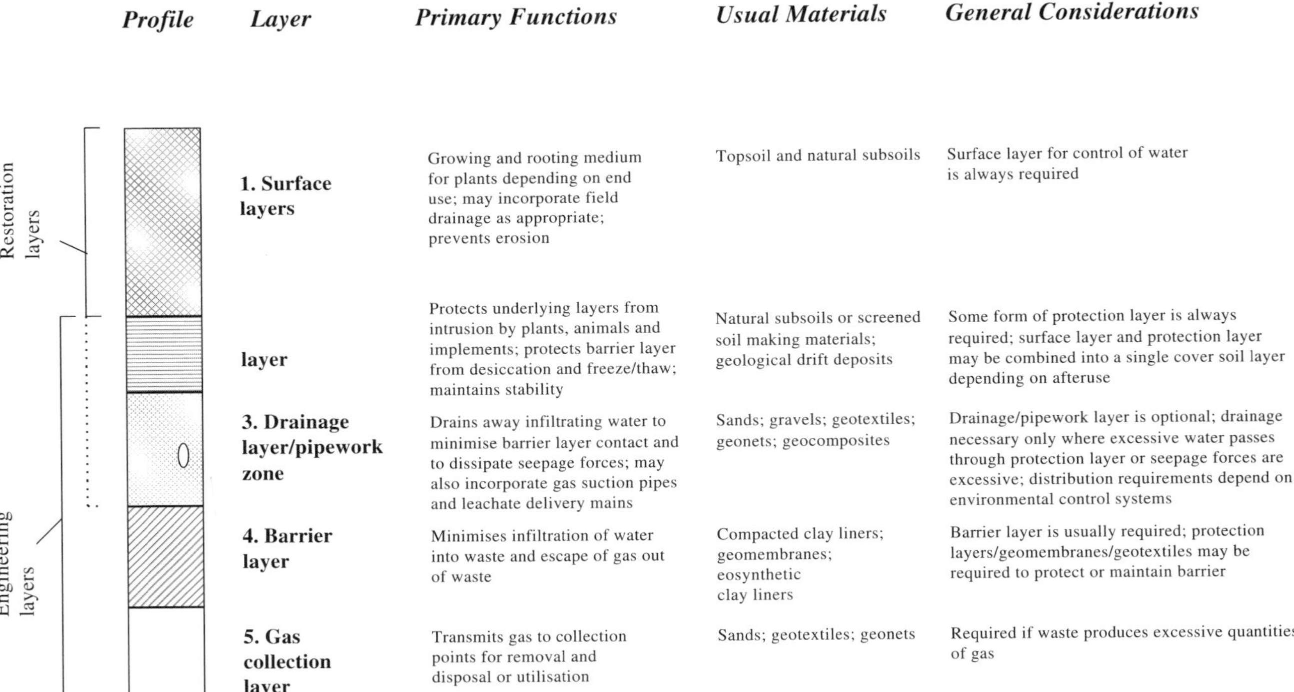

Profile	Layer	Primary Functions	Usual Materials	General Considerations
	1. Surface layers	Growing and rooting medium for plants depending on end use; may incorporate field drainage as appropriate; prevents erosion	Topsoil and natural subsoils	Surface layer for control of water is always required
	layer	Protects underlying layers from intrusion by plants, animals and implements; protects barrier layer from desiccation and freeze/thaw; maintains stability	Natural subsoils or screened soil making materials; geological drift deposits	Some form of protection layer is always required; surface layer and protection layer may be combined into a single cover soil layer depending on afteruse
	3. Drainage layer/pipework zone	Drains away infiltrating water to minimise barrier layer contact and to dissipate seepage forces; may also incorporate gas suction pipes and leachate delivery mains	Sands; gravels; geotextiles; geonets; geocomposites	Drainage/pipework layer is optional; drainage necessary only where excessive water passes through protection layer or seepage forces are excessive; distribution requirements depend on environmental control systems
	4. Barrier layer	Minimises infiltration of water into waste and escape of gas out of waste	Compacted clay liners; geomembranes; eosynthetic clay liners	Barrier layer is usually required; protection layers/geomembranes/geotextiles may be required to protect or maintain barrier
	5. Gas collection layer	Transmits gas to collection points for removal and disposal or utilisation	Sands; geotextiles; geonets	Required if waste produces excessive quantities of gas

Figure 5.17 *Components of a landfill capping system. Source: Waste Management Paper 26B, Landfill Design, Construction and Operational Practice. Department of the Environment, HMSO, London, 1995. Crown copyright is reproduced with the permission of the Controller of Her Majesty's Stationery Office*

to the public or the local environment. When a site is deemed complete, post-closure pollution controls and leachate and landfill gas control systems would no longer be required. Stabilisation is defined in terms of the quantity and composition of the leachate and landfill gas produced at the site. In the UK, a Certificate of Completion is required from the Environment Agency before a site licence can be relinquished (Waste Management Paper 26A, 1994). To obtain the Certificate of Completion, a demonstration of leachate and landfill gas volumes and concentrations requires a monitoring and sampling programme throughout the operation and the post-closure period of the landfill. In addition, the future potential of the site to generate leachate and landfill gas would be a criterion in deciding whether a completion certificate would be granted. A further deciding factor is the settlement of the site and the possibility of physical instability of the waste or retaining structures.

Assessment of completion depends on the type of landfill site. For example, sites which have taken only inert wastes pose a low risk to human health and the environment since only low or zero levels of leachate and landfill gas are likely to be generated. For biodegradable wastes such as municipal solid waste, then a full assessment of the leachate composition and gas volume and the potential future generation rates, together with an assessment of the waste settlement, would be required (Waste Management Paper 26A, 1994).

To assess whether stabilisation has occurred and completion has been reached, a sampling and analysis programme for leachate and landfill gas is implemented. The monitoring programme then identifies whether completion conditions have been reached, which equates to the point where the composition of the leachate and the volume of gas have reached defined criteria of low levels. Assessment of the settlement of the waste is carried out by engineers; this is to assess the settlement of the waste under its own weight due to consolidation as degradation takes place.

Table 5.27 shows an example of the completion criteria which should be reached for leachate where there is a likelihood of the leachate entering groundwater (Waste Management Paper 26A, 1995). The completion criteria for a remote site where there is no chance of the leachate causing an environmental hazard or entering groundwater may be different from those listed in Table 5.27. In terms of landfill gas, completion is deemed to have been reached when the maximum gas concentrations from waste degradation remain less than 1% by volume for methane and for less than 1.5% by volume for carbon dioxide (Waste Management Paper 26A, 1994; Waste Management Paper 27, 1994). This condition should have been demonstrated over a 2-year period on at least four separate occasions in all the monitoring boreholes.

The need for a Certificate of Completion of the landfill operation before the operator can relinquish the site licence represents a major liability for the operator. The stabilisation of the landfill may take decades to complete, representing decades of responsibility and liability for leachate and landfill gas management. Financial liabilities go hand-in-hand with technical liabilities and the provision of monitoring, collection and treatment systems for gas and leachate long after the site has accepted the final load of waste (Griffiths 1996).

Landfill site restoration and post-closure management of landfills is discussed in detail in Waste Management Paper 26E (1996). The range of options available for

Table 5.27 Example of completion criteria for landfill leachate (mg/l)

Parameter	Concentration
pH	6.5–8.5
Conductivity	4000
Chloride	2000
Sulphate	2500
Calcium	1000
Magnesium	500
Sodium	1500
Potassium	120
Aluminium	2
Nitrate	500
Nitrite	1
Ammonia	5
Total organic carbon	10
Iron	2
Manganese	0.5
Copper	1
Zinc	1
Phosphorus	10
Fluoride	10
Barium	1
Silver	0.1
Arsenic	0.5
Cadmium	0.05
Cyanides	0.5
Chromium	0.5
Mercury	0.01
Nickel	0.5
Lead	0.5
Antimony	0.1
Selenium	0.1
Mineral oils	0.1
Phenols	0.005
Organo chlorine compounds (other than pesticides)	0.01
Pesticides	
individually	0.001
collectively	0.005
Polycyclic aromatic hydrocarbons	0.002

Source: Waste Management Paper 26A, Landfill Completion. Department of the Environment, HMSO, London, 1995. Crown copyright is reproduced with the permission of the Controller of Her Majesty's Stationery Office.

the post-closure after-use of a landfill site includes agriculture, woodland, amenity/conservation and built developments, structures and hard standing areas. The choice of option is dictated by several factors. These include whether there is an agreement to restore to a particular after-use by the landowner, or a requirement of the local planning authority. In addition, the options may be restricted by the soil characteristics available in the area, and the type of waste in the landfill.

5.10 Energy Recovery

The development of larger and larger landfill sites throughout many countries has provided for economies of scale and the economic viability of utilisation of landfill gas. The modern site is seen in this context as a 'bio-reactor', used to stabilise waste and produce landfill gas for energy recovery. Therefore, whilst landfill sites exist which are used for disposal without energy recovery, the modern purpose-built site would normally incorporate a landfill gas extraction system for the recovery of energy. It is estimated that by 1995 there were 75 operational energy recovery schemes utilising landfill gas in the UK, with a further 20 either under construction or in the planning stage (DTI 1996). In the USA there were approximately 137 landfill gas utilisation projects by 1994 (Thorneloe et al 1995). Other countries developing landfill gas utilisation are Germany with about 80 schemes, Canada with 9 schemes and the Netherlands with about 10 schemes (Willumsen 1996). Gendebien et al (1992) have extensively reviewed the utilisation of landfill gas in energy recovery schemes and have presented several case studies in detail.

Estimates of the amount of landfill gas generated throughout the lifetime of a landfill site are highly variable, with estimates of between 39 and 500 m^3/tonne (McBean et al 1995). Annual rates of gas production have been estimated for a typical municipal solid waste landfill at between 6 and 8 m^3/tonne/year, but much higher rates of over 20 m^3/tonne/year have been recorded. Figure 5.18 shows the results of a survey of 100 UK landfill sites with gas generation rates which range between 2.5 and 27.5 m^3/tonne/year (DOE 1995b). The results were mainly for biodegradable-type landfills, but also included co-disposal sites. This allows the potential amount of energy which could be generated from the site to be calculated, knowing that undiluted landfill gas can have a calorific value of between 15 and 21 MJ/m^3, compared with the calorific value of natural gas at about 37 MJ/m^3 (Waste Management Paper 27, 1994). The calorific value of the gas depends on the percentage composition of combustible gases such as methane and non-combustible gases such as carbon dioxide. Figure 5.19 shows the percentage of methane and carbon dioxide in extracted landfill gas in a survey of 100 UK landfill sites (DOE 1995b).

Figure 5.20 shows the development of landfill gas schemes in the UK (Brown and Maunder 1994). Early energy recovery from landfill gas schemes in the UK was based on using landfill gas as a replacement fuel for kilns, boilers and furnaces located close to the landfill site. The disadvantage is that a suitable end-user for the landfill gas should be located close to the landfill site. Later schemes developed via landfill gas-powered engines to provide power and electricity generation. The introduction of the non-fossil fuel obligation (NFFO) and premium electricity prices further stimulated the recovery of energy from landfill gas for the generation of electricity and export to the national grid. The majority of energy recovery schemes associated with landfill gas utilisation recover energy in the form of electricity generation, currently representing over 100 MW capacity in the UK with a further 90 MW planned (DTI 1996). A further development is to use the waste heat from the power generation phase to produce combined heat and power systems.

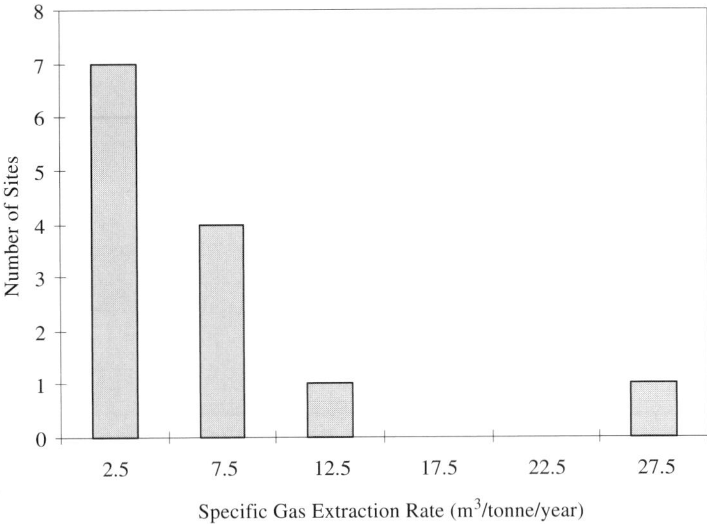

Figure 5.18 *Landfill gas extraction rate in the survey of 100 UK landfills (m^3/tonne/year). Source: DOE, Characterisation of 100 UK Landfill Sites. Report CWM 015/90, Wastes Technical Division, Department of the Environment, London, 1995. Crown copyright is reproduced with the permission of the Controller of Her Majesty's Stationery Office*

In the USA, a similar growth in landfill gas utilisation schemes developed through use of landfill gas in the early 1980s mainly through direct use of gas in boilers and furnaces. By 1994, approximately 53% of schemes were electricity generation from gas use in engines, 20% via power generation using gas turbines, 24% via direct use, for example, in boilers and furnaces, and the remainder through other energy recovery systems (Thorneloe et al 1995). Total electrical output from landfill gas utilisation in the USA has been estimated as almost 350 MW per year (Thorneloe et al 1995). New regulations proposed in the USA under the Clean Air Act 1991 on the limit of gases from large landfill sites (over 2.25 million tonnes of waste) are expected to encourage the use of landfill gas in energy recovery projects.

Figure 5.21 shows a schematic diagram of a landfill gas energy recovery project (Brown and Maunder 1994). The energy recovery technology is based around the gas collection system and the pre-treatment and power generation technology. Gas collection is via either vertical gas wells or horizontal well collection systems, depending on the type of site, site-filling techniques, depth of waste and leachate level. The gas is collected in a series of perforated gas pipelines connected to a central pipeline. The spacing of the wells for optimum gas collection depends on a number of factors, including the rate of gas generation, but they would typically be between 20 and 50 m apart. A condensate removal system is required since the gas is at temperatures above ambient and is saturated with water vapour and organic vapours. As the gas cools, the water vapour condenses to form water in the pipe which reduces the efficiency of gas collection and transport. The condensate system used to remove the water vapour consists of baffled or expansion chambers which cool and condense the water. Condensate systems both below and above ground may be required to

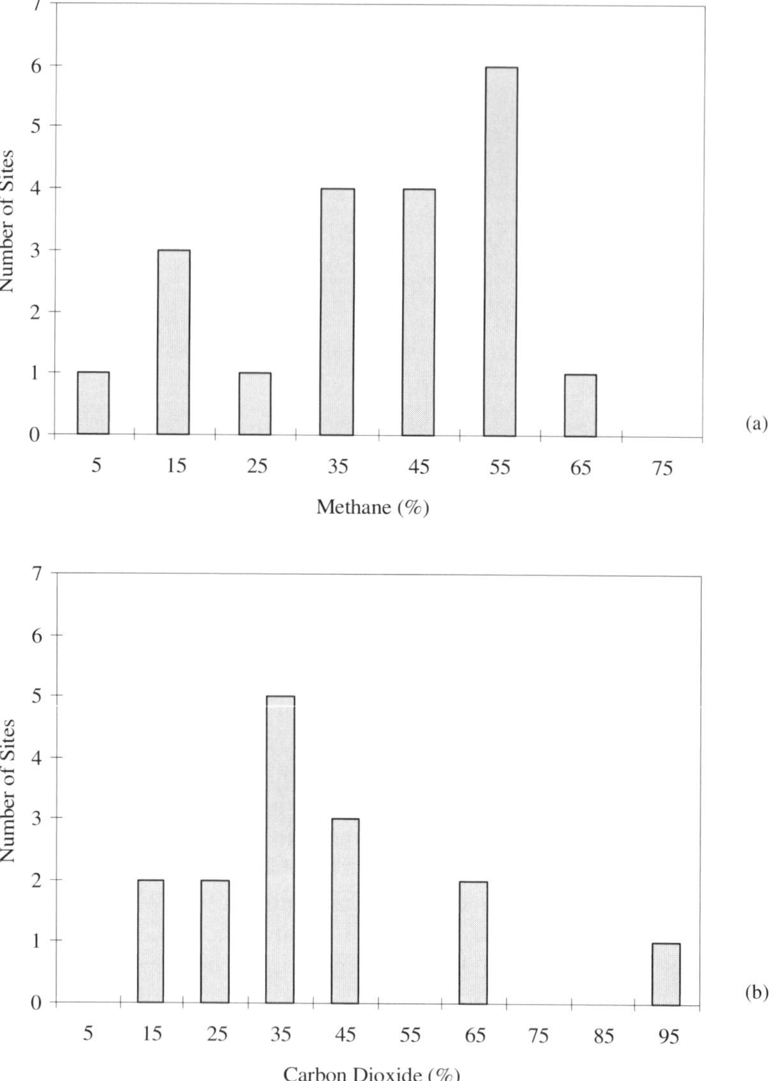

Figure 5.19 *Percentage of (a) methane and (b) carbon dioxide in extracted landfill gas in the survey of 100 UK landfills. Source: DOE, Characterisation of 100 UK Landfill Sites. Report CWM 015/90, Wastes Technical Division, Department of the Environment, London, 1995. Crown copyright is reproduced with the permission of the Controller of Her Majesty's Stationery Office*

dewater the gas. A filter would also be included to remove fine particulate material from the gas flow. The gas is then compressed and possibly passed to a pre-treatment if a greater degree of clean-up is required, for example, to remove corrosive trace gases and vapours from the gas stream. Such possible pre-treatments may include further filtration, gas chilling to condense certain constituents, absorption and adsorption systems to scrub the gases, and other gas clean-up systems such as

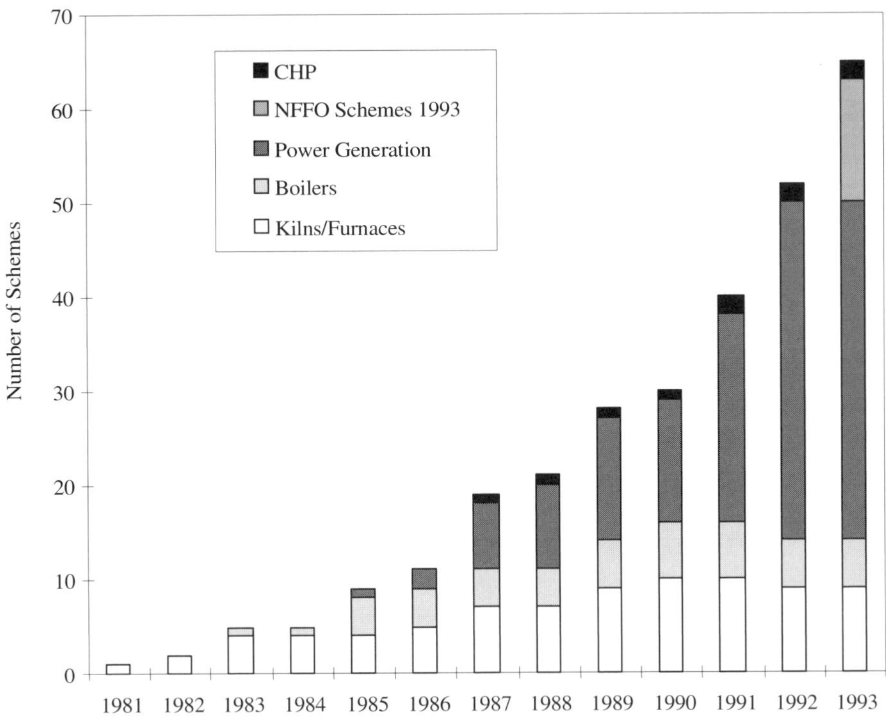

Figure 5.20 *Development of the UK landfill gas industry. Source: Reprinted from Brown K.A. and Maunder D.H., Exploitation of landfill gas: A UK perspective. Water Science and Technology, 30, 143–151, 1994 with kind permission from Elsevier Science Ltd., The Boulevard, Langford Lane, Kidlington OX5 1GB, UK*

membranes and molecular sieves to remove trace contaminants. A large proportion of the landfill gas consists of carbon dioxide, which is non-combustible and therefore reduces the overall calorific value of the gas. Therefore, for utilisation systems requiring a high-specification gas or a high calorific value, then clean-up systems to remove carbon dioxide may be required. Such systems include water scrubbing, absorption on zeolites and membrane separation, and are expensive to install and maintain (Brown and Maunder 1994; Stegmann 1996).

The utilisation of the landfill gas is via direct use as substitute fuel in boilers, kilns and furnaces, for electricity generation, or by upgrading to produce CNG, LNG substitute natural gas or for use as a chemical feedstock. Direct utilisation in boilers, kilns and furnaces close to the landfill site represents the easiest and cheapest option, since minimal modifications to the burner system of the combustion unit are required and transport costs are minimised. Power generation is produced from spark ignition engines, diesel engines and gas turbines, with availabilities of about 95% and load factors of over 80%. The engines are used with pure landfill gas or co-fuelled with natural gas. In some cases modifications to the engines are required before operation on landfill gas. The modifications are to take account of the

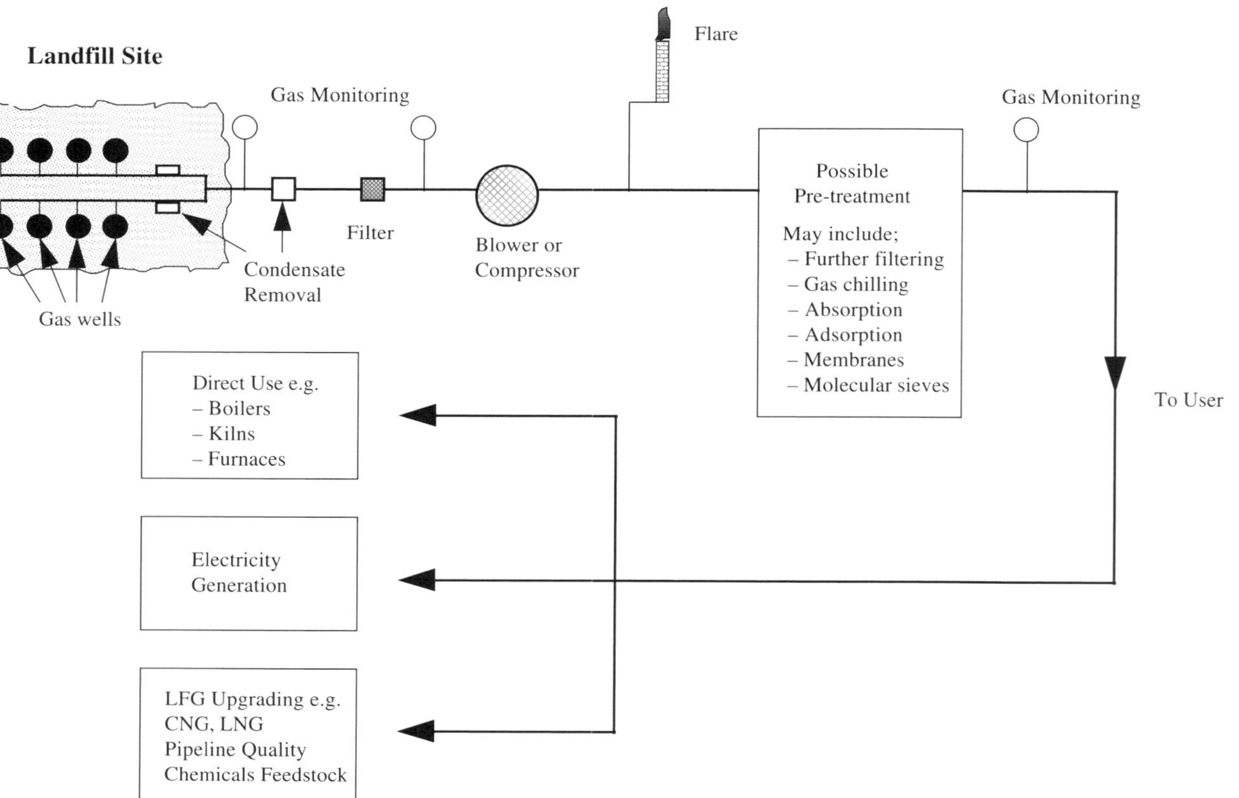

Figure 5.21 *Schematic diagram of a landfill gas energy recovery project. Source: Reprinted from Brown K.A. and Maunder D.H., Exploitation of landfill gas: A UK perspective. Water Science and Technology, 30, 143–151, 1994, with kind permission from Elsevier Science Ltd., The Boulevard, Langford Lane, Kidlington OX5 1GB, UK*

presence of carbon dioxide, which lowers the calorific value and ignitability of the gas compared to natural gas. Upgraded landfill gas has also been used as a fuel for vehicles used on the landfill site itself. This requires upgrading of the gas for the high specifications required of a vehicle. If the gas is to be used as substitute natural gas for direct input into natural gas pipelines, the gas has to be thoroughly cleaned up to comply with natural gas industry specifications. For example, a minimum calorific value would be specified, which would mean removal of carbon dioxide to a predetermined low level. In addition, fine particulate material, trace components, hydrogen sulphide etc. would be required to be removed and the gas should attain a consistent composition. It is also technically possible to use the gas as a chemical feedstock, with a wide range of products being potentially available from the methane. Table 5.28 shows the major landfill gas end-uses, their limitations and the landfill gas treatments required (Gendebien et al 1992; Stegmann 1996). In the UK, of the 75 current (1995) energy recovery schemes involving landfill gas, 62 recover energy in the form of electricity generation, including two combined heat and power schemes, and the remainder use the gas directly in boilers, furnaces and kilns (DTI 1996). The main advantage of electricity generation is that the end-user does not have to be located close to the landfill site, since the electricity can be transported via the national grid.

A further consideration for combustion of landfill gas is the presence of chlorinated organic compounds at trace level in the gas. The combustion of such chlorinated compounds may lead to the formation of dioxins and furans in the exhaust from the system. High combustion temperatures with long residence times and rapid quenching of the combustion products would be recommended to minimise the formation of dioxins and furans. High chlorine contents in landfill gas may also lead to high levels of hydrogen chloride gas which condenses to hydrochloric acid, resulting in corrosion down-stream of the combustion section of the system. Similarly, high sulphur contents may cause problems due to sulphuric acid formation.

The legislation covering exhaust emissions from power stations generating electricity from landfill gas with an installed capacity of less than 20 MW (the majority of schemes in the UK would be less than 20 MW) is not subject to air pollution control legislation under the Environmental Protection Act 1990, Part I. However, the power station would be covered by the Act under nuisance regulations (DTI 1996).

In many cases landfill gas is flared without energy recovery to destroy the methane and organic micro-pollutants as a means of gas hazard and odour control. In addition, the flare may be required to burn off any excess gas or to act as a standby for any plant shutdowns. The flare may be an exposed open flame, usually on a pedestal or enclosed in a ceramic furnace. The open-type flare has to maintain a flame even under extremes of weather conditions. Enclosed flares have greater control of the combustion conditions and also allow longer residence times to burn out completely the organic compounds in the landfill gas. The stability of the flame is related to the gas composition, weather conditions, burner design etc. However, the flame will be stable at methane concentrations of between 30 and 60%.

The typical composition of landfill gas was shown in Table 5.19, and is mainly methane and carbon dioxide with lower concentrations of hydrocarbons.

Table 5.28 *Major landfill gas end-uses, pre-treatments and advantages/disadvantages*

Landfill gas application	Required pre-treatment	Advantages/disadvantages
1. Direct use, e.g., boilers, kilns, furnaces	Removal of condensate Particulate removal Dehydration (raw)	Changes required to burner design Must be consumed at or close to the landfill site or transported short distances only
2. Power generation, e.g., spark ignition engines, diesel engines, gas turbines	Removal of condensate Particulate removal Dehydration (raw) Removal of halocarbons (if in high concentration)	Can be transported via pipeline, moderate distances Relatively high maintenance costs for spark ignition engines Gas turbines used mainly for large gas throughputs exceeding 2500 m^3/h
	Removal of condensate Particulate removal Partial removal of CO_2 Thorough dehydration Removal of halocarbons (if in high concentration)	Can be transported via pipeline, moderate distances and mixed with natural gas at low ratios Relatively high maintenance costs for spark ignition engines Gas turbines used mainly for large gas throughputs exceeding 2500 m^3/h
3. Vehicle fuel	Removal of CO_2 Particulate removal Removal of halocarbons (if in high concentration) Compression of the gas	Limited to on-site use
4. Chemical feedstock	Removal of condensate Particulate removal Removal of CO_2 Removal of H_2S Removal of halocarbons (if in high concentration)	Expensive pre-treatment
5. Injection to the national gas grid	Removal of condensate Particulate removal Removal of CO_2 Removal of H_2S Removal of halocarbons (if in high concentration)	Expensive pre-treatment

Sources: [1] Gendebien A., Pauwels M., Constant M., Ledrut-Damanet M.J., Nyns E.J., Willumsen H.C., Butson J., Fabry R. and Ferrero G.L., Landfill Gas: From Environment to Energy. Commission of the European Communities, Contract No. 88b-7030-11-3-17, Commission of the European Communities, Luxembourg, 1992; [2] DTI, Landfill Gas: Development Guidelines. Energy Technology Support Unit, Harwell, Department of Trade and Industry, HMSO, London, 1996; [3] Stegmann R., Landfill gas utilisation: An overview. In: Christensen T.H., Cossu R. and Stegmann R. (Eds), Landfilling of Waste: Biogas. E. and F.N. Spon, London, 1996.

Table 5.29 *Pollutant emissions from the combustion of landfill gas from three UK landfill sites in different types of power plant*

Emission	Dual fuel diesel engine[1] (mg/m^3)	Spark ignition engine[2] (mg/m^3)	Gas turbine[3] (mg/m^3)
Particulate matter	4.3	125	9
Carbon monoxide	800	~10 000	14
Unburnt hydrocarbons	22	> 200	15
Nitrogen oxides	795	~1170	61
Hydrogen chloride	12	15	38
Sulphur dioxide	51	22	6
Dioxins (ng/m^3)	0.4	0.6	0.6
Furans (ng/m^3)	0.4	2.7	1.2

Gas pre-treatments:
1 Drying, filtering, compression and cooling of gas prior to use.
2 Drying, filtering, compression and cooling of gas prior to use.
3 Wet scrubbing, compression, cooling, filtering and heating to 70 °C prior to use.

Sources: [1] Blakey N.C., Emissions from power generation plants fuelled by landfill gas. In: Fourth International Landfill Symposium, Sardinia, October 1993; [2] Young C.P. and Blakey N.C., Emissions from landfill gas power generation plants. In: Christensen T.H., Cossu R. and Stegmann R. (Eds.) Landfilling of Wastes: Biogas. E. and F.N. Spon, 1996.

Combustion of the methane in the landfill gas will produce mainly carbon dioxide and water vapour, and other minor pollutants. Table 5.29 shows the concentration of pollutants produced from the combustion of landfill gas in a dual-fuelled diesel engine, spark ignition engine and gas turbine (Young and Blakey 1993). The emissions from the combustion of landfill gas will vary depending not only on the type of combustion system, but also the composition of the gas used.

The accurate assessment of landfill gas generation from a site is a major factor in deciding whether the site will be developed for the recovery of energy via landfill gas. Assessment of the landfill gas generation curve over the lifetime of the site then becomes a basis for financial investment in a landfill gas utilisation project. The assessment would include both predictive computer modelling and physical site assessments. Difficulties arise in the assessment of the potential for landfill gas utilisation using physical assessment methods due to the heterogeneous nature of waste and often poor records of waste emplaced. A sample well or probe at one part of the site may give completely different results from one at another part of the site, even at sample points close to each other. Table 5.30 summarises the physical assessment techniques to estimate landfill gas generation (DTI 1996). Other factors influencing the generation of landfill gas include the site size and geometry. Landfill gas utilisation would normally only be considered for large sites with a minimum of between 200 000 and 500 000 tonnes of waste. In addition, higher gas recovery rates are obtained from large, deeper sites rather than shallower sites.

Estimation of landfill gas production involves not only the total amount of gas potentially available, but also the rate and duration of gas production (Cossu et al 1996). Modelling techniques used to estimate the gas generation rates from a landfill site are based on the assumption that a certain unit mass of biodegradable waste will

Table 5.30 *Physical assessment techniques to estimate landfill gas generation*

Technique	Basis	Advantages	Disadvantages
Waste sampling from trial pits and boreholes	Qualitative assessment of nature of waste by inspection	Rapid, low cost. Can provide useful information about the nature and location of wastes in the absence of records	Small samples taken from a large site may not be representative of the whole
Waste analysis	Chemical analysis of waste samples. Volatile solids or organic carbon content indicate landfill gas potential. Samples incubated in the laboratory can be used for gas generation	Rapid, low cost. Can provide useful information about the nature and location of wastes in the absence of records	Small samples taken from a large site may not be representative of the whole
Measurement of passive landfill gas emissions	Measurement of flow of landfill gas from boreholes or into flux boxes installed within the landfill	Measures landfill gas generated under field conditions	Variable pressure fields in landfills can introduce errors in the scale-up computation. Atmospheric pressure influences flow
Pumping trials	Pumping of landfill gas from boreholes until steady state is achieved, with an estimate of the volume of waste affected by measuring the pressure drop in the landfill	Measures landfill gas generated and extracted under field conditions. Provides early indications of the problems likely to be encountered later (e.g., low permediability waste, air leakage)	More costly than other methods. Measurement of the radius of influence can lack precision.[1]

[1] Landfill gas is drawn towards the test well under the influence of the pressure gradient within the landfill. The pressure difference decreases with increasing distance from the well until a point is reached where the applied suction at the test well has no influence on the pressure in the landfill. The horizontal distance to this point is termed the radius of influence.

Source: DTI, Landfill Gas Development Guidelines. Energy Technology Support Unit, Harwell. Department of Trade and Industry, HMSO, London, 1996. Crown copyright is reproduced with the permission of the Controller of Her Majesty's Stationery Office.

produce a certain quantity of landfill gas. As the waste is further and further biodegraded, the rate of landfill gas production will decrease in proportion to the quantity remaining to be processed. Figure 5.22 shows the most commonly applied equation to describe the rate of landfill gas generation (DTI 1996). Cossu et al (1996) have reviewed landfill gas production models and the factors influencing those models.

$$\text{Rate} = kL_0e^{-kt}$$

Rate = Rate of landfill gas production
k = rate constant, represents the decay value or half life of the waste.
L_0 = ultimate yield of landfill gas
t = time

Figure 5.22 *Equation to describe the production of landfill gas. Source: DTI, Landfill Gas Development Guidelines. Energy Technology Support Unit, Harwell. Department of Trade and Industry, HMSO, London, 1996*

Table 5.31 *Problems associated with old waste landfill sites*

Parameter	Problem
Waste	Data on waste quantities and composition inputs are usually scarce
Siting	Old landfills tend to be sited for convenience rather than on grounds of geological and hydrogeological criteria
Surrounding environment	Old landfills are often sited close to conurbations, or new developments have been close to old landfills
Waste degradation	The waste degradation process is slow, resulting in long-term generation of leachate and landfill gas after the site is closed
Leachate and gas migration	Leachate and landfill gas may be migrating from the site

Source: Jefferis S.A. In: Sarsby R.W. (Ed.) Waste Disposal by Landfill. Balkema, Rotterdam, 1995.

5.11 Old Landfill Sites

Since landfilling of wastes has occurred throughout history, there are innumerable old landfill sites. For the majority of old landfill sites, very little is known of the input of the waste, monitoring of leachate and landfill gas. Table 5.31 shows some of the problems associated with old waste landfill sites (Jefferis 1995). Early landfill designs were based on the attenuate and disperse type, and whilst few incidents of leachate migration causing problems have occurred, old landfill sites may have the potential for future environmental hazards. Gas emission and migration from landfills is also potentially environmentally hazardous. There have been several incidents of landfill gas migration causing explosions, including the Loscoe incident in the UK (see Box 5.5).

A survey undertaken in 1993 of 4000 UK landfill sites (Figure 5.23) found that 230 had suffered a serious pollution incident requiring substantial remedial action (The Waste Manager 1994b). The majority of problems were associated with ground water and surface water pollution and landfill gas migration. Other failures were attributed to subsidence and fires. Clean-up of the sites was estimated at between

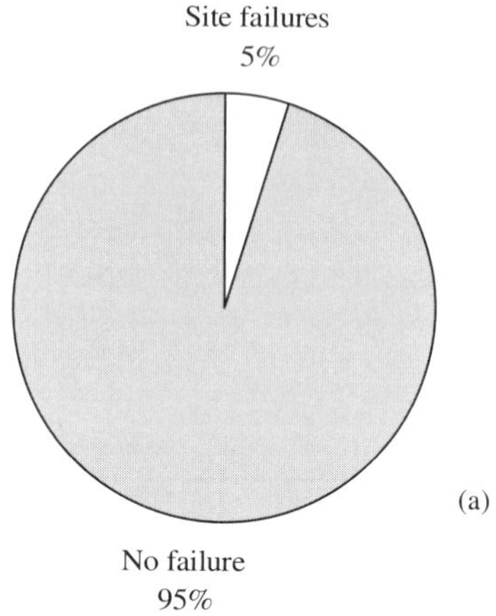

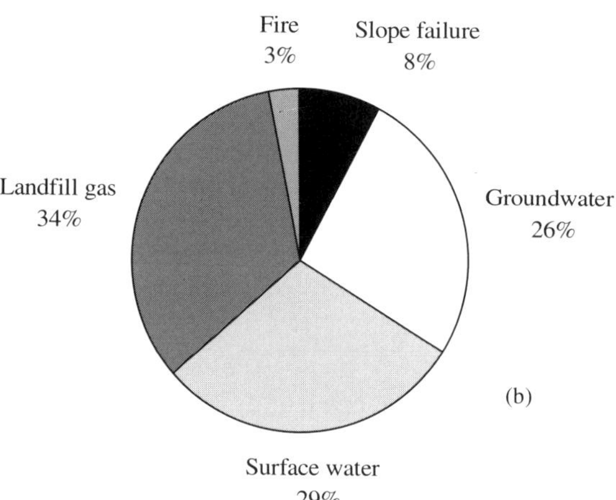

Figure 5.23 *Incidence of (a) significant landfill failures and (b) types of failure. Source: Rose J., Landfill clean-ups: A booming business. The Waste Manager, 1994. Reproduced from The Waste Manager, 154 Buckingham Palace Road, London*

£100 000 and £500 000. Interestingly, almost a third of the sites were modern containment sites, and more than 10 of the problem sites had been started after 1990, suggesting that site failure is not associated solely with the older, poorer designed sites.

Therefore, for old landfills the main problems are associated with the uncontrolled migration of leachate and gas into the surrounding environment. The control of leachate and landfill gas migration from old sites is via slurry trench cut-off walls (Jefferis 1995). A trench is excavated in the ground around the boundaries of the landfill and a slurry of bentonite clay, ground granulated blast furnace slag, Portland cement and water is poured in. The mixture sets to form a low-permeability barrier. The permeabilities of such barrier walls are of the order of 10^{-9} m/s hydraulic conductivity. In some cases a polymeric plastic membrane may also be added to reach lower permeabilities.

The creation of a barrier beneath an old landfill is more complicated, and requires drilling and high-pressure injection of the slurry. The slurry is injected in a series of overlapping 'V' cuts to create an interlocking layer of low-permeability barriers. The gas and leachate collection and control systems are introduced by excavating wells or boreholes for collection, extraction and treatment of the leachate and landfill gas. Finally, in many cases old landfills are not capped to the high standards of current landfills. However, the site is much more easily accessible than the sides and base of the landfill and can easily be capped with suitable barrier systems and top soils (Jefferis 1995).

A new development associated with old landfill sites is landfill mining and reclamation, where previously deposited wastes after stabilisation are excavated for processing to obtain recyclable materials, a combustible fraction and soil, and also to create new void space for further landfilling (The World Resource Foundation 1995b). In addition, landfill mining and reclamation aids the remediation of poorly designed landfills, or upgrades landfills that do not meet current environmental standards or operational practices. A number of landfill mining and reclamation schemes have been developed throughout the World, for example in the USA. Figure 5.24 shows a possible landfill mining operation involving extensive processing (The World Resource Foundation 1995b). Excavation involves similar practices to open-cast mining using excavators, clamshell grabs and lorries or conveyor belts to transport the waste to the processing facility, which may be either on-site or off-site. Processing involves the sizing of the material using a coarse screen to remove large items which cannot be processed. A further fine screen such as a trommel screen allows the separated fine material to be recovered as soil. The material passing through the screen is processed via magnetic separation to remove ferrous metals. Air classification of the non-ferrous metal fraction separates the light organic materials which may be used to produce a refuse derived fuel, and the residual heavy material such as non-ferrous metals, glass and wood may be further fractionated.

The factors involved in landfill mining and reclamation schemes involve the recovery efficiency of recyclable materials from the site and their quality. Table 5.32 shows the typical recovery efficiencies of the available product and the purities of the product streams from landfill mining and reclamation (The World Resource Foundation 1995b). The majority of the material recovered from landfill mining and reclamation is the soil fraction, which can be typically 50–60% of the material mined. Table 5.32 shows that whilst a high proportion of the available recyclable materials in the landfill can be recovered, the data on product purity shows that impurities in the product may limit the market demand for these materials.

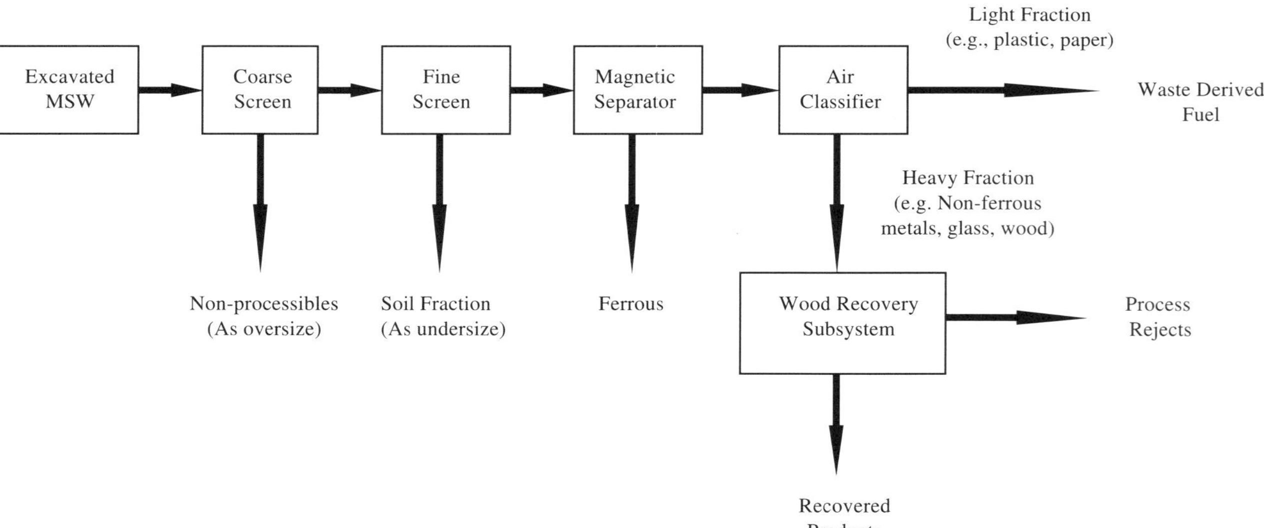

Figure 5.24 *An example of an extensive landfill mining and reclamation operation. Source: The World Resource Foundation, Technical Brief; Landfill Mining. The World Resource Foundation, Tonbridge, 1995. Reproduced with the permission of the World Resource Foundation*

Table 5.32 Typical recovery efficiencies and purities of the product streams from landfill mining and reclamation

Product	Recovery efficiency (%)	Purity (%)
Soil	85–95	90–95
Ferrous metal	70–90	80–95
Plastic	50–70	70–90

Source: The World Resource Foundation, Technical Brief. Landfill Mining. The World Resource Foundation, Tonbridge, UK, 1995.

5.12 Economics

There have been only a limited number of economic appraisals of landfill which have been published. Table 5.33 shows a summary of landfill costs for a municipal solid waste landfill site (Burnett 1993; Royal Commission on Environmental Pollution 1993). The data shown in Table 5.33 are estimated for a typical 200 000 tonne per year facility with a 20-year operational project life. A typical landfill operates in phases of development, waste emplacement and restoration throughout the lifetime of the site. Consequently the costs of each stage are approximately evenly distributed throughout the 20-year project and analysed as such for the investment appraisal. The site acquisition cost of £8 million is based on the throughput of waste contracted for disposal of 200 000 tonnes per year for 20 years. Therefore a 4 million tonne waste disposal site would be required. With an estimated waste density of 1 m^3/tonne, a 4 million m^3 site is required. An estimated cost of £2/m^3 void space was in put into the economic appraisal.

Site assessment costs are estimated at £159 000, which is made up of reconnaissance, preliminary site investigation, market survey, full site investigation, planning and environmental assessment costs. The major capital expenditure and development costs include the site lining materials, leachate collection and treatment systems, and landfill gas management systems, which together comprise about 90% of the development costs. For the assessment outlined in Table 5.33, the liner system used for estimation was a polymeric plastic liner plus natural clay. Other development costs include the weighbridge, wheel cleaning system, land excavation, security fencing, offices etc. Operating costs represent the greatest costs of the landfill project, and include employee salaries, maintenance of equipment and vehicles, environmental monitoring etc., for the 20 years of the project. Restoration costs occur throughout the project as each phase is completed, and include the capping material of impervious clay and top soil, and landscaping. Aftercare costs include monitoring of leachate, gas and settlement for a period after site closure, which is represented in the estimate in Table 5.33 as 30 years (Burnett 1993; Royal Commission on Environmental Pollution 1993). Figure 5.25 shows a percentage breakdown of the various stages of costs (Royal Commission on Environmental Pollution 1993).

Table 5.33 *Summary of landfill costs*

Operating conditions	
Project life	20 years
Municipal solid waste throughput	200 000 tonnes/year
Discount factor for discounted cash flow investment appraisal	10%
Gross expenditure	
Stage	Total expenditure (£)
Site acquisition	8 000 000
Site assessment	159 000
Capital expenditure and development	6 516 000
Operational costs (£852 200/year)	17 044 000
Site restoration	2 472 000
Aftercare	1 600 000
Total	35 791 000
Discounted cost per tonne	£9.83/tonne
Energy recovery via electricity generation	
Power output from landfill gas combustion and electricity generation from 200 000 tonnes/year waste	2 MW
Annual revenue (pool price 2.5 p/kwh)	35 000
Net cost per tonne	£9.55

Sources: [1] Royal Commission on Environmental Pollution, 17th Report, Incineration of Waste. HMSO, London, 1993; [2] Burnett S., The comparative costs of incineration. In: Incineration: An Environmental Solution for Waste Disposal. IBC Conference, Manchester, February, 1993.

The total cost estimate for the 20-year project is £35.79 million. The estimate was prepared for a landfill site with biodegradable waste in the form of municipal solid waste. The estimate was also extended to incorporate the disposal of hazardous and special wastes in a co-disposal landfill. The additional requirements of a co-disposal site in terms of increased environmental assessment costs, added personnel for operation and environmental monitoring, additional leachate treatment, and a higher standard lining and capping system gave an estimated £5.5 million increase in overall costs, representing about 15% increase (Burnett 1993; Royal Commission on Environmental Pollution 1993).

A further factor involved in the assessment of landfill costs is the potential to recover energy in the form of landfill gas combustion to generate either electricity or provide heat for district heating. Table 5.33 shows the estimation of the disposal costs/tonne of waste taking into account the generation of electricity from the combustion of landfill gas to generate electricity at a pool price of 2.5 p/kwh. The generation of electricity from landfill gas allows operators to bid for a non-fossil fuel obligation (NFFO) contract, which allows a negotiated fee for the electricity at rates above the general pool price (Chapter 2). The generation of electricity from a landfill site does not guarantee that a NFFO contract will be granted, and therefore the general electricity pool price would then be obtained. However, if a NFFO contract

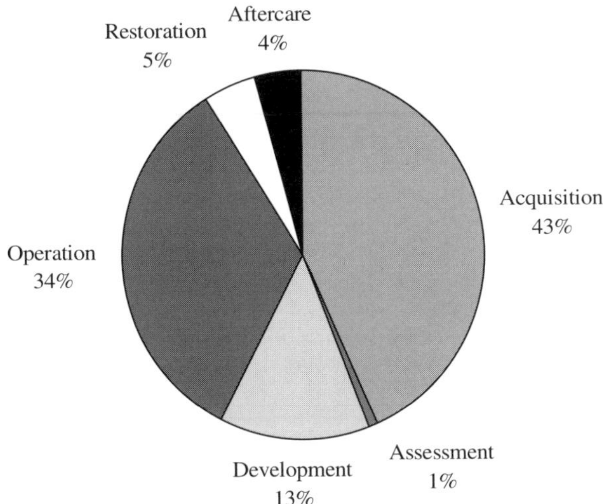

Figure 5.25 *Percentage distribution of landfill costs. Source: Royal Commission on Environmental Pollution. 17th Report, Incineration of Waste. HMSO, London, 1993. Crown copyright is reproduced with the permission of the Controller of Her Majesty's Stationery Office*

were to be obtained, then the cost of landfill would be further reduced. The introduction of NFFO to encourage the development of renewable energy schemes such as energy from waste was seen as a method to develop the technologies. It was envisaged by the UK Government that eventually they would be able to compete with fossil-fuel-derived electricity generation and combined heat and power schemes. For landfill gas schemes that situation has virtually arrived. The NFFO price paid to electricity generators using landfill gas as currently just above the fossil-fuel-generated pool price for electricity. Some electricity generating schemes have recently been developed outside the NFFO scheme by direct sales of the generated electricity to the end-user, thereby avoiding the electricity pool operated by the National Grid Company and also avoiding the fossil fuel levy which fossil-fuel-generated electricity pays as part of the NFFO scheme. For example, the Wingmoor landfill in Bishop's Cleeve, Cheltenham, generates electricity from landfill gas, with a capacity of 3 MW. The electricity is sold direct to the Cheltenham and Gloucester College of Higher Education (Cosslett 1996).

Landfill gas energy recovery projects can have a significant influence on the economics of waste landfills. However, in the UK, landfill gas utilisation for recovery of energy accounts for only about 75 landfill sites out of a total of over 4000 licensed sites (Digest of Environmental Protection and Water Statistics 1994; DTI 1996). Direct use of landfill gas in furnaces, boilers and kilns are the lowest-cost options since gas transport costs and modifications to the burner of the combustion system are minimal. Examples include the firing of a kiln using landfill gas from the Cheadle landfill with an output of 2.54 $MW_{thermal}$ (DTI 1996). The capital costs of such direct-use schemes are lower than those for power production, for example, the capital cost of the Cheadle landfill gas utilisation scheme, including

Table 5.34 *Typical installed costs (1996) for plant generating electricity from landfill gas in relation to size of plant output*

Power plant	Output of 1 MW_e Cost £/$kW_{electrical}$	Output of 2 MW_e Cost £/$kW_{electrical}$	Output of 5 MW_e Cost £/$kW_{electrical}$
Gas turbine	Not applicable	Not applicable	400–500
Spark ignition engine	350–450	300–400	Not applicable
Dual-fuel engine	450–550	400–500	350–400

N.B. Costs include the power plant, generator and associated electrical equipment.

Source: DTI, Landfill Gas Development Guidelines. Energy Technology Support Unit, Harwell. Department of Trade and Industry, HMSO, London, 1996. Crown copyright is reproduced with the permission of the Controller of Her Majesty's Stationery Office.

costs of wells, pipework, blowers, compressors and modifications to the kiln, was approximately £147/kW for a 2.54 $MW_{thermal}$ scheme, representing an installed cost of £370 000. Electricity generating schemes relying on spark ignition, diesel engines and gas turbine plant require investment in the power plant, generation units and associated electricity. Table 5.34 shows typical installed costs for plant generating electricity from landfill gas (DTI 1996). From Table 5.34, a 2 MW spark ignition power plant operating on landfill gas would cost between £0.6 and £0.8 million for the installed costs of the power plant, generator and associated electrical equipment. In addition, the scheme would also involve costs associated with the gas extraction system, wells, well heads, pipework, pumps, compressors, gas monitoring equipment, flares and costs for connection to the national electricity grid. The impact of such costs on the overall capital costs of a full scheme are illustrated in Table 5.35, which shows the capital, operating and maintenance costs of several operating schemes in the UK (DTI 1996). The operating and maintenance costs of such schemes are low.

Economies of scale in development and operation of landfill sites indicate that smaller sites, below 50 000 tonnes/year throughput, are significantly more expensive than larger sites. Figure 5.26 shows the economies of scale for landfill in terms of the estimated cost per tonne of municipal solid waste in relation to annual throughput of waste (Royal Commission on Environmental Pollution 1993).

The influence of transport costs is also an important parameter in assessing the costs of landfill, particularly where landfill sites close to the point of waste generation are becoming scarce. Where the round-trip distance increases, progressively increased costs are involved. Figure 5.27 shows the increase in disposal costs with increased transport distance of the waste to the landfill site. Costs are shown for waste disposal via landfill where for longer distances a waste transfer facility would be essential (Royal Commission on Environmental Pollution 1993).

A review of landfill costs reflects only the *costs* involved, whereas the price or gate-fee which can be charged for accepting waste at the site is a different figure, which is based on a variety of factors. Such factors include the normal criteria of the market place, competition such as the availability of alternative landfill void space in the area or alternative disposal options, and the profit margin of the operator.

Table 5.35 *Examples of capital, and operating and maintenance costs for electricity generating schemes from landfill gas in the UK*

Site	Opening date	Rated output ($MW_{electrical}$)	Capital cost (£/kWh)	Operating and maintenance cost (£/kWh)
Stewartby	1987	0.74	550[1]	0.98
Packington	1987	3.03	654[1]	1.0
Middleton Broom	1993	1.2	920[1]	1.4[2]
Un-named landfill	1992	2.0	850[1]	2.0
Calvert	1992	2.7	890	2.0[3]
Brogborough				
Phase 1	1991	4.0	525	2.0[3,4]
Phase 2	1992	6.7	642	1.8[3,4]

N.B. Costs relate to year of opening.
[1] Excluding the costs of the gas collection system.
[2] Excluding consumables.
[3] Excluding rates.
[4] Excluding insurance.

Source: DTI, Landfill Gas Development Guidelines. Energy Technology Support Unit, Harwell. Department of Trade and Industry, HMSO, London, 1996. Crown copyright is reproduced with the permission of the Controller of Her Majesty's Stationery Office.

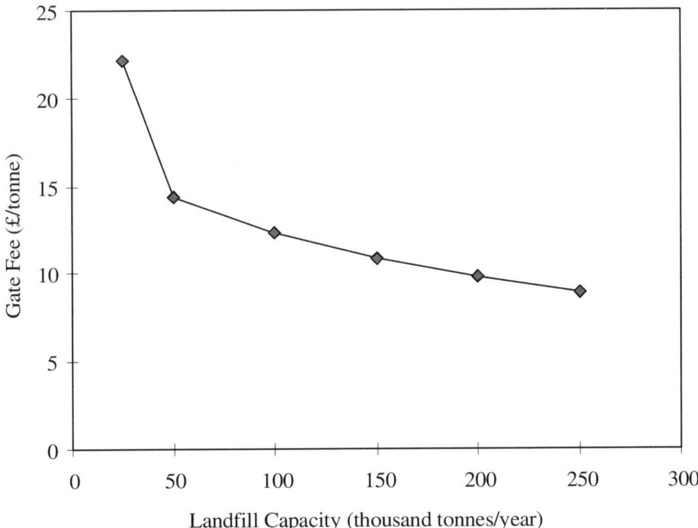

Figure 5.26 *Influence of economies of scale on the costs of landfilling municipal solid wastes. Source: Royal Commission on Environmental Pollution. 17th Report, Incineration of Waste. HMSO, London, 1993. Crown copyright is reproduced with the permission of the Controller of Her Majesty's Stationery Office*

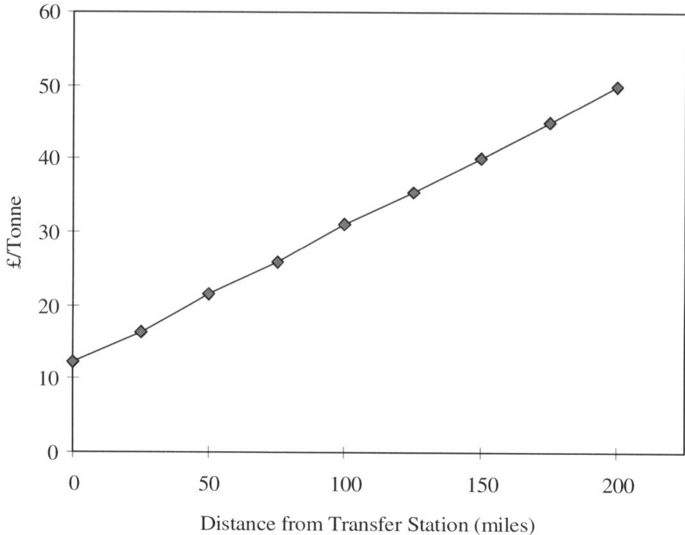

Figure 5.27 Influence of round-trip transport distance on the costs of waste disposal by landfill. Source: Royal Commission on Environmental Pollution. 17th Report, Incineration of Waste. HMSO, London, 1993. Crown copyright is reproduced with the permission of the Controller of Her Majesty's Stationery Office

The economic data presented show only the costs involved, and no account is taken of the Landfill Tax at £7/tonne for biodegradable waste and £2/tonne for inert wastes (Chapter 2). Further impacts on the costs of landfill in the future are centred on the proposed EC Waste Landfill Directive, which is discussed in the next section.

Hirshfield et al (1992) have suggested that the costs of landfill should also include some estimate of social cost due to the siting of the landfill. Such costs would include the depreciation of house prices due to their proximity to the landfill site. In addition, opportunity costs for the loss of alternative uses for the land and the land surrounding the site should also be estimated. They conclude that such costs are significant, and represent about 18% of the normally estimated costs of land, construction, operation, closure and post-closure activities, and leachate and gas monitoring.

5.13 Proposed European Waste Landfill Directive

The proposed European Waste Landfill Directive (Commission of the European Communities, COM(97) 105 Final, 1997) contains a number of provisions which, if adopted, would have major implications throughout Europe on the management of waste. The UK in particular, with its heavy reliance on landfilling, would see a drastic change in the treatment and disposal of waste. The European Waste Landfill Directive was first proposed by the European Commission in 1990, and after intense debate, delays and defeat in the European Parliament, a revised proposal has been submitted for adoption by the states of the European Union. The basis of the Directive is to harmonise standards throughout the EU for waste management facilities on the basis of a high level of environmental protection. It is clear that the European Commission regards landfilling of waste as the least favourable option due to the fact that 'landfilling does not make use of waste as a resource' and may result in 'substantial negative impacts on the environment'. The most important of these have been highlighted by the Commission, and include 'emissions of hazardous substances to soil and groundwater, emissions of methane into the atmosphere, dust, noise, explosion risks and deterioration of land' (Commission of the European Communities, COM(97) 105 Final, 1997). Against this background of antagonism towards landfill, the waste Landfill Directive proposed in March 1997 contains stringent measures to limit the environmental impact of landfills, and will have a major impact on waste treatment and disposal in the UK.

The proposal introduces provisions to reduce the landfilling of biodegradable waste and to ensure that the gases produced in existing, as well as new, landfills are collected, treated and used. Therefore a limit on the disposal of biodegradable waste has been introduced with the aim of encouraging the separate collection of biodegradable waste and reducing landfilling of waste in general. The proposal requires Member States to reduce landfilling of biodegradable solid waste to 75% of the total produced by 2002, 50% by 2005 and 25% by 2010 from a baseline level taken at 1993 (ENDS 1997b). This approach allows flexibility within the principles of subsidiarity, in that Member States are allowed to decide how the reduction targets for biodegradable waste are to be met. Several European countries, including Germany, Austria, Finland, France and the Netherlands, have already introduced limits or guidelines for biodegradable wastes going to landfill (Commission of the European Communities, COM(97) 105 Final, 1997). The reduction in biodegradable wastes going to landfill has a central aim of reducing emissions of methane as a major greenhouse gas in line with a further European Community Strategy in relation to climate change and reduction of methane emissions. To comply with this Strategy, reducing the contribution of methane to the atmosphere by reducing the biodegradable waste going to landfill is seen as the most effective and cheapest option.

However, the removal of biodegradable waste from the waste stream going to landfill is contrary to the adoption of the concept of the 'flushing bioreactor' promoted in the UK by the Department of the Environment. The flushing bioreactor is founded on a continuation of landfilling of biodegradable waste as a disposal option, with a projected stabilisation of the waste within a period of 30–50 years using leachate recirculation to accelerate gas generation and flush out pollutants.

One further factor liable to different interpretations may be the definitions of waste used in the proposed Directive (ENDS 1997b). In the UK, 'municipal waste' is usually defined as including household waste, waste from civic amenity sites, street sweepings and park waste, and waste collected from trade premises by local authorities. However, the proposed Directive has a broader definition of municipal waste as including household waste together with commercial, industrial, institutional and other waste which, because of its nature or composition, is similar to waste from households. Consequently it is not clear which wastes in the UK will count towards the biodegradable waste reduction targets.

Waste landfills are to be classified as those for hazardous waste, non-hazardous waste and inert waste, with each category used only for the specified purpose. A further measure in the proposal of direct consequence to the UK is that there should be no joint disposal of hazardous and non-hazardous waste in landfills. This would prevent the co-disposal of hazardous/industrial or special waste with municipal waste, which has been used extensively in the UK. The practice in the UK has been accepted since it was shown that the physical, chemical and biological processes that degrade municipal waste also help to degrade the hazardous/industrial or special waste and thereby reduce the hazard.

Another requirement for waste set out in the proposal is that waste be pre-treated before it is landfilled. Pre-treatment is defined in the proposal as physical, chemical or biological processes, including sorting, that change the characteristics of the waste in order to reduce its volume or hazardous nature, facilitate its handling or enhance recovery (Commission of the European Communities, COM(97) 105 Final, 1997).

The landfilling of tyres, whether they be shredded or whole is prohibited under the proposal. Tyres have been designated as a Priority Waste Stream by the Commission, requiring special treatment and disposal, and the ban on landfill is designed to encourage recovery and recycling schemes. In addition, it has been shown that tyres placed in landfills make the sites unstable and also increase the risk of fire. Other types of waste banned by the proposal are liquid waste, infectious hospital or clinical waste, flammable or highly flammable waste, and explosive or oxidising waste (Commission of the European Communities, COM(97) 105 Final, 1997).

Finally, the proposed Directive outlines the licensing system which in the UK has already been implemented under the 1990 Environmental Protection Act and Waste Management and Licensing Regulations 1994 (Waste Management Paper 4, 1996). These set out the licence provisions, which include types and quantities of waste, a geological and hydrogeological description of the site, methods for pollution monitoring and control, an environmental impact assessment, post-closure and after-care plan, financial security of the licence holder etc.

Recent Federal Regulations introduced in the USA covering landfill include a range of siting, design and operational requirements similar to those proposed by the European Waste Landfill Directive (Solid Waste Disposal Facility Criteria 1991). For example, all landfills are required to be built with a composite liner system consisting of 60 cm of clay and a 0.6-cm-high-density polyethylene liner. Other requirements are a leachate collection and removal system, landfill gas control system, groundwater monitoring, final capping system equal in hydraulic conductivity to the landfill liner system, and that 30 years of monitoring for groundwater and landfill gas and cover maintenance are required after final closure (Hickman 1996).

Bibliography

Bagchi A. 1994. Design Construction and Monitoring of Landfill. Wiley, New York.

Beaven R.P. 1996. A hydrogeological approach to sustainable landfill design. In: Harwell Waste Management Symposium: Progress, Practice and Landfill Research. Harwell, May.

Bishop D.J. and Carter G. 1995. Waste disposal by landfill lining systems. In: Sarsby, R.W. (Ed.) Waste Disposal by Landfill. Balkema, Rotterdam.

Blakey N.C. 1993. Emissions from power generation plants fuelled by landfill gas. In: Christensen T.H., Cossu R. and Stegmann R. (Eds.) Sardinia 93, Fourth International Landfill Symposium (Cagliari, Sardinia, October), 1993. CISA-Environmental Sanitary Engineering Centre, Cagliari.

Blakey N., Archer D. and Reynolds P. 1995. Bioreactor landfill: A microbiological review. In: Christensen T.H., Cossu R. and Stegmann R. (Eds.) Sardinia 95, Fifth International Landfill Symposium (Cagliari, Sardinia, October 1995), CISA-Environmental Sanitary Engineering Centre, Cagliari.

Brown K.A. and Maunder D.H. 1994. Exploitation of landfill gas: A UK perspective. Water Science and Technology, 30, 143–151.

Burnett S. 1993. The comparative costs of incineration. In: Incineration: An Environmental Solution for Waste Disposal. IBC Conference, Manchester.

Chaiampo F., Conti R. and Cometto D. 1996. Morphological characterisation of MSW landfills. Resources, Conservation and Recycling, 17, 37–45.

Christensen T.H., Kjeldsen P. and Lindhardt B. 1996a. Gas generating processes in landfills. In: Christensen, T.H., Cossu R. and Stegman R. (Eds.) Landfilling of Waste: Biogas. E. and F.N. Spon, London.

Christensen T.H., Cossu R. and Stegmann R. 1996b. Landfilling of Waste: Barriers. E. and F.N. Spon, London.

Cosslett G. 1996. Live wires in landfill gas. The Waste Manager, May, 15–17.

Cossu R., Andreottola G. and Muntoni A. 1996. Modelling landfill gas production. In: Christensen T.H., Cossu R. and Stegmann, R. (Eds.) Landfill of Waste: Biogas. E. and F.N. Spon, London.

De Rome L. and Gronow J. 1995. Leachate recirculation in the UK: An overview of research projects. In: Christensen T.H., Cossu R. and Stegmann R. (Eds.) Sardinia 95, Fifth International Landfill Symposium (Cagliari, Sardinia, October 1995), CISA-Environmental Sanitary Engineering Centre, Cagliari.

Digest of Environmental Protection and Water Statistics, 1991. Department of the Environment, HMSO, London.

Digest of Environmental Protection and Water Statistics, 1994. Department of the Environment, HMSO, London.

DOE, 1989. A Review of the Technical Aspects of Co-disposal. Department of the Environment, Wastes Technical Division, Report CWM/007/89.

DOE, 1995a. Analysis of Industrial and Commercial Waste going to Landfill in the UK. Department of the Environment, HMSO, London.

DOE, 1995b. Characterisation of 100 UK Landfill Sites. Report CWM 015/90, Wastes Technical Division, Department of the Environment, HMSO, London.

DOE, 1995c. Environmental Impacts from Landfills Accepting Non-Domestic Wastes. Report CWM 036/91, 1991, Wastes Technical Division, Department of the Environment, London.

DOE, 1995d. Landfill Monitoring Investigations at Gorsethorpe Landfill, Sherwood Sandstone, Notts, 1978–1992. Report CWM/034/94, 1994. Wastes Technical Division, Department of the Environment, London.

DOE, 1995e. A review of the composition of leachates from domestic wastes in landfill sites. Department of the Environment Research Report No. CWM 072/94. In: Waste

Management Paper 26B, Landfill Design, Construction and Operational Practice. Department of the Environment, HMSO, London.

DTI, 1996. Landfill Gas Development Guidelines. Energy Technology Support Unit, Department of Trade and Industry, HMSO, London.

EC Directive 80/68/EEC, 1980. Council Directive of 17 December 1979 on the protection of groundwater against pollution from certain dangerous substances. Official Journal of the European Communities L020, January, Brussels.

ENDS, 1997a. Report 265, Environmental Data Services Ltd., Bowling Green Lane, London, 12–14.

ENDS, 1997b. Report 265, Environmental Data Services Ltd., Bowling Green Lane, London, 40–41.

Energy from Waste: Best Practice Guide, 1996. Department of Trade and Industry, HMSO, London.

Fifield F.W. and Haines P.J., 1995. Environmental Analytical Chemistry. Blackie Academic and Professional, London.

Gendebien A., Pauwels M., Constant M., Ledrut-Damanet M.J., Nyns E.J., Willumsen H.C., Butson J., Fabry R. and Ferrero G.L. 1992. Landfill Gas: From Environment to Energy. Commission of the European Communities, Contract No. 88b-7030-11-3-17, Commission of the European Communities, Luxembourg.

Griffiths B. 1996. Financial liabilities: A case for equitable treatment. In: Harwell Waste Management Symposium: Progress, Practice and Landfill Research. Harwell, May.

Ham R.K. 1993. Overview and implications of US sanitary landfill practice. Air and Waste, 43, 187–190.

Hickman L.H. 1996. MSW Management: State of Practice in the USA. International Solid Waste Association, Yearbook, Paris.

Hirshfield S., Vesilind P.A. and Pas E.I. 1992. Assessing the true cost of landfills. Waste Management and Research, 10, 471–484.

Hjelmar O., Johannessen L.M., Knox K., Ehrig H.J., Flyvbjerg J., Winther P. and Christensen, 1995. Composition and management of leachate from landfills within the EU. In: Christensen T.H., Cossu R. and Stegmann R. (Eds.) Sardinia 95, Fifth International Landfill Symposium (Cagliari, Sardinia, October 1995), CISA-Environmental Sanitary Engineering Centre, Cagliari.

Ikeguchi T. 1994. Progress in sanitary landfill technology and regulations in Japan: A review. Waste Management and Research, 12, 109–127.

Jefferis S.A. 1995. Old landfills: Perception and remediation of problem sites. In: Sarsby R.W. (Ed.) Waste Disposal by Llandfill. Balkema, Rotterdam.

Joseph J.B. and Mather J.D. 1993. Landfill: Does current containment practice represent the best option? In: Christensen T.H., Cossu R. and Stegmann R. (Eds.) Sardinia 93, Fourth International Landfill Symposium (Cagliari, Sardinia, October 1993), CISA-Environmental Sanitary Engineering Centre, Cagliari, 99–107.

LaGrega M.D., Buckingham P.L. and Evans J.C. 1994. Hazardous Waste Management. McGraw-Hill, New York.

Lee F.G. and Jones-Lee A. 1993. Landfill and groundwater pollution issues: 'Dry tomb' vs. wet-cell landfills. In: Christensen T.H., Cossu R. and Stegmann R. (Eds.) Proceedings Sardinia 93, Fourth International Landfill Symposium, Cagliari, Sardinia, 11–15 October 1993, 1777–1796.

Maier T.B., Steinhauser E.S., Vasuki N.C. and Pohland F.G. 1995. Integrated leachate and landfill gas management. In: Christensen T.H., Cossu R. and Stegmann R. (Eds.) Sardinia 95, Fifth International Landfill Symposium (Cagliari, Sardinia, October 1995), CISA-Environmental Sanitary Engineering Centre, Cagliari.

Making Waste Work, 1995. Department of the Environment and Welsh Office, HMSO, London.

McBean E.A., Rovers F.A. and Farquhar G.J. 1995. Solid Waste Landfill Engineering and Design. Prentice Hall, Englewood Cliffs, NJ.

Moss H. 1997. Dynamotive Technologies UK Ltd., Bedford (personal communication).

Pavelka C., Loehr R.C. and Haikola B. 1993. Hazardous waste landfill leachate characteristics. Waste Management, 13, 573–580.
Pescod, M.B. (Ed.) 1991–93. Urban Solid Waste Management. World Health Organisation, Copenhagen.
Petts J. and Eduljee G. 1994. Environmental Impact Assessment for Waste Treatment and Disposal Facilities. Wiley, Chichester.
Porteous A. 1992. Dictionary of Environmental Science and Technology. Wiley, Chichester.
Proposal for a Council Directive on the Landfill of Waste (Catalogue No. CB-CO-97-095-EN-C) 1997. Brussels 5.3.97, COM(97) 105 Final, Brussels.
Reinhart D.R. 1996. Full-scale experiences with leachate recirculating landfills: Case studies. Waste Management and Research, 14, 347–365.
Reinhart D.R. and Al-Yousfi A.B. 1996. The impact of leachate recirculation on municipal solid waste landfill operating characteristics. Waste Management and Research, 14, 337–346.
Rettenberger G. and Stegmann R. 1996. Landfill gas components. In: Christensen T.H., Cossu R. and Stegmann R. (Eds.) Landfilling of Wastes: Biogas. E. and F.N. Spon, London.
Royal Commission on Environmental Pollution, 1993. 17th Report, Incineration of Waste. HMSO, London.
Rugge K., Bjerg P.L. and Christensen T.H. 1995. Distribution of organic compounds from municipal solid waste in the groundwater downgradient of a landfill (Grinsted, Denmark). Environmental Science and Technology, 29, 1395–1400.
Sarsby R.W. (Ed.) 1995. Waste Disposal by Landfill. Balkema, Rotterdam.
Sharma H.D. and Lewis S.P. 1994. Waste Containment Systems, Waste Stabilisation, and Landfills: Design and Evaluation. Wiley, New York.
Solid Waste Disposal Facility Criteria, 1991. Final Rule, 40 CFR Parts 257 and 258, United States Environmental Protection Agency, Federal Register Vol. 56, No. 196, 9 October, Washington, DC.
Stegmann R. 1996. Landfill gas utilisation: An overview. In: Christensen T.H., Cossu R. and Stegmann R. (Eds) Landfilling of Waste: Biogas. E. and F.N. Spon, London.
Tchobanoglous G. and Burton F.L. 1991. Waste Water Engineering, Treatment, Disposal and Re-Use. McGraw-Hill (Metcalf and Eddy Inc.), New York.
Tchobanoglous G. and O'Leary P.R. 1994. Landfilling. In: Kreith F. (Ed.) Handbook of Solid Waste Management. McGraw-Hill, New York.
The Waste Manager, 1994a. February, p. 22.
The Waste Manager, 1994b. Landfill clean-ups: A booming business. (Data from Frank Graham Consulting Engineers.) Rose J., The Waste Manager, 20–22 March.
The World Resource Foundation, 1995a. Technical Brief: Landfill Techniques. The World Resource Foundation, Tonbridge.
The World Resource Foundation, 1995b. Technical Brief: Landfill Mining. The World Resource Foundation, Tonbridge.
Thorneloe S.A., Pacey J.G. and Doorn M. 1995. Technical and non-technical issues regarding landfill gas to energy: Impact on the US landfill gas industry. In: Christensen T.H., Cossu R. and Stegmann R. (Eds.) Sardinia 95, Fifth International Landfill Symposium (Cagliari, Sardinia, October 1995), CISA-Environmental Sanitary Engineering Centre, Cagliari.
Town and Country Planning Regulations, 1988. Assessment of Environmental Effect. HMSO, London.
Townsend T.G., Miller W.L., Lee H.-J. and Earle J.F.K. 1996. Acceleration of landfill stabilisation using leachate recycle. Journal of Environmental Engineering, April, 263–268.
Van den Brook B., Lambropoulos N.A. and Haggett, K. 1995. Biorector landfill research in Australia. In: Christensen T.H., Cossu R. and Stegmann R. (Eds.) Sardinia 95, Fifth International Landfill Symposium (Cagliari, Sardinia, October 1995), CISA-Environmental Sanitary Engineering Centre, Cagliari.
Ward R.S., Williams G.M. and Hills C.C. 1996. Changes in major and trace components of landfill gas during subsurface migration. Waste Management and Research, 14, 243–261.

Warmer Bulletin 44, 1995. Information Sheet. Journal of the World Resource Foundation, Tonbridge.
Waste Management in Japan, 1995. Department of Trade and Industry, Overseas Science and Technology Expert Mission Visit Report, Institute of Wastes Management, Northampton.
Waste Management Licencing Regulations, 1994. HMSO, London.
Waste Management Paper 26, 1986. Landfilling Wastes. HMSO, London.
Waste Management Paper 28, 1992. Recycling. Department of the Environment, HMSO, London.
Waste Management Paper 26A, 1995. Landfill Completion. Department of the Environment, HMSO, London.
Waste Management Paper 26F, 1994. Landfill Co-Disposal. Draft for external consultation. Wastes Technical Division, Department of the Environment, London.
Waste Management Paper 27, 1994. Landfill Gas. Department of the Environment, HMSO, London.
Waste Management Paper 26B, 1995. Landfill Design, Construction and Operational Practice. Department of the Environment, HMSO, London.
Waste Management Paper 4, 1996. Licencing of Waste Management Facilities. Department of the Environment, HMSO, London.
Waste Management Paper 26E, 1996. Landfill Restoration and Post-Closure Management. Consultation draft, August 1996. Environment Agency, Department of the Environment, London.
Westlake K. 1995. Landfill Pollution and Control. Albion, Chichester.
White P.R., Franke M. and Hindle P. 1995. Integrated Solid Waste Management: A Lifecycle Inventory. Blackie Academic and Professional, London.
Williams A. (Ed.) 1994. Methane Emissions. Watt Committee Report 28, The Watt Committee on Energy, London.
Williams G.M. and Aitkenhead N. 1991. Lessons from Loscoe: The uncontrolled migration of landfill gas. Quarterly Journal of Engineering Geology, 24, 191–207.
Willumsen H.C. 1996. Landfill gas utilisation: Statistics of existing plants. In: Christensen T.H., Cossu R. and Stegmann R. (Eds) Landfilling of Wastes: Biogas. E. and F.N. Spon, London.
Young C.P. and Blakey N.C. 1996. Emissions from landfill gas power generation plants. In: Christensen T.H., Cossu R. and Stegmann R. (Eds.) Landfilling of Wastes: Biogas. E. and F.N. Spon, London.
Yuen S.T.S., Styles J.R. and McMahon T.A. 1995. An active landfill management by leachate recirculation: A review and an outline of a full-scale project. In: Christensen T.H., Cossu R. and Stegmann R. (Eds.) Sardinia 95, Fifth International Landfill Symposium (Cagliari, Sardinia, October 1995), CISA-Environmental Sanitary Engineering Centre, Cagliari.

6
Incineration

Summary

This chapter is concerned with incineration, the second major option for waste treatment and disposal in many countries throughout the world. The various incineration systems are discussed. Concentration is made on mass burn incineration of municipal solid waste, following the process through waste delivery, the bunker and feeding system, the furnace, and heat recovery systems. Emphasis on emissions formation and control is made with discussion of formation and control of particulate matter, heavy metals, toxic and corrosive gases, products of incomplete combustion such as polycyclic aromatic hydrocarbons (PAH), dioxins and furans. The contaminated wastewater and contaminated bottom and flyash arising from waste incineration is described. The dispersion of emissions from the chimney stack are discussed. Energy recovery via district heating and electricity generation are described. Other types of incineration, including fluidised bed incinerators, starved air incinerators, rotary kiln incinerators, cement kilns, and liquid and gaseous waste incinerators, and the types of waste incinerated in the different types, are discussed. The economic aspects of waste incineration are reviewed. Current examples of the different types of incinerator are described throughout.

6.1 Introduction

Incineration of waste in the UK accounts for about 5% of household waste, 7.5% of commercial waste and less than 2% of industrial waste (Making Waste Work 1995). Municipal solid waste incineration has historically been seen in terms of simply a means of waste disposal. However, modern incinerators would now include a means of energy recovery as an economic necessity. Energy recovery is usually by the generation of electricity from high temperature steam turbines or through district heating schemes. Incineration of some commercial and industrial wastes which are

Table 6.1 *Waste incineration in the UK*

Type of incinerator	1991		1996	
	Number	Throughput (1000 tonnes/y)	Number	Throughput (1000 tonnes/y)
Municipal waste	30	2900	6	1900[1]
Clinical waste	700	300	40	200
Sewage sludge	6	75[3]	10[2]	110[3]
Hazardous waste	4	125	4	125
Chemical companies[4]	40	Unknown	50	Unknown

[1] Includes 0.37 million tonnes new incinerator capacity.
[2] Estimated.
[3] Dry weight.
[4] In-house.

Sources: [1] Holmes J., The UK Waste Management Industry, 1995. Institute of Waste Management, Northampton, 1995; [2] Making Waste Work, Department of the Environment and Welsh Office, HMSO, London, 1995; [3] Cosslett G., Going up in smoke. The Waste Manager, July/August. The Environmental Services Association, London, 1996; [4] Paterson B., Clinical waste treatment. In: Institute of Wastes Management, Annual Conference Proceedings, Torbay, June. Institute of Wastes Management, Northampton, 1997.

hazardous and have low throughputs use incineration as a means of disposal and energy recovery is often a secondary objective. In the UK there are a number of incineration plants, ranging from large scale mass burn municipal waste incinerators, through in-house and merchant industrial and commercial incinerators of specialised wastes, to clinical waste incinerators (Table 6.1) (Holmes 1995; Making Waste Work 1995; Cosslett 1996).

The majority of the incinerators in the UK in the early 1990s were clinical waste incinerators, the vast majority of which were associated with hospitals and were of small scale, with low throughputs in the range 1–2 tonnes per hour. However, in 1991 clinical waste incinerators in the UK lost their 'Crown Immunity', which exempted National Health Service hospitals from emissions legislation. In addition, with the introduction of UK legislation on emissions from clinical waste incinerators which was implemented in 1995 under the Environmental Protection Act of 1990, the hospital incinerators became subject to emission levels which they could not meet. As a result, many of the incinerators were closed down, some incinerators were retrofitted with extensive gas clean-up systems, or entirely new plants were commissioned. The current trend for clinical waste incineration is for fewer, larger plants serving two or more hospitals on a regional basis. The incinerators are often built and operated by the private sector, or as private sector plus Hospital Trust joint ventures.

Similar to the decline in numbers of clinical waste incinerators are large scale mass burn municipal solid waste incinerators. In 1991 there were approximately 30 municipal solid waste incinerators operating in the UK. The majority were designed and constructed in the late 1960s and early 1970s with the prime objective of waste disposal, and only six were equipped with some form of energy recovery. These older designed plants had minimal pollution control. However, introduction of EC emissions legislation (Council Directive 89/429/EEC, 1989) which took effect in 1996

Table 6.2 Existing and planned municipal solid waste incineration plant in the UK

Location	Throughput (tonnes/year)	Comments
New plant		
SELCHP, S.E. London	420 000	Completed 1994, 32 MW electricity and district heating for 7500 homes and local facilities
Tyseley, Birmingham	350 000	Completion 1996, 25 MW output electricity generation, total cost £95 million, includes clinical waste incineration
Billingham, Cleveland	220 000	Completion 1997, 20 MW output electricity generation, estimated cost £40 million
Dudley	90 000	Completion 1998, estimated cost £30 million
Stoke	180 000	Completion 1997, estimated cost £50 million, 12.5 MW electricity generation
Wolverhampton	105 000	Completion 1998, estimated cost £30 million
Upgraded plant		
Coventry	280 000	Originally opened 1975
Nottingham	150 000	Originally opened 1973
Sheffield	138 000	Originally opened 1976
Edmonton, N. London	500 000	Originally opened 1969

Source: Cosslett G., Going up in smoke. The Waste Manager, July/August. The Environmental Services Association, London, 1996.

meant closure of many of these plants. However, four incinerators will continue to operate after retrofitting of extensive gas clean-up systems, and a number of new modern design incinerators are under construction or have recently been commissioned (Table 6.2) (Cosslett 1996). A feature of the new plants is the involvement of the private sector through joint venture companies with Local Authorities. Upgrading to meet EC emissions limits for heavy metals, acidic gases, particulates and organic hydrocarbons such as dioxins and furans has involved expensive, sophisticated flue gas clean-up systems, which have resulted in costs of between £10 million and £15 million for the four incinerators upgraded so far. New plants are even more expensive, with capital costs estimated at between £30 million and £95 million. At the core of the new plants is energy recovery to generate income from district heating or electricity generation.

Sewage sludge incineration is predicted to increase because the disposal of sludge at sea, which was the disposal route of 30% of sewage sludge in the UK, is to be phased out by 1998 (Making Waste Work 1995). The main alternative to sea dumping is landspreading of the sludge, which is predicted to increase to account for 66% of UK sewage sludge by 2005. Incineration of the sludge with energy recovery is also predicted to increase, and by the year 2005 to account for about 28% of UK sewage sludge treatment and disposal, equivalent to about 300 000 tonnes of dry solids per year (Making Waste Work 1995).

Hazardous waste incineration in the UK is centred mainly on four private sector incinerators located in Pontypool, Sheffield, Fawley and Ellesmere Port, which together account for 125 000 tonnes per year throughput. In addition, there are

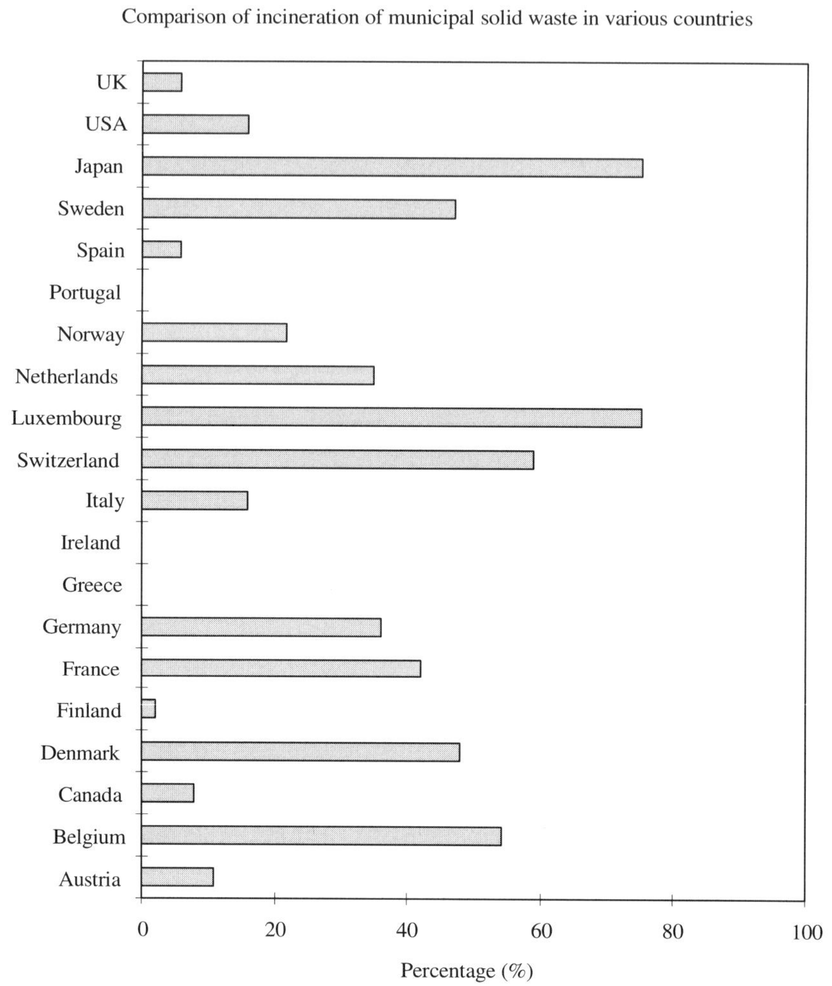

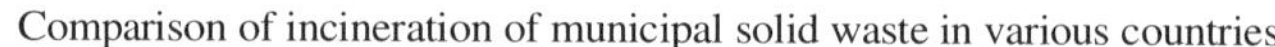

Figure 6.1 *Comparison of municipal solid waste incineration in various countries. Sources: [1] Warmer Bulletin 44, Journal of the World Resource Foundation, Tonbridge, 1995; [2] Making Waste Work, Department of the Environment and Welsh Office, London, 1995*

many in-house incinerators and over 50 in-house hazardous waste and chemical waste incinerators which exist throughout the UK (Holmes 1995).

Comparison of the use of incineration as a waste treatment and disposal option in other countries shows that generally landfill remains the first option, with incineration a distant second (Chapter 5). For example, municipal solid waste incineration accounts for only 19% and hazardous waste incineration 8% of the options used in OECD Europe (Stanners and Bourdeau 1995). However, in some countries such as Japan, Luxembourg, Switzerland, Belgium and Sweden incineration of municipal solid waste accounts for over 50% of the waste disposal generated (Figure 6.1, Warmer Bulletin 44, 1995). Similarly for other wastes such as hazardous waste, sewage sludge and clinical waste, incineration is a significant option in their treatment and disposal in many countries.

Incineration of waste has a number of advantages over landfill.

- Incineration can usually be carried out near the point of collection. The number of landfill sites close to the point of waste generation are becoming scarcer, resulting in transport of waste over long distances.
- The waste is reduced into a biologically sterile ash product, which for municipal solid waste is approximately one-tenth of its pre-burnt volume and one-third of its pre-burnt weight.
- Incineration produces no methane, unlike landfill. Methane is a 'greenhouse gas' and is a significant contributor to global warming.
- Waste incineration can be used as a low cost source of energy to produce steam for electric power generation, industrial process heating, or hot water for district heating, thereby conserving valuable primary fuel resources.
- The bottom ash residues can be used for materials recovery or as secondary aggregates in construction.
- Incineration is the best practicable environmental option for many hazardous wastes such as highly flammable, volatile, toxic and infectious waste.

However, incineration also has disadvantages.

- It generally entails much higher costs and longer pay back periods due to the high capital investment
- There is sometimes a lack of flexibility in choice of waste disposal options once the incineration route is chosen; because of the high capital cost the incinerator must be tied to long-term waste disposal contracts.
- The incinerator is designed on the basis of a certain calorific value for the waste. Removal of materials such as paper and plastics for recycling may reduce the overall calorific value of the waste and consequently may affect incinerator performance. (However, see p. 335)
- Whilst modern incinerators comply with existing emissions legislation there is some public concern that the levels emitted may still have an adverse effect on health.
- The incineration process still produces a solid waste residue which requires management.

6.2 Incineration Systems

The modern incinerator is an efficient combustion system with sophisticated gas clean-up which produces energy and reduces the waste to an inert residue with minimum pollution. Incineration plants may be classified on a variety of criteria, for example, their capacity, the nature of the waste to be combusted, the type of system etc. However, a broad division can be made between mass burn incineration and other types.

Mass burn incineration Large scale incineration of municipal solid waste in a single-stage chamber unit in which complete combustion or oxidation occurs. Typical throughputs of waste are between 10 and 50 tonnes per hour.

Other types of incineration Other types of incineration involve smaller scale throughputs of between 1 and 2 tonnes per hour of wastes such as clinical waste, sewage sludge and hazardous waste. Typical examples of such systems include fluidised bed, cyclonic, starved air or pyrolytic, rotary kiln, rocking kiln, cement kiln, and liquid and gaseous incinerators.

6.2.1 Mass Burn Incineration

Mass burn incineration is used for the treatment and disposal of municipal solid waste throughout the world. The economic viability of incineration as a waste treatment and disposal route for municipal solid waste depends on the recovery of energy from the process to offset the high costs involved in incineration. The properties and composition of municipal solid waste were discussed in Chapter 3. Of particular importance are the 'fuel' properties of the waste, the proximate analysis (ash, moisture, volatile contents) and the ultimate (elemental) analysis, which can be used to assess how the waste will burn in the incinerator and the emissions which are likely to result. Moisture content is obviously important since ignition will not occur if the material is wet, and moisture also diminishes the gross calorific value of a fuel. Volatile matter contains the combustible fraction of the waste and consists of gases such as hydrogen, carbon monoxide, methane, ethane etc., a more complex organic hydrocarbon fraction and an aqueous phase derived by decomposition of water-bound compounds. The ash content is important since a high ash percentage will lower the calorific value of the waste and must be removed and disposed of after combustion. Waste ash is highly heterogeneous and contains inert non-combusted material such as glass and metal cans. The composition of waste may be generally represented in a ternary diagram; Figure 6.2 shows the range of analyses acceptable to the combustion system (Buekens and Patrick 1985; Hall and Knowles 1985). The shaded area represents the typical composition of municipal solid waste which can sustain combustion without the requirement for auxiliary fuel. The area encloses the minimum acceptable calorific value, and the maximum permissible moisture content. The elemental analysis allows emissions from combustion such as sulphur dioxide to be calculated.

A typical modern municipal waste incineration plant with energy recovery is shown in Figure 6.3 (DTI 1996). The incinerator may be divided into five main areas:

1. waste delivery, bunker and feeding system;
2. furnace;

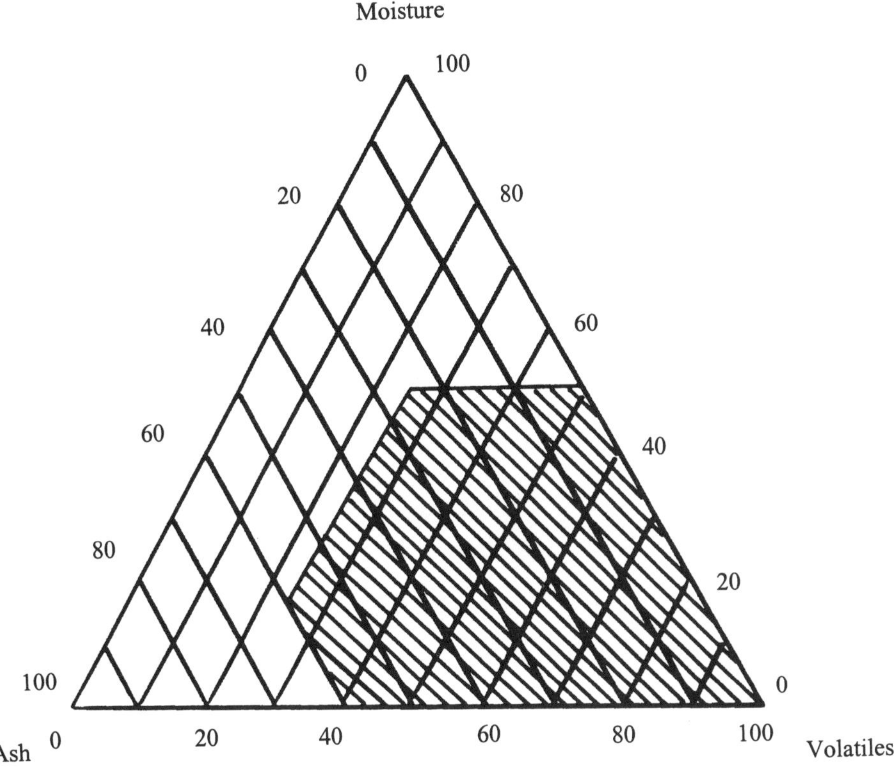

Figure 6.2 *Suitability of municipal solid waste composition for incineration. Sources: [1] Buekens A. and Patrick P.K., Incineration. In: Solid Waste Management: Selected Topics. Suess M.J. (Ed.) World Health Organisation, Copenhagen, 1985; [2] Hall G.S. and Knowles M.J.D., Good design and operation reaps benefits on UK refuse incineration boilers over the last ten years. In: Energy Recovery from Refuse Incineration, Institution of Mechanical Engineers, London, 1985*

3. heat recovery;
4. pollution control;
5. energy recovery via district heating and electricity generation.

6.2.1.1 *Waste Delivery, Bunker and Feeding System*

The waste is usually delivered by collection vehicles, although in some European incinerators, barges or trains may be used. The collection vehicles are weighed on arrival and departure to provide accurate weights of the waste throughput for determining the fees to be charged for disposal and for incinerator operational control. The incinerator may handle a variety of wastes from households, commercial sites and industry, and these would be monitored not only to differentiate the fees charged, but also since they may have very different combustion properties which would influence incinerator performance. Odour may result from the waste and its handling, and therefore plants are normally kept under a slight negative

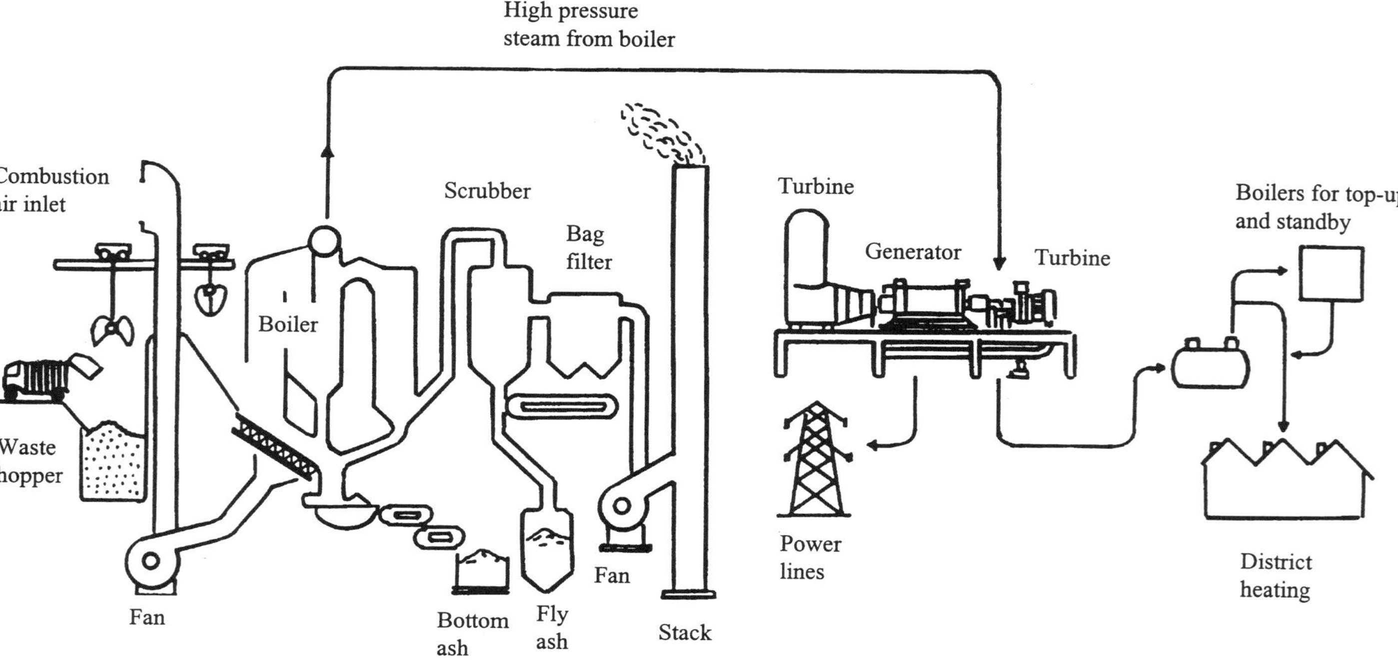

Figure 6.3 *Schematic diagram of a typical mass-burn municipal solid waste incinerator with energy recovery. Source: DTI, Energy from Waste: Best Practice Guide. Department of Trade and Industry, London, 1996. Crown copyright is reproduced with the permission of the Controller of Her Majesty's Stationery Office*

pressure, because the combustion air is taken from the waste storage area, which prevents escape of odour.

The bunker is large enough to allow for storing the waste to ensure a balance between the uneven delivery of the waste and the continuous operation of the plant. Therefore, the bunker would be designed to hold about 2–3 days equivalent weight of waste, which would typically be 1000–3000 tonnes of waste. Longer periods of storage are undesirable due to the rotting of the waste and consequent bad odours. The waste is delivered to the bunker, which may be divided into different sections in separate unloading bays to allow for the mixing of the waste of different calorific values and combustion properties by the crane operator. The crane is of a travelling type, and the crane operator will not only mix the wastes but also extract any bulky or dangerous items from the refuse for separate treatment. The operator then loads the waste into the feeding system. The crane grab can hold up to 6 m^3 of waste.

The feeding system is a steel hopper where the waste is allowed to flow into the incinerator under its own weight without bridging or blocking. The hoppers are kept partly filled with waste to minimise air leakage into the furnace and to ensure there is no interruption of feed to the grate; monitors are used to measure the level of waste in the hopper. To prevent the fire in the furnace burning back up into the feeding hopper, hydraulic shutters are used to seal the hopper at the furnace entrance. Also the feed chute may be water-cooled or refractory-lined to prevent fire.

6.2.1.2 Furnace

Each incinerator may have several furnaces fed by the operator from the waste bunker. For example, a typical 50 tonne/hour incinerator might have five separate 10 tonne/hour furnaces. The use of multiple furnaces allows for down time of the furnace for repair and regular maintenance. The waste is fed into the furnace either by an independently controlled ram, or by the action of the first part of the grate which draws the waste into the furnace (Figure 6.4, Clayton et al 1991). In the furnace the waste undergoes three stages of incineration:

1. drying and devolatilisation;
2. combustion of volatiles and soot;
3. combustion of the solid carbonaceous residue.

In practice the various stages merge, since the components of the waste stream differ in moisture content, thermal degradation temperature, volatile composition, ignition temperature and carbon (fixed) content.

As the waste enters the hot furnace the waste is heated up via contact with hot combustion gases, pre-heated air or radiated heat from the incinerator walls, and initially moisture is driven off in the temperature range 50–100 °C. The water content of waste is very important since heat is required to evaporate the moisture; therefore more of the available calorific value of the waste is lost in heating up the wet waste and less energy is available. In addition, the rate of heating up of the waste and therefore the rate of thermal decomposition will also be affected by the water content of the waste. Water contents of municipal solid waste can vary

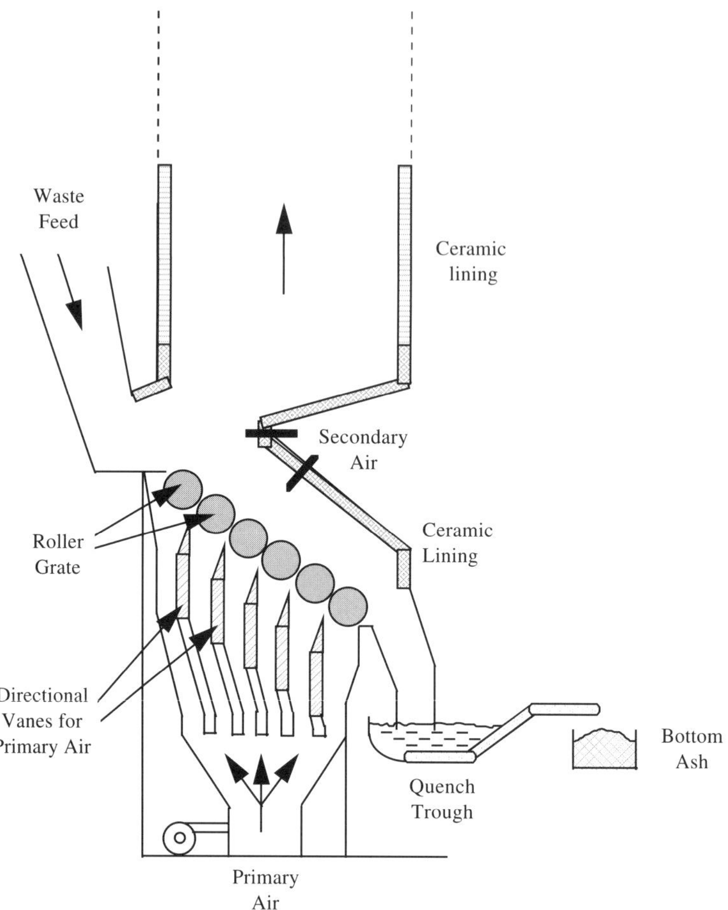

Figure 6.4 *Municipal solid waste incinerator furnace and combustion chamber (roller grate). Source: Adapted from Clayton P., Coleman P., Leonard A., Loader A., Marlowe I., Mitchell D., Richardson S., Scott D. and Woodfield M., Review of Solid Waste Incineration in the UK. Warren Spring Laboratory (National Environmental Technology Centre) Report No., LR776 (PA), Department of Trade and Industry, London, 1991*

between 25 and 50%. After moisture release, the waste then undergoes thermal decomposition and pyrolysis of any organic material such as paper, plastics, food waste, textiles etc., in the waste which generates the volatile matter, the combustible gases and vapours. The volatile components of organic material in municipal solid waste comprise typically between 70 and 90%, and are produced in the form of hydrogen, carbon monoxide, methane, ethane and other higher molecular weight hydrocarbons. Devolatilisation takes place over a wide range of temperatures from about 200 °C to 750 °C, with the main release of volatiles between 425 and 550 °C. Thermal decomposition of the waste and volatile release will also be dependent on the different components present in the waste. For example, polystyrene decomposes over the temperature range 450–500 °C and yields almost 99% volatiles, whereas wood decomposes over the temperature range 280–500 °C and produces about 70% volatiles. In addition to composition, the physical state of the waste will influence the rate of thermal decomposition. For example, cellulosic material in thin form such as paper will decompose in a few seconds, whereas in the form of a large piece of wood may take several minutes to decompose totally.

The combustion of volatiles to produce the flames of the fire takes place immediately above the surface of the waste on the grate and in the combustion chamber above the grate. Complete combustion of the gases and vapours requires a sufficiently high temperature, adequate residence time and excess turbulent air to ensure good mixing. The volatile gases and vapours released immediately ignite in the furnace since the furnace gas temperature will be between 750 and 1000 °C. The ignition temperature of the volatiles derived from the waste are well below these temperatures. Combustion chamber temperatures above about 1200 °C are avoided, as above this temperature ash fusion is likely to occur leading to a build-up of slag on refractory material. Typical mean residence times of the gases and vapours in the combustion chamber are 2–4 s, which compares with typical burn-out times for volatile hydrocarbons of the order of milliseconds. Secondary air is blown in through nozzles above the grate to ensure excess air for combustion and provide turbulence. Excess secondary air is required to avoid areas of zero oxygen levels which serve to pyrolyse rather than combust the hydrocarbons, since these can produce potentially hazardous high molecular weight hydrocarbons and soot. Therefore the distribution and turbulence characteristics of the secondary air are important factors in minimising the formation of pollutants in the combustion chamber.

After the drying and devolatilisation stages, the residue consists of a carbonaceous char and the inert material. The carbonaceous char, which is defined as the fixed carbon rather than the volatile carbon contained in the volatile gases such as methane, ethane and other hydrocarbons, combusts on the grate and may take between 30 and 60 min for complete burn-out. The ash and metals residue is discharged continuously at the end of the last grate section into a water trough and quenched. The handling equipment is subject to heavy wear due to the moist and abrasive nature of the material. The residue should be completely burnt out and biologically sterile. The ash, known as bottom ash, is removed continuously or periodically via a conveyor, and usually disposed of in landfill sites. However, increasingly the bottom ash is being considered for further use to obtain maximum use of the waste, and uses have included secondary aggregate for construction projects such as road building and concrete production. The lighter flyash is transported through the system as particulate material, which will also adsorb metals and organic material as it cools through the incinerator heat recovery and gas clean-up system. The flyash is collected in the cyclones, electrostatic precipitators and bag filters of the gas clean-up system. Because the flyash contains heavy metals, polycyclic aromatic hydrocarbons and dioxins and furans it has no recyclable value and is therefore landfilled.

At the heart of the incinerator is the grate, and a number of different types of furnace grate exist for municipal waste incineration, for example the roller (Dusseldorf) system, reciprocating systems, the reverse reciprocating (Martin)

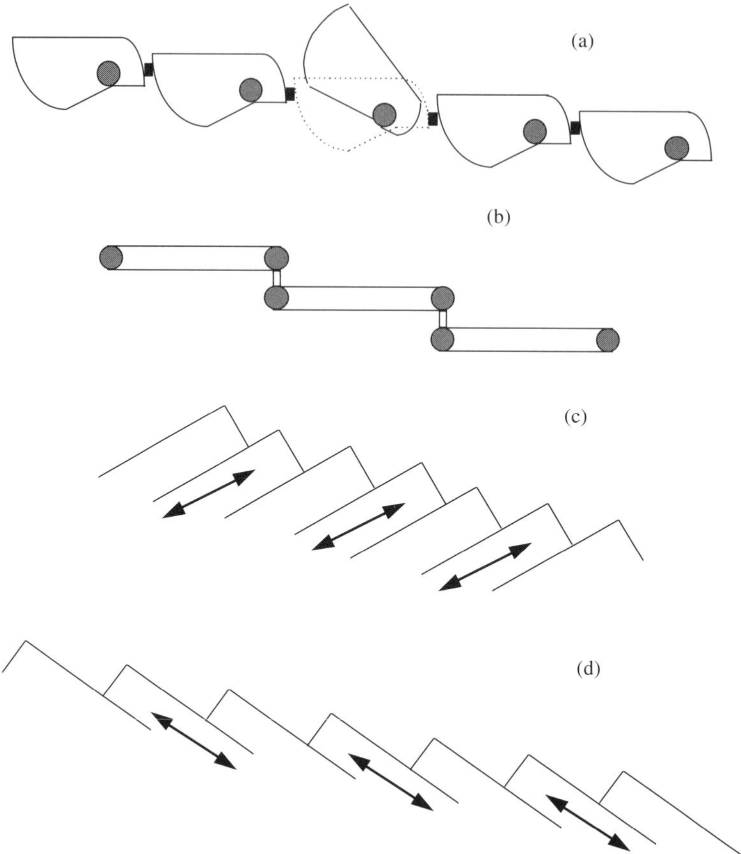

Figure 6.5 *Types of municipal solid waste incinerator grate. (a) Rocker grate, (b) L-stoker grate, (c) Forward-acting reciprocating grate and (d) Reverse-acting reciprocating grate. Source: Adapted from Clayton P., Coleman P., Leonard A., Loader A., Marlowe I., Mitchell D., Richardson S., Scott D. and Woodfield M., Review of Solid Waste Incineration in the UK. Warren Spring Laboratory (National Environmental Technology Centre) Report No., LR776 (PA), Department of Trade and Industry, London, 1991*

systems, the rocker (Nichols) system and continuous (L stoker) systems have all been used in the UK (Figures 6.4 and 6.5, Clayton et al 1991).

The grates are automatic and serve to move the waste from the inlet hopper end to the discharge end, whilst providing agitation or tumbling of the waste. The grate has a variable speed drive to adjust the residence time of the waste in the combustion zone to allow for changes in composition. For example, the roller grate system has a roller diameter of 1.5 m and a circumferential speed of between 5 and 15 m/h. The grates in the roller system are inclined at about 30°, which assists in the movement of the burning waste through the furnace. Horizontal grates are generally arranged in sections of drying, ignition and burn-out, and also assist the distribution and control of primary air. Reciprocating grates consist of three or more sections with a step of about 0.5–1 m between sections. Each section consists of a series of fixed bars and moveable bars in a staircase-like arrangement. The movement of the mobile bars serves to agitate and move the waste down the grate. Control of the air supply to the furnace and combustion chamber is essential for efficient combustion. Primary air is blown evenly through the waste bed via the underside of the grate through slits in the grate which assists in combustion and in cooling the grate. Overgrate or secondary air is introduced through nozzles above the fuel bed, and in some plants tertiary air is added to cool the flue gases before gas cleaning treatment.

The size and shape of the combustion chamber itself are important in determining optimum combustion efficiency, and a number of different designs exist (Figure 6.6, Woodfield 1987). The size determines the mean residence time of the volatiles, and affects their burn-out. The shape affects the heating pattern of the incoming waste from hot flue gases and furnace-wall heat radiation, which influences drying time, ignition time and burn-out time. Shape also influences the gross flow patterns within the chamber, including recirculation and bulk mixing, which in turn influence combustion. The initial drying and devolatilisation of municipal solid waste can be very malodorous, and the flow pattern of the gases and vapours should be through the hottest part of the furnace to combust completely, and therefore destroy, the odorous organic compounds.

The furnace and combustion chamber are lined throughout with refractory materials within a steel outer casing. Between the outer casing and the refractory of the combustion chamber may also be the water tubes of the boiler which generate the steam for energy recovery. The main boiler tubes are located in the main boiler chamber above the combustion chamber through which the hot flue gases flow. The refractory material of the furnace and combustion chamber essentially contains the combustion process in an area which will not fail due to thermal stress or degradation from high temperature corrosion and abrasion. Refractories also re-radiate heat to accelerate drying, ignition and combustion of the incoming waste. The type of refractories used vary with the different parts of the furnace and combustion chamber. This is because at each stage the temperatures and fluctuations in temperature, oxidation and reduction conditions, abrasion from hard objects, erosion from dust-laden flue gases, and corrosion from gases and slags will require different properties from the refractories. For example, the alumino-silicates with high alumina refractories, or silicon carbide bricks are used in the hotter, grate-level part of the furnace, and the upper walls of the combustion chamber would be lower specification alumino-silicate firebricks (Turner 1996).

6.2.1.3 Heat Recovery

The potential for heat recovery from the incineration process is due to the fact that the combustion gases must be cooled before they can be discharged through the flue gas cleaning system. The temperature of the gases leaving the combustion chamber, between 800 and 1100 °C, is too high for direct discharge since gas temperatures below 250–300 °C are required for the gas cleaning equipment such as electrostatic precipitators, scrubbers and bag filters. Cooling is by the integral boiler and boiler chamber system in the modern municipal waste incinerator, although older incinerators where heat recovery was not practised used water injection and air cooling. The

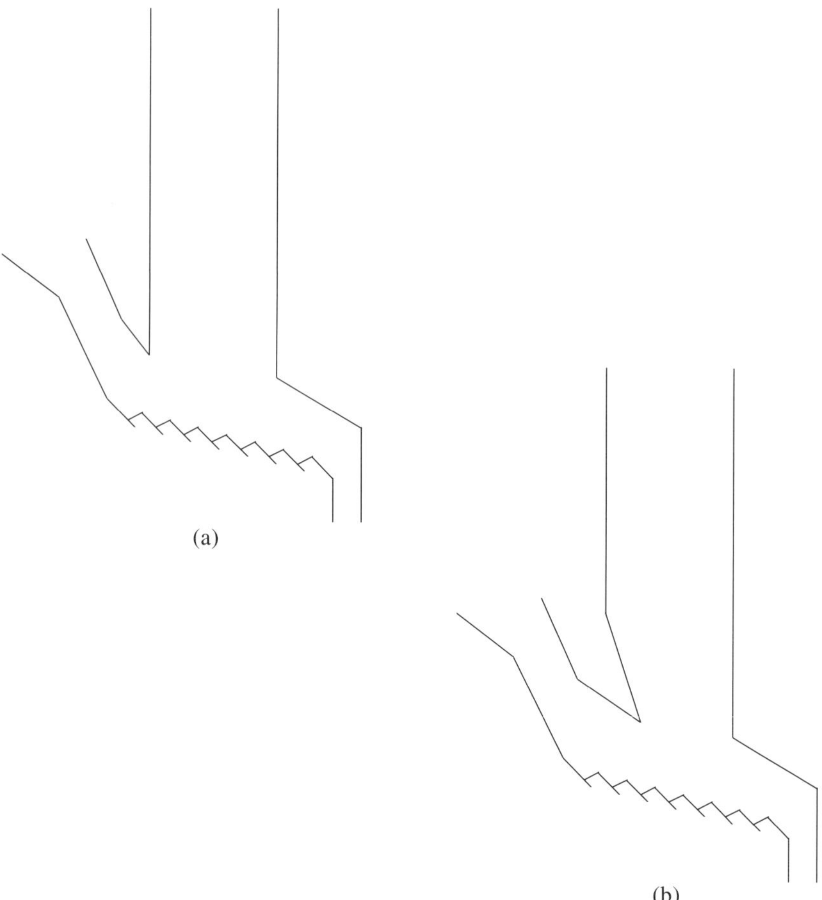

Figure 6.6 *Types of municipal solid waste incinerator combustion chamber. Source: Woodfield M., The Environmental Impact of Refuse Incineration in the UK. Warren Spring Laboratory (National Environmental Technology Centre) Report Department of Trade and Industry, London, 1987*

boiler consists of banks of steel tubes through which water flows to generate steam from the heat generated in the furnace. The integral type or water-wall boilers are constructed around, and integrated with, the combustion chamber; the main boiler is located in a separate boiler chamber above the combustion chamber and the tubes are heated by the hot flue gases. After the boiler there may be an economiser, which is a heat exchange system to heat water in a tube bank to produce further hot water from the flue gases before they enter the flue gas cleaning system. The hot water or steam produced in the boiler may also be used within the plant to provide power and space heating.

In an incineration plant, waste is burned at a more or less constant rate, generally near to the design capacity, and therefore the output of energy cannot be varied to meet fluctuating demand. For space or district heating, utilisation of this energy may

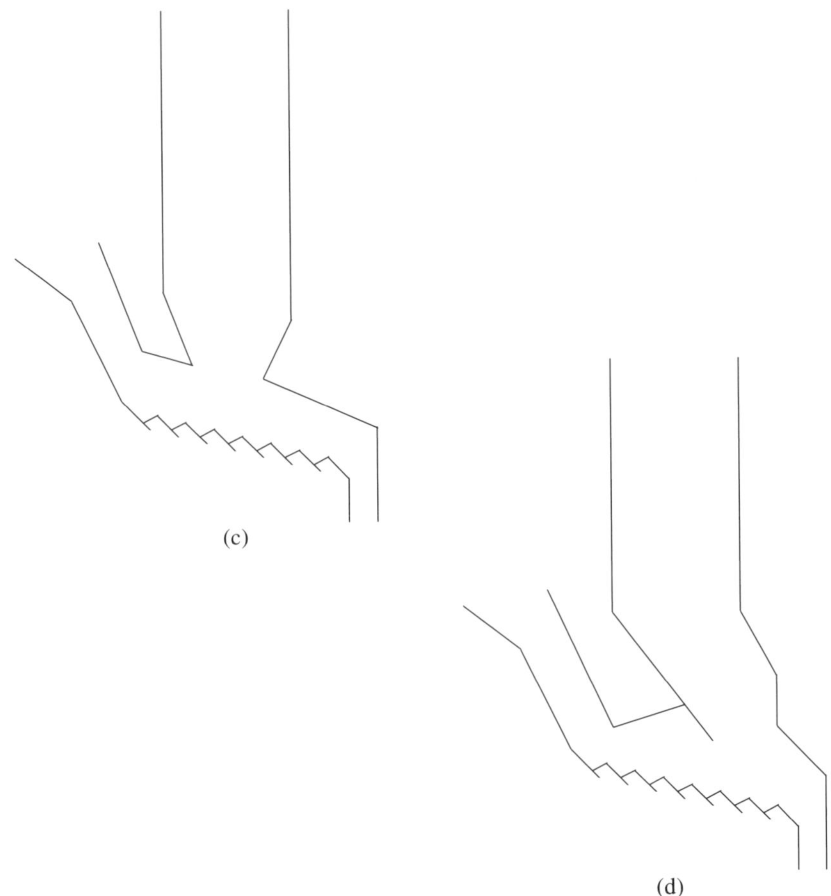

Figure 6.6 *(continued)*

be a problem, whereas electricity generated may be sold to the mains grid. Therefore if continuous output of heat is required, a back-up furnace is required with consequent additional investment and maintenance costs. Similarly, if heat demand is reduced, for example, in the summer months, alternative uses for the heat or a system of flue gas cooling is required so that waste incineration may be continued. The boiler is designed to ensure good heat transfer with the optimum circulation of the water without the occurrence of excessive fouling, to allow for the cleaning of the boiler surfaces, and to be mechanically stable in the operating conditions. Figure 6.7 shows a typical boiler configuration with four passes of the hot gases over the boiler tube bundles (Darley 1996). The temperature of the flue gases is reduced progressively from about 1200 °C just above the grate until eventually the gases leave the fourth boiler/economiser stage at about 250 °C.

A major factor in the efficient operation of the boiler is that the tubes should not be fouled with deposits from the flue gases, which contain flyash, soot, volatilised

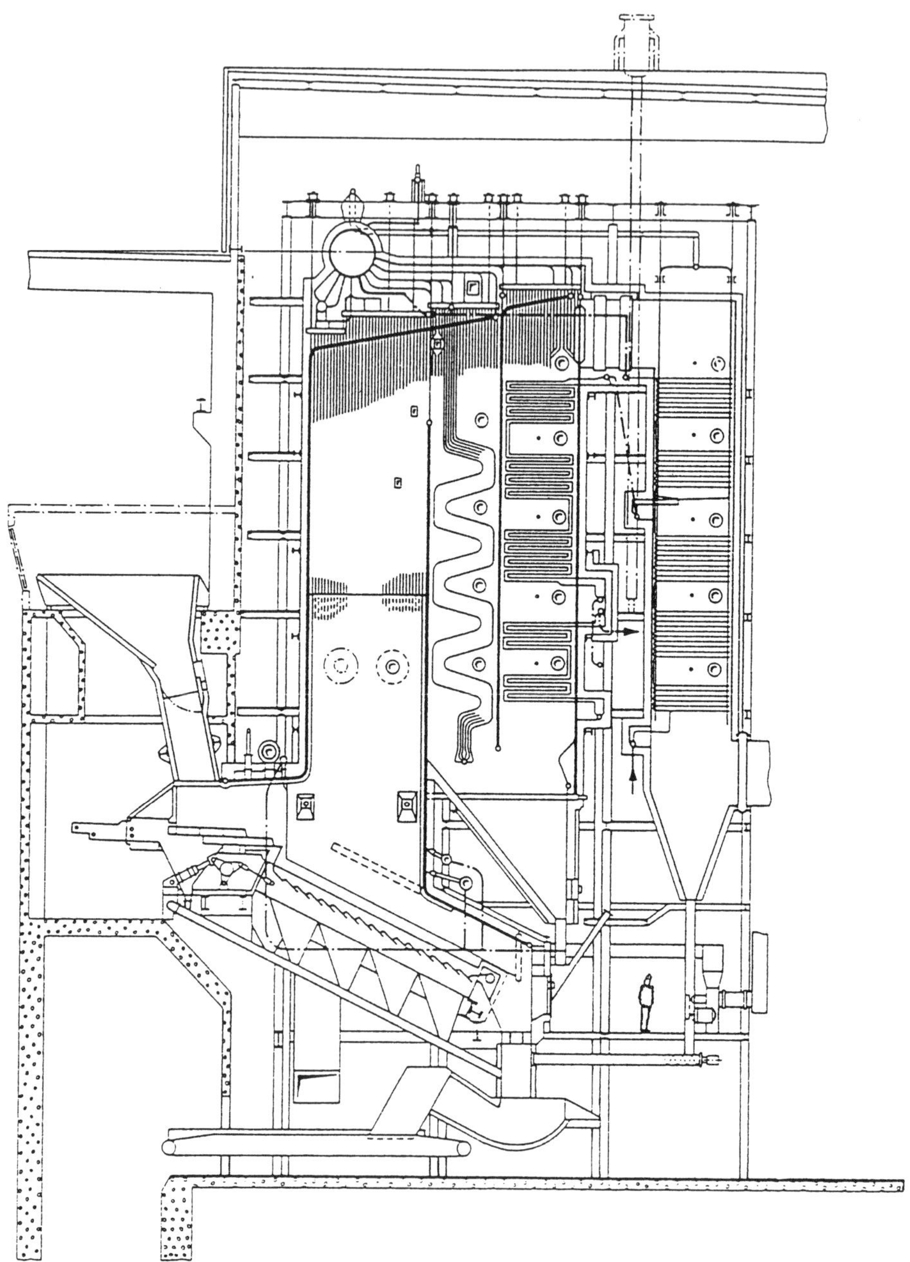

Figure 6.7 *Typical boiler configuration for energy recovery. Source: Darley P. (MAN GHH Ltd., London), Energy Recovery. In: University of Leeds Short Course, Incineration of Municipal Waste with Energy Recovery. Department of Fuel and Energy, University of Leeds, 1996. Reproduced by permission from EVT Energie – und Verfahrenstechnik GmbH*

metal compounds etc. The deposits stick to the boiler tubes and therefore reduce the transfer of heat from the hot flue gases to the water in the steel tubes, and hence the generation of steam and recovery of energy. The rate at which tube fouling deposits build up depends on the dust loading of the flue gases, the stickiness of the flyash, which in turn depends on temperature, the flue gas velocity and the tube bank geometry. The adherence of flyash to boiler tubes is mainly determined by the presence of molten salts such as calcium, magnesium and sodium, sulphates, oxides, bisulphates, chlorides, pyrosulphates etc. in the flyash, and the presence of SO_3 and HCl (Buekens and Schoeters 1984). Scale deposits can be partially removed by means of soot blowers (using superheated steam), by shot cleaning (dropping cast iron shot on the tubes to knock off the deposits) or by rapping the tubes (rapping the tube banks to knock off the deposits). Soot blowers are the most common method and are usually operated once per shift. When the outlet temperature of the flue gases reaches a pre-determined maximum value, the operation has to be halted for a thorough mechanical or wet cleaning of the boiler.

Corrosion is another primary consideration in the design and operation of incinerator boilers (Krause 1991). The formation of HCl by the combustion of chlorine-containing wastes such as paper and board and plastics such as PVC may cause serious corrosion of the tubes due to low-temperature acid corrosion. Precise control of temperature is critical to prevent high-temperature and low-temperature corrosion of the boiler. High-temperature corrosion involves superheater boiler tubes in the boiler chamber at temperatures above 500 °C, and involves a series of chemical interactions between tube metal, tube scale deposits, slag deposits and flue gases. The rate of corrosion is influenced by temperature, the presence of low melting phases such as alkali bisulphates and pyrosulphates, and acid gases such as HCl and SO_3, the nature of the tube metal and the periodic occurrence of reducing conditions. Low-temperature corrosion is due to condensation of acid gases such as HCl and H_2SO_4 formed as the temperature falls below the dew point. The dew point for H_2SO_4 is between about 40 and 155 °C, and for HCl it is between 27 and 60 °C depending on the gas concentration and water content in the flue gas (Krause 1991). Therefore gas temperatures of more than 200 °C are required to minimise down-stream dew-point corrosion.

6.2.1.4 Pollution Control

Following the introduction of EC legislation on emissions from incinerators, which applied to all incinerators throughout Europe from 1996, control of pollution from an incinerator became a major part of the process (Council Directive 89/369/EEC 1989; Council Directive 89/429/EEC 1989). It now constitutes a major proportion of the cost, technological sophistication and space requirement of an incinerator.

Of the pollutant emissions arising from the incineration of waste, those emitted to the atmosphere have received most attention from environmentalists and legislators. Table 6.3 shows emission limits for municipal waste, mass burn incineration for a number of countries (Clayton et al 1991; Sakai et al 1996).

There are a wide variety of emissions limits, but it is clear that the emissions of most concern are total particulate matter or dust, acidic gases such as hydrogen

Table 6.3 *Comparison of municipal solid waste mass-burn incineration emission limits (mg/m³)*

	Germany	Sweden	Netherlands	USA	European Union	
	11% O_2	10% O_2	11% O_2	7% O_2	11% O_2 or 9% CO_2	11% O_2
	1991	1986	1989	1995	1989	Draft
Particulate	10	20	5	24	30	10
HCl	10	100	10	25 p.p.m.	50	10
HF	1	1	1		2	1
SO_2	50	200	40	30 p.p.m. (or 80% removal)	300	50
NOx	200	400	70	150 p.p.m.		
CO	50		50	–	100	100
Total C	10		10		20	20
Dioxin (ng/m³)	0.1 (TEQ)	0.1 (TEQ)	0.1 (TEQ)	13 (0.14–0.21 TEQ at 11% O_2)	850 °C, 2 s $\geq$6% O_2	0.1 (TEQ)
Heavy metals						
Total class I					0.2	
Cadmium	0.05	0.02	0.05	0.02		0.05
Mercury	0.05	0.08	0.05	0.08 (or 85% removal)		0.05
Total Class II					1.0	0.5
Arsenic						
Nickel						
Total Class III		0.05			5.0	
Lead		(Pb + Zn)			(Pb,Cr,Mn,Cu)	
Chromium						
Sample average	24 h	1 month	1 h	4 h		1 month

Sources: [1] Clayton et al., Review of Solid Waste Incineration in the UK. Warren Spring Laboratory (National Environmental Technology Centre), Report LR776 (PA), Department of Trade and Industry, HMSO, London, 1991; [2] Sakai et al., Waste Management, 16, 341–350, 1996.

chloride, hydrogen fluoride and sulphur dioxide, and heavy metals such as mercury, cadmium and lead. In addition, the combustion efficiency is controlled by limits on the emission of carbon monoxide and organic carbon, and in some countries there are also limits on the emission of dioxins. The 1989 European Directive controls the emission of dioxins by setting the combustion conditions of a minimum gas temperature, the residence time and a minimum oxygen level to ensure efficient burn-out of organic compounds. However, the 1993 draft legislation sets a limit of 0.1 ng TEQ/m^3 (TEQ = toxic equivalent, see p. 234). Table 6.3 (and also Table 6.5) shows emissions in terms of reference conditions. All legislative emission limits and plant emissions data are related to a set of reference conditions, such as 9% O_2, 9% CO_2, 11% O_2, 11% CO_2 etc., so that emissions can be compared from plants which may actually be operated under very different conditions. This makes the emission limit data for most countries very similar. For example, Table 6.3 shows USA data

Table 6.4 Typical concentration ranges of emissions from municipal solid waste mass-burn incineration before gas clean-up (mg/m^3)

Emissions	Minimum	Maximum	Mean
Particulate	1500	8000	3000
Hydrogen chloride	400	2200	1150
Hydrogen fluoride	5	20	9
Sulphur oxides	200	2000	500
Nitrogen oxides	150	650	250
Lead	6	55	30
Cadmium	0.3	3.6	1.8
Mercury	0.1	1.1	0.5

Source: Warmer Bulletin 18, Journal of the World Resource Foundation, Tonbridge, June, 1988.

in terms of 7% oxygen; for dioxins the legislative limit is 13 ng/m^3 at 7% oxygen, which equates to about 0.14 ng/m^3 at 11% oxygen.

Table 6.4 shows typical concentration ranges for emissions *before* any gas clean-up treatment (Warmer Bulletin 18, 1988). The emissions are very much higher than are legally permitted and this emphasises the need for efficient and sophisticated gas clean-up to reduce the emissions to below legislated values. The layout of a hypothetical gas clean-up system for a municipal waste incinerator is shown in Figure 6.8. The particulate material is first removed by an electrostatic precipitator and pre-collector, and the acid gases are removed by a lime scrubber which may be of the dry lime or wet lime type. The lime scrubber is followed by the addition of additives such as activated carbon and lime to adsorb mercury and dioxins and furans. This is followed by a fabric filter to remove any fine particulate matter, including flyash and the activated carbon and lime containing the adsorbed pollutants. Finally, the oxides of nitrogen are removed by the addition of ammonia to form inert nitrogen.

The main emissions control equipment used in the incineration of waste is described in Boxes 6.1–6.6. Box 6.1, cyclones; Box 6.2, electrostatic precipitators; Box 6.3, fabric filters; Box 6.4, wet scrubbers; Box 6.5, dry and semi-dry scrubbers; Box 6.6, de-NOx systems.

Table 6.5 shows a comparison of emissions to the atmosphere from older plant and modern plant throughout the world (Clayton et al 1991; Atkins 1996). Older plant built before about 1985 represents older technology where the gas clean-up consisted mainly of particulate removal systems such as electrostatic precipitators. There was little public or legislative concern over acidic gas emissions, heavy metals such cadmium and mercury, or organic micropollutants such as dioxins and furans and polycyclic aromatic hydrocarbons. Consequently, municipal waste incinerators deservedly developed a poor image as pollution sources. However, the modern incinerator built in the late 1980s and 1990s is an efficient combustion plant with a gas clean-up system complying with the highest environmental standards. The older polluting incinerator plants have largely been decommissioned as the introduction of stringent emissions legislation has been enacted throughout the world.

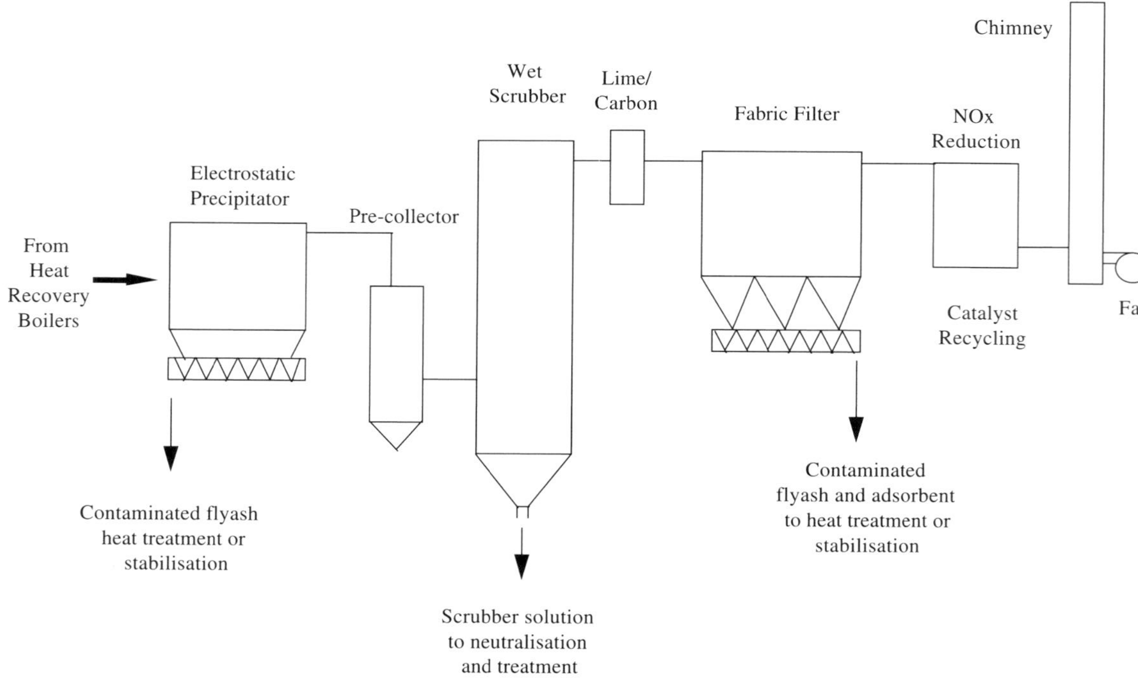

Figure 6.8 *Typical advanced gas clean-up system for a municipal solid waste incinerator. Source: Adapted from Wade J. (ABB Environmental Systems Ltd., Staines), Flue gas emissions control. In: University of Leeds Short Course, Incineration of Municipal Waste with Energy Recovery. Department of Fuel and Energy, University of Leeds, 1996*

Box 6.1
Cyclones

Cyclones are most effective in removing particles larger than 15 μm, but they are much less efficient when it comes to the finer particles. Incinerator particulates have a significant proportion of particles which are less than 15 μm size. In addition, it is the finer grain size that has a concentration of heavy metals and organic micropollutants. Therefore, cyclones are only used as a preliminary collector prior to an electrostatic precipitator or fabric filter to improve the collection efficiency of these more effective trapping systems. The principle of the cyclone is that the dust-laden gas stream enters the cyclone tangentially or axially, forms a vortex and rotates in a helical path down the tube. The particles are flung to the inner wall of the cyclone by centrifugal force and then drop down to the bottom of the cyclone where they are collected. The cleaned gas forms a second vortex which flows up the middle of the cyclone and out through the central inner cylinder (a). Cyclones may be operated in banks of smaller sized cyclones as the collection efficiency for smaller particles is improved with smaller inlet-orifice cyclones (b).

Sources: [1] Porteous A., Dictionary of Environmental Science and Technology. John Wiley and Sons, Chichester, 1992. Figure reproduced by permission from Professor A. Porteous; [2] Clayton et al., Review of Municipal Solid Waste Incineration in the UK. Warren Spring Laboratory (National Environmental Technology Centre), Report LR776 (PA), HMSO, London, 1991. Crown copyright (figure) is reproduced with the permission of the Controller of Her Majesty's Stationery Office.

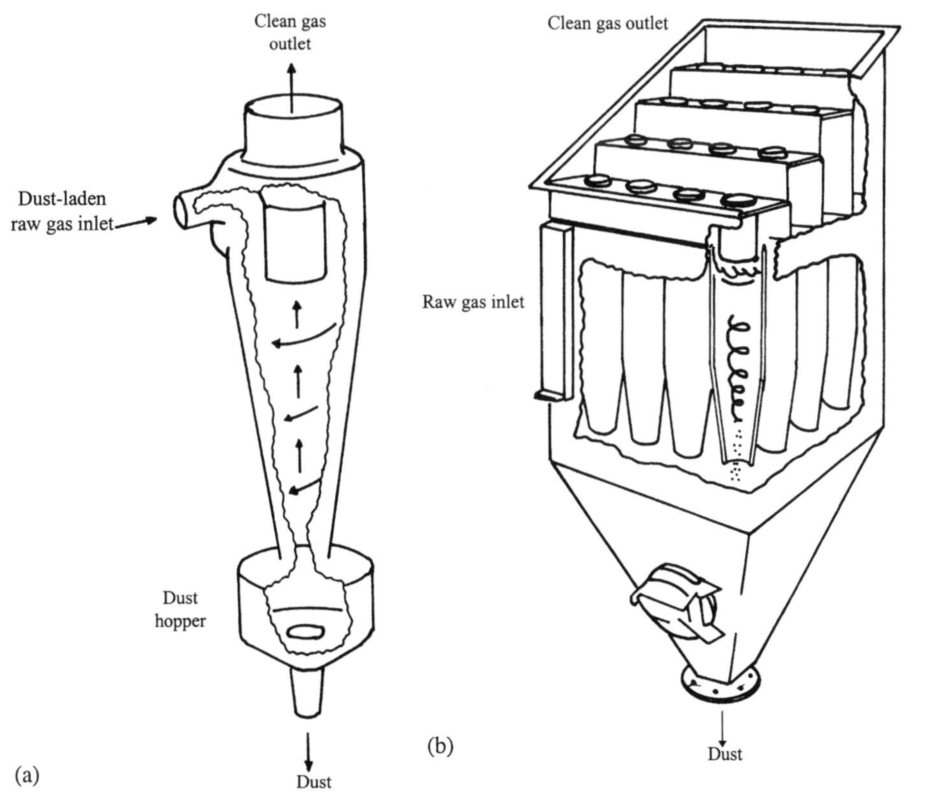

Box 6.2
Electrostatic Precipitators

Electrostatic precipitators are the most common form of particulate removal system for municipal waste incinerators. The dust-laden gas stream enters the electrostatic precipitator where the particles are first charged by negative ions produced by corona discharge in an intense electrostatic field. Typical voltages used are about 50 kV. The particles entering the field become negatively charged by the free electrons. The charged particles are electrically attracted to the collector electrode plates which have an opposite charge, and the particles accumulate. The plates are regularly cleaned by 'rapping' with a rotating hammer system which dislodges the layers of accumulated particles. The dust from the plates falls to the bottom of the precipitator where it collects in hoppers. Typically, the electrostatic precipitator consists of an array of wires or thin metal rods which form the charging electrodes, with the collector plates running between them and separated by a distance of about 25 cm. The size of a typical electrostatic precipitator would be about 7 m^3. Electrostatic precipitators can remove 97–99.5% of the particulates in a gas stream and are extremely efficient down to the sub-micron size range. In addition, the mechanism of electrostatic precipitation applies to liquid and tar droplets, and therefore they can also remove other pollutants in addition to particulates. Electrostatic precipitators must be operated under design conditions, since their efficiency can be markedly affected by flue gas temperature, humidity and the layers of particulates accumulated on the collector electrodes.

Sources: [1] Porteous A., Dictionary of Environmental Science and Technology. John Wiley and Sons, Chichester, 1992. Figure reproduced by permission from Professor A. Porteous; [2] Clayton et al., Review of Municipal Solid Waste Incineration in the UK. Warren Spring Laboratory (National Environmental Technology Centre), Report LR776 (PA), HMSO, London, 1991. Crown copyright (figure) is reproduced with the permission of the Controller of Her Majesty's Stationery Office.

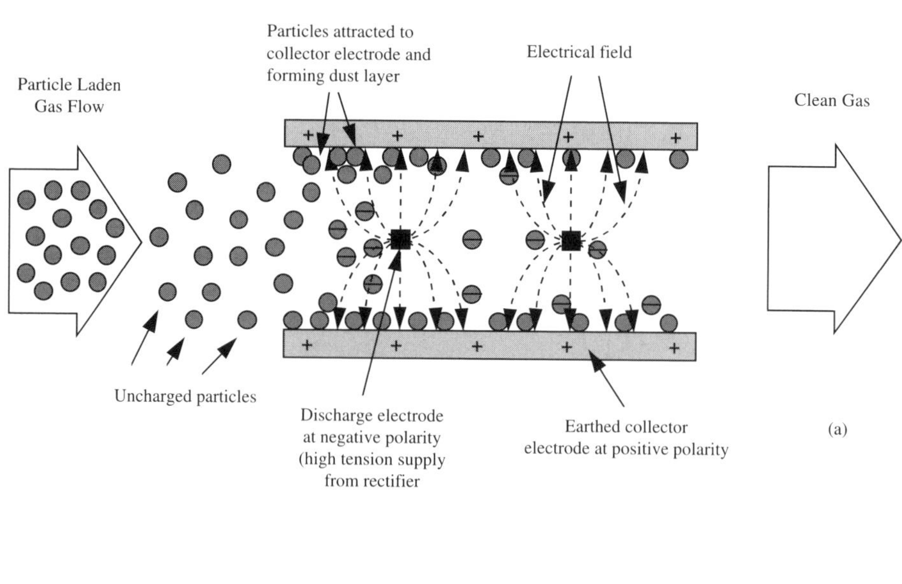

(a)

Box 6.2
(continued)

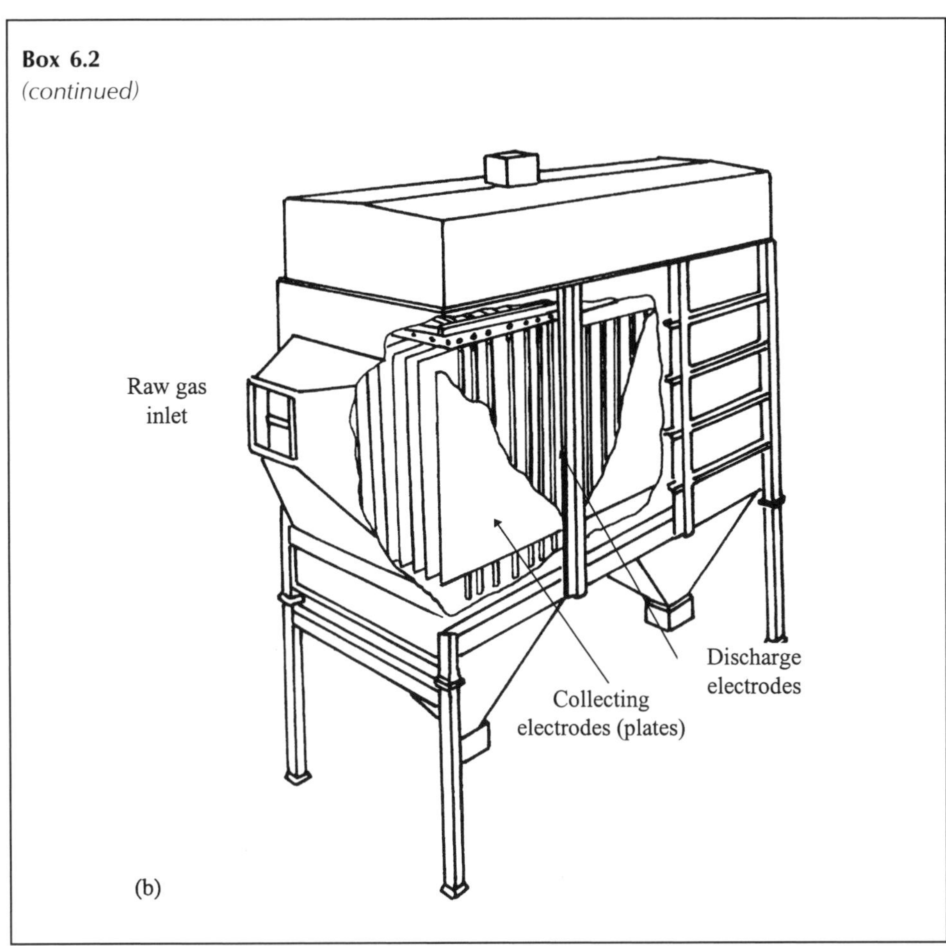

(b)

The emissions to the environment of most concern in relation to mass burn, municipal waste incinerators are those covered by the legislation, and are listed below.

1. Particulate matter.
2. Heavy metals such as mercury, cadmium, lead, arsenic, zinc, chromium, copper, nickel etc.
3. Acidic and corrosive gases such as hydrogen chloride, hydrogen fluoride, sulphur dioxide and nitrogen oxides.
4. Products of incomplete combustion such as carbon monoxide, dioxins, furans and polycyclic aromatic hydrocarbons.
5. Contaminated wastewater.
6. Contaminated ash.

Each will be discussed in turn in relation to their formation, environmental impact and control.

Box 6.3
Fabric Filters

Fabric filters consist of a series of elongated, permeable fabric bags through which the particulate-laden gas flows. The fine fabric filters out the particles from the gas stream. The fabric can be designed with a range of pore sizes, although the filtering process will also include impaction, diffusion and electrostatic attraction in addition to simple filtration. The build-up of particulate matter will also aid removal. The bags, which may number up to 100, are housed in a casing. Because of the fabric construction, the operation temperature of fabric filters is usually low and acid gases can cause damage. They are effective for the removal of particulates down to submicron size and can remove particles from the gas stream to levels of less than 10 mg/m^3, which easily meet current EC legislation. Increasingly, fabric filters are used as the final gas clean-up after the electrostatic precipitator and acid gas scrubber, with a pre-addition of additives such as lime and activated carbon to remove heavy metals such as mercury and organic micropollutants such as dioxins and furans. The fabric filter collects the fine particulates escaping from the electrostatic precipitator and scrubber and the solid additives. Alternatively, the bags themselves can be coated with additives where chemical adsorption or absorption would then take place. The bags are cleaned regularly to remove the accumulated particulate matter by an air pulse, which rapidly expands the bag and releases the dust into a hopper at the base of the baghouse.

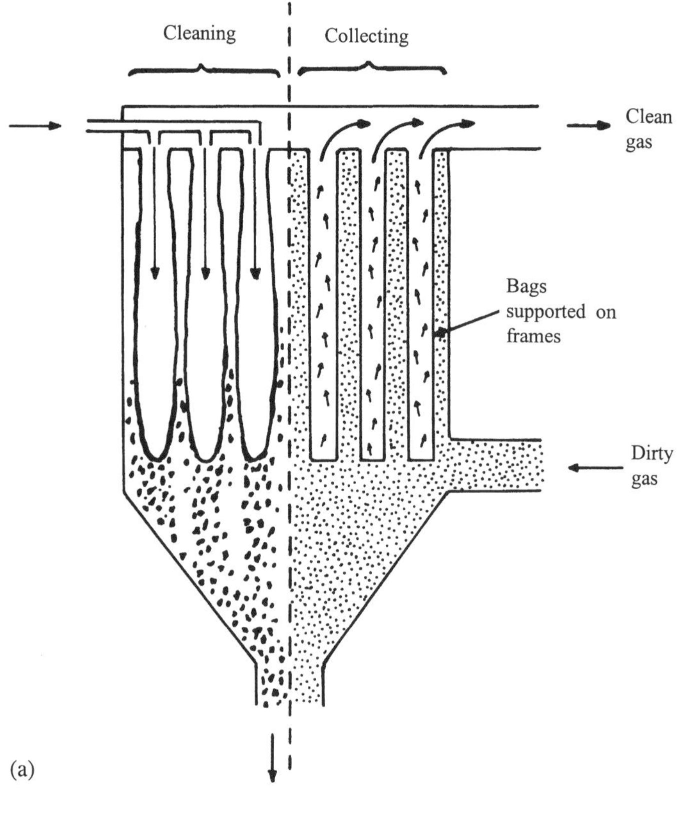

Box 6.3
(continued)

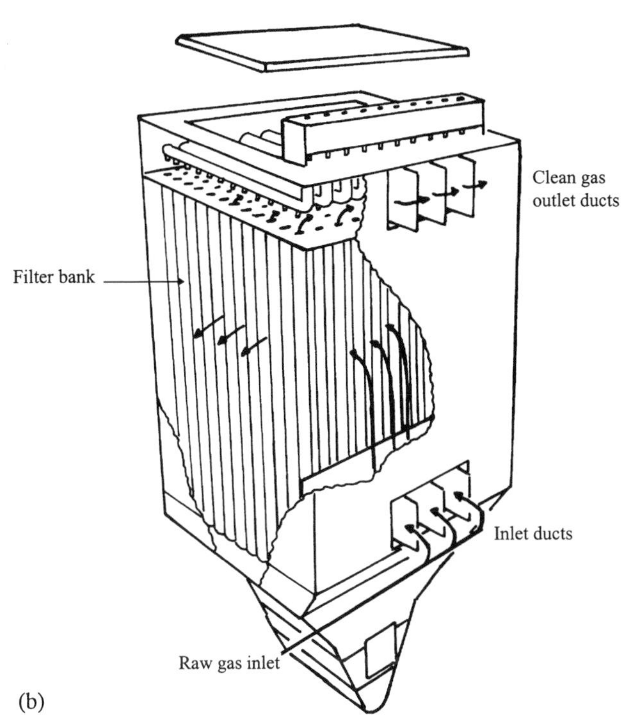

(b)

Sources: [1] Clayton et al., Review of Municipal Solid Waste Incineration in the UK. Warren Spring Laboratory (National Environmental Technology Centre), Report LR776 (PA), HMSO, London, 1991. Crown copyright (figure) is reproduced with the permission of the Controller of Her Majesty's Stationery Office; [2] Wade J. (ABB Environmental Systems Ltd., Staines), Flue gas emissions control. In: University of Leeds Short Course, Incineration of Municipal Waste with Energy Recovery. Department of Fuel and Energy, University of Leeds, 1996. Figure reproduced with the permission of ABB Flakt Industri AB, Vaxjo, Sweden.

Particulate matter The combustion of waste is a very dusty process. The agitation of the waste as it tumbles down the grate, the blowing of primary air through the bed, the high ash content of the waste, and the heterogeneous nature of the waste all serve to produce a high particulate loading in the flue gases (Buekens and Patrick 1985). The design of the incinerator also influences the particulate loading in the flue gases. Such design factors include the size of the incinerator, the grate type, and the combustion chamber design. Particulate emissions from incinerators are the most visual to the public and require efficient and high levels of removal so that complaints

Box 6.4
Wet Scrubbers

Wet scrubbers are used in incinerator clean-up systems to remove soluble acid gases such as hydrogen chloride, hydrogen fluoride and sulphur dioxide. The gases are passed countercurrent through thin films or sprays of liquid in a tower, thus providing a very large surface area for the reaction of the gas. The most common liquid used to trap the acid gases is an alkaline solution such as lime (calcium hydroxide) or sodium hydroxide. Wet scrubbers can also remove other pollutants such as particulates and heavy metals. For example, a typical scrubber design would have a quench unit before the scrubber tower to cool the gases to about 60 °C. The gases are passed to the scrubber tower with a water spray in the first stage, which would absorb the hydrogen chloride and hydrogen fluoride gas to produce hydrochloric and hydrofluoric acid. The acid solution passes down through the tower, removing heavy metals which are soluble in acid solution. In the second stage, which may be in the upper portion of the tower or in a separate tower, the alkaline solution is used to remove any remaining hydrogen chloride and sulphur dioxide. After a de-mister stage to remove liquid carryover, the gases leave the scrubber. Acid

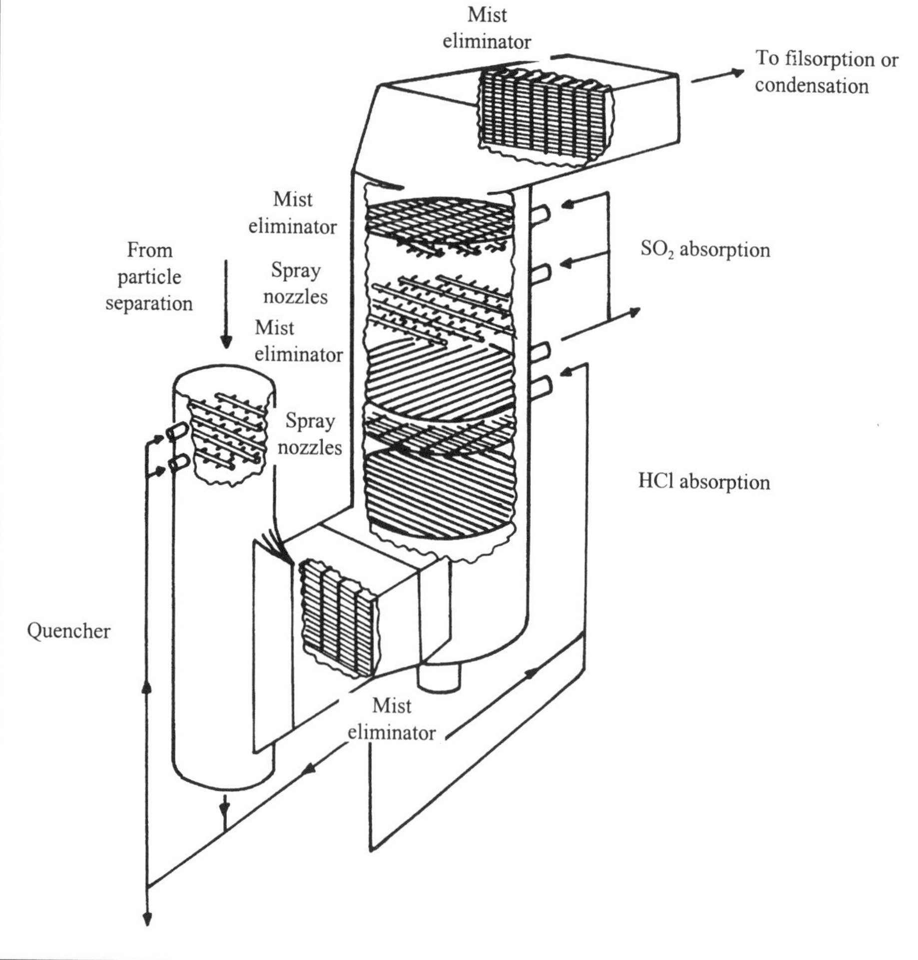

Box 6.4
(continued)

gas removal efficiencies are very high at over 95% for hydrogen chloride, and efficiencies for removing heavy metals are also high, of the order of 99% for lead and 92% for cadmium. The disadvantages of wet scrubbers include the fact that the gases are cool and saturated with water vapour and consequently have to be re-heated prior to discharge from the stack to prevent the formation of a visible plume. A further major disadvantage is that the derived liquids are very polluted and acidic and have to undergo extensive and expensive clean-up before discharge to the sewer.

Source: Wade J. (ABB Environmental Systems Ltd., Staines), Flue gas emissions control. In: University of Leeds Short Course, Incineration of Municipal Waste with Energy Recovery. Department of Fuel and Energy, University of Leeds, 1996. Figure reproduced with the permission of ABB Flakt Industri AB, Vaxjo, Sweden.

do not arise. The particulate matter is largely composed of ash, but in addition, pollutants of a more toxic nature such as heavy metals and dioxins and furans are associated with particulate matter, either as individual solid particles or adsorbed on the surface of the particles.

The particulate matter may also contain carbon and adsorbed acidic gases such as hydrochloric, sulphuric or even hydrofluoric acid to produce corrosive acid 'smuts'. Soot is formed when carbon-containing wastes are combusted in conditions of high temperature and low oxygen content. Polycyclic aromatic hydrocarbons (PAH) have been cited as chemical intermediaries in soot formation, and an alternative proposed mechanism is via acetylene radicals which build to form large soot molecules (A. Williams 1990). The control of soot formation is via adequate residence time for the combustion process completely to burn out any soot being formed, with good mixing of the primary and secondary combustion air.

The emission of untreated flue gases would give rise to a dark plume and the deposition of dust downwind of the incinerator stack. The size range of incinerator particulates found in the flue gases is from $<1\ \mu m$ to $75\ \mu m$, the larger particles tending to settle out before they reach the flue (Niessen 1978). It is the ultrafine particles that are of particular concern in assessment of health effects, since they contain ash and adsorbed acid gases, heavy metals and organic micropollutants which, because of their size, can pass deep into the respiratory system of humans. There is currently some concern that the important factor in determining the deleterious effects of fine particles is not particularly their composition but their ultrafine nature and the fact that they can penetrate deep into the lungs. A separate size category of particulate matter of environmental concern, PM_{10} (particulate matter), has been designated for particulate matter of less than $10\ \mu m$ in size. A large fraction of municipal waste incinerator particulates are of such small size (Niessen 1978). In addition, their small size promotes both short- and long-range dispersion from the chimney stack into the environment (Denison and Silbergeld 1988).

Particulate emissions from mass burn incinerators are controlled by a range of possible equipment which is effective in removing particulate material. These include

Box 6.5
Dry and Semi-dry Scrubbers

The main disadvantage of wet scrubbers is the need for extensive treatment of the derived highly polluted liquid, which increases the costs of gas clean-up. Therefore, the dry and semi-dry scrubbers were developed for the treatment of acid gases. The dry system (a) utilises dry fine-grained powder, for example, dry calcium hydroxide, which is sprayed onto the flowing gases. The incoming gases to a dry scrubber tower are cooled to about 160 °C. The reaction of the acid gases takes place in the dry state, to produce, for example, calcium chloride and calcium sulphate with hydrogen chloride and sulphur dioxide, respectively, which pass to the bottom of the tower for treatment or recycling. The system is used in conjunction with a down-stream fabric filter which traps the particulate material. In addition, the dry alkaline powder trapped on the fabric filter presents an additional surface for reaction of the acid gases as they pass through the filter. The dry system is used for acid gases, but can also be designed to include heavy metals such as cadmium and mercury and organic micropollutants such as

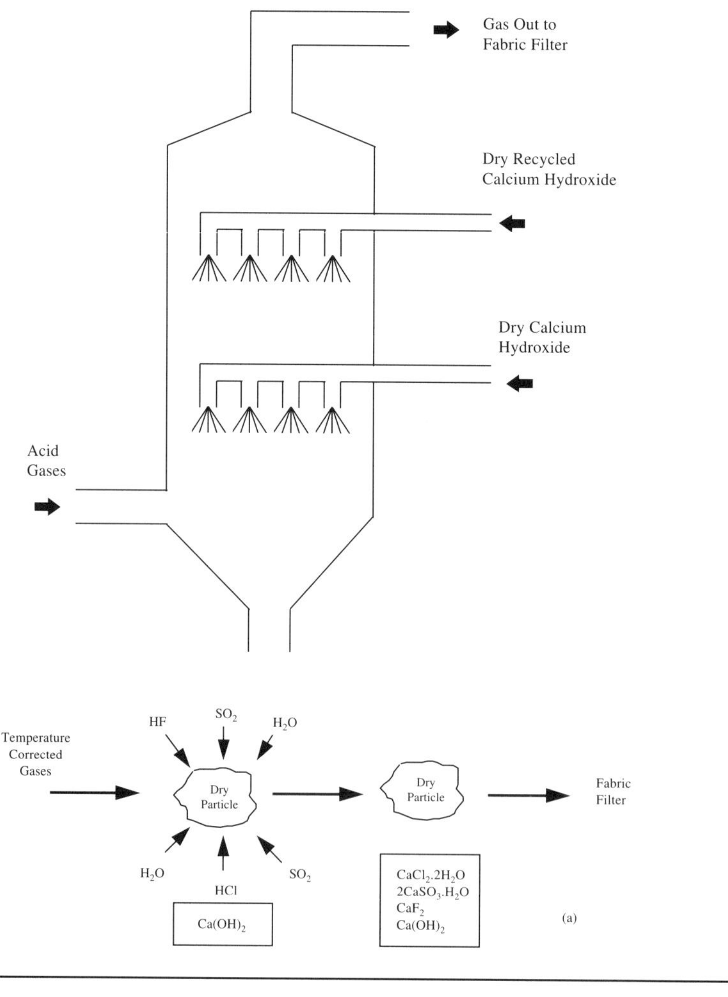

Box 6.5
(continued)

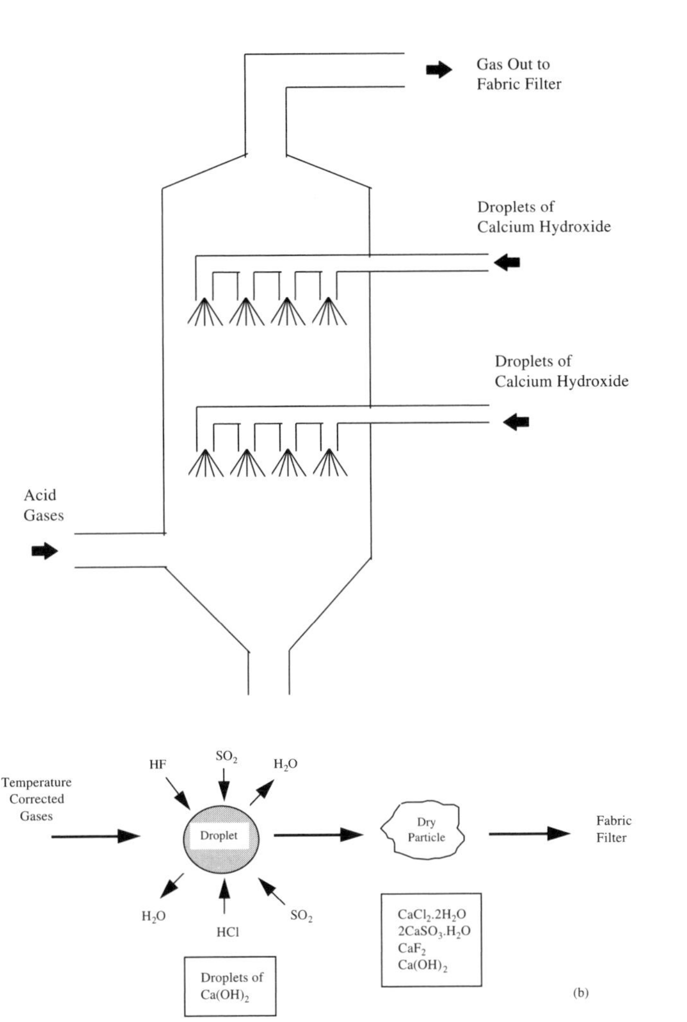

dioxins by the addition of activated carbon to the calcium hydroxide. The semi-dry scrubber (b) utilises a spray of droplets of calcium hydroxide which react with the acid gases, but as the droplets pass through the scrubber tower the water in the droplets evaporates to form fine-grained solid particles. Again the semi-dry scrubber is used in conjunction with a down-stream fabric filter. Dry and semi-dry scrubbers used in conjunction with fabric filters are very effective in the removal of acid gases and, when used with an activated carbon additive, volatile heavy metals and dioxins and furans.

Sources: [1] Clayton et al., Review of Municipal Solid Waste Incineration in the UK. Warren Spring Laboratory (National Environmental Technology Centre), Report LR776 (PA), HMSO, London, 1991; [2] Wade J. (ABB Environmental Systems Ltd., Staines), Flue gas emissions control. In: University of Leeds Short Course, Incineration of Municipal Waste with Energy Recovery. Department of Fuel and Energy, University of Leeds, 1996. Figures adapted from Wade, 1996.

Box 6.6
De-NOx Systems

Nitrogen oxides (NOx) are formed in combustion processes, and because municipal waste incinerators are 'large combustion plant' they are subject to NOx regulation. NOx emissions in flue gases are most common as NO (nitrogen oxide, typically about 90%) and NO_2 (nitrogen dioxide, typically about 10%). NOx forms from nitrogen in the air used for combustion and in the fuel itself, with higher levels of NOx being formed at higher combustion temperatures and higher fuel nitrogen. NOx contribute to acid rain and act as a photochemical oxidant in the atmosphere. Reduction of NOx can be achieved by restricting its formation by control of combustion conditions. Lower temperatures and lower oxygen levels reduce NOx formation. Non-catalytic reduction of NOx can be achieved by the introduction of ammonia in a narrow temperature range of 870–900 °C. The ammonia reduces the NOx to nitrogen and water. At higher temperatures the ammonia itself breaks down to produce NOx, and at lower temperatures the reaction is too slow to be effective. Flue gas NOx are more often controlled by a selective catalytic reduction (SCR) process in the presence of added ammonia, which reproduces the ammonia reduction reaction but at a lower temperature and wider temperature range. Typical catalysts are platinum, palladium, vanadium oxide and titanium oxide. The ammonia which is added up-stream of the catalyst bed reacts with the NOx in the presence of the catalyst at 300–400 °C to produce nitrogen and water. Since the catalyst can be deactivated by heavy metals, the de-NOx system is located after the fabric filter. Selective catalytic reduction can reduce NOx levels by over 90%.

Source: Wade J. (ABB Environmental Systems Ltd., Staines), Flue gas emissions control. In: University of Leeds Short Course, Incineration of Municipal Waste with Energy Recovery. Department of Fuel and Energy, University of Leeds, 1996.

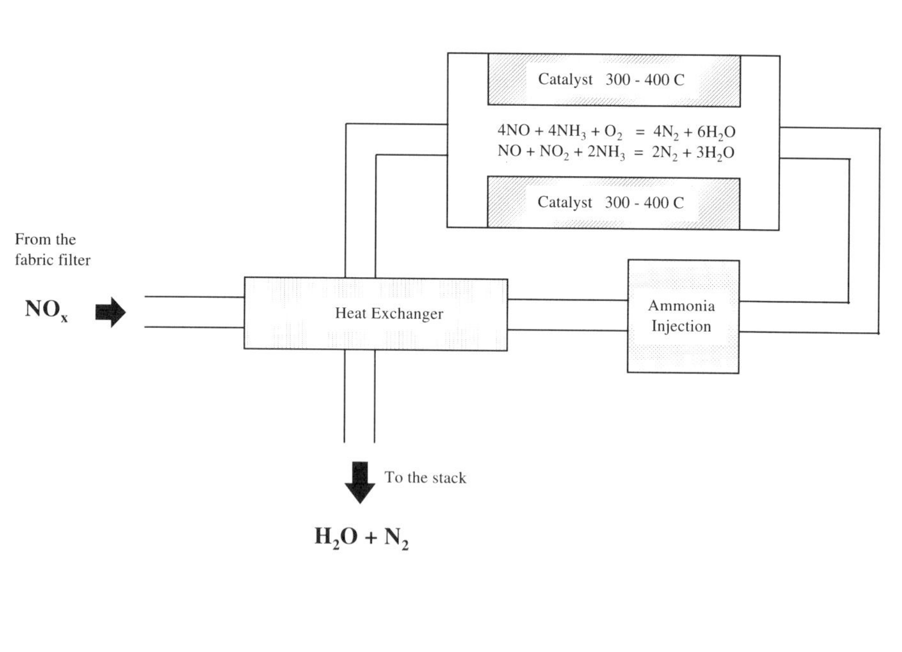

Table 6.5 *Emissions to the atmosphere comparing older plant with typical modern municipal solid waste incinerators*

Emission (mg/m^3)	UK		Sweden		Canada modern plant[5]	Germany modern plant[6]
	Older plant[1] (range)	Modern plant[2]	Older plant[3] (range)	Modern plant[4]		
Particulates	16–2800	1.4	1–90	1.2	–	15
CO	6–640	9	–	–	–	–
HCl	345–950	11	450–900	25	–	<2
SO_2	180–670	8	90–360	17	–	–
HF	–	0.01	4.5–9	<2	–	–
NOx	–	370	180–360	–	–	–
Pb	0.1–50	0.05[a]	0.45–2.7	0.06	0.055	0.358
Cd	<0.1–3.5	0.001	0.045–0.9	0.002	0.004	0.026
Hg	0.21–0.39	0.005	0.27–0.36	0.09	0.02	0.067
TCDD ng/m^3	0.73–1215	0.02[b]	–	–	0.0	–
TCDF ng/m^3	6.84–1425	–	–	–	0.1	–
PAH $\mu g/m^3$	–	–	0.9–90	–	0.1	–

Reference conditions:
[1] STP dry gas, 273 K, 101.3 kPa.
[2] 11% O_2, 273 K, 101.3 kPa, dry gas.
[3] 9% CO_2, 273 K, 101.3 kPa, dry gas.
[4] 9% CO_2, 273 K, 101.3 kPa, dry gas.
[5] 9% CO_2, 273 K, 101.3 kPa, dry gas.
[6] 11% CO_2, 273 K, 101.3 kPa, dry gas.
[a] Total of the heavy metals arsenic, chromium, copper, lead, manganese, nickel and tin.
[b] Toxic equivalent.

Sources: [1] Clayton et al., Review of Municipal Solid Waste Incineration in the UK. Warren Spring Report (National Environment Technology Centre), LR776 (PA), Department of Trade and Industry, HMSO, London, 1991; [2] Atkins G. (SELCHP Ltd., London), Energy from waste: The best practicable environmental option. In: University of Leeds Short Course, Incineration of Municipal Waste with Energy Recovery. Department of Fuel and Energy, University of Leeds, 1997.

cyclones, electrostatic precipitators and fabric filters (Boxes 6.1–6.3). However, because of the need to have sophisticated equipment, such as wet scrubbers and fabric filters, further down-stream to trap heavy metals and organic micropollutants which are in the solid phase or adsorbed on fine ash particles, the most common initial particulate removal apparatus is an electrostatic precipitator.

Heavy metals Metals and metal compounds are present in the components of raw waste. For example, municipal refuse may contain lead from lead-based paints, mercury and cadmium from batteries, aluminium foil, lead plumbing, zinc sheets, volatile metal compounds etc. Table 6.6 shows the range of trace components found in municipal solid waste illustrated by examples taken from Europe and the USA (Law and Gordon 1979; Lorber 1985). High levels occur and the concentrations are very variable. Metals and metal compounds may evaporate in the furnace and condense eventually in the colder parts of the flues and generate an aerosol of submicron particles, or they may become adsorbed onto flyash particles through a

Table 6.6 *The range of trace components in municipal solid waste (g/tonne)*

Trace component	USA[1]	Europe[2]
Iron (Fe)	1000–3500	25000–75000
Chromium (Cr)	20–100	100–450
Nickel (Ni)	9–90	50–200
Copper (Cu)	80–900	450–2500
Zinc (Zn)	200–2500	900–3500
Lead (Pb)	110–1500	750–2500
Cadmium (Cd)	2–22	10–40
Mercury (Hg)	0.7–1.9	2–7

Sources: [1][1] Law S.L. and Gordon G.E., Environmental Science and Technology, 13, 432–438, 1979; [2][2] Lorber K.E., In: Sorting of Household Waste and Thermal Treatment of Waste. Ferranti M.P. and Ferrero G.L. (Eds.) CEC, Elsevier Applied Science, London, 1985.

range of processes (Figure 6.9, Barton et al 1990). The extent of evaporation of these metals and metal compounds in the furnace depends on complex and interrelated factors such as operating temperature, oxidative or reductive conditions, and the presence of scavengers, mainly halogens such as chlorine (Buekens and Patrick 1985). The volatility of these metals and salts is low, for example: Cd, 765 °C; Hg, 357 °C; As, 130 °C; $PbCl_2$, 950 °C; $HgCl_2$, 302 °C. However, for some compounds the volatility temperatures are not known.

The distribution of the metals in the various outputs from municipal waste incinerators have been investigated by a number of workers. Figure 6.10 shows the distribution of heavy metals as a mass balance into and out of an incinerator equipped with an electrostatic precipitator as the only gas clean-up measure, in terms of that fraction either emitted to the flue gas, or captured in the electrostatic precipitator or the slag from the furnace (Brunner and Monch 1986). It is suggested that the partitioning is a function of the physico-chemical properties of the elements and their derived compounds, so that volatile mercury and cadmium compounds with high vapour pressures and low boiling points are most likely to be found in the flue gas. Metals with a medium vapour pressure and boiling point, such as lead and zinc, are retained better in the slag and are less concentrated in the electrostatic precipitator dust. Other metals with low vapour pressure and high boiling points, such as iron and copper, are almost completely trapped in the slag. The speciation of these metals in the incinerator off-gas is strongly influenced by the presence of compounds of chlorine, sulphur, carbon, nitrogen, fluorine and others during combustion and gas cooling. The off-gases containing metals and chlorine species, particularly hydrogen chloride, lead to the formation of metal chlorides. For example, cadmium is easily volatilised during incineration and is oxidised in the presence of hydrogen chloride to give cadmium chloride as the main product (Vogg et al 1986). Mercury has also been shown to be present largely in the halogenated form, predominantly mercury (II) chloride and to a lesser extent mercury (I) chloride. Whilst initially mercury is vaporised as the metal in the furnace, it quickly

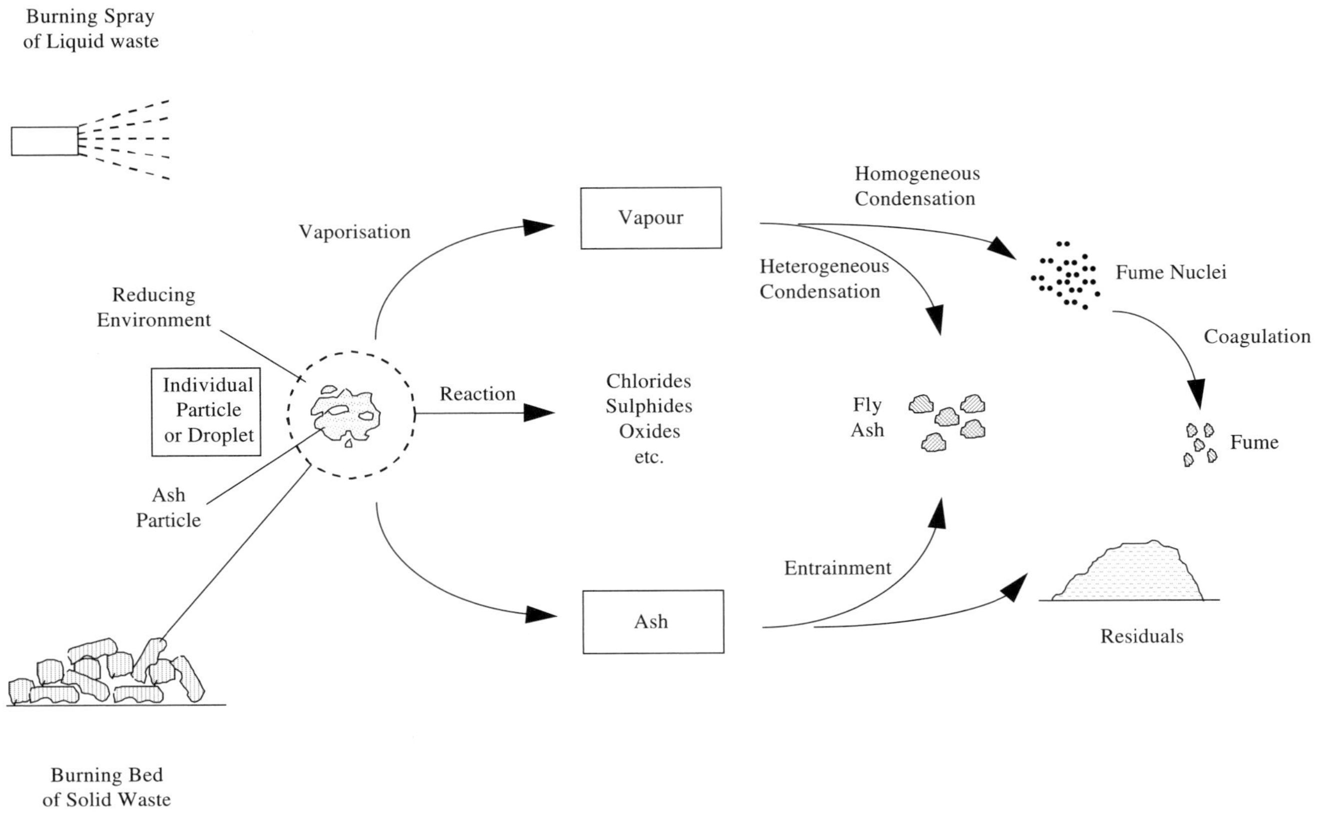

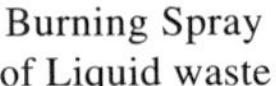

Figure 6.9 *Metals behaviour during incineration. Source: Barton R.G., Clark W.D. and Seeker W.R., Combustion Science and Technology, 74, 327–342, 1990. Reproduced by permission from Gordon and Breach Science Publishers, Lausanne, Switzerland*

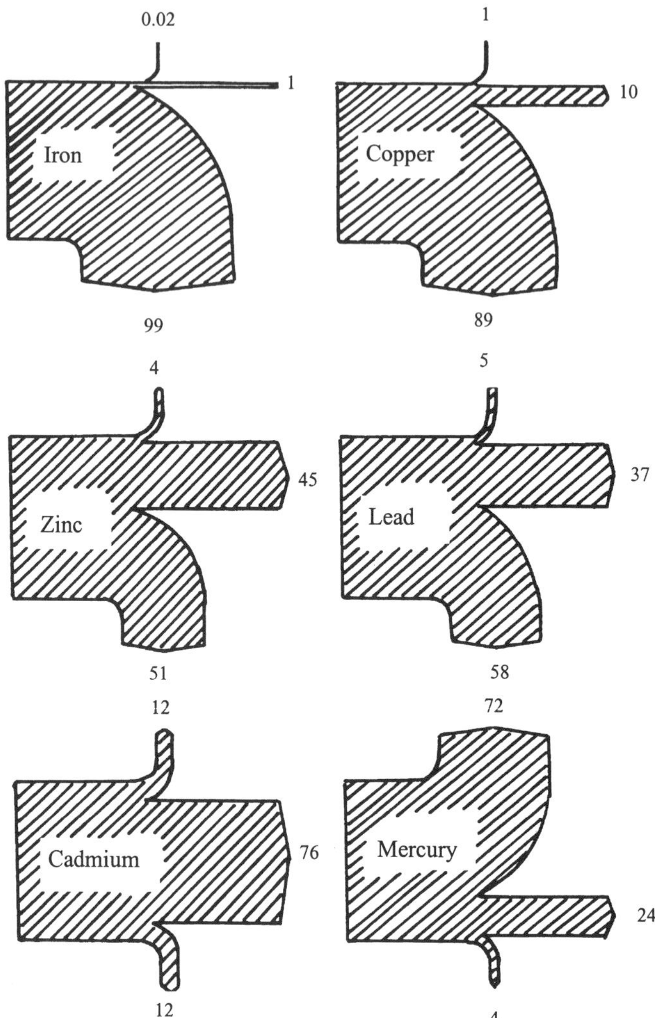

Figure 6.10 The partitioning of metals during municipal solid waste incineration to flue gas, electrostatic precipitator and bottom ash. Source: Brunner P.H. and Monch H., Waste Management and Research, 4, 105–119, 1986. Reproduced by permission from ISWA – International Solid Waste Association

becomes oxidised to the halogenated form and only a small percentage is present as metal vapour (Vogg et al 1986).

The gas clean-up systems required to control heavy metals are dependent on the metal's volatility. Measures used to control total particulate emissions, such as electrostatic precipitators and fabric filters, will collect the associated heavy metals which are in the flyash, are adsorbed to the surface, or are discrete heavy metal particles. The heavy metals are associated with the particulate matter, because of the volatilisation of metals during the combustion of the waste, subsequent condensation

Table 6.7 *Emission levels and removal efficiencies for heavy metals for different gas cleaning systems*

	Wet scrubber + electrostatic precipitator		Dry injection + fabric filter	
Heavy metal	Emission level (mg/m^3)	Removal efficiency (%)	Emission level (mg/m^3)	Removal efficiency (%)
Cadmium	0.020	98.5	0.00014	99.98
Lead	0.240	98.8	0.012	99.9
Zinc	1.300	98.3	0.090	99.8
Mercury	0.150	50.0	0.050	85.0

Source: Carlsson K., Waste Management and Research, 4, 15–20, 1986.

at lower temperatures and adsorption onto the fine particulates in the flue gas. The heavy metals tend to concentrate in the finer grained size fraction of the particulates (Greenberg et al 1978; Bouscaren 1988).

The more volatile heavy metals, particularly mercury and to some extent cadmium, require more sophisticated emissions control equipment. Table 6.7 shows a comparison of different gas clean-up systems from two municipal solid waste incinerators in Europe, with particular regard to the heavy metal emissions (Carlsson 1986). The two clean-up systems consisted of a wet scrubber with a spray tower of hydrated lime (calcium hydroxide) slurry coupled with an electrostatic precipitator, and a dry scrubber of injection of hydrated lime coupled with a fabric filter. Both systems produced emission levels below the regulated EC limits. The highest collection efficiency for the metals under consideration was the dry injection of hydrated lime with a fabric filter system, with over 99.8% collection efficiency for most heavy metals. The exception is the more volatile mercury, where only 85% removal occurred because of the high vapour pressure of the derived mercury and mercury compounds.

To meet the legislative requirements for mercury emissions it may in some cases be necessary for some form of reagent addition upstream of the scrubber system and fabric filter. Additives which have proven effective are sodium sulphide, TMT 15 (trimercapto-s-triazine) and activated carbon. The additive is added at concentration levels of between 0.1 and 0.5 g/m^3 of waste gas, and high removal efficiencies have been reported (Wade 1996). The additives add to the cost of gas clean-up, with sodium sulphide being the most cost effective, followed by activated carbon which is approximately three times the cost and TMT 15 at seven times the cost of sodium sulphide. Activated carbon also has the advantage of removing dioxins and furans from the gas stream. An alternative to gas clean-up of heavy metals is to eliminate them from the raw waste material; the recycling of batteries for the removal of cadmium and mercury has been shown to be effective in reducing the emissions of these metals (WHO 1990).

Heavy metals exert a range of toxic health effects including carcinogenic, neurological, hepatic, renal and hematopoietic (Denison and Silbergeld 1988). Major episodes of ill health have been reported among populations acutely and chronically exposed to heavy metals, particularly cadmium, mercury and lead (Friberg et al 1986). The presence of cadmium represents a health risk via accumulation in living tissue leading to respiratory ailments, kidney damage, hypertension and in extreme circumstances damage to bones and joints (Probert et al 1987). Mercury and mercury compounds give rise to toxic effects associated with the central nervous system, the major areas affected being associated with the sensory, visual and auditory functions as well as those concerned with coordination (WHO 1990). Lead exposure has been associated with dysfunction in the haematological system and central nervous system. Decreases in intelligence and behaviour have been reported in children subject to exposure to increased levels of lead (WHO 1990). The primary route for human exposure to heavy metals released by incineration is the food chain. Of the heavy metals, cadmium, mercury and lead are deemed of most importance in relation to municipal waste incinerators since, whilst other metals occur, their toxicities or emission levels are much lower. The health effects of heavy metals arising from incineration is increased because they are readily available to the body since they are concentrated on the finer size fraction, tend to be adsorbed to the surface of particles, and their fine size means they are more easily ingested (Bouscaren 1988; Greenberg et al 1978; Buekens and Schoeters 1984). Whilst the effects of heavy metals on human health are clearly devastating, the gas clean-up measures required to meet legislative limits of heavy metal emissions from incinerators means that they do not pose a problem to the environment. Indeed, a recent Royal Commission on Environmental Pollution (1993) concluded that no effects on health have been linked to the release of heavy metals from incineration plants.

Toxic and corrosive gases Municipal waste contains a range of compounds which contain chlorine, fluorine, sulphur, nitrogen and other elements which may result in the generation of toxic or corrosive gases. Nitrogen oxides also result from the nitrogen in the combustion air formed by reaction with oxygen at the high temperatures of the combustion zone. Typical waste contains about 7000–8000 mg/kg chlorine, 100–200 mg/kg fluorine and 2700–5000 mg/kg sulphur (Reimann 1986). The waste chloride and fluoride are in the form of waste plastics, for example PVC and PTFE; chlorides are also found in paper and board, rubber and leather, and as sodium chloride, for example, from street sweepings after road salting in icy conditions (Buekens and Schoeters 1984). The sulphur content is low compared with coal sulphur contents, although some waste oils may contain up to 5% sulphur and 1–2% chlorine (Probert et al 1987).

Normally, because the combustion of the hydrocarbon volatile fraction in an incinerator is almost complete, the flue gases consist mainly of nitrogen, oxygen, water vapour and carbon dioxide. However, the combustible waste compounds which contain the chlorine, fluorine, sulphur or nitrogen during combustion generate gaseous contaminants such as hydrogen chloride, hydrogen fluoride, sulphur oxides and nitrogen oxides:

$$C, H, Cl, F, S, N + O_2 \Rightarrow CO_2 + H_2O + HCl + HF + SO_2 + NOx$$

At the high temperatures of the combustion zone carbon dioxide and water vapour may partially dissociate, but the resulting carbon monoxide, hydrogen and oxygen recombine when the temperature decreases. Many of the other products, including hydrogen chloride, hydrogen fluoride and sulphur dioxide, are stable. Normally chlorine is not detectable in the furnace emissions since it is reduced by numerous gases or solid-reducing agents to hydrogen chloride (Buekens and Schoeters 1984).

The origin of hydrogen chloride (HCl) in incinerator flue gas has been the subject of much research due to its corrosive nature at low temperature, i.e., dew point corrosion, and high-temperature corrosion, and also its implication in dioxin formation. One of the major sources of HCl is regarded as being PVC plastic, and a direct relationship between HCl in the flue gas and the PVC content of the waste for a municipal waste incinerator has been demonstrated (Buekens and Schoeters 1984). However, other sources, such as metal chlorides like NaCl or $CaCl_2$ from paper and board, rubber, leather and vegetable matter, are also regarded as significant sources of HCl (Uchida et al 1988). PVC emits HCl by a gradual process of thermal decomposition which takes place between 180 °C and 600 °C. Chlorides are also implicated in the formation of dioxins and furans. However, the link between the PVC plastic content of municipal solid waste and dioxin formation is disputed. High concentrations of organic chlorides are found in rubber and leather products. Hydrogen fluoride is even more reactive and corrosive than HCl, and arises from combustion of fluorinated hydrocarbons such as plastics like PTFE.

Municipal waste incinerators are regarded as only a minor source of sulphur dioxide (SO_2) emission when compared with power plants and industrial boilers firing heavy fuel oil or coal (Clayton et al 1991). About 1% of the SO_2 may be further oxidised to sulphur trioxide, SO_3, which reacts with water vapour to form highly corrosive sulphuric acid, H_2SO_4, in the flue gas. The higher HCl and lower SO_2 emissions from incinerators have prompted manufacturers to suggest that a better measure of incinerator emissions would be total acidity, combining the acidic gases and thus representing a better comparison with other forms of fuel.

Nitrogen oxides arise from the nitrogen in the waste (fuel NOx) and by direct combination of the nitrogen and oxygen present in the combustion air which occurs more rapidly at high temperature (thermal NOx). In practice, thermal NO is formed almost exclusively at high temperatures in the flame, particularly under oxidising conditions; in reducing conditions little NO is formed (Buekens and Patrick 1985). Fuel NOx is formed from the nitrogen in the waste, but may also form nitrogen. NOx generation is reduced by factors such as burner design, the level of excess air, the temperature of combustion, and recirculation of the flue gases (Buekens and Patrick 1985). At low temperatures, i.e., below 200 °C, NO is slowly oxidised into NO_2. This reaction continues after emission of the flue gases into the atmosphere.

The potential problem of acid gases in the back-end of the incinerator plant is due to their low dew point, which results in corrosive damage to metals. The emission of the pollutant gases to the atmosphere produces the well-documented acid rain, with its associated environmental damage, whilst NO after atmospheric oxidation to NO_2 is active in the generation of photochemical smog. HCl, HF, SO_2 and NOx all produce acids in the atmosphere which contribute to acid rain, forming hydrochloric, hydrofluoric, sulphuric and nitric acids, respectively. Increased acidification of the atmosphere has resulted in damage to buildings, acidification of lakes, respiratory problems, die-back of forests etc.

Table 6.8 *Effectiveness of gas cleaning systems (% removal)*

Control system	Pollutant			
	HCl	HF	SO_2	NOx
Wet scrubber + electrostatic precipitator	>95	–	>70	–
Dry scrubber + fabric filter	>90	92	>75	–
Semi-dry scrubber + fabric filter	44–99	86	70–98	–
Selective catalytic reduction	–	–	–	80–90

Source: Brna T.G., Combustion Science and Technology, 74, 83–98, 1990.

The level of emissions of the toxic and corrosive acid gases HCl, HF, SO_2 and NOx are all regulated and require extensive gas clean-up measures to comply with the regulations. Brna (1990) has reviewed the clean-up of flue gases, including acid gas control from municipal waste incinerators. Wet, dry and semi-dry processes are used to remove the acid gases produced by waste combustion. Wet-scrubbing systems use slurries and solutions at relatively low temperatures and produce a liquid or wet solid/sludge reaction product. The adsorbents used include calcium oxide, calcium hydroxide and sodium hydroxide. The liquid or sludge product is highly polluted and difficult and expensive to treat. In addition, the slurry or solution used requires to be made up to certain specifications which again adds to the cost (Wade 1996). Consequently, developments have centred on new methods of control which generate a solid residue which is easier to handle. Dry systems use a dry powder and possibly up-stream humidification to improve gas/sorbent reaction. Semi-dry processes use an alkaline sorbent slurry or solution which is atomised into fine droplets into the flue gas; the droplets react and dry in the hot flue gases to produce a dry powder. For both the dry and semi-dry systems adsorption is improved by the use of a down-stream fabric filter, which increases the contact time between the gases and the alkaline filter cake formed on the filter by the adsorbent (see Boxes 6.4 and 6.5).

Nitrogen oxide, the main oxide of nitrogen found in flue gases, cannot be reduced by scrubbing because of its low solubility. The addition of ammonia to form nitrogen and water has been used to reduce NOx, but is only effective in a very narrow temperature range of 870–900 °C. Therefore, flue gas reduction of NOx for municipal waste incinerators is usually achieved with selective catalytic reduction (SCR) (see Box 6.6).

Table 6.8 shows the effectiveness of different gas cleaning systems (Brna 1990).

Odours from incineration of waste may not pose a health hazard as such since the concentration levels of compounds where the nose can detect odour are extremely low. However, odours cause nuisance to the local population and are often the most common cause of complaint. Odours associated with incineration of waste are usually complex mixtures of organic compounds and result from incomplete combustion. Threshold limit values, above which the odour can be detected, can be very low, for example, many organic compounds which have been associated with waste incinerators have a threshold limit value between 0.001 and 10 p.p.m. (Brunner 1985). The odours released into the atmosphere will be influenced by the efficiency of the combustion process and the gas clean-up system, and peaks of emission may last only a few seconds. Emission to the atmosphere is influenced by dilution in the surrounding atmosphere and the dispersion of the stack plume, which in turn will be influenced by meteorological conditions. Brunner (1985) has reviewed the odours from incineration of waste, the threshold limit values of common odours and their control.

Products of incomplete combustion: Polycyclic aromatic hydrocarbons (PAH), dioxins and furans The volatile matter arising from the thermal degradation of waste is normally completely combusted by providing adequate residence time, post-combustion temperature and turbulent mixing. However, it is a consequence of the incineration process that there will be some areas in the incinerator which allow incomplete combustion of the gases and vapours. These incompletely combusted vapours may contain CO, volatile organic compounds such as polycyclic aromatic hydrocarbons (PAH), dioxins and furans, tar and soot particles. Incomplete combustion may occur when the incinerator is improperly operated, for example, operation at excessively low temperatures, i.e., below 800 °C, or overloading of the plant. The occurrence of incomplete combustion can be detected by monitoring the flue gas composition. For efficient waste combustion, the concentration level of carbon monoxide must consistently remain below 0.1 volume %.

There are potentially a wide range of waste-derived thermally degraded organic compounds which may be found in incinerator flue gases. Of these organic micro-pollutants there is most public concern associated with the emissions of polycyclic aromatic hydrocarbons, dioxins and furans from the incineration of waste.

Polycyclic aromatic hydrocarbons (PAH) Polycyclic aromatic hydrocarbons (PAH) are compounds based on aromatic benzene rings which are fused to form two or more polycyclic rings. Figure 6.11 shows examples of PAHs. PAHs are known to occur naturally in the environment, for example in sediments, fossil fuels and by natural combustion in forest fires (Lee et al 1981). The major sources of PAHs, however, are anthropogenic; examples include oil- and coal-fired power generation plant, coke production, residential furnaces, diesel and gasoline engines and in waste combustion (Lee et al 1981; P.T. Williams 1990). Concern over the emission of PAHs to the environment is centred on the associated health hazard, because PAHs comprise the largest group of carcinogens among the environmental chemical groups (Lee et al 1981). PAHs adsorbed onto air-borne particles are believed to be a major contributory reason why death rates from lung cancer are higher in urban than in rural areas (Lave and Seskin 1970). Cancers of the lung, stomach, kidneys, scrotum and liver have been associated with exposure to PAHs (Gelboin and Tso 1978; Lee et al 1981). In addition, PAHs have been suggested as precursors to the formation of soot in combustion systems (A. Williams 1990).

Table 6.9 shows the PAH emission from incinerators firing municipal solid waste and refuse derived fuel (RDF) (Davies et al 1976; Parker and Roberts 1985; Colmsjo et al 1986). The RDF incinerator is the Eksjo plant in Sweden, which uses RDF pellets in a fluidised bed. A large percentage of the PAHs were reported to be

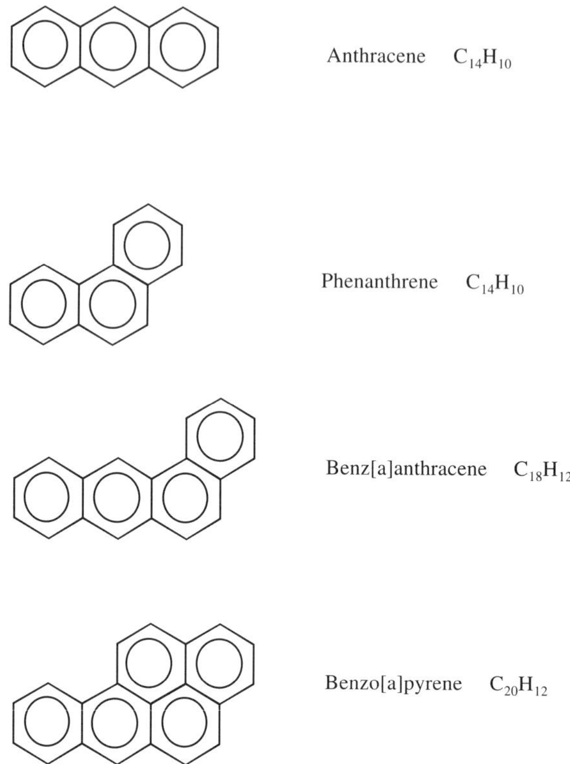

Figure 6.11 *Examples of polycyclic aromatic hydrocarbons*

adsorbed on the surface of the fly ash. Table 6.9 also shows the distribution of PAH from the stack gases from a UK incinerator (Davies et al 1976). It was also reported that the largest emission was associated with the solid residues associated with the flyash; the concentration in water was low, reflecting the low solubility of PAH in water. Also shown in Table 6.9 are PAH emissions during the cold start-up of a municipal solid waste incinerator in Sweden (Colmsjo et al 1986). The emissions are clearly much higher, reflecting the less than optimum combustion efficiency during start-up. However, under normal operation the concentration of each individual PAH never exceeded 10 ng/m^3. Also detected were a number of halogenated PAH and chlorobenzenes and chlorophenols, which are known to act as precursors for the formation of dioxins and furans.

The PAH reported from municipal waste incineration include some species known to be biologically active in human and bacterial cell tests, for example benzo[a]pyrene, benzo[e]pyrene, phenanthene, methylphenanthenes, fluoranthene and the methylfluorenes (Barfnecht et al 1980; Lee et al 1981; Longwell 1983). The levels of PAH reported for incinerator emissions are similar to those reported for coal- and oil-fired power stations and lower than those emitted from older diesel engines, reflecting the combustion process in all its forms as a source of PAH

Table 6.9 *Emissions of polycyclic aromatic hydrocarbons (PAH) from municipal waste incineration (μg/m³)*

PAH	UK MSW[1]	Sweden RDF[2]	Sweden MSW
Fluorene	–	0.2	12.0
Methylfluorenes	–	0.09	–
Phenanthrene	–	–	43.0
Fluoranthene	0.58	0.2	11.0
Pyrene	1.58	0.09	6.8
Benzo[a]anthracene	0.72 (Benzo[a]anthracene + Chrysene)	0.1	1.1
Chrysene		0.5	3.0
Benzofluoranthenes	0.32	0.04	4.0
Benzo[a]pyrene	0.02 (Benzo[a]pyrene + Benzo[e]pyrene)	0.04	–
Benzo[e]pyrene		0.04	0.7
Perylene	0.18	0.04	–
Benzo[ghi]perylene	0.42	0.04	–
Coronene	0.04	–	–

[1] MSW incinerator emissions during plant start-up.
[2] RDF, refuse derived fuel.

Sources: [1] Davies I.W., Harrison R.M., Perry R., Ratnayaka D. and Wellings R.A., Environmental Science and Technology, 10, 451–453, 1976; [2] Parker C. and Roberts T., Energy from Waste: An Evaluation of Conservation Technologies. Elsevier Applied Sciences, London, 1985; [3] Colmsjo A.C., Zebuhr Y.U. and Ostman C.E., Atmospheric Environment, 20, 2279–2282, 1986.

(Williams 1994). The physical and chemical properties of PAHs suggests that control mechanisms introduced for the control of dioxins and furans would easily control PAH emissions from incinerators. For example, gas clean-up systems incorporating sprayers and electrostatic precipitators are highly efficient at removing PAHs (Davies et al 1976).

Dioxins and furans Polychlorinated dibenzo-ρ-dioxins (PCDD), or 'dioxins', and the closely related polychlorinated dibenzofurans (PCDF), or 'furans', constitute a group of chemicals that have been demonstrated to occur ubiquitously in the environment. They have been detected in soils and sediments, rivers and lakes, chemical formulations and wastes, herbicides, hazardous waste site samples, landfill sludges and leachates (Tiernan 1983). PCDD and PCDF have a number of recognised sources, among which are their formation as by-products of chemical processes such as the manufacture of wood preservatives and herbicides, the smelting of copper and scrap metal, the recovery of plastic-coated wire and natural combustion such as forest fires (Steisel et al 1987; DOE 1989). More contentiously, they are found in combustion products, i.e., the ash, stack effluents, water and other process fluids from the combustion of municipal waste, coal, wood and industrial waste (Tiernan 1983). The concern over dioxins and furans arises from a number of animal studies which show that for some species they are highly toxic at very low levels of exposure (Tosine 1983; Oakland 1988). The extrapolation of these animal data to man – though contentious – has led to dioxins and furans acquiring their notoriety as one of the most toxic chemicals known to man. Boxes 6.7 and 6.8 discuss the conflicting arguments relating to the toxicity of dioxins and furans.

PCDD and PCDF are highly stable environmentally and present difficult sampling and analytical problems because of interferences, their low concentration and their perceived toxicity (Tosine 1983; Oakland 1988; Williams 1992). PCDDs and PCDFs have been involved in a number of incidents in recent years which give them their notoriety. For example, the Seveso accident in Italy in 1976, the herbicide spraying programme of Agent Orange in Vietnam in the late 1960s, and the Times Beach Missouri land poisoning of 1982 (British Medical Association 1991).

The generalised molecular structures of PCDD and PCDF are shown in Figure 6.12; they are tricyclic aromatic compounds containing two (dioxin) or one (furan) oxygen atoms. Each of these structures represents a whole series of discrete compounds having between one and eight chlorine atoms attached to the ring; for example, Figure 6.12 shows the tetra isomers with four chlorine atoms in the 2,3,7,8 positions, i.e., 2,3,7,8 tetrachloro dibenzo-ρ-dioxin (2,3,7,8 TCDD) and 2,3,7,8 tetrachloro dibenzofuran (2,3,7,8 TCDF). Since each chlorine atom can occupy any of the eight available ring positions, it can be calculated that there are 75 PCDD isomers and 135 PCDF isomers. All the PCDDs and PCDFs are solids with high melting and boiling points, and with limited solubility in water. Many of these isomers have not been prepared in pure form, and hence their toxicology has not been assessed and their identification becomes difficult.

The potential threat of PCDD and PCDF to humans should be assessed bearing in mind that the 75 PCDD isomers and 135 PCDF isomers have differing toxicities and are often present in multiple mixtures. The toxicities of the PCDF isomers generally parallel those of PCDD (Tiernan 1983). The assessment of the toxicity of PCDD and PCDF mixtures has led to the development of the toxic equivalent (TEQ) scheme. This uses the available toxicological and biological data to generate a set of weighting factors each of which expresses the toxicity of a particular PCDD or PCDF in terms of an equivalent amount of the most toxic and most analysed PCDD, which is 2,3,7,8 TCDD. Thus in the most widely accepted method 2378 TCDD would have a TEQ of 1.0 and OCDD, for example, would have a TEQ of 0.001 (DOE 1989).

Table 6.5 shows the emissions of the TCDD and TCDF isomers from a range of municipal solid waste incinerators throughout the world, comparing modern plants with the older type incinerators with minimal pollution control.

A number of theories have been proposed for the formation of PCDD and PCDF during combustion, and their formation route may be a combination of processes depending on prevailing conditions (Lustenhouwer et al 1980).

1. PCDD and PCDF occur as trace constituents in the waste and because of their thermal stabilities they survive the combustion process. Waste material has been shown to contain PCDD and PCDF at trace levels (Oakland 1988), but mass balances have shown that higher concentrations have been found in the emissions than are found in the input (Commoner et al 1987). However,

Box 6.7
The Health Hazard Associated with Dioxins and Furans

The case for:

A number of comprehensive reviews have been published dealing with the health effects of PCDD and PCDF, by far the majority of work being on animal tests. It is clear that this group of compounds is very toxic. Toxicity depends on the number and position of the chlorine substituents; 2,3,7,8-tetra-chlorinated dibenzo-p-dioxin is the most toxic dioxin and has been shown to cause lethal effects in certain laboratory animals at very low levels, for example the LD_{50} (lethal dose where 50% of the species tested dies) for guinea pigs is 0.6 μg/kg body weight. There is less information with respect to the toxic effects on humans, and most existing data have been derived from occupational exposure or industrial accident victims. The effects attributed to PCDD and PCDF exposure include a persistent skin acne condition known as chloracne, and systemic effects such as digestive disorders and muscle and joint pains, neurological disorders such as headaches and loss of hearing, and psychiatric effects such as depression and sleep disturbance. Also potentially connected to PCDD and PCDF exposure are long-term health risks such as chromosome damage, heart attacks and cancer. Recent work reported in the Lancet examined workers exposed to TCDD and suggested that there was an increased risk of cancer in exposed workers. The researchers were able to separate workers exposed to TCDD separately from those exposed to other chlorinated hydrocarbons. Risk assessments of various US incinerators in relation to PCDD and PCDF emissions range from 1 to 270 additional cancers per million of the exposed population. The 1994 US–EPA report on dioxins has received considerable publicity. The report presents a number of conclusions.

(i) TCDD and related compounds are potent toxicants in animals with the potential to produce a spectrum of effects. Some of these effects may be occurring in humans at very low levels, and some may be resulting in adverse impacts on human health. However, the report also states that despite this potential there is currently no clear indication of increased disease in the general population attributable to dioxin-like compounds.
(ii) The primary mechanism by which dioxins enter the food chain is via atmospheric deposition.
(iii) There is adequate evidence to support the inference that humans are likely to respond with a broad spectrum of effects from dioxins if exposures are high enough.
(iv) The margin of exposure between background levels and levels where effects are detectable in humans is considerably smaller than previously estimated, within an order of magnitude of current background levels.
(v) Dioxins are likely to present a cancer hazard to humans.

Sources: [1] Cattabeni F. et al., Dioxin: Toxicological and Chemical Aspects. SP Medical and Scientific Books, London, 1978; [2] Choudhary G., Keith L.M. and Rappe C. (Eds.) Occupational Exposure. Chlorinated Dioxins and Dibenzofurans in the Total Environment. Section V. Butterworth, London, 1983; [3] Hay A.W.M., in Hutzinger O., Frei R.W., Merian E. and Pocchiari F. (Eds.), Chlorinated Dioxins and Related Compounds: Impact on the Environment. Pergamon, New York, 1982; [4] Skene S.A., Dewhurst I.C. and Greenberg M., Human Toxicology, 8, 173–203, 1989; [5] Buser H.R. and Rappe C., Analytical Chemistry, 56, 442–448, 1984; [6] Baker P.G., Analytical Processes, November, 478–480, 1981; [7] Manz A. et al., Lancet, 338, 959–964, 1991; [8] Saracci R. et al., Lancet, 338, 1027–1032, 1991; [9] Commoner B., Shapiro K. and Webster T., Waste Management and Research, 5, 327, 1987; [10] Dioxin risk: Are we sure yet? EPA Draft Report Review. Environmental Science and Technology, 29, 24A–35A, 1995.

Box 6.8
The Health Hazard Associated with Dioxins and Furans

The case against:

Reports on a number of human accidental and occupational exposures to PCDD and PCDF and their effect on health have shown that human epidemiological studies are difficult to interpret since there have been problems in controlled methodologies, inadequate information on intake, and exposure mode and level. Animal tests have shown a very species-dependent toxicological response, for example, whilst the LD_{50} for guinea pigs is 0.6 μg/kg body weight, for mice it is 284 μg/kg body weight. Therefore, the extrapolation of animal tests to humans is contentious. Occupational and accidental exposures have often been to mixtures of PCDDs and/or PCDFs and also in conjunction with other related and possibly hazardous compounds. Data suggest that the effects of PCDD and PCDF on humans were inconclusive and required further study. Indeed an international steering group reporting on the Seveso incident in Italy where an uncontrolled release of 2,3,7,8-TCDD occurred from a plant manufacturing 245-trichlorophenol, concluded that no clear-cut adverse health effects attributable to 2,3,7,8-TCDD besides chloracne could be observed. In addition, it has been reported that no human deaths have been attributed conclusively to 2,3,7,8-TCDD or any other dioxin, that there is no convincing evidence of a link between exposure to dioxins and cancer, and that no major adverse health effects on human reproduction have been detected as a result of low-level exposure to dioxins. Where chronic symptoms from exposure to PCDD and PCDF have been observed in the human population, it has been shown that after removal of the source, the symptoms have generally cleared within a few months and have always disappeared within two years. The conclusions of the 1994 US–EPA report on dioxins have also been disputed. The main contention is that evidence related to animal tests cannot be linked with humans with present knowledge. Where human studies have been carried out, the isolation of dioxins as the main carcinogen is also disputed since in such studies workers were often exposed to other toxic substances, and the high influence of cigarette smoking is often ignored. Also, animal effects have only been found with high dose levels. A recent report by the UK Department of the Environment has assessed the risk to health posed by emissions of dioxins from a large scale municipal waste incinerator and characterised the extent of exposure. The report concluded that incinerators operating within legislative limits posed no health risk to individuals exposed to plant emissions of dioxins or to the population in the surrounding environment. Whilst the evidence for a clear link between exposure to PCDD and PCDF and long-term adverse health effects is inconclusive, the perceived risk is still of some concern to the public.

Sources: [1] Ahlborg U.G. and Victorin K., In: Waste Management and Research, 5, 203, 1987; [2] Department of the Environment, Pollution Paper No. 27. Dioxins in the Environment. HMSO, London, 1989; [3] Steisel N., Morris R. and Clarke M.J., Waste Management and Research, 5, 381, 1987; [4] Dioxin risk: Are we sure yet? EPA Draft Report Review. Environmental Science and Technology, 29, 24A–35A, 1995; [5] Baker P.G., Analytical Processes, November, 478–480, 1981; [6] Risk Assessment of Dioxin Releases from Municipal Waste Incineration Processes. Environment Agency (HMIP), Department of the Environment, HMSO, London, 1996.

conditions undoubtedly exist for the thermally stable PCDD and PCDF to survive the combustion process, particularly at the lower combustion temperatures that prevail in certain zones of some incinerators.

2. PCDD and PCDF are produced during the incineration process from precursors such as polychlorinated biphenyls (PCB), chlorinated benzenes, pentachlorophenols etc. The in-situ synthesis of PCDD and PCDF therefore occurs via re-arrangement, free-radical condensation, dechlorination and other molecular

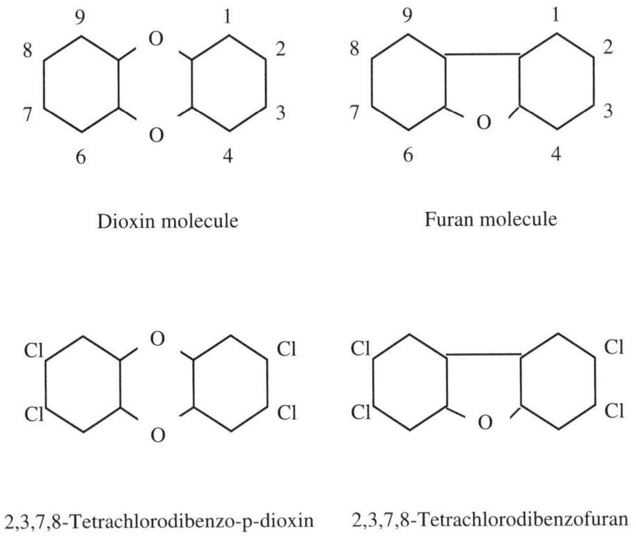

Dioxin molecule

Furan molecule

2,3,7,8-Tetrachlorodibenzo-p-dioxin

2,3,7,8-Tetrachlorodibenzofuran

Figure 6.12 *Dioxin and furan molecules and the 2,3,7,8-tetra isomers*

reactions (Hagenmaier et al 1988). The precursor theory to the formation of PCDD and PCDF has arisen from laboratory studies and incinerator waste input–emission output studies which show that these compounds can be formed from chlorophenols, chlorobenzenes, PCBs and brominated diphenylethers (Buser and Rappe 1979; Hutzinger et al 1982). These precursors may be present in the waste or formed by the combustion of chlorinated plastics and other chlorinated organic materials in the waste (Marklund et al 1987; Oakland 1988; Probert et al 1987).

3. PCDD and PCDF are produced as a result of elementary reactions of the appropriate elements, i.e., carbon, hydrogen, oxygen and chlorine atoms. This reaction is called a de-novo synthesis of PCDD and PCDF. PCDD and PCDF have been shown to form on flyash containing residual carbon collected within a combustion system at temperatures in the region of 300–400 °C in the presence of flue gases containing HCl, O_2 and H_2O (Hagenmaier et al 1988). It is thought that the reaction is catalysed by various metals, metal oxides, silicates etc., particularly copper chloride present in the flyash. This theory is borne out by the observation that low levels of PCDD have been observed in the furnace exit of incinerators, but levels 100 times greater were found in the electrostatic precipitator ash of the same plant (Commoner et al 1987; Hagenmaier et al 1988). Figure 6.13 shows a schematic diagram of the mass balance input–output of dioxins and furans for a municipal waste incinerator (Commoner et al 1987).

The control of PCDD and PCDF emissions may be approached by either restricting their formation, efficient combustion control and/or by clean up of the flue gases after they have formed.

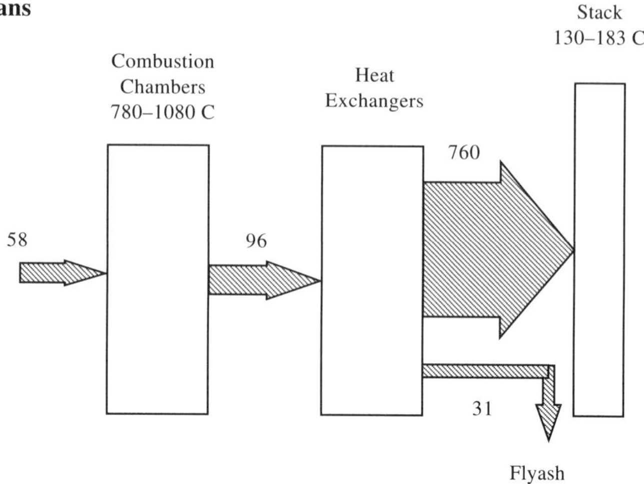

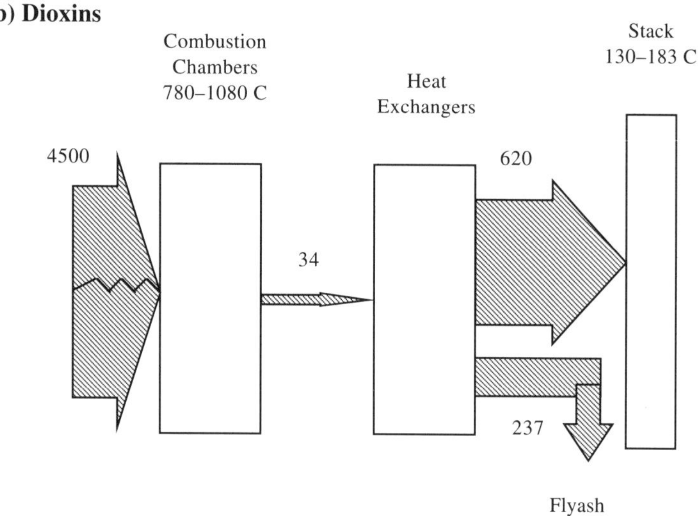

Figure 6.13 Diagrammatic representation of the mass balance of dioxins and furans for a municipal solid waste incinerator (a) furans; (b) dioxins (units μg/h). Source: Commoner B., Shapiro K. and Webster T., Waste Management and Research, 5, 327, 1987. Reproduced by permission from ISWA – International Solid Waste Association

The removal of the chlorine- and hydrogen chloride-producing plastic components from the waste prior to incineration has been suggested as a mechanism of PCDD and PCDF control. However, experiments on hydrogen chloride formation in incinerators have shown that even when all the plastic is removed from the waste significant concentrations of hydrogen chloride are still produced in the flue gases from other sources of chlorine such as paper and board (Buekens and Schoeters

1984; Visalli 1987). In addition, no correlation has been reported between PVC plastic in the waste stream with PCDD and PCDF emissions from incinerators (Uchida et al 1983; Visalli 1987). Also, wood burning has been shown to produce chlorophenols and chlorobenzenes which can combine to form PCDDs and PCDFs without the presence of chlorine, hydrogen chloride or chlorinated plastic.

Combustion control has centred on the destruction of PCDD and PCDF at high temperatures. Consequently, recommended conditions are temperatures above 1000 °C, residence times of >1 s, and turbulence to ensure good mixing, with excess air. Decomposition increases exponentially with temperature, for example, at 1200 °C a residence time of only 1 ms is required for destruction. Incinerators have been shown to be particularly effective in destroying dioxins (Commoner et al 1987). The emission of carbon monoxide from incinerators is used as a measure of efficient combustion control, so that minimum carbon monoxide correlates with efficient combustion and therefore also with minimum PCDD and PCDF emission. However, de-novo synthesis has been shown to be the dominant formation route to PCDD and PCDF in many municipal waste incinerators (Hagenmaier et al 1988; Commoner et al 1987; Vogg et al 1987). Therefore PCDD and PCDF may be destroyed in the high temperature of the furnace with efficient combustion control, but the overall emission of PCDD and PCDF from the incinerator may not be affected by this destruction since formation of these compounds takes place in the cooler parts of the incinerator system, down-stream of the furnace. Furnace conditions are optimised for efficient combustion, however, since this also influences the production of the products of incomplete combustion, carbon content, chloride content and heavy metal content, as well as the surface activity of the ash particles which are essential for the de-novo synthesis and therefore should be minimised. In addition, since the temperature of formation by the de-novo route is maximised in the 300–400 °C range, incinerator flue gases are rapidly cooled from the furnace outlet to below this 'temperature window' before entering the clean-up system to minimise their formation. De-novo formation of PCDD and PCDF will still occur to some extent since within the economisers, or electrostatic precipitator or fabric filter of the clean-up system, ideal conditions for the synthesis will occur.

Post-combustion control of PCDD and PCDF has centred on the efficient collection of particulates since they are shown to be found mainly on flyash, either adsorbed or formed in-situ; they also exist at lower levels in the gas phase. Gas-phase removal of PCDD and PCDF by wet-scrubbing systems is not so effective, since dioxins such as TCDD have very low solubilities in water. However, wet/dry scrubbers with lime slurry as the active scrubbing agent have been shown to be effective in the removal of PCDD and PCDF (Nielsen et al 1986). In addition, calcium oxide and calcium hydroxide addition used to remove hydrogen chloride will also influence PCDD and PCDF emissions by removing the chlorine as a building block of the PCDD and PCDF molecule. The most effective system for removal of PCDD and PCDF has been shown to be the addition of small quantities of activated carbon as a fine powder injected independently or added to the calcium oxide adsorbent prior to a fabric filter (Atkins 1996; Wade 1996). The activated carbon adsorbs the PCDD and PCDF, and the build-up of the activated carbon on the fabric also provides a surface for further adsorption. PCDD and PCDF

Table 6.10 *UK inventory of annual PCDD and PCDF emissions to the atmosphere*

Process	Emissions to the atmosphere (g TEQ[1]/year)	
	1994 estimate	Future estimate
MSW incineration	460–580	15
Chemical waste incineration	1.5–8.7	0.3
Clinical waste incineration	18–88	5
Sewage sludge incineration	0.7–6	0.9
Crematoria	1–35	1–35
Coke production	2	2
Industrial coal combustion	5–67	5–67
Industrial wood combustion	1.4–2.9	1.4–2.9
Domestic coal combustion	20–34	20–34
Domestic wood combustion	2–18	2–18
Straw combustion	3.4–10	3.4–10
Tyre combustion	1.7	1.7
Landfill gas combustion	1.6–5.5	1.6–5.5
Sinter plants	29–54	29–47
Iron and steel	3–41	14
Non-ferrous metals	5–35	10
Traffic	1–45	1–45

[1] TEQ = toxic equivalent.

Source: DOE, Review of Dioxin Emissions in the UK. Department of the Environment, HMSO, London, 1995.

emissions from incinerators using such gas clean-up systems have reported levels well below legislated limits (Atkins 1997).

Comparison of emission levels of PCDD and PCDF from various sources in the UK is shown in Table 6.10 (DOE 1995a). The 1994 data show that by far the highest emissions were from municipal waste combustion. However, with the implementation of European Directive emission limits on plant after December 1996, the emissions are estimated to fall by a factor of between 30 and 40. Other waste combustion systems such as chemical waste incineration and clinical waste incineration are also predicted to have large decreases in the emissions of PCDD and PCDF as again emission limit legislation takes effect. Significantly, other major sources of PCDD and PCDF in the UK are from domestic coal combustion, industrial sources and traffic.

Contaminated wastewater Water pollution from incinerators is not generally regarded as an important problem, because the limited amount of wastewater generated is of the order of 2.5 m^3 per tonne of municipal waste incinerated. However, the wastewater from municipal waste incinerator plants has been shown to be contaminated with heavy metals and inorganic salts, and have high temperatures and high acidities or alkalinities (Reimann 1987).

The main sources of wastewater from incinerators are from flue gas treatment as flue gas scrubber water, alkaline scrubbing of the gases to remove acid gases, and the quenching of incinerator ash. Other minor sources include, for example, scrubber water pre-treatment and the purification of boiler feedwater where a boiler plant is installed. Where the gases are scrubbed or cooled with water, the absorbed acid gases will make the water very acidic and will consequently also contain significant quantities of heavy metals, which are soluble in the acidic solution. Where the gases are scrubbed with an alkaline solution such as sodium hydroxide or calcium hydroxide to remove acid gases, the scrubber water will be very alkaline.

The bottom ash from the incinerator grate is removed in a unit which serves to cool the ash and also maintain a partial vacuum in the incinerator chamber. Approximately 0.3–0.8 m^3 of water is used per tonne of waste (Reimann 1987). Bottom ash wastewater is alkaline and contains only low levels of dissolved heavy metals, below permitted sewerage discharge levels.

The main pollutants in incinerator wastewaters are heavy metals. Clean-up of the heavy metals is usually through neutralisation via precipitation with calcium hydroxide with the additive TMT15 (trimercaptotriazine) (Reimann 1987). The calcium hydroxide causes the major part of the pollutants to be precipitated as hydroxide sludges. However, more than 60% of the mercury and other heavy metals remain in the wastewater and the addition of TMT15 is required, which precipitates the mercury and heavy metals down to levels well below the permissible limits (Reimann 1987).

Whilst there is most concern over the presence of heavy metals in wastewater, the presence of organic pollutants such as PAH and dioxins and furans should also be suspected. However, reported levels of PAH and PCDD and PCDF are very low as dissolved compounds in wastewater (Ozvacic et al 1985; Reimann 1987), but have been detected in significant concentrations in suspended particles, which therefore are required to be filtered out (Bumb et al 1980; Ozvacic et al 1985).

Contaminated ash If the incinerator is operating correctly, the residue or ash should be completely burnt out and biologically sterile. Bottom ash from the furnace grate represents the bulk (75–90%) of total ash and is composed mainly of mineral oxides. Its heavy metal content is generally lower than 1.5% by weight, but is highly variable. Table 6.11 shows concentrations of heavy metals in bottom ash and flyash from two incinerators (Buekens and Schoeters 1984; Reimann 1989). The high heavy metal concentrations present in the ash residues from incineration become of more significance when they are placed in landfill sites where leaching of the pollutants may be a source of groundwater contamination. Generally the flyash is more readily leached than the clinker fraction, since the heavy metals largely occur in the smallest size fraction of less than 10 μm and are concentrated at or near the surface of the particles (Buekens and Schoeters 1984). In addition, the high chlorine content of the waste results in the majority of the metal species being present as the metal chlorides, which are generally more soluble in water than other species (Brunner and Monch 1986; Denison and Silbergeld 1988). It has been shown that up to 32.5% of the available zinc, 1.75% of lead, 5.7% of manganese and 94% of the available cadmium can be leached from flyash (Buekens and Schoeters 1984). Whilst water in contact with flyash produces alkaline solutions rather than acidic, copper, lead, zinc and cadmium show increased solubilities at high alkalinities, that is, they are

Table 6.11 *Concentration of heavy metals and halides in incinerator bottom ash and flyash (μg/g)*

Heavy metal/halide	Bottom ash USA	Flyash USA	Bottom ash Germany	Fly ash Germany
Cadmium (Cd)	41	64	3.8	–
Chromium (Cr)	520	1160	655	450
Copper (Cu)	450	510	1520	1050
Mercury (Hg)	0.4	0.9	0.7	8
Manganese (Mn)	3100	1500	–	–
Nickel (Ni)	210	1800	260	145
Lead (Pb)	1700	7200	1010	4900
Zinc (Zn)	5500	10000	4570	16600
Chloride (Cl)	–	–	1970	38000
Fluoride (F)	–	–	215	105

Sources: [1] Buekens A. and Schoeters J., Thermal methods in waste disposal: Incineration. Study performed for EEC under contract number ECI 1011/B7210/83/B, Free University of Brussels, Brussels, 1984. [2] Reimann D.O., Waste Management and Research, 7, 57–62, 1989.

amphoteric in nature, showing significant solubilities at both low and high pH values (Denison and Silbergeld 1988).

The ash is usually disposed of on land in the UK, although recycling the bottom ash for aggregate use in the construction industry and road building is common in Europe (Sakai et al 1996). For example, in Germany approximately 50% of bottom ash is utilised; in Denmark and the Netherlands about 90% is utilised (Warmer Bulletin Fact Sheet 1995). The main use in Denmark is for the development of a granular sub-base for car parking, bicycle paths and paved and un-paved roads etc. In Germany, utilisation has been through, for example, sub-base paving applications, and in the Netherlands bottom ash has been used in the construction industry as a granular base or in-fill road base, and for embankments and noise and wind barriers. In the Netherlands, bottom ash has also been used as aggregate in asphalt and concrete.

The presence of PCDD and PCDF has been shown in flyashes derived from the incineration of solid waste – Table 6.12 shows total PCDD and PCDF and certain isomers found in flyashes from German, Canadian and Dutch incinerators (Olie et al 1982; Hagenmaier et al 1987; Shaub 1989). The German data represent the average of 52 flyash samples from 10 incinerators, and the Canadian data an average of 8 samples. The significant concentrations of PCDD and PCDF found in flyash samples illustrates the de-novo synthesis route to their formation, a route catalysed by the flyash itself. Whilst flyash contains significant concentrations of PCDD and PCDF, bottom ash concentrations are negligible since bottom ashes are quickly quenched before significant PCDD and PCDF production can take place (Ozvacic et al 1985; Shaub 1989).

The contamination of flyash with heavy metals and PCDD and PCDF may result in the ash attaining the status of special or hazardous waste and consequently requiring special permits for landfill disposal (Mulder 1996; Sakai et al 1996). Other treatment methods used or under investigation to stabilise incinerator flyash are solidification, chemical stabilisation, ash melting or vitrification and extraction/recovery processes (Sakai et al 1996).

Table 6.12 Concentration of PCDD and PCDF in municipal solid waste incinerator flyash (ng/g)

PCDD/PCDF	Germany[1]	Canada[2]	Netherlands[3]
PCDD			
TCDD	11	13	93
PeCDD	34	23	254
HxCDD	50	26	604
HpCDD	57	15	760
OCDD	65	6	345
Total PCDD	210	83	2056
PCDF			
TCDF	72	–	173
PeCDF	95	–	312
HxCDF	82	–	459
HpCDF	56	–	314
OCDF	13	–	51
Total PCDF	275	–	1309

Sources: [1][1] Hagenmaier H., Brunner H., Haag R., Kraft M. and Lutzke K., Waste Management and Research, 5, 239, 1987; [2][2] Olie K., Lustenhouwer J.W.A. and Hutzinger O., In: Hutzinger O., Frei R.W., Merian E. and Pocchiari F. (Eds.) Chlorinated Dioxins and Related Compounds: Impact on the Environment. Pergamon, Oxford, 1982; [3][3] Shaub W.M., An overview of what is known about dioxin and furan formation, destruction and control during incineration of MSW. Coalition on Resource, Recovery and the Environment (CORRE), US EPA, MSW Technology Conference, San Diego, CA, Jan.–Feb., 1989.

Dispersion of emissions from the chimney stack A serious consideration in relation to incinerators of all types is the height of the chimney stack and the dispersion of the plume of exhaust gases. The dispersion of the plume involves an initial rise followed by a horizontal spreading about the plume centre line (Figure 6.14a, Clarke 1986). The concentration profiles in the vertical and horizontal directions are assumed to approach a Gaussian distribution form (Figure 6.14b, Clarke 1986). The dispersion of the plume will influence the down-wind ground concentrations of pollutants. Such considerations are vitally important in the assessment of the impact of an incinerator on the local environment. The rate of dispersion and hence ground concentrations are influenced by meteorological considerations, wind speed and rate of plume emission. Unstable conditions promote dispersion, whilst stable conditions such as fog result in very slow-spreading plumes. The ground-level concentration close to the stack is zero, but eventually the plume spreads out and reaches ground level at a point known as the radius of maximum effect. For incinerators this point could be several kilometres from the incinerator. Further from the stack, the ground concentration becomes reduced as the plume becomes more diluted in the atmosphere. Figure 6.14c (Clarke 1996) shows plume dispersal in stable, unstable and neutral meteorological conditions. Unstable conditions give a higher ground concentration closer to the stack than stable conditions.

Dispersion is also influenced by the height of the stack, higher stacks promoting increased rates of dispersion. The use of dispersion models and guidelines give recommended stack heights for incinerators. Guidance notes for incinerators require

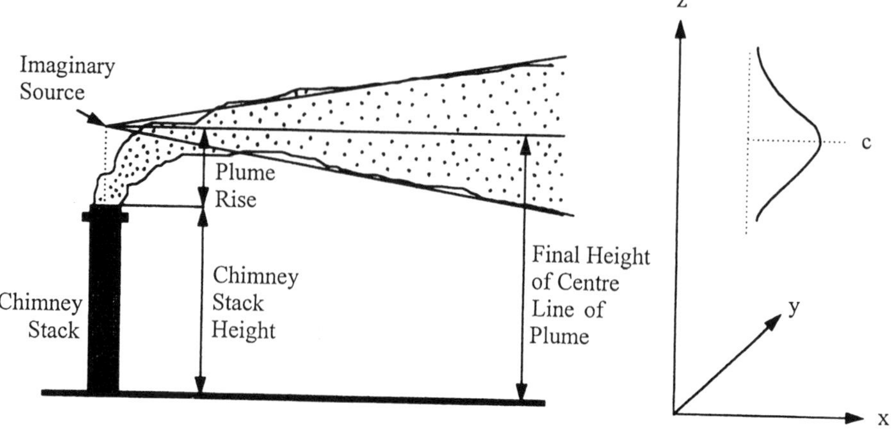

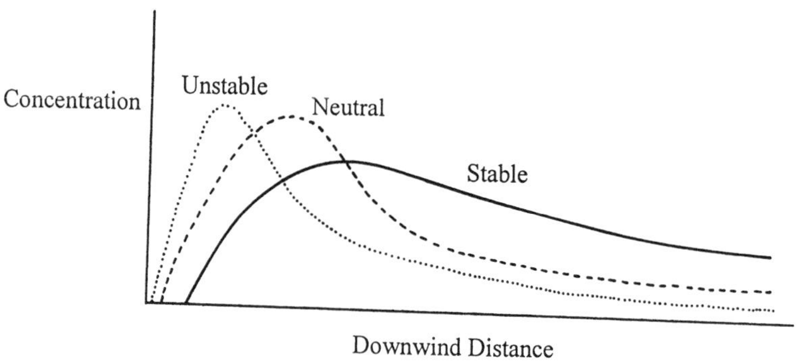

Figure 6.14 *Plume dispersion from chimney stacks and ground-level concentration in relation to atmospheric stability. Sources: [1] Clarke, A.G., The Air. In: Hester R.E. (Ed.) Understanding our Environment. The Royal Society of Chemistry, London, 1986; [2] Clarke A.G., Chimney design. In: University of Leeds Short Course, Incineration of Municipal Waste with Energy Recovery. Department of Fuel and Energy, University of Leeds, 1996*

that no plume should be visible from the stack for 90% of the time, with exemption during start-up. Where there is a significant water content in the flue gases, for example, where a wet scrubber has been used, then some form of re-heating of the gases would be required (Clarke 1996).

6.2.1.5 *Energy Recovery via District Heating and Electricity Generation*

The modern municipal waste incinerator relies on the production of steam for electricity generation or district heating to ensure the cost-effectiveness of the process. Some schemes may incorporate both electricity generation and district heating as combined heat and power (CHP) systems. Electricity is generated from the steam produced in the boilers via a steam-condensing turbine. The high pressure, high temperature steam enters the turbine and passes through the various stages of the turbine. As it does so it expands and reaches high velocity, turning the blades of the turbine and hence the turbine shaft, which generates the electricity (Gilpin 1982; Porteous 1992). Where district heating is the objective, the high temperature and pressure steam passes through heat exchangers which generate hot water under pressure for distribution to homes, offices and institutions. CHP systems would use a different type of steam turbine which would generate a lower amount of electricity, but the steam effluent from the turbine would be at a higher temperature, allowing district heating to be incorporated (Porteous 1992). Evaluation of which option is the best is very much a site-specific issue. Contracts to sell the electricity or heat supply to the local district should be secured. In addition, contracting for waste to fuel the plant on a long-term basis should also be secured. Electricity generation income may be enhanced if a non-fossil fuel obligation (NFFO) contract can be obtained from the regional electricity company. District heating schemes rely on a market for the heat, which in the case of domestic and commercial premises may be seasonal.

There has been some concern that recycling schemes may reduce the calorific value of municipal solid waste and adversely affect the performance and energy recovery potential of large scale municipal solid waste incineration. However, a recent study (Atkinson et al 1996) has shown that source separation of waste does not significantly affect the calorific value of the derived waste used in the incinerator. The study examined a range of 'bring' and 'kerbside' collection systems for recovering recyclable materials. In each case the level of recovery for the four types of recycled materials examined – newspapers and magazines, glass bottles and jars, metal cans and plastic bottles – was estimated to be between 4.4% and 13% of the household waste stream (see Table 4.16). However, the reduction in the calorific value of the waste was estimated at only between 5 and 10% because the calorific value of the resulting derived waste is increased as the recycled materials are removed. This was due to the fact that non-combustible materials such as metal cans and glass are removed, but also a decrease in the moisture and ash content of the derived waste. It was concluded that such variations in calorific value were not significant and would make little difference to the energy recovery performance of a municipal solid waste incinerator.

Box 6.9 describes the Sheffield municipal solid waste incinerator which utilises energy recovery through a district heating system, and Box 6.10 describes the SELCHP municipal solid waste incinerator which utilises energy recovery through electricity generation (Lawrence 1995; Atkins 1997).

6.2.2 Other Types of Incineration

There are a wide variety of incineration types used to incinerate a wide variety of wastes. In this section a number of different design types will be discussed in relation to their technologies and application to different types of waste:

Box 6.9
The Sheffield MSW Incinerator

The Sheffield incinerator was completed in 1975 and operated with energy recovery via district heating. The incinerator represents an older design, but has recently been upgraded throughout to meet EC emissions legislation. The Sheffield incinerator is operated by Sheffield Heat and Power. The company has a joint shareholding between Sheffield City Council, British Gas and EKONO, of Finland, a specialist CHP company. The incinerator consists of two 10 tonne per hour boilers fired by municipal solid waste, giving a total annual throughput of 120 000 tonnes. The waste is combusted to produce about 34 MW of heat energy. The furnace grates are of the roller type. The heat produces about 3.2 tonnes of steam per tonne of waste throughput, which for the two 10 tonne per hour boilers is equivalent to a total of approximately 65 tonnes of high pressure steam per hour. In addition, a separate clinical waste incinerator which is located on the same site handles 4000 tonnes per year of waste generated from Sheffield hospitals. The clinical waste incinerator is also equipped with energy recovery boilers which produce 4 MW of energy to the district heating system.

The generated steam is passed to a heat exchanger to produce hot water under pressure at up to 16 bar and a temperature up to 120 °C for transfer to the district heating system, which provides heat for 3500 homes and over 50 commercial and institutional buildings. These include university buildings, hospitals, theatres, banks, cinemas, libraries, local government offices and commercial premises. Two 15 MW boilers operated by fossil fuels are available for back-up in the event of high heat demands for the district heating system. When there is low demand for the generated heat, dump condensers are capable of condensing all the generated steam raised in the furnaces. Upgrading of the plant to meet the 1996 EC emissions legislation has led to the installation of a dry scrubber and fabric filter system in addition to the existing electrostatic precipitators. A new system for collecting and storing the ash and dust has also been installed. In addition, a new 7.6 MW steam turbine has been added to generate electricity. The hot exhaust from the turbine will be used to provide the heat for the district heating system, and consequently Sheffield will operate a true CHP system. The Sheffield plant successfully applied for a non-fossil fuel obligation (NFFO) contract with the local Regional Electricity Company. The electricity and heat output from the incinerator are optimised to the demand for heat on the district heating system.

The district heating system is based on a network of pre-insulated pipes of between 80 and 300 mm internal diameter, which stretch for over 34 kilometres throughout the city. The pipes consist of an inner pipe, foam insulation and an outer casing bonded together, with an in-built monitoring system to detect leakages. The heat energy from the hot water district system is transferred to the consumer's internal heating and hot water system via heat exchangers. The rate of heat transfer is governed by water flow rate and heat exchanger surface area; charging is via electronic heat meters. In addition to energy recovery, the Sheffield plant also recovers ferrous metals from the bottom ash.

Source: Lawrence D., Incineration of solid municipal waste with energy recovery by district heating: The Sheffield MSW Incineration Plant. Incineration of Municipal Waste with Energy Recovery. University of Leeds, Industrial Short Course, Department of Fuel and Energy, University of Leeds, 1995.

Box 6.10
The SELCHP Incinerator

The South East London Combined Heat and Power (SELCHP) limited incinerator is a new-generation incinerator completed in 1994, with energy recovered in the form of electricity generation. There are plans to extend the energy recovery to include a district heating system to provide heat for 7500 houses. SELCHP is owned by a consortia of local authorities and private companies, including the London Boroughs of Lewisham and Greenwich, private waste-to-energy plant developers, designers and operating companies, and the local electricity company.

The plant has a disposal capacity of 420 000 tonnes/year of municipal and commercial waste. The plant has two furnaces streams, each capable of 29 tonnes/hour throughput. The furnace grate is of the reciprocating counterflow Martin type, comprising alternate steps of fixed and moving grate bars, and is inclined at 26° to the horizontal. This type of grate continually rotates the burning bed of waste which is pushed back up the grate, resulting in combustion throughout the grate, including the front portion. Consequently, the drying, ignition and combustion stages take place simultaneously. The waste is fed to the grate via ram feeders from the feed hopper. Forced draught fans provide primary air via the under-grate air zones to the burning waste bed. The boilers generate 144 tonnes/hour steam at 395 °C and 46 bar pressure. The boiler is of the four-pass variety, incorporating a superheater through to an economiser. The flue gas clean-up system comprises a semi-dry lime scrubber and fabric filter. Emissions have been shown to be well below the EC legislated limits.

The steam produces 32 MW of electrical energy from a steam turbine. The turbine is currently used only for electricity generation, and the hot exhaust gases are presently just condensed, but the turbine is also designed to utilise the exhaust heat for the planned district heating system. Electricity is generated at 11 kV and, by the use of transformers, up-rated to 132 kV for export to the National Grid. The SELCHP incinerator was awarded a non-fossil fuel obligation (NFFO) contract and the electricity is sold to the Non-Fossil Fuel Purchasing Agency at a premium price. The NFFO places an obligation on the Regional Electricity Companies to purchase a certain capacity of electricity generated from non-fossil fuels such as waste. In addition to income from electricity sales, the incinerator also recycles ferrous metals from the ash residues, which are separated by a magnetic separator. The ash residues are then landfilled, although there are plans to recycle the ash for construction applications.

Source: Atkins G., Incineration of solid municipal waste with energy recovery via electricity generation: The SELCHP scheme. Incineration of Municipal Waste with Energy Recovery. University of Leeds, Industrial Short Course, Department of Fuel and Energy, University of Leeds, 1997.

- fluidised bed incinerators;
- starved air incinerators;
- rotary kiln incinerators and cement kilns;
- liquid and gaseous waste incinerators.

6.2.2.1 Fluidised Bed Incinerators

Fluidised bed incinerators have been used for a wide variety of wastes, including municipal solid waste, sewage sludge, hazardous waste, liquid and gaseous wastes, and those wastes with difficult combustion properties (Neissen 1978; Buekens and Patrick 1985; Waste Incineration 1996). Fluidised beds are mainly of the bubbling, turbulent or circulating bed types, although some pressurised fluidised beds have been built for coal combustion for power generation. Figure 6.15 shows schematic diagrams of bubbling bed, turbulent and circulating fluidised beds (Buekens and Schoeters 1984).

Fluidised beds consist of a bed of sand particles contained in a refractory-lined chamber through which the primary combustion air is blown from below; the sand particles are hence fluidised by adjusting the air flow. Increasing the air flow produces a turbulent flow of solids, and to prevent elutriation of the bed material out of the freeboard, cyclones are placed within the freeboard to re-circulate the solids back into the bed. Further increase in air flow produces a circulating fluidised bed, where the solids are intentionally elutriated out of the bed into a cyclone and the material is re-circulated back to the bed. However, in the circulating fluidised bed combustion also takes place in the cyclone. Such beds are much longer and produce longer residence times of the solid particles of waste in the hot zone, resulting in higher burn-out of the products of combustion and reduced organic emissions.

The fluidised bed reactor promotes the dispersion of incoming waste, with rapid heating to ignition temperature, and promotes sufficient residence time in the reactor for their complete combustion. Secondary functions include the uniform heating of excess air, good heat transfer for heat exchange surfaces within the bed, and the ability to reduce gaseous emissions by control of temperature or the addition of additives directly to the bed to adsorb pollutants, for example, the addition of lime to reduce sulphur dioxide emissions (Probert et al 1987; Waste Incineration 1996). The fluidised bed reactor greatly increases the burning rate of waste since the rate of pyrolysis of the solid waste material is increased by direct contact with the hot, inert bed material. Gases in the bed are continuously mixed by the bed material, thus enhancing the flow of gases to and from the burning solid surface and enhancing the completeness and rate of gas-phase combustion reaction. This factor becomes more pronounced with circulating fluidised beds with their longer residence times in the hot zone. In addition, the charred surface of the burning solid material is continuously abraded by the bed material, enhancing the rate of new char formation and the rate of char oxidation. Fluidised beds are compact and have high heat storage with fast dynamic response to throughput or demand. They also have high heat transfer rates, and thus allow faster ignition of low combustible waste. Because of the high heat transfer rates found in fluidised beds they are very good for heat recovery processes, and the heat transfer surfaces may be placed within the bed (Buekens and Schoeters 1984).

Fluidised beds, by the nature of their combustion, are able to incinerate a range of wastes. Municipal solid waste may be incinerated in a fluidised bed incinerator, but this is best achieved by some form of pre-screening and shredding or the production of RDF pellets (Ballantyne et al 1980). The solid waste is then fed by screw gravity feeding or pneumatically. Processing routes have involved the production of refuse derived fuel in shredded or pelletised form. Large particles in fluidised beds can cause problems due to agglomeration which prevents fluidisation and the bed

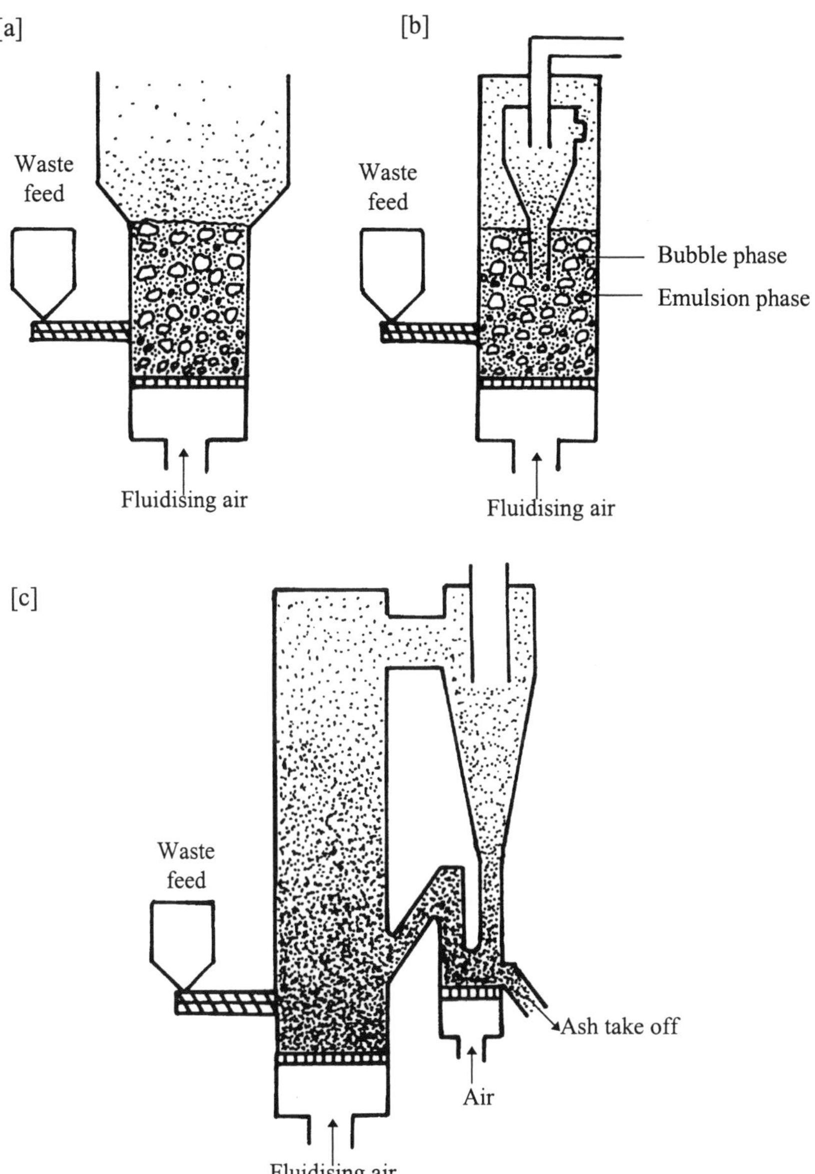

Figure 6.15 *Schematic diagram of (a) a bubbling bed, (b) a turbulent and (c) a circulating fluidised bed. Source: Adapted from Buekens A. and Schoeters J., Thermal methods in waste disposal. Study performed for the EEC under contract EC1 1011/B7210/83/B, Free University of Brussels, Brussels, 1984*

consequently slumps. The agglomerated masses then have to be removed. Fusion of ash particles in very hot zones in the bed can also cause agglomerates to form. Examples of fluidised beds include the Fayetteville, North Carolina, USA, plant used for the incineration of shredded MSW via a fluidised bed (DTI OSTEMS Mission 1996). The shredded MSW is derived from a 245 000 tonne/year throughput, integrated recycling and energy from waste facility in which materials are extracted from the waste stream to leave a combustible fraction for the fluidised bed incinerator. Sweden also has well-developed fluidised bed technology for incineration of municipal solid waste. For example, Lidkoping utilises a 50 000 tonne/year bubbling fluidised bed for shredded refuse derived fuel, where the energy recovered is used for a district heating scheme (DTI OSTEMS Mission 1996). An example of a circulating fluidised bed occurs in Sundsvall, Sweden, where a mixture of shredded refuse derived fuel, peat and wood chips is used. Japan also has a large fluidised bed industry to incinerate MSW, with over 100 bubbling-type incinerators (Waste Management in Japan 1995; Whiting 1996). The incinerators are fed by waste which has already been pre-sorted by the householder to produce a combustible fraction. Consequently there is a lower requirement for pre-processing, only shredding generally being required. The range of incinerator sizes are from 35 000 tonnes/year to over 150 000 tonnes/year. A schematic diagram of a typical refuse derived fuel circulating fluidised bed incinerator is shown in Figure 6.16 (Urban Waste and Power 1988). In this diagram limestone is included, which can be added to the bed to remove sulphur dioxide, also optional fuels such as coal or wood to aid in the combustion of difficult wastes.

Fluidised beds have been successfully developed for the incineration of sewage sludge, and in the UK, fluidised bed technology is the preferred option (Frost 1992; Waste Incineration 1996). Sewage sludge has a high water content, typically 96%. The dried solids have a relatively high calorific value of about 20–24 MJ/kg, but a high ash content of between 20 and 50% (Bruce et al 1989; Frost 1992). Dewatering the sludge is expensive, so that a balance is struck between dewatering and raising the calorific value of the wet sludge sufficiently to allow combustion to take place. Lower levels of dewatering can be sufficient if a supplementary fuel is used with the sludge. Normally the sludge entering the fluidised bed would have been thickened and dewatered to some extent using mechanical dewatering. The limiting dry solids content of a sludge which, if fed to the furnace, would require no supplementary fuel is termed the ‘autothermic solids content’ (Frost 1992). A typical bubbling fluidised bed for sewage sludge incineration is shown in Figure 6.17 (Oppelt 1987; Frost 1992). Figure 6.18 shows two sewage sludge fluidised bed incinerator designs (Frost 1992). Figure 6.18a shows a simple mechanical dewatering system to produce 24% dry solids in the sludge. Such a sludge would not be autothermic and therefore supplementary fuel would be required. Figure 6.18b shows that after the mechanical dewatering stage, flue gas heat from the combustion process is used to dry the sludge and increase the solids content to 45%. Supplementary fuel is not required here except in the initial start-up process. The hot flue gases may be passed to a boiler system for energy recovery. Typical throughputs of sewage sludge are between 15 000 and 25 000 tonnes of dry solids per year.

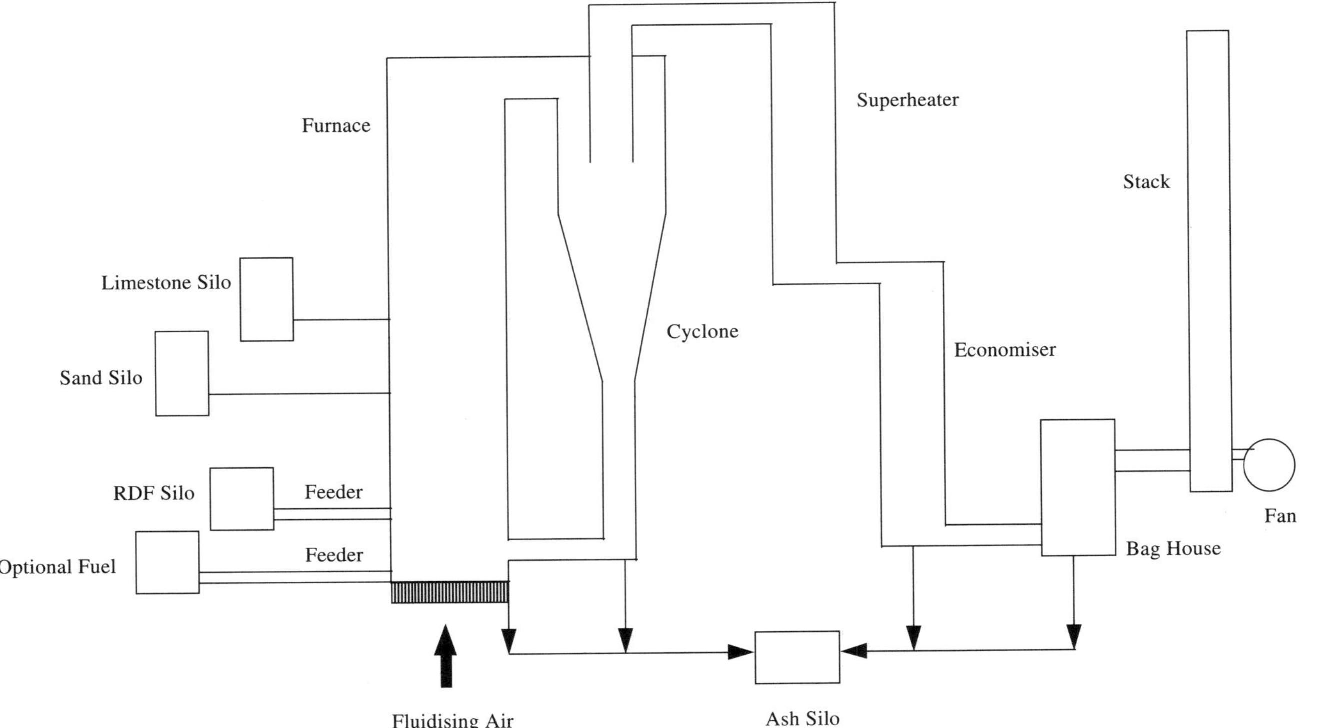

Figure 6.16 *Schematic diagram of a typical circulating fluidised bed incinerator for refuse derived fuel. Source: Adapted from Urban Waste and Power, Power from Waste, Ashford, 1988*

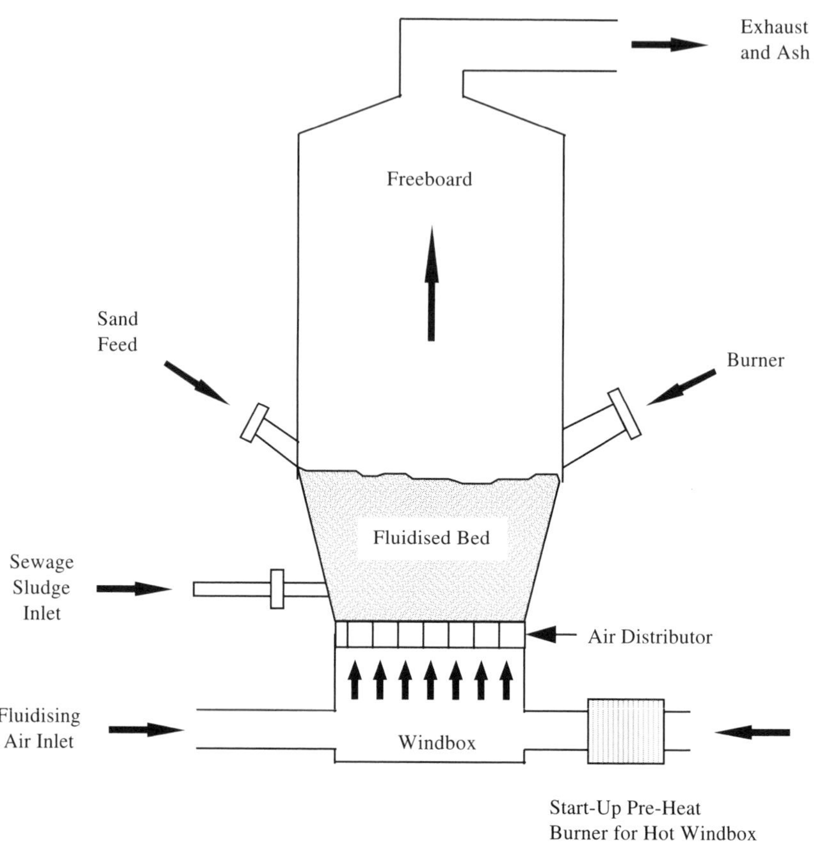

Figure 6.17 *Typical bubbling bed fluidised bed incinerator for sewage sludge. Sources: [1] Frost R., Incineration of Sewage Sludge. University of Leeds Short Course 'Incineration and Energy from Waste'. University of Leeds, 1992; [2] Oppelt H.M., Journal of the Air Pollution Control Association, 37, 558–586, 1987*

Sewage sludge incinerators of the multiple-hearth type have been used in the UK and are also used extensively throughout the USA and Japan (Figure 6.19, Oppelt 1987; Brunner 1991). The incinerator has between 5 and 12 hearths and is designed to handle wastes with high moisture contents. The sludge, which requires dewatering to at least 15% dry solids content, is fed into the top of the incinerator. The sludge moves down through the hearths by movement of the rabble arms, which move the sludge alternately through the centre and edge of the hearths. Flue gases pass up through the furnace. The upper hearths act as drying hearths utilising the hot flue gases from the burning sludge in the middle and lower hearths. Furnace temperatures are up to 900 °C. Ash residues exit at the base of the incinerator (Bruce et al 1989; Frost 1992; Hall, 1992).

Box 6.11 describes an example of a sewage sludge fluidised bed incinerator (Frost 1992; Hudson 1993).

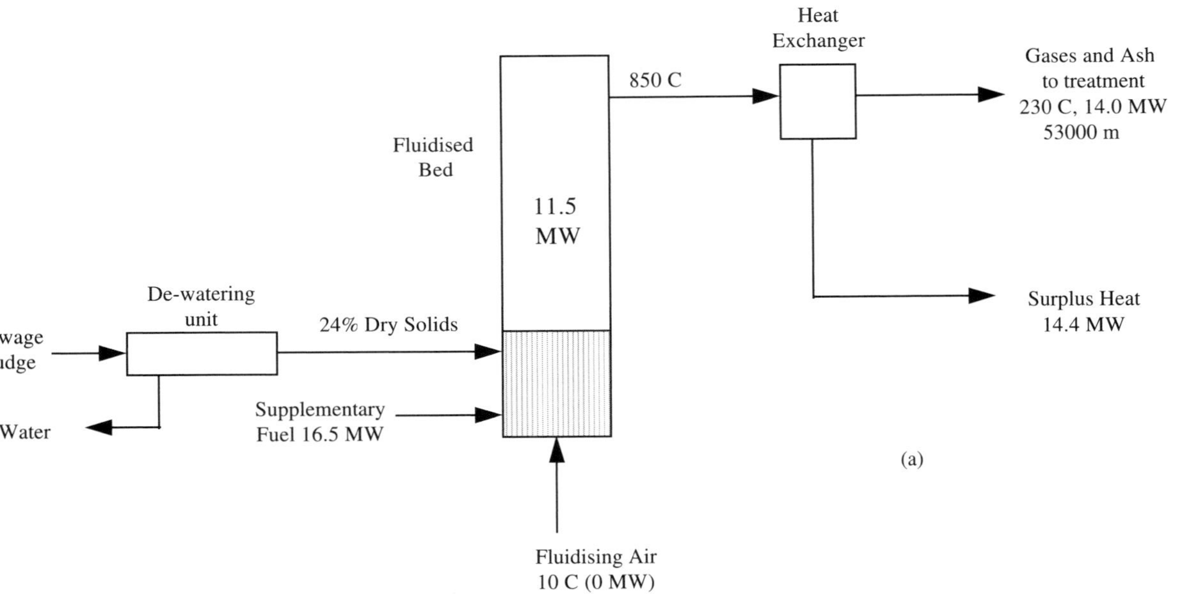

Figure 6.18 *Schematic diagrams of bubbling bed fluidised bed designs for incineration of sewage sludge. Source: Adapted from Frost R., Incineration of Sewage Sludge. University of Leeds Short Course 'Incineration and Energy from Waste'. University of Leeds, 1992*

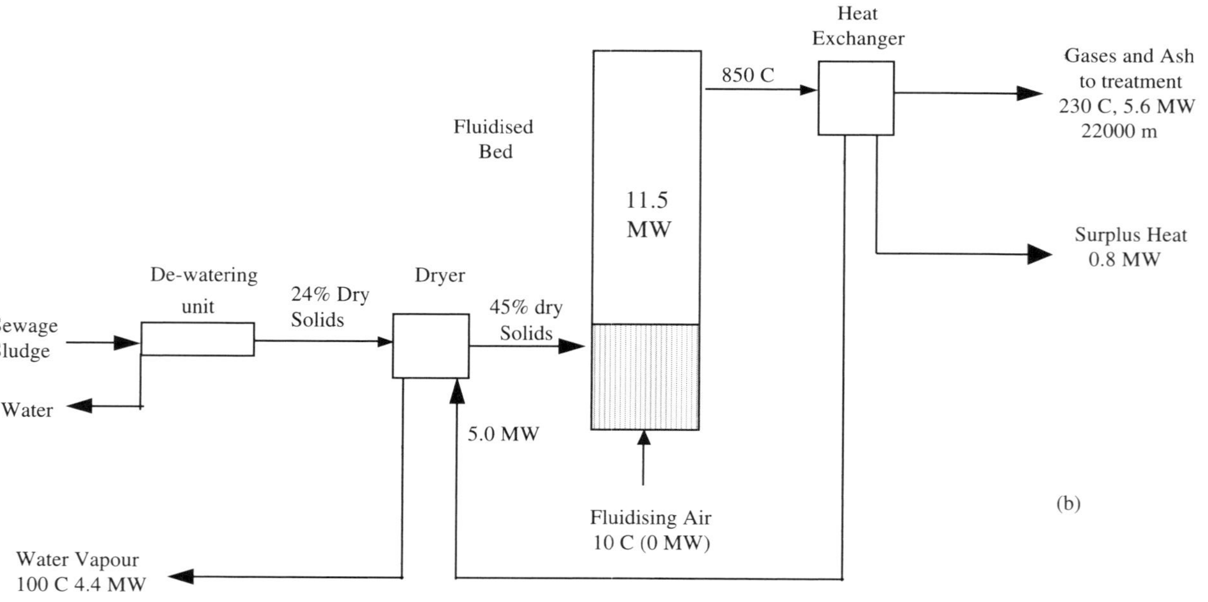

Figure 6.18 *(continued)*

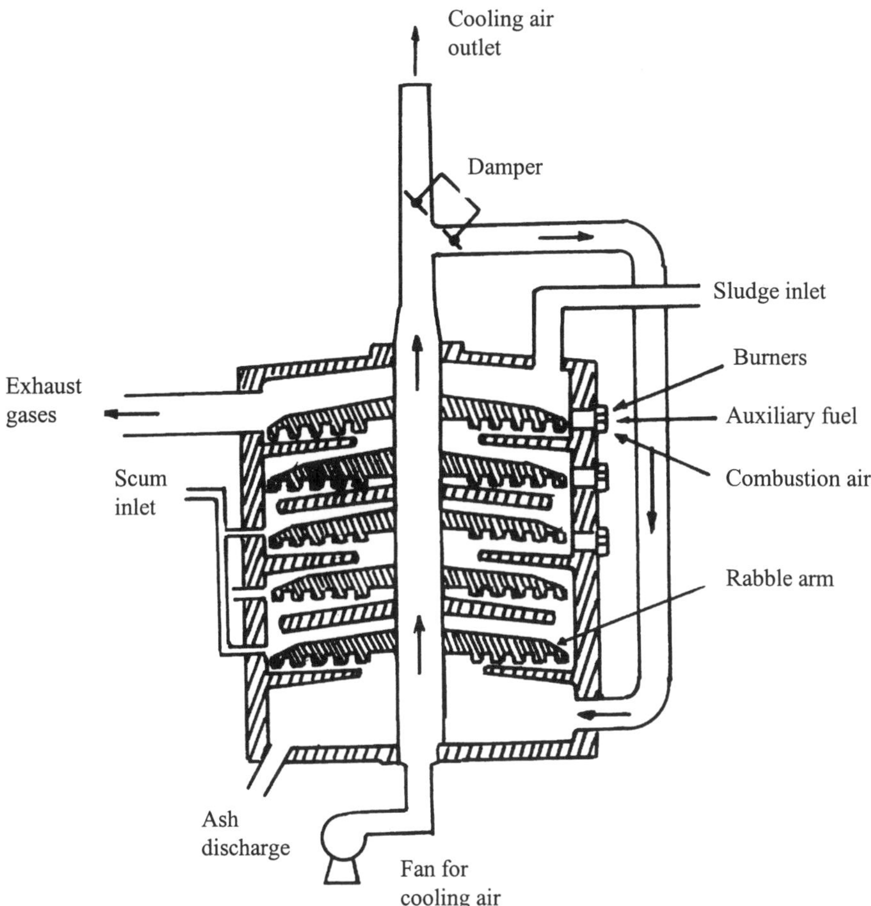

Figure 6.19 *Schematic diagram of a multiple-hearth incinerator used for sewage sludge. Sources: [1] Oppelt H.M., Journal of the Air Pollution Control Association, 37, 558–586, 1987; [2] Brunner C.R., Handbook of Incineration Systems. McGraw-Hill, New York, 1991*

6.2.2.2 Starved Air Incinerators

Starved air or pyrolytic incinerators are two-stage combustion type incinerators which are widely used for clinical waste incineration and also for some industrial wastes. The two stages consist of a pyrolytic stage and a combustion stage. The system is used mainly for solid waste, although it can be designed for liquid sludges and gaseous waste. A typical two-stage solid waste incinerator with a waste heat recovery system is shown in Figure 6.20 (Oppelt 1987). The advantages of the starved air incinerator are a more controlled combustion process leading to lower releases of volatile organic compounds and carbon monoxide (Waste Incineration 1996). In addition, the low combustion air flow results in low entrainment of

Box 6.11
Fluidised Bed Sewage Sludge Incinerator: Esholt, Yorkshire

The incinerator is a bubbling fluidised bed type which has a design operating temperature of about 850 °C with a flue gas residence time of 2 s to ensure efficient burn-out of organic products. Gas cleaning is by electrostatic precipitation followed by wet scrubbing. The calorific value of the dry solids in the sludges are typically between 26 000 and 28 000 KJ/kg, with a high ash content of about 35%. The Esholt incinerator does not have a drying stage, and consequently a high efficiency mechanical dewatering stage is required to produce sludge with a dry solids content of above 30%. The incinerator therefore operates autothermically, but with pre-heated combustion air using exhaust flue gas heat.

Emissions data

	IPR 5/11 legislated emission limits[1] (mg/m^3)	Esholt emissions (1990–92)
Particulate matter	20	3.6
CO	50	15
NOx	650	210
HCl	30	3.6
H_2S	2	0.6
SO_2	300	139
Metals		
Hg	0.1	0.06
Cd	0.1	0.14
Ni + As + Pb + Cr + Cu + Sn + Mn	1.0	0.49
Dioxins and furans (TEQ ng/m^3)	1.0	<0.12

[1] Environmental Protection Act 1990. Process Guidance Note IPR 5/11. Sewage sludge incineration.

Sources: [1] Hudson J,. Sewage sludge incineration. In: Incineration: An Environmental Solution for Waste Disposal. IBC Technical Services Conference, 16/17 February, Manchester, 1993; [2] Frost R., Incineration of sewage sludge, Incineration of Municipal Waste with Energy Recovery. University of Leeds, Industrial Short Course, Department of Fuel and Energy, University of Leeds, 1992.

particulate matter in the flue gases, which also reduces other particulate-borne pollutants such as heavy metals and dioxins and furans.

Pyrolysis is defined as the chemical decomposition of the waste by the action of heat. Heating the waste in an inert atmosphere produces a gas which, when ignited, is self-supporting in air. In practice the two-stage combustor relies on semi-pyrolysis, where the heat for the thermal decomposition or gasification of the waste is produced by sub-stoichiometric combustion of the waste. The waste is combusted under sub-stoichiometric conditions, i.e., where there is insufficient air to provide complete combustion and therefore there is a high proportion of the products of incomplete combustion which pass through to the second stage (Priest 1985; Brunner 1991).

The two-stage process ensures that gas velocities are relatively low and particulate matter is largely retained in the first stage However, with the stringent legislative requirements on emissions, the full range of gas clean-up systems may be required to control other emissions such as acid gases, heavy metals and dioxins and furans.

Box 6.11
(continued)

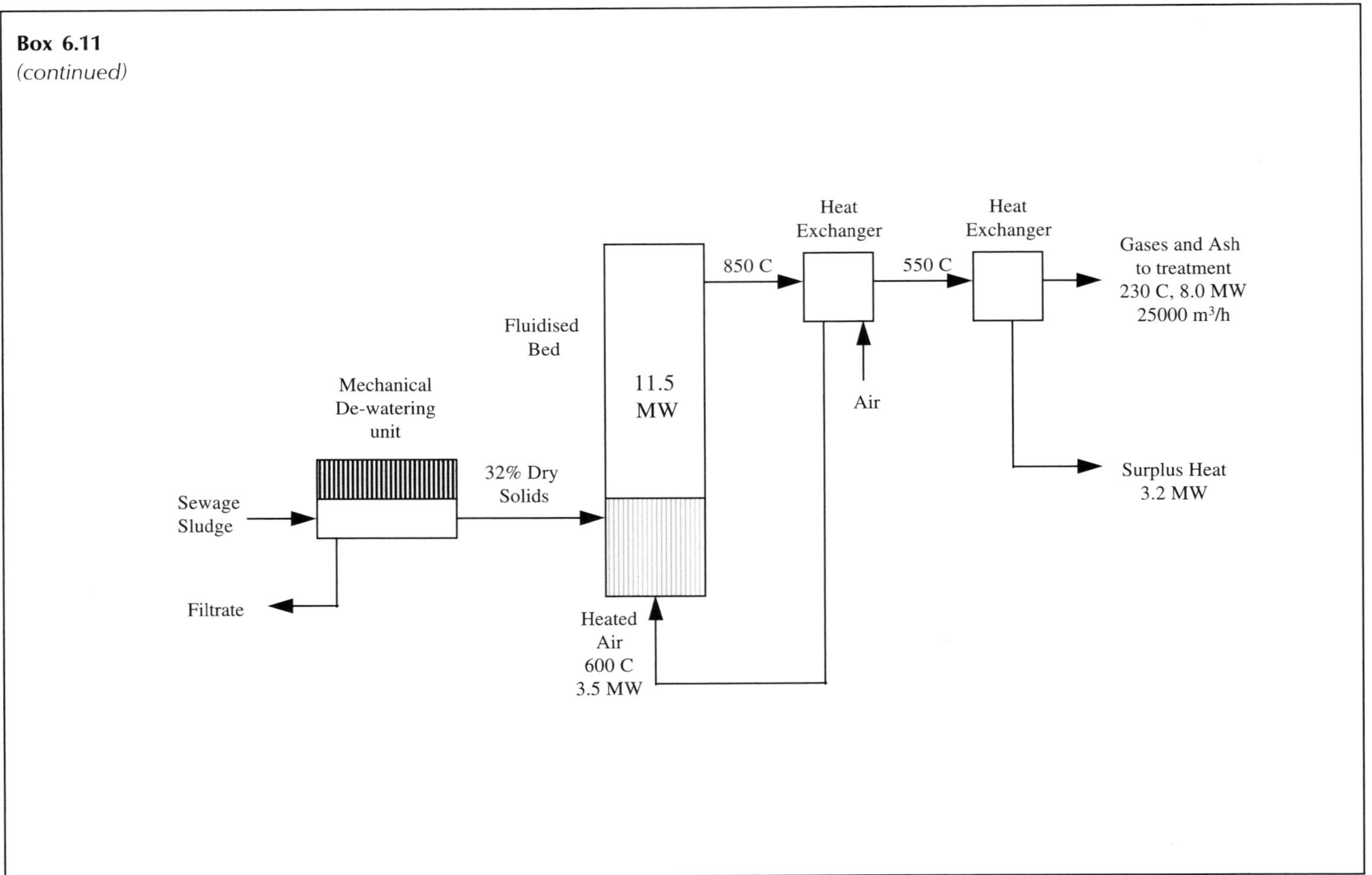

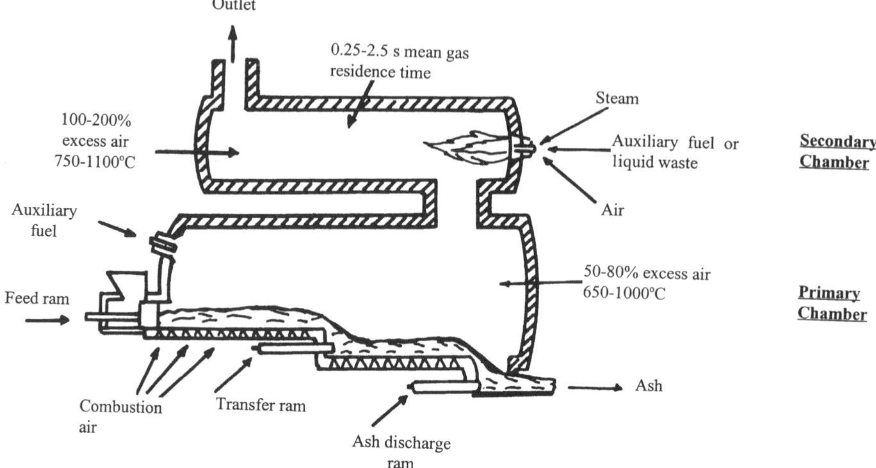

Figure 6.20 *Schematic diagram of a starved air two-stage incinerator. Source: Oppelt H.M., Journal of the Air Pollution Control Association, 37, 558–586, 1987. Reproduced with the permission of the Air and Waste Management Association*

The pyrolytic/gasification reactions that take place are numerous and complex. Figure 6.21 shows the reactions for a typical hydrocarbon $(CH_2)n$ (Priest 1985). The sub-stoichiometric conditions produce a reducing atmosphere within the primary chamber, and the heat generated breaks down the hydrocarbon in the pyrolysis zone into carbon and hydrogen. The carbon reacts with the CO_2 and H_2O generated earlier to give CO and H_2, which passes to the second-stage combustion zone where complete combustion takes place.

The temperature of the gases leaving the pyrolytic section are of the order of 700–800 °C, since a high proportion of the heat generated is used in the endothermic pyrolytic process. These gases will then pass to the secondary section, where secondary excess air, approximately 200% stoichiometric, is added to give a temperature of 1000–1200 °C which completes the combustion process by combusting the hydrogen, carbon monoxide and hydrocarbons. The two-stage combustion process inhibits the formation of NOx (Waste Incineration 1996).

The relatively long residence time within the secondary chamber plus the high temperature of over 1000 °C will destroy any dioxins and PCBs that are contained in the secondary gases. This does not preclude any de-novo formation of dioxins, and consequently clean-up procedures of, for example, additive activated carbon and/or lime plus a fabric filter are also required In practical terms the smaller two-stage incinerator of up to 0.75 tonnes/hour capacity tends to be made of vertical units operating on a batch basis. The larger units of 0.75–5 tonnes/hour capacity are designed as horizontal units and have automatic feeding and de-ashing (Priest 1985).

In most cases it is not necessary to pre-treat the waste prior to loading into the furnace. Charging is normally done using a hydraulic ram system which pushes the

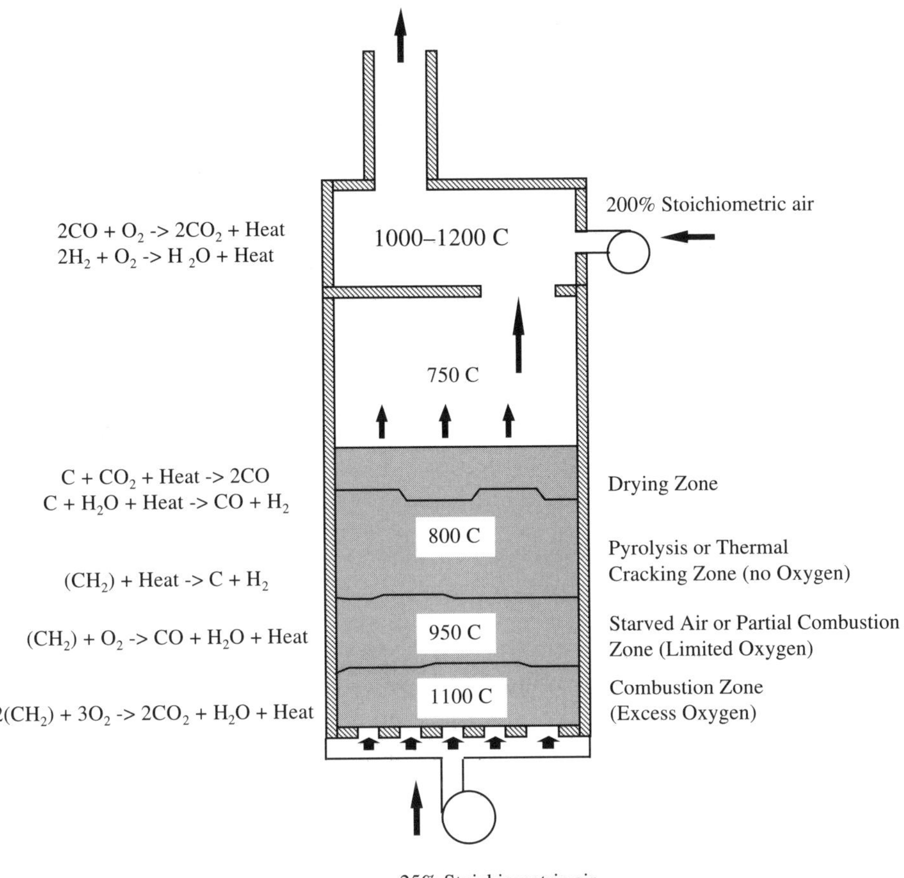

Figure 6.21 *Starved air, two-stage semi-pyrolytic combustion reaction for a typical hydrocarbon. Source: Priest G.M., Producing steam from the combustion of solid waste. In: Energy Recovery from Refuse Incineration. Seminar, Institution of Mechanical Engineers, Strathclyde, August 1985. Reproduced by permission from Mechanical Engineering Publications Ltd.*

waste via a refractory-lined guillotine door to the pyrolysis chamber. The primary chamber process consists of combustion, partial combustion, pyrolysis and drying. Primary air is fed through the grate to give even combustion at the base of the waste. The residence time of the waste in the primary chamber will be dependent on the hearth area and the characteristics of the waste, and should produce an almost carbon-free ash; the time may be anything from 6 to 12 hours. The ash is finally discharged at the rear of the incinerator into a water trough.

The gases entering the secondary chamber have a sufficiently high calorific value to be self-sustaining in combustion. Secondary air is introduced to provide the excess air conditions with a high degree of turbulence to create sufficient mixing to sustain combustion without the use of support fuel at typical operating temperatures of between 1000 and 1200 °C. Long residence times in the secondary chamber also

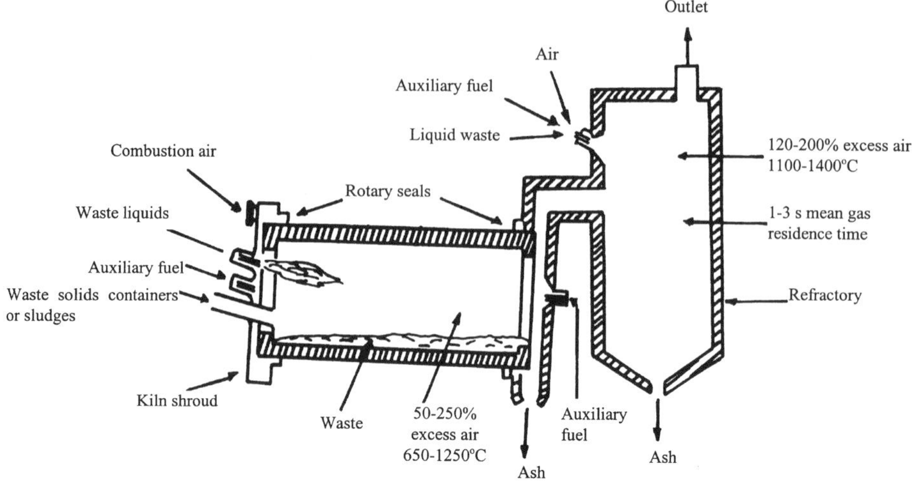

Figure 6.22 *Schematic diagram of a rotary kiln incinerator. Source: Oppelt H.M., Journal of the Air Pollution Control Association, 37, 558–586, 1987. Reproduced with the permission of the Air and Waste Management Association*

allow complete burn-out of the combustible gases, vapours, tars and soot. Auxiliary burners are also employed for the initial start-up and then combustion is sustained by the gases from the pyrolytic stage, when the burner may be switched off (Priest 1985; Brunner 1991).

6.2.2.3 Rotary Kiln Incinerators

The rotary kiln is a two-stage incineration type, but the first stage is usually operated in the oxidative mode, i.e., with about 50–200% excess air, rather than the semi-pyrolytic mode. Rotary kilns have been used for a wide variety of wastes, including municipal solid waste, sewage sludge, industrial waste and hazardous waste, and for clean-up of contaminated soils (Waste Incineration 1996). However, they are most common for the treatment of hazardous and industrial wastes, where in some cases whole drums of waste are fed to the rotary kiln to be completely destroyed (Waste Incineration 1996). Figure 6.22 shows a schematic diagram of a rotary kiln incinerator (Oppelt 1987).

The rotary kiln is the primary chamber consisting of an inclined cylinder lined with ceramic material which rotates at rates which can vary between two revolutions per minute to six revolutions per hour depending on the type of waste and type of rotary kiln. The size of the rotary kiln can be from 1 to 6 m diameter and from 4 to 20 m in length. The kiln is rotated by a series of rollers on which the kiln is located. The waste is fed to the front end and ignited by a burner. The combusting wastes are tumbled and agitated by the rotation of the kiln and move down the kiln to reach the end as ash. Internal baffles may be used to increase the mixing and turning of the waste.

The kiln normally operates at between 900 and 1100 °C, but the 'slagging type' of rotary kiln operates at temperatures up to 1500 °C. The high temperature of the 'slagging type' rotary kiln allows the formation of a molten slag of the ash, and whole drums of waste can be incinerated as the metal drums will melt (Waste Incineration 1996). The presence of the molten slag absorbs particulate matter, including heavy metals. The ash or molten slag exits from the kiln into a quench pit. The molten ash forms a glass-like material which is less susceptible to leaching of the dissolved metals (Brunner 1991).

The gases from the primary rotary kiln pass to the secondary chamber where excess air conditions with auxiliary burners serve to burn out completely the combustible gases, vapours, tars and soot. A secondary chamber is particularly necessary for hazardous wastes, where the time, temperature and turbulence may be insufficient to guarantee the complete combustion of all the organic components of the waste (Oppelt 1987). Typical temperatures in the secondary chamber would be 1400 °C, with residence times of between 1 and 3 s and up to 200% excess air levels. The rotary kiln combines a long residence time and a high temperature, which allows the complete combustion of complex hazardous wastes (Oppelt 1987).

Box 6.12 describes an example of a slagging rotary kiln incinerator used for the incineration of hazardous waste.

Cement kilns used for the production of cement are used in some countries, including the UK, to dispose of a variety of wastes including municipal solid waste, industrial waste, tyres and hazardous wastes. The waste is blended with other fuels in cement kilns, which utilise rotary kiln technology. The rotary kiln cylinder is up to 250 m in length and 4 m diameter and is lined with alumina bricks. Normally coal or oil is used as the fuel, but this is sometimes supplemented with waste. Chalk or limestone, plus clay or shale are mixed with water to form a slurry which is passed through the high temperature furnace to form cement clinker. After processing through the kiln, the cement clinker is ground and gypsum is added to produce cement. The combustion temperatures within the cement kiln are very high, typically more than 1400 °C (Gilpin 1982). The process is very energy-intensive, and the use of waste material offsets the costs of fuel. The high temperatures and long residence times used in the process serve to destroy the waste. In addition, when chlorinated or fluorinated wastes are combusted, the large mass of alkaline clinker from the process absorbs and neutralises the acidic stack gases (Holmes 1995; Benestad 1989).

6.2.2.4 *Liquid and Gaseous Waste Incinerators*

Liquid and gaseous waste incinerators pass the waste into a burner which mixes the (combustible) waste with air to form a flame zone which burns the waste. Figure 6.23 shows a typical liquid waste incinerator (Oppelt 1987). Supplementary fuel may be required depending on the calorific value of the waste, or else the liquid waste is pumped directly into a flame generated by the burner fired by a conventional fuel. The flame is fired into a ceramically lined combustion chamber which radiates heat back into the exhaust gases, thus providing an extended hot zone to burn out completely the products from the combustion of the waste. Very high temperatures

Box 6.12
Slagging Rotary Kiln Incineration of Hazardous Waste, Rechem International Ltd., Fawley, Hampshire

The slagging rotary kiln incinerator was commissioned in 1990. It has a thermal capacity of 22.2 MW and allows a nominal throughput of 4.4 tonnes/hour. The plant handles hazardous waste and clinical wastes. The liquid wastes are delivered in tankers or drums and may be blended to produce a composition suitable for the incinerator. High hazard classification liquid wastes are stored and incinerated in isolation from other wastes. Drummed solid wastes are passed to the incinerator via a shredder/blender continuous feed system for the majority of the solid wastes. Those drummed solid wastes unacceptable to the shredder/blender system are fed directly to the incinerator by a drum feeding system. The gas clean-up system comprises the secondary combustion chamber which operates at 1100 °C to burn out any hydrocarbon gases. Further gas clean-up is by a quench tower to partially remove particulates and acid gases under acidic quench liquor conditions. There follows a packed absorber tower for almost complete removal of particulates and acid gases by recirculation of alkaline liquor. Finally, two wet electrostatic precipitators in series with pre-conditioning water sprays are used to remove particulates. The table shows Environment Agency Guidance limits (IPR 5/1) for hazardous waste incinerators and the actual emissions of the Fawley plant over a 5-month period.

	IPR 5/1 Hazardous Waste Guidance Limits (mg/m^3)	Fawley plant actual data[1] (mg/m^3)
Particulates	20	3.4
Volatile organic compounds	20	5.6
Hydrogen chloride	10	<1
Hydrogen fluoride	2	<1
Sulphur dioxide	50	<1
Nitrogen oxides	350	76
Mercury	0.1	0.003
Cadmium	0.1	0.0005
Other heavy metals	1.0	0.115
Carbon monoxide[2]	100	<1
Dioxins (TEQ, ng/m^3)	1.0/0.1	0.14[3]

NB All measurements at reference conditions: 273 K; 101.3 kPa; 11% O_2 dry gas.
[1] Average of 14 measurements, January to May 1995.
[2] Hourly average.
[3] Average of 4 measurements.

Source: Pritchard M.L. and Turner G., Technical and Economic Study of Incineration Processes. DOE Report DOE/HMIP/RR/95/027, The Environment Agency, Department of the Environment, Bristol, 1996.

occur in the flame, of the order of 1400–1650 °C, and the furnace chamber temperatures are between 820 and 1200 °C. The chamber may be horizontal or vertical. Liquid waste incinerators are used extensively for the combustion of hazardous wastes. The key section of the incinerator is the burner, which essentially serves to atomise the waste to form a fine spray of droplets and vapour which ignites to form the flame. Several different designs of burner nozzle exist to cope with the wide range of properties found with liquid wastes and sludges (Brunner 1991). The liquid waste is

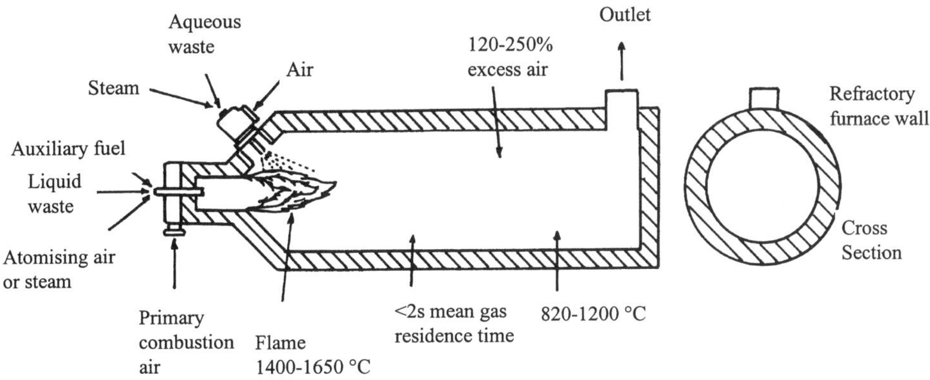

Figure 6.23 *Typical liquid waste incinerator. Source: Oppelt H.M., Journal of the Air Pollution Control Association, 37, 558–586, 1987. Reproduced with the permission of the Air and Waste Management Association*

pumped under high pressure through the burner nozzle, which produces a fine spray of atomised droplets of size range typically between 10 and 150 μm. The smaller the droplet size, the easier vaporisation becomes, and consequently the burn-out of each droplet takes place in a much shorter time (A. Williams 1990).

Gaseous waste incinerators operate on a similar system to liquid waste incinerators but the difficulties of producing a fine spray or vapour for combustion have already been overcome. Gaseous wastes usually consist of organic hydrocarbons, which are combustible. The gases or vapours may be of low concentration and consequently are not autothermic, and therefore the gases or vapours are passed into a burner with either the supplementary fuel gas or combustion air, or may be passed directly into the flame zone. The combustion chamber provides a long residence time for complete burn-out of the gaseous waste (Brunner 1991).

6.3 Economics of Incineration

Incineration is mainly capital-intensive, the major cost factors being interest paid on borrowed capital, equipment and building costs, maintenance, wages, power and services and ash disposal. Incineration may also generate revenue from the sale of heat, power or steam.

It is often difficult to assess the cost of incinerators in general terms, particularly for large scale municipal incinerators where the costs are very much site-dependent. Also, for other wastes such as industrial and chemical wastes the incineration cost will depend on a variety of factors, including the physical form of the waste, the handling of the waste, calorific value, the amount of gas clean-up, the throughput, the requirement for the energy generated etc. There have been a number of recent estimates of the costs involved in incineration projects (Royal Commission on Environmental Pollution 1993; W.S. Atkins 1993; DOE 1995b). Such estimates are,

Table 6.13 *Estimated capital costs and reported costs for large scale mass-burn MSW incinerators in the UK*

(a) Estimated hypothetical costs

Estimator	Throughput (tonnes/year)	Comments
ETSU[1]	200 000	Estimated cost £40–55 million
	300 000	Estimated cost £50–72 million
	400 000	Estimated cost £60–80 million
W.S. Atkins Ltd.[2]	200 000	Estimated cost £49.9–52.4 million
	400 000	Estimated cost £69.5–82.8 million
Aspinwall and Co. Ltd.[2]	200 000	Estimated cost £38.8 million

(b) Existing, commissioned and proposed plant costs

Location	Throughput (tonnes/year)	Comments
SELCHP, S.E. London[3]	420 000	Completed 1994, cost £85 million
Tyseley, Birmingham[4]	350 000	Completion 1996, estimated cost £95 million
Billingham, Cleveland[4]	220 000	Completion 1997, estimated cost £40 million
Dudley[4]	90 000	Completion 1998, estimated cost £30 million
Stoke[4]	180 000	Completion 1997, estimated cost £50 million
Wolverhampton[4]	105 000	Completion 1998, estimated cost £30 million

[1] 1991 prices; [2] 1993 prices; [3] 1992 prices; [4] 1996 prices.

Sources: [1]*[1] Chappell, P., In: Energy from Waste: Clean, Green and Profitable, Institute of Energy seminar, London, October, Institute of Energy, London, 1991;* [2]*[2] W.S. Atkins Consultants Ltd., ETSU Report B RI/00341/REP, ETSU, Harwell, 1993;* [2]*[3] Royal Commission on Environmental Pollution. 17th Report, Incineration of Waste. Aspinwall and Co. Ltd., HMSO, London, 1993;* [3]*[4] SELCHP, Technical Literature. SELCHP, London, 1995;* [4]*[5] Cosslett G., Going up in smoke. The Waste Manager, July/August. The Environmental Services Association, London, 1996.*

however, no substitute for the real costs involved in the construction and operation of a real incinerator. However, such costs, and indeed income streams, are often proprietary information and consequently not available.

Table 6.13 shows estimated capital costs and reported costs for large scale mass-burn municipal solid waste incinerators in the UK (Chappell 1991; Royal Commission on Environmental Pollution 1993; W.S. Atkins 1993; SELCHP 1995; Cosslett 1996).

The 'estimated hypothetical costs' were based on commercial quality tenders for a hypothetical mass-burn incineration plant but with an unspecified location. As such, the estimates are realistic. The 'existing plant costs' represented by SELCHP are clearly the most accurate since this plant is already operating. Similarly the 'commissioned and proposed plant costs' represent costs associated with specific locations.

Differences in costs are represented by a range of different factors, which emphasises the difficulties in estimating costs for incineration since they are very

Table 6.14 Operating costs mass-burn MSW incineration

Estimator	Throughput (tonnes/year)	Comments
W.S. Atkins Ltd	200 000	Estimated cost £3.9–4.0 million/year
	400 000	Estimated cost £6.1–6.5 million/year
Aspinwall and Co. Ltd.	200 000	Estimated cost £3.5 million/year

Sources: W.S. Atkins Consultants Ltd., ETSU Report B RI/00341/REP, ETSU, Harwell, 1993; [2] Royal Commission on Environmental Pollution. 17th Report, Incineration of Waste. Aspinwall and Co. Ltd., HMSO, London, 1993.

much site-specific. For example, mass-burn incineration costs will vary due to differences in manufacturer's preferred equipment and designs, and the degree to which there is any involvement of the operation of the plant or shareholding in the plant by the equipment manufacturer, or whether they are merely supplying the equipment. The furnace, boiler and gas cleaning involved for emissions clean-up are the major capital costs of a modern municipal solid waste incinerator, but variations in the preferred technologies by each manufacturer will significantly influence costs. In addition, variations occur due to whether foreign technology, and hence currency exchange rates are involved, the price of land, site clearance costs, the level of architectural design, whether standby equipment is required etc. The key area of costing between the hypothetical and existing, commissioned and proposed costs is that specific location costs are included. In addition, the date of the estimates for the hypothetical plants was 1993, whilst those of the existing, commissioned and proposed plants are 1996.

Operating costs are also difficult to obtain from existing plant since they are again variable and sometimes proprietary, and consequently not available. Table 6.14 shows estimated operating costs for a mass-burn municipal solid waste incineration project (Royal Commission on Environmental Pollution 1993; W.S. Atkins 1993). Operating costs include labour, maintenance, disposal costs of the ash residues and wastewater, rates, electrical costs, consumables and other operating costs.

Comparisons of costs for waste incineration are usually based on a 'cost per tonne' comparison with suitable base-case assumptions. For example, the majority of cost comparisons have a base-case situation of 20 years project lifetime and estimated waste throughput. In addition, the discount rate used to calculate the discounted cash flow is usually set at 10%, although some institutions or local authorities may use higher or lower figures for their discounted cash flow calculations. The discount rate represents the minimum expected rate of return for the project, representing return on investment. Income from energy recovery is used in the estimates to offset the costs of the plant and plant operation. Overall costs and income are then used to calculate a 'cost per tonne' of waste treatment and disposal for comparison of technologies and options. Table 6.15 shows the cost per tonne of municipal solid waste incineration from various estimates.

Economies of scale lead to a reduction in the cost/tonne of incineration. Figure 6.24 shows the estimated reduction in gate fee in relation to economies of scale for a

Table 6.15 *Cost per tonne for MSW incineration*

Estimator	Throughput (tonnes/year)	Estimated cost/tonne
W.S. Atkins Ltd. (1993)	200 000	Energy recovery via electricity generation pool price income at 2.5 p/kwh, £39/tonne
		Energy recovery via electricity generation NFFO income at 6.5 p/kwh[1], £29/tonne
	400 000	Energy recovery via electricity generation pool price income at 2.5 p/kwh, £26/tonne
		Energy recovery via electricity generation NFFO income at 6.5 p/kwh[1], £17/tonne
Aspinwall and Co. Ltd. (1993)		No energy recovery – £37/tonne
		Energy recovery via electricity generation pool price income at 2.5 p/kwh, £29/tonne
		Energy recovery via electricity generation NFFO income at 4.5 p/kwh[2], £23/tonne
Comparative costs (estimator Aspinwall and Co. Ltd.)		
MSW landfill (1993)	200 000	No energy recovery, £9.8/tonne
		Energy recovery via landfill gas with electricity generation, £8.8–9.5/tonne
Sewage sludge incineration		£40–70/tonne
Hazardous waste incineration		£120–150/tonne

[1] Assumes a premium price for the electricity at 6.5 p/kwh under a NFFO contract for 7 years followed by pool price electricity.
[2] Assumes a premium price for the electricity at 4.5 p/kwh under a NFFO contract for 10 years followed by pool price electricity.

Sources: [1] W.S. Atkins Consultants Ltd., ETSU Report B RI/00341/REP, ETSU, Harwell, 1993; [2] Royal Commission on Environmental Pollution. 17th Report, Incineration of Waste. Aspinwall and Co. Ltd., HMSO, London, 1993.

municipal solid waste incinerator with energy recovery via electricity generation and a market pool price of 2.5 p/kwh (Royal Commission on Environmental Pollution 1993).

On a comparative cost basis, including energy recovery, incineration has a cost per tonne of between £17 and £39 depending on the size of plant, market electricity price, and whether a non-fossil fuel obligation contract exists, and at what price the electricity is secured. Compared with landfill costs, estimated from Chapter 5 at £9.8/tonne and with energy recovery at £8.8/tonne, landfill is obviously the better economic option. However, further considerations are pertinent to such comparisons. One element of incineration income is the disposal cost charged to the customer, which should also be taken into account. Specialised waste disposal charges to waste producers can generate high incomes, further offsetting the capital and operating costs. In addition, the landfill estimates do not include the landfill tax at £7/tonne, or £2/tonne for inactive waste. Also, the proposed EC Landfill Directive discussed in Chapter 5 places obligations on the landfill operator in relation to site

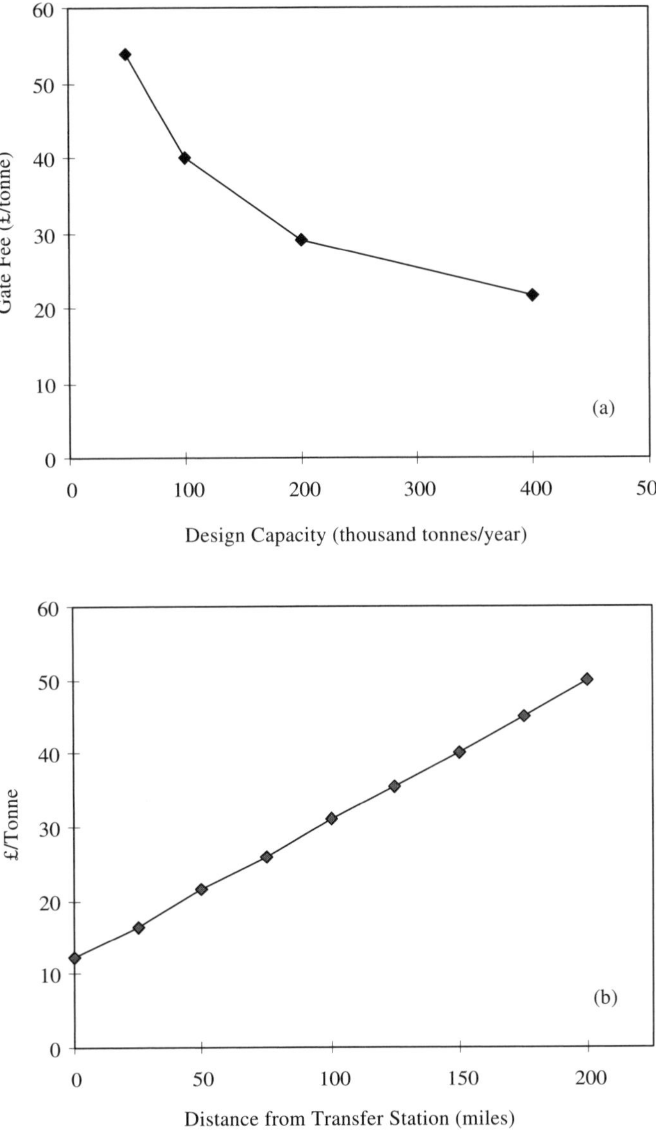

Figure 6.24 *Economies of scale in relation to municipal solid waste incineration. Source: Royal Commission on Environmental Pollution. 17th Report, Incineration of Waste. Data from Aspinwall and Co. Ltd., HMSO, London, 1993. Crown copyright is reproduced with the permission of the Controller of Her Majesty's Stationery Office*

control, emissions monitoring and aftercare. All these factors lead to an increase in the costs of landfill.

A further consideration are transport costs of the waste to the disposal facility. Transport costs associated with incineration are low in comparison with those associated with a landfill site which may be located far from the source of waste generation.

Estimated costs of other types of incineration Estimating costs for other types of incineration are extremely difficult, not only because in many cases the information is confidential, but also because of the wide range of incinerator designs available, different throughputs of waste, different gas clean-up systems, the provision for energy recovery etc. Examples of estimated costs have been presented by Pritchard and Turner (1996). For example, capital costs for clinical waste incineration have been estimated at between £2 million and £4 million depending on the incinerator design and gas clean-up system for a clinical waste incinerator with a throughput of between 1.25 and 1.5 tonnes/hour. It was estimated that between 20% and 28% of the capital cost was due to the pollution control system.

Capital costs of a typical merchant chemical waste incinerator plant with a throughput of 30 000 tonnes/year have been estimated at £20 million, of which 25% were attributed to the pollution control system (Pritchard and Turner 1996). Operating costs were estimated at £5 million over the 20-year lifetime of the project. The average cost of merchant chemical waste incineration was estimated at £500/tonne.

A 20 000 tonnes dry solids/year sewage sludge incinerator would have capital costs of £30 million, of which approximately 7% are attributed to the pollution control system (Pritchard and Turner 1996). The operating costs were estimated at between £2 million and £6 million over the 10-year lifetime of the project. The average cost of sewage sludge incineration was estimated at £370 per tonne of dry solids.

Bibliography

Ahlborg U.G. and Victorin K. 1987. Impact on health of chlorinated dioxins and other base organic emissions. Waste Management and Research 5, 203.

Atkins G. (SELCHP Ltd., London), 1997. Energy from waste; the best practicable environmental option. In: University of Leeds Short Course, Incineration of Municipal Waste with Energy Recovery. Department of Fuel and Energy, The University of Leeds, Leeds.

Atkinson W., New R., Papworth R., Pearson J., Poll J. and Scott D. 1996. The impact of recycling household waste on down stream energy recovery systems. ETSU Report B/RI/00286/REP, Energy Technology Support Unit, Department of Trade and Industry, London.

Baker P.G. 1981. Determination of polychlorodibenzo-p-dioxins. Analytical Processes, November, 478–480.

Ballantyne W.E., Huffman W.J., Curran L.M. and Stewart D.H. 1980. Energy recovery from municipal solid waste and sewage sludge using multisolid fluidised bed combustion technology. In: Jones J.L. and Radding S.B. (Eds.) Thermal Conversion of Solid Wastes and Biomass. ACS Symposium Series 130, American Chemical Society, Washinghton, DC.

Barfnecht T.R., Andon B.M., Thilley W.G. and Hites R.A. 1980. Soot and mutation in bacteria and human cells. In: Cocke M. and Dennis A.J. (Eds) Proceedings of the Fifth International Symposium on PAH. Battelle, Columbus, OH, pp. 231–242.

Barton R.G., Clark W.D. and Seeker W.R. 1990. Fate of metals in waste combustion systems. Combustion Science and Technology, 74, 327–342.

Benestad C. 1989. Incineration of hazardous waste in cement kilns. Waste Management and Research, 7, 351–361.

Bouscaren B. 1988. Reduction of the emission of heavy metals in incineration of wastes. In: Brown A., Evemy P. and Ferrero G.L. (Eds.) Energy Recovery through Waste Combustion. Elsevier Applied Science, London, pp. 144–153.

British Medical Association, 1991. Hazardous Waste and Human Health. Oxford University Press, Oxford.

Brna T.G. 1990. Cleaning of flue gases from waste combustors. Combustion Science and Technology, 74, 83–98.

Bruce A.M., Colin F. and Newman P.J. 1989. Treatment of Sewage Sludge. Elsevier Applied Science, London.

Brunner C.R. 1985. Hazardous Air Emissions from Incineration. Chapman and Hall, New York.

Brunner C.R. 1991. Handbook of Incineration Systems. McGraw-Hill, New York.

Brunner P.H. and Monch H. 1986. The flux of metals through municipal solid waste incinerators. Waste Management and Research, 4, 105–119.

Buekens A. and Patrick P.K. 1985. Incineration. In: Seuss M.J. (Ed.) Solid Waste Management: Selected Topics. World Health Organisation, Copenhagen.

Buekens A. and Schoeters J. 1984. Thermal methods in waste disposal. Study performed for the EEC under contract EC1 1011/B7210/83/B. Free University of Brussels, Brussels.

Bumb R.R., Crummett W.B., Cutie S.S., Gledhill J.R., Hummel R.H., Kagel R.O., Lamparski L.L., Luomoa E.V., Miller D.L., Nestrick L.A., Shadoff L.A., Stehl R.H. and Woods J.S. 1980. Trace chenistries of fire: A source of chlorinated dioxins. Science, 210, 385.

Buser H.R. and Rappe C. 1979. Formation of polychlorinated dibenzoflurans (PCOFs) from the pyrolysis of individual PCB isomers. Chemosphere, 8, 157.

Buser H.R. and Rappe C. 1984. Isomer-specific separation of 2,3,7,8-substituted polychlorinated dibenzo-p-dioxins by high resolution gas chromatography mass spectrometry. Analytical Chemistry, 56, 442–448.

Carlsson K. 1986. Heavy from 'Energy from Waste' plants – comparison of gas-cleaning systems. Waste Management and Research, 4, 15–20.

Cattabeni F., Cavallaro A. and Galli G. 1978. Dioxin: Toxicological and Chemical Aspects. SP Medical and Scientific Books, London.

Chappell, P. 1991. A review of municipal waste combustion technology. In: Energy from Waste: Clean, Green and Profitable. Institute of Energy Seminar, London, October. Institute of Energy, London.

Choudhary G., Keith L.M. and Rappe C. (Eds.) 1983. Occupational Exposure. Chlorinated Dioxins and Dibenzofurans in the Total Environment. Section V. Butterworth, London.

Clarke A.G. 1986. The air. In: Hester R.E. (Ed.) Understanding our Environment. Royal Society of Chemistry, London.

Clarke A.G. 1996. Chimney design. In: University of Leeds Short Course, Incineration of Municipal Waste with Energy Recovery. Department of Fuel and Energy, The University of Leeds, Leeds.

Clayton P., Coleman P., Leonard A., Loader A., Marlowe I., Mitchell D., Richardson S., Scott D. and Woodfield M. 1991. Review of Solid Waste Incineration in the UK. Warren Spring Laboratory (National Environmental Technology Centre) Report No. LR776 (PA), Department of Trade and Industry, London.

Colmsjo A.C., Zebuhr Y.U. and Ostman C.E. 1986. Polynuclear aromatic compounds in flue

gases and ambient air in the vicinity of a muncipal incineration plant. Atmospheric Environment, 20, 2279–2282.
Commoner B., Shapiro K. and Webster T. 1987. The origin and health risks of PCDD and PCDF. Waste Management and Research, 5, 327.
Cosslett G. 1996. Going up in smoke. The Waste Manager, July/August, Environmental Services Association, London.
Council Directive 89/369/EEC, 1989. Reduction of air pollution from new municipal waste incinerators. Official Journal of the European Communities, Brussels.
Council Directive 89/429/EEC, 1989. Reduction of air pollution from existing municipal waste incinerators. Official Journal of the European Communities, Brussels.
Darley P. 1996. (MAN GHH Ltd., London) Energy recovery. In: University of Leeds Short Course, Incineration of Municipal Waste with Energy Recovery. Department of Fuel and Energy, The University of Leeds, Leeds.
Davies I.W., Harrison R.M., Perry R., Ratnayaka D. and Wellings R.A. 1976. Municipal incinerator as source of polynuclear aromatic hydrocarbons in environment. Environmental Science and Technology, 10, 451–453.
Denison R.A. and Silbergeld E.K. 1988. Risks of municipal solid waste incineration: An environmental perspective. Risk Analysis, 8, 343–355.
Dioxin risk: Are we sure yet, 1995. EPA Draft Report Review. Environmental Science and Technology, 29, 24A–35A.
DOE, 1989. Pollution Paper 27, Dioxins in the Environment. Department of the Environment, HMSO, London.
DOE, 1995a. A Review of Dioxin Emissions in the UK. Department of the Environment, HMSO, London.
DOE, 1995b. Technical and Economic Study of Incineration Processes. Department of Environment Research Report No. DOE/HMIP/RR/95/027, Environment Agency, Bristol.
DTI, 1996. Energy from Waste: Best Practice Guide. Department of Trade and Industry, London.
DTI OSTEMS Mission, 1996. Energy from Waste. Waste Incineration and Anaerobic Digestion with Energy Recovery. Institution of Civil Engineers, London.
ENDS, 1996. Report 263. Environmental Data Services Ltd., December, Bowling Green Lane, London.
ENDS, 1997. Report 257. Environmental Data Services Ltd., June, Bowling Green Lane, London.
Eurostat Year Book, 1995. Office for Official Publications of the European Communities, Luxembourg.
Friberg L., Nordberg G.F. and Vouk V.B. 1986. Handbook of the Toxicology of Metals. Vol. I. Elsevier Applied Science, Amsterdam.
Frost R. 1992. Incineration of Sewage Sludge. University of Leeds Short Course 'Incineration and Energy from Waste'. University of Leeds, Leeds.
Gelboin H.V. and Tso P.O.P. (Eds.) 1978. Polycyclic Hydrocarbons and Cancer. Academic Press, New York.
Gilpin A. 1982. Dictionary of Energy Technology. Butterworth Scientific and Ann Arbor Science, London.
Greenberg R.R., Zoller W.H. and Gordon G.E. 1978. Composition and size distributions of particles released in refuse incineration. Environmental Science and Technology, 12, 566–573.
Hagenmaier H., Brunner H., Haag R., Kraft M. and Lutzke K. 1987. Problems associated with the measurement of PCDD and PCDF emissions from waste incineration plants. Waste Management and Research, 5, 239.
Hagenmaier H., Kraft M., Haag R. and Brunner H. 1988. In: Brown A., Evemy P. and Ferrero G.L. (Eds.) Energy Recovery through Waste Combustion. Elsevier Applied Science, London.
Hall J.E. 1992. Treatment and use of sewage sludge. In: Bradshaw A.D., Southwood R. and Warner F. (Eds.) The Treatment and Handling of Wastes. Chapman and Hall, London.

Hall G.S. and Knowles M.J.D. 1985. Good design and operation reaps benefits on UK refuse incineration boilers over the last ten years. In: Energy Recovery from Refuse Incineration. Institution of Mechanical Engineers, London.

Hay A.W.M. 1982. In: Hutzinger O., Frei R.W., Merian E. and Pocchiari F. (Eds.) Chlorinated Dioxins and Related Compounds: Impact on the Environment. Pergamon, New York.

Holmes J. 1995. The UK Waste Management Industry, 1995. Institute of Waste Management, Northampton.

Hudson, J. 1993. Sewage slude incineration. In: Incineration: An Environmental Solution for Waste Disposal. IBC Technical Services Conference 16/17 February, Manchester.

Hutzinger O., Frei R.W., Meriam E. and Pocchiari F. 1982. Chlorinated Dioxins and Related Compounds: Impact on the Environment. Pergamon, New York.

Krause H.H. 1991. Corrosion by chlorine in waste fueled boilers. In: Bryers R.W. (Ed.) Incinerating Municipal and Industrial Waste. Hemisphere, New York, pp. 145–159.

Law S.L. and Gordon G.E. 1979. Sources of metals in municipal incinerator emissions. Environmental Science and Technology, 13, 432–438.

Lave L.B. and Seskin E.P. 1970. Air pollution and human health. Science, 169, 723.

Lawrence D. 1995. Incineration of solid municipal waste with energy recovery by district heating: The Sheffield MSW incineration plant. Incineration of Municipal Waste with Energy Recovery. University of Leeds, Industrial Short Course, Department of Fuel and Energy. University of Leeds, Leeds.

Lee M.L., Novotny M. and Bartle K.D. 1981. Analytical Chemistry of Polycyclic Aromatic Compounds. Academic Press, New York.

Longwell J.P. 1983. Polycyclic aromatic hydrocarbons and soot from practical combustion systems. In: Lahaye J.L. and Prado G. (Eds.) Soot in Combustion Systems and its Toxic Properties. Plenum Press, New York, pp. 37–56.

Lorber K.E. 1985. In: Ferranti M.P. and Ferrero G.L. (Eds.) Sorting of Household Waste and Thermal Treatment of Waste. CEC, Elsevier Applied Science, London.

Lustenhouwer J.W.A., Olie K. and Hutzinger O. 1980. Chlorinated dibenzo-p-dioxins and related compounds in incinerator effluents. Chemosphere, 9, 501.

Making Waste Work, 1995. Department of the Environment and the Welsh Office, HMSO, London.

Manz A., Berger J., Dwyer J.H., Hesch-Janys D., Nagel S. and Walsgott H. 1991. Cancer mortality among workers in chemical plant contaminated with dioxin. Lancet, 338, 959–964.

Marklund S., Rappe C., Tysklind M. and Egeback K.E. 1987. Identification of polycholorinated dibenzofurans and dioxins in exhausts from cars run on leaded gasoline. Chemosphere, 16, 29.

Mulder E. 1996. Pre-treatment of MSWI flyash for useful application. Waste Management, 16, 181–184.

Nielsen K.K., Moeller J.T. and Rasmussen S. 1986. Reduction of dioxins and furans by spray dryer absorption from incinerator flue gas. Chemosphere, 15, 1247.

Niessen W.R. 1978. Combustion and Incineration Processes. Marcel Dekker, New York.

Oakland D. 1988. Dioxins: Sources, combustion, theories and effects. Filtration and Separation, Jan./Feb., 39.

Olie K., Lustenhouwer J.W.A. and Hutzinger O. 1982. Polychlorinated dibenzo-b-dioxins and related compounds in incinerator effluents. In: Hutzinger O., Frei R.W., Merian E. and Pocchiari F. (Eds.) Chlorinated Dioxins and Related Compounds: Impact on the Environment. Pergamon, Oxford, p. 227.

Oppelt H.M. 1987. Incineration of hazardous waste: A critical review. Journal of the Air Pollution Control Association, 37, 558–586.

Ozvacic V., Wong G., Tosine H., Clement R.E. and Osborne J. 1985. Emissions of chlorinated organics from two municipal incinerators in Ontario. Journal of the Air Pollution Control Association, 35, 849.

Parker C. and Roberts T. 1985. Energy From Waste: An Evaluation of Conservation Technologies. Elsevier Applied Science, London.

Paterson B. 1997. Clinical waste treatment. In: Institute of Wastes Management, Annual Conference Proceedings, Torbay, June, Institute of Wastes Management, Northampton.

Porteous A. 1992. Dictionary of Environmental Science and Technology. Wiley, Chichester.

Priest G.M. 1985. Producing steam from the combustion of solid waste. In: Energy Recovery from Refuse Incineration. Seminar, Institution of Mechanical Engineers, Strathclyde, August.

Pritchard M.L. and Turner G. 1996. Technical and Economic Study of Incineration Processes. DOE Report DOE/HMIP/RR/95/027, The Environment Agency, Department of the Environment, Bristol.

Probert S.D., Kerr K. and Brown J. 1987. Harnessing energy from domestic, municipal and industrial refuse. Applied Energy, 27, 89–168.

Rappe C., Marklund S., Bergqvist A. and Hansson M. 1983. Polychlorinated dibenzo-p-dioxins, dibenzofurans and other polynuclear aromatics during incineration and polychlorinated fires. In: Choudhary G., Keith L.M. and Rappe C. (Eds.) Chlorinated Dioxins and Dibenzofurans in the Total Environment. Butterworth, London.

Reimann D.O. 1986. Mercury output from garbage incineration. Waste Management and Research, 4, 45–56.

Reimann D.O. 1987. Treatment of waste water from refuse incineration plants. Waste Management and Research, 5, 147.

Reimann D.O. 1989. Heavy metals in domestic refuse and their distribution in incinerator residues. Waste Management and Research, 7, 57–62.

Royal Commission on Environmental Pollution, 1983. 17th Report, Incineration of Waste. HMSO, London.

Sakai S., Sawell S.E., Chandler A.J., Eighmy T.T., Kosson D.S., Vehlow J., van der Sllot H.A., Hartlen J. and Hjelmar O. 1996. World trends in municipal solid waste management. Waste Management, 16, 341–350.

Saracci R., Kogevinas M., Bertazzi P.-A., Mesquita B.H., Coggon D., Green L.M., Kauppinen T., L'Abbe K.A., Littorin M., Lynge E., Mathews J.D., Neuberger M., Osman J., Pearce N. and Winkelmann R. 1991. Cancer mortality in workers exposed to chlorophenoxy herbicides and chlorophenols. Lancet, 338, 1027–1032.

SELCHP, 1995. Technical Literature. SELCHP, London.

Shaub W.M. 1989. An overview of what is known about dioxin and furan formation, destruction and control during incineration of MSW. Coalition on Resource, Recovery and the Environment (CORRE), US EPA, MSW Technology Conference, San Diego, CA, Jan.–Feb.

Skene S.A., Dewhurst I.C. and Greenberg M. 1989. Polychlorinated dibenzo-p-dioxins and polychlorinated dibenzofurans: The risks to human health. A review. Human Toxicology, 8, 173–203.

Stanners D. and Bourdeau P. 1995. Europe's Environment. The Dobris Assessment, European Environment Agency, Copenhagen.

Steisel N., Morris R. and Clarke M.J. 1987. The impact of the dioxin issue on resource recovery in the United States. Waste Management and Research, 5, 381.

Tiernan T.O. 1983. Analytical chemistry of polychlorinated dibenzo-p-dioxins and dibenzofurans: A review of the current status. In: Choudhary G., Keith L.M. and Rappe C. (Eds.) Chlorinated Dioxins and Dibenzofurans in the Total Environment. Butterworth, London.

Tosine H. 1983. Dioxins: A Canadian perspective. In: Choudhary G., Keith L.M. and Rappe C. (Eds.) Chlorinated Dioxins and Dibenzofurans in the Total Environment. Butterworth, London, p. 3.

Turner B. (Hepworth Refractories Ltd., Sheffield), 1996. Refractory materials for incinerators. In: University of Leeds Short Course, Incineration of Municipal Waste with Energy Recovery. Department of Fuel and Energy, University of Leeds, Leeds.

Uchida S., Kamo H., Kubota H. and Kanaya K. 1983. Reaction kinetics of formation of HCl

in municipal refuse incineration. Industrial Engineering Chemistry, Process Design and Development, 22, 144.
Uchida S., Kamo H. and Kubota H. 1988. The source of HCl emission from municipal refuse incinerators. Industrial Engineering Chemistry Research, 27, 2188.
Urban Waste and Power 1988. Power from Waste (company technical literature). Urban Waste and Power, Ashford.
Visalli J.R. 1987. A comparison of dioxin, furan and combustion gas data from test programs at three MSW incinerators. Journal of the Air Pollution Control Association, 37, 1451.
Vogg H., Braun H., Metzger M. and Schneider J. 1986. The specific role of cadmium and mercury in municipal solid waste incineration. Waste Management and Research, 4, 65–74.
Vogg H., Metzger M. and Steiglitz L. 1987. Recent findings on the formation and decomposition of PCDD/PCDF in municipal solid waste incineration. Waste Management and Research, 5, 285.
Wade J. (ABB Environmental Systems Ltd., Staines), 1996. Flue gas emissions control. In: University of Leeds Short Course, Incineration of Municipal Waste with Energy Recovery. Department of Fuel and Energy, University of Leeds, Leeds.
Warmer Bulletin Vol. 18, 1988. Journal of the World Resource Foundation, June, Tonbridge.
Warmer Bulletin 44, 1995. Information Sheet. The World Resource Foundation, Tonbridge.
Warmer Bulletin Fact Sheet, 1995. Ash handling from waste combustion. Journal of the World Resource Foundation, Tonbridge.
Waste Incineration, 1996. IPC Guidance Note S2 5.01. Processes Subject to Integrated Pollution Control: Waste Incineration. Environment Agency, Bristol.
Waste Management in Japan, 1995. Department of Trade and Industry, Overseas Science and Technology Expert Mission Visit Report. Institute of Wastes Management, Northampton.
Whiting, K. (Consultant), 1996. Large scale MSW incineration technologies. In: University of Leeds Short Course, Incineration of Municipal Waste with Energy Recovery. Department of Fuel and Energy, University of Leeds, Leeds.
WHO, 1990. Heavy Metal and PAH Compounds from Municipal Incinerators. Environmental Health Series No. 32, World Health Organisation, Copenhagen.
Williams A. 1990. Combustion of Liquid Fuel Sprays. Butterworths, London.
Williams P.T. 1990. Sampling and analysis of polycyclic aromatic compounds from combustion sources: A review. Journal of the Institute of Energy, 63, 22–30.
Williams P.T. 1992. The sampling and analysis of dioxins and furans from combustion sources. Journal of the Institute of Energy, 65, 46.
Williams P.T. 1994. Pollutants from incineration: An overview. In: Hester R.E. and Harrision R.M. (Eds.) Waste Incineration and the Environment. Issues in Environmental Science and Technology. Royal Society of Chemistry, London.
Woodfield M. 1987. The Environmental Impact of Refuse Incineration in the UK. Warren Spring Laboratory (National Environmental Technology Centre), Report Department of Trade and Industry, London.
W.S. Atkins Consultants Ltd. 1993. ETSU Report B RI/00341/REP, ETSU, Harwell.

7

Other Treatment Technologies: Pyrolysis, Gasification, Composting, Anaerobic Digestion

Summary

This chapter discusses other options for waste treatment and disposal. Pyrolysis of waste, the types of products formed during pyrolysis and their utilisation, and the different pyrolysis technologies are discussed. Gasification of waste, gasification technologies and utilisation of the product gas are described. Composting of waste is described, including the composting process and the different types of composter. Anaerobic digestion of waste, the degradation process, and the operation and technology for anaerobic digestion are discussed. Current examples of different types of pyrolysis, gasification, composting and anaerobic digestion are described throughout.

7.1 Introduction

The hierarchy of waste management and the concept of sustainable waste management have led to the development of alternative waste treatment and disposal options other than the traditional reliance in the UK on the options of landfill and incineration. Alternatives which have a minimal environmental impact with an image of recycling or energy recovery with low pollution have received particular attention. Amongst such technologies are pyrolysis; gasification; composting and anaerobic digestion.

7.2 Pyrolysis

Pyrolysis is the thermal degradation of organic waste in the absence of oxygen to produce a carbonaceous char, oils and combustible gases. Relatively low temperatures are used, in the range 400–800 °C. The application of pyrolysis to waste materials is a relatively recent development. In particular, the production of oils from the pyrolysis of waste has been investigated with the aim of using the oils directly in fuel applications or with upgrading to produce refined fuels. The pyrolysis oils derived from a variety of wastes have also been shown to be complex in composition and contain a wide variety of chemicals which may be used as chemical feedstock. The oil has a higher energy density, that is, a higher energy content per unit weight, than the raw waste. The solid char can be used as a solid fuel or as a char-oil (char-water slurry for fuel); alternatively the char can be used as carbon black or upgraded to activated carbon. The gases generated have medium to high calorific values and may contain sufficient energy to supply the energy requirements of a pyrolysis plant (Bridgwater and Bridge 1991).

The process conditions are altered to produce the desired char, gas or oil end-product, with pyrolysis temperature and heating rate having the most influence on the product distribution. The heat is supplied by indirect heating such as the combustion of the gases or oil, or directly by hot gas transfer. Pyrolysis systems for municipal solid waste, tyres, plastics, sewage sludge and biomass have been investigated, and some have been taken through towards commercial exploitation (Bridgwater and Evans 1993; Roy 1994; Lee 1995; Williams et al 1995).

In the UK, pyrolysis of waste as a route to produce energy has received a further boost from the non-fossil fuel obligation, which provides a premium price for electricity generated from the oils, char or gases produced from waste. The product oil from pyrolysis has the advantage that it can be used in conventional electricity generating systems such as diesel engines and gas turbines. However, the properties of the pyrolysis oil fuel may not match the specifications of a petroleum-derived fuel and may require modifications to the power plant or upgrading of the fuel.

Pyrolysis produces char, gas and oil, the relative proportions of which are dictated by the pyrolysis technology used and the process parameters, of which temperature and heating rate have the most influence (Bridgwater and Bridge 1991). In some cases the oil product is described as a liquid, but depending on the feedstock and the pyrolysis process conditions, it may represent either a true oil, an oil/aqueous phase, or separated oil and aqueous phases.

Very slow heating rates coupled with a low final maximum temperature maximises the yield of char, for example, the production of char from wood in the form of charcoal involves a very slow heating rate to moderate temperatures. The process of carbonisation of waste results in reduced concentrations of oil/tar and gas product and are regarded as by-products of the main charcoal-forming process.

Moderate heating rates in the range of about 20 °C/min to 100 °C/min and maximum temperatures of 600 °C give an approximately equal distribution of oils, char and gases. This is referred to as conventional pyrolysis or slow pyrolysis. Because of the slow heating rates and generally slow removal of the products of pyrolysis from the hot pyrolysis reactor, secondary reactions of the products can take place. Generally, a more complex product slate is found.

Very high heating rates of about 100 °C/s to 1000 °C/s at temperatures below 650 °C and with rapid quenching lead to the formation of a mainly liquid product; this is referred to as fast or flash pyrolysis. Liquid yields up to 70% have been reported for biomass feedstocks using flash pyrolysis. In addition, the carbonaceous char and gas production are minimised. The primary liquid products of pyrolysis are rapidly quenched and this prevents breakdown of the products to gases in the hot reactor. The high reaction rates also cause char-forming reactions from the oil products to be minimised (Bridgwater and Bridge 1991; Bridgwater and Evans 1993).

At high heating rates and high temperatures, the oil products quickly breakdown to yield a mainly gas product. Typical yield of gas from the original feedstock hydrocarbon is 70%. This process differs from gasification, which is a series of reactions involving carbon and oxygen in the form of oxygen gas, air or steam to produce a gas product consisting mainly of CO, CO_2, H_2 and CH_4. Table 7.1 shows the typical characteristics of different types of pyrolysis (Bridgwater and Bridge 1991; Bridgwater and Evans 1993).

Pyrolysis process conditions can be optimised to produce either a solid char, a gas or a liquid/oil product. Table 7.2 shows the yields of char, oil/liquid and gas from various waste feedstocks (Kaminsky and Sinn 1980; Rampling and Hickey 1988; Williams and Besler 1992, 1996; Williams et al 1995; Horne and Williams 1996; Williams and Williams 1997).

The solid char product from carbonisation or slow pyrolysis of wood has been used for centuries as the process to produce charcoal for use as fuel, and charcoal product yields of between 30 and 40% are common. Pyrolysis of waste materials also produces a char product, the percentage production depending on process conditions. Pyrolysis of municipal solid waste produces a 35% char product which has a high ash content of up to 37%, tyre pyrolysis under low heating rate conditions produces a char of up to 50% with an ash content of about 10%. The chars may be used directly as fuels, briquetted to produce solid fuels, used as adsorptive materials such as activated carbon, upgraded to produce a higher grade activated carbon, or crushed and mixed with the pyrolysis oil product to produce a slurry for combustion.

The calorific values of the chars are relatively high, for example, char derived from municipal solid waste has a calorific value of about 19 MJ/kg, tyre char about 29 MJ/kg and wood waste produces a char of calorific value about 33 MJ/kg (Rampling and Hickey 1988; Williams and Besler 1992, 1996; Williams et al 1995). These figures compare well with a typical bituminous coal calorific value of 30 MJ/kg.

The significance of a high ash content in the chars means that the value of the char as a fuel is reduced. In addition, the use of pyrolysis chars as substitutes for activated carbon is greatly diminished if they have a high ash content. Chars from wood have very low ash contents and have been successfully upgraded using steam activation to produce activated carbon. However, the upgrading of tyre chars to activated carbon requires an additional processing step of de-ashing to make the product acceptable to the activated carbon industry. In addition, the specifications of activated carbon derived from traditional routes such as coconut shell are well established, and as

Table 7.1 *Typical characteristics of different types of pyrolysis*

Pyrolysis	Residence time	Heating rate	Reaction environment	Pressure (bar)	Temperature (°C)	Major product
Carbonisation	hours or days	Very low	Combustion products	1	400	Charcoal
Conventional	10 s–10 min	Low–moderate	Primary/secondary products	1	< 600	Gas, char, liquid
Flash–liquid	< 1 s	High	Primary products	1	< 600	Liquid
Flash–gas	< 1 s	High	Primary products	1	> 700	Gas
Ultra	< 0.5 s	Very high	Primary products	1	1000	Gas, chemicals
Other pyrolysis types						
Vacuum	2–30 s	Medium	Vacuum	< 0.1	400	Liquid
Hydropyrolysis	< 10 s	High	H_2 + primary	~20	< 500	Liquid, chemicals
Methanolysis	0.5–1.5 s	High	CH_4 + primary products	~3	1050	Benzene, toluene, xylene + alkenes

Sources: [1] Bridgwater A.V. and Bridge S.A., In: Biomass Pyrolysis Liquids Upgrading and Utilisation. Bridgwater A.V. and Grassi G. (Eds.) Elsevier Applied Science, London, 1991; [2] Bridgwater A.V. and Evans G.D., An assessment of thermochemical conversion systems for processing biomass and refuse. Energy Technology Support Unit, Harwell. Report ETSU/T1/00207/REP, Department of Trade and Industry, 1993.

Table 7.2 *Product yields from the pyrolysis of waste*

Waste	Pyrolysis process	Temperature (°C)	Heating Rate	Char (%)	Liquid (%)	Gas (%)
Wood[1]	Moderate (batch)	600	20 °C/min	22.6	50.4	27.0
Wood[2]	Fast (fluidised bed)	550	~300 °C/s	17.3	67.0	14.9
Tyre[3]	Moderate (batch)	600	20 °C/min	39.2	54.0	6.8
Tyre[3]	Slow/moderate (batch)	850	~5 °C/min	49.5	32.5	18.0
Tyre[4]	Fast (fluidised bed)	640	–	38	40	18
RDF[5]*	Moderate (batch)	600	20 °C/min	35.2	49.2	18.8
RDF[6]*	Moderate (batch)	700	–	30	49	22
Plastic[7] (mixed)	Moderate	700	25 °C/min	2.9	75.1	9.6

* Refuse derived fuel from municipal solid waste.

Sources: [1] Williams P.T. and Besler S., Influence of temperature and heating rate on the slow pyrolysis of biomass. Renewable Energy, 7, 233–250, 1996; [2] Horne P.A. and Williams P.T., Influence of temperature on the products from the flash pyrolysis of biomass. Fuel, 75, 1051–1059, 1996; [3] Williams P.T., Besler S., Taylor D.T. and Bottrill R.P., Pyrolysis of automotive tyre waste. Journal of the Institute of Energy, 68, 11–21, 1995; [4] Kaminsky W. and Sinn H., Pyrolysis of plastic waste and scrap tyres using a fluidised bed process. In: Jones J.L. and Radding S.B. (Eds.) Thermal Conversion of Solid Wastes and Biomass. ACS Symposium Series 130, American Chemical Society, Washington, DC, 1980; [5] Williams P.T. and Besler S., The pyrolysis of municipal solid waste. Journal of the Institute of Energy, 65, 192–200, 1992; [6] Rampling T.W. and Hickey T.J., The laboratory characterisation of refuse derived fuel. Warren Spring Laboratory Report, LR643, Department of Trade and Industry, HMSO, London, 1988; [7] Williams E.A and Williams P.T., Journal of Chemical Technology and Biotechnology, 70, 9–20, 1997.

with most new products it is difficult for an alternative product to break into an established market. Even though the waste-derived chars may be cheaper, the specifications, quality and maintenance of quality have to be guaranteed.

The advantages of producing an oil product from waste are that the oil can be transported away from the pyrolysis process plant, and this therefore de-couples the processing of the waste from the product utilisation. The oil may be used directly as a fuel, added to petroleum refinery stocks, upgraded using catalysts to a premium grade fuel or used as a chemical feedstock. The composition of the oil is dependent on the chemical composition of the feedstock and the processing conditions. For example, oils derived from biomass have a high oxygen content, of the order of 35% by weight, due to the content of cellulose, hemicellulose and lignin in the biomass. These are large polymeric structures containing mainly carbon, hydrogen and oxygen. Similarly, oils derived from municipal solid waste have a high oxygen content due to the presence of cellulosic components in the waste such as paper, cardboard and wood. Biomass and municipal solid waste pyrolysis oils derived from flash pyrolysis processes tend to have a lower viscosity and consist of a single water/oil phase. The oils are therefore high in water, which markedly reduces their calorific value. Slow pyrolysis produces a liquid product with higher viscosities which tend to have two phases due to the more extensive degree of secondary reactions which occur. Oils derived from scrap tyre pyrolysis and plastics, on the other hand, are composed mainly of carbon and hydrogen reflecting the composition of their source. Figure 7.1 shows the chemical structures of cellulose, hemicellulose and lignin, and Figure 7.2 shows the structures of a typical tyre rubber and several common plastics (Graham et al 1984; Menachem and Goklestein 1984).

The oils have significant calorific values, ranging from 25 MK/kg for oils derived from municipal solid waste to 42 MJ/kg for oils derived from scrap tyres, while a typical petroleum-derived fuel oil has a calorific value of 46 MJ/kg (Table 7.3). Table 7.3 shows the properties of oils derived from the pyrolysis of tyres, municipal solid waste and wood (Pober and Bauer 1977; Rick and Vix 1991; Williams et al 1995). Comparison with petroleum-derived diesel fuel shows that in many respects the oils derived from waste are quite similar. However, the direct use of such fuels in combustion systems designed and optimised on fuels refined from petroleum may be difficult. For example, biomass and municipal solid waste pyrolysis oils can be viscous and highly acidic due to the organic acids present in the oils, and they readily polymerise. In addition, pyrolysis oils may contain solid char particles due to carry-over from the pyrolysis reactor. Consequently, their use in liquid spray or atomisation combustion systems such as diesel engines, furnaces and boilers may result in the spray or atomisation system becoming blocked and/or corroded.

In addition, performance guarantees for the use of non-standard fuels in combustion systems may invalidate the manufacturers' warranties, which would be based around standard, i.e., petroleum-refined, fuels. Emission limits from the combustion system, set at National and European level, would also have to be met irrespective of the fuel being used. However, the fuels derived from waste materials such as tyres, wood and municipal solid waste have been successfully combusted in a variety of systems, and further research on combustion performance, long-term effects and emissions is being undertaken (Bridgwater and Evans 1993).

The oils derived from the pyrolysis of waste materials tend to be chemically very complex due to the polymeric nature of the wastes and the range of potential primary and secondary reactions. Biomass and municipal solid waste pyrolysis oils contain hundreds of different chemical compounds, including organic acids, phenols, alcohols, aldehydes, ketones, furans etc. (Desbene et al 1991) Tyre pyrolysis oils consist mainly of alkanes, alkenes and monoaromatic and polycyclic aromatic compounds (Williams et al 1995). Oils derived from mixed plastic waste at typical pyrolysis temperatures of 500 °C are highly viscous and are better described as waxes. They consist largely of alkanes, alkenes and aromatic compounds. Where single plastics are pyrolysed the wax is similar to the basic structure from which the plastic was formed. Consequently, for example, polyethylene and polypropylene will produce mainly alkane and alkene waxes, whilst polystyrene will produce an alkane, alkene and aromatic product (Williams and Williams 1997).

Because of the range of compounds found in pyrolysis oils, there is some interest in using the oils as chemical feedstocks for speciality chemicals. For example, wood

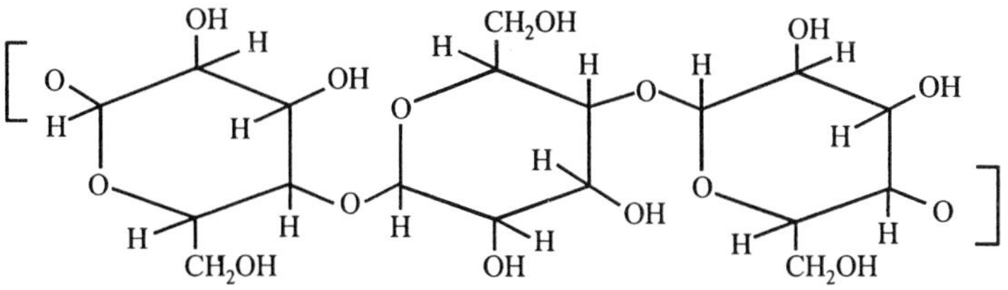

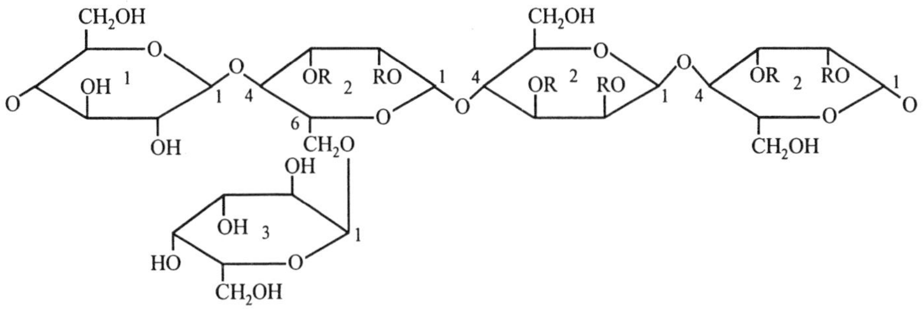

Hemicellulose for softwoods

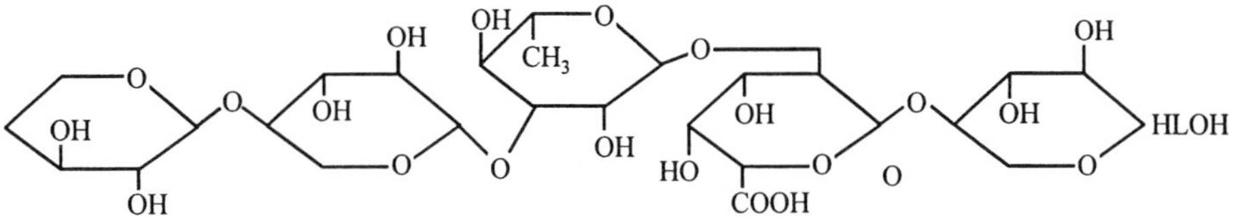

Hemicellulose for hardwoods

Figure 7.1 *Chemical structures of cellulose, hemicellulose and lignin as components of biomass. Sources: [1] Menachem L. and Goklestein I., International Fibre Science, Vol. 11. Wood Structure and Composition. Marcel Dekker, New York, 1984; [2] Lignin figure reprinted from Graham R.G., Bergougnou M.A. and Overend R.P., Fast pyrolysis of biomass. Journal of Analytical and Applied Pyrolysis, 6, 95–135, 1984, with kind permission from Elsevier Science – NL, Sara Burgerhartstraat 25, 1055 KV Amsterdam, The Netherlands*

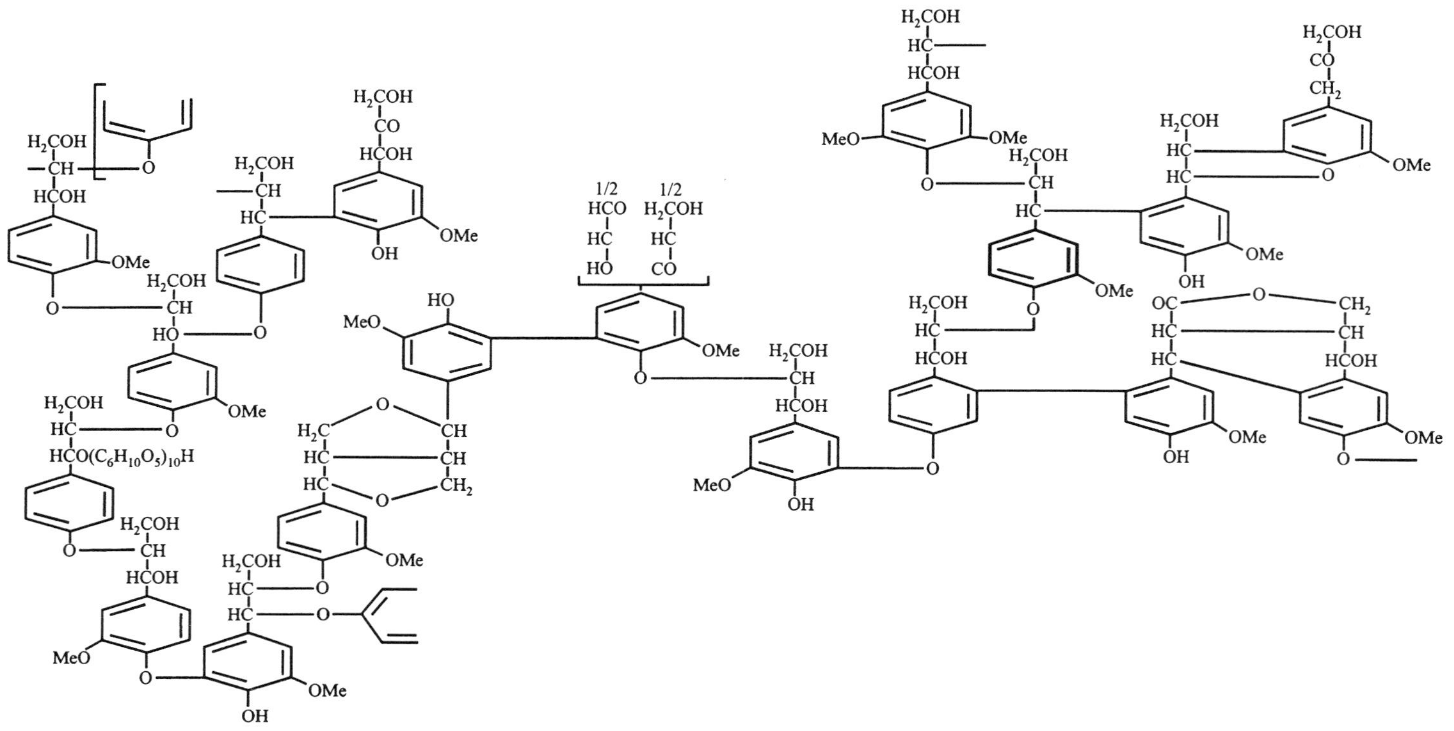

Figure 7.1 *(continued)*

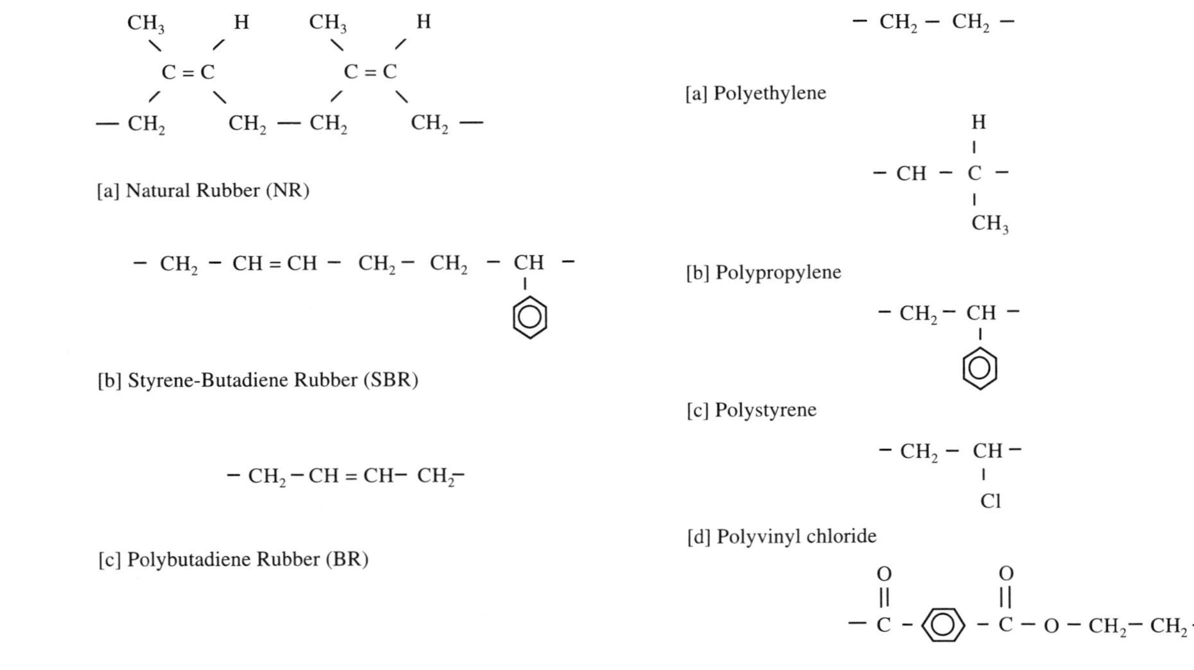

Figure 7.2 *Chemical structures of rubbers used in tyres and common plastics*

Table 7.3 *Typical fuel properties of waste derived pyrolysis oils*

Parameter	Tyre oil	MSW oil	Biomass oil	Diesel oil
Carbon residue (%)	0.7	–	–	<0.35
Mid B.Pt. °C	230	–	–	300
Viscosity (cSt)	2.12 °60 (C)	–	17 (100 °C)	1.3 (60 °C)
	3.50 (40 °C)	–	90 (50 °C)	3.3 (40 °C)
Density (kg/m^3)	0.91	1.3	1.2	0.78
API gravity	20.41	–	–	31
Flash Point (°C)	24	56	110–120	75
Hydrogen (%)	9.98	7.6	7–8	12.8
Carbon (%)	87.0	57.5	50–67	–
Nitrogen (%)	0.4	0.9	0.8–1	–
Oxygen (%)	0.7	33.4	15–25	–
Initial B.Pt. (°C)	80	–	–	180
10% B.Pt. (°C)	140	–	–	–
50% B.Pt. (°C)	230	–	–	300
90% B.Pt. (°C)	340	–	–	–
CV (MJ/kg)	42.0	24.4	24.7 (lower)	46.0
Sulphur (%)	1.5	0.1–0.3	<0.01	0.9

Sources: [1] Williams P.T., Besler S., Taylor D.T. and Bottrill R.P., Pyrolysis of automotive tyre waste. Journal of the Institute of Energy, 68, 11–21, 1995; [2] Pober K.W. and Bauer H.F., The nature of pyrolytic oil from municipal solid waste. In: Anderson L.L. and Tillman D.A. (Eds.) Fuels from Waste. Academic Press, London, 1977; [3] Rick F. and Vix U., Product standards for use as fuel in industrial firing plants. In: Bridgewater A.V. and Grassi G., Biomass Pyrolysis Liquids: Upgrading and Utilisation. Elsevier Applied Science, London, 1991.

pyrolysis oils contain oxygenated compounds such as methylphenols (cresol), methyoxyphenol (guaiacol), furaldehyde (fufural) and methoxypropenylphenol (isoeugenol), which have applications in the pharmaceutical, food and paint industries (Stoikos 1991; Bridgwater and Evans 1993). Tyre oil contains dl-limonene, which is used in the formulation of industrial solvents, resins and adhesives and as a replacement for chlorofluorocarbon for cleaning electronic circuit boards (Pakdel et al 1991). The oils also contain significant concentrations of benzene, xylenes, styrene and toluene, which are used extensively in the chemical and pharmaceutical industries (Williams et al 1995). The wax-like product derived from pyrolysis of mixed plastic waste has been successfully re-processed in petroleum refineries using catalyst cracking to produce gasoline or plastics (Lee 1995).

To overcome some of the problems of high oxygen content, high viscosity, acidity and polymerisation associated with oils derived from waste materials containing high oxygen contents, e.g. biomass and municipal solid waste, research has been undertaken to upgrade these oils (Stoikos 1991). The research has concentrated on the use of catalysts to produce a premium quality fuel or high value chemical feedstock. Two main routes to catalytic upgrading have been investigated: high pressure catalytic hydrotreatment, and low pressure catalysis using shape-selective catalysts of the zeolite type. Catalytic hydrotreatment of the oils with hydrogen or hydrogen and carbon monoxide under high pressure, and/or in the presence of hydrogen donor solvents using transition metal catalysts, has produced oils similar in composition to gasoline and diesel. The upgrading takes place through deoxygenation and hydrocracking of the heavy fractions in the oil. Zeolite ZSM-5 catalysts have a strong acidity, high activities and shape selectivities, which convert the oxygenated oil to a light hydrocarbon mixture in the C_1–C_{10} range by dehydration and deoxygenation reactions. The oxygen in the oxygenated compounds of biomass pyrolysis oils is converted largely to CO, CO_2 and H_2O, and the resultant oil is highly aromatic with a dominance of single-ring aromatic compounds, and is similar in composition to gasoline (Bridgwater and Bridge 1991; Williams and Horne 1994).

The gases produced from waste pyrolysis are mainly carbon dioxide, carbon monoxide, hydrogen, methane and lower concentrations of other hydrocarbon gases. In addition, the gas contains a significant proportion of uncondensed pyrolysis oils. The oxygenated waste materials such as biomass and municipal solid waste have higher concentrations of carbon dioxide and carbon monoxide, whilst scrap tyre and mixed plastics waste pyrolysis produces higher concentrations of hydrogen, methane and other hydrocarbon gases. The gases have a significant calorific value, for example, the gas produced from the conventional pyrolysis of municipal solid waste has a calorific value of the order of 18 MJ/m^3, and wood waste produces a gas of calorific value 16 MJ/m^3 (Williams and Besler 1992, 1996). Tyre pyrolysis produces a gas of much higher calorific value, about 40 MJ/m^3 depending on the process conditions (Williams et al 1995). The high calorific value is due to the high concentrations of hydrogen and other hydrocarbons. By comparison, the calorific value of natural gas is about 37 MJ/m^3. The high calorific value of pyrolysis gases has prompted suggestions that the gas could be used to provide the energy requirements for the pyrolysis process plant.

The gases are produced from the thermal degradation reactions of the waste constituents as they break down, and also through secondary cracking reactions of the primary products. Consequently, higher gas yields are found where the products of pyrolysis spend a relatively longer time in the hot zone of the reactor rather than undergoing rapid quenching, which produces higher oil yields and lower char yields. Also, higher gas yields are found at pyrolysis temperatures above about 700 °C, where accelerated cracking of the pyrolysis products occurs.

A wide variety of pyrolysis technologies have been investigated for the pyrolysis of waste materials. Examples of technologies which have been used for waste pyrolysis include fluidised beds, fixed-bed reactors, ablative pyrolysis at hot surfaces, rotary kilns, entrained flow reactors and vacuum pyrolysis (Bridgwater and Evans 1993). The design is dictated by the type of pyrolysis being undertaken, for example, fast or slow, to produce the targeted end-product. Many are still at the pilot-scale stage, whilst others are at the commercial or near commercial stage. Box 7.1 shows an example of a pyrolysis system for wastes.

7.3 Gasification

Gasification differs from pyrolysis in that oxygen in the form of air, steam or pure oxygen is reacted at high temperature with the available carbon in the waste to

Box 7.1
Ensyn Engineering Associates Inc., Canada, Biomass Pyrolysis System

The Ensyn Engineering Associates Inc. Canada, biomass pyrolysis system shown in the figure below is a rapid-heating fast pyrolysis system to produce an oil product for use as a chemical feedstock and fuel oil. The materials pyrolysed include wood waste, agricultural waste, heavy petroleum oils and tyre crumb. The system utilises either a solid heat carrier of sand or a gas heat carrier such as nitrogen to carry heat into a turbulent vertical reactor. The sand material used to transfer heat to the feedstock is heated by external furnaces. Very rapid interaction occurs with the waste feedstock to produce fast pyrolysis primary products, which are then

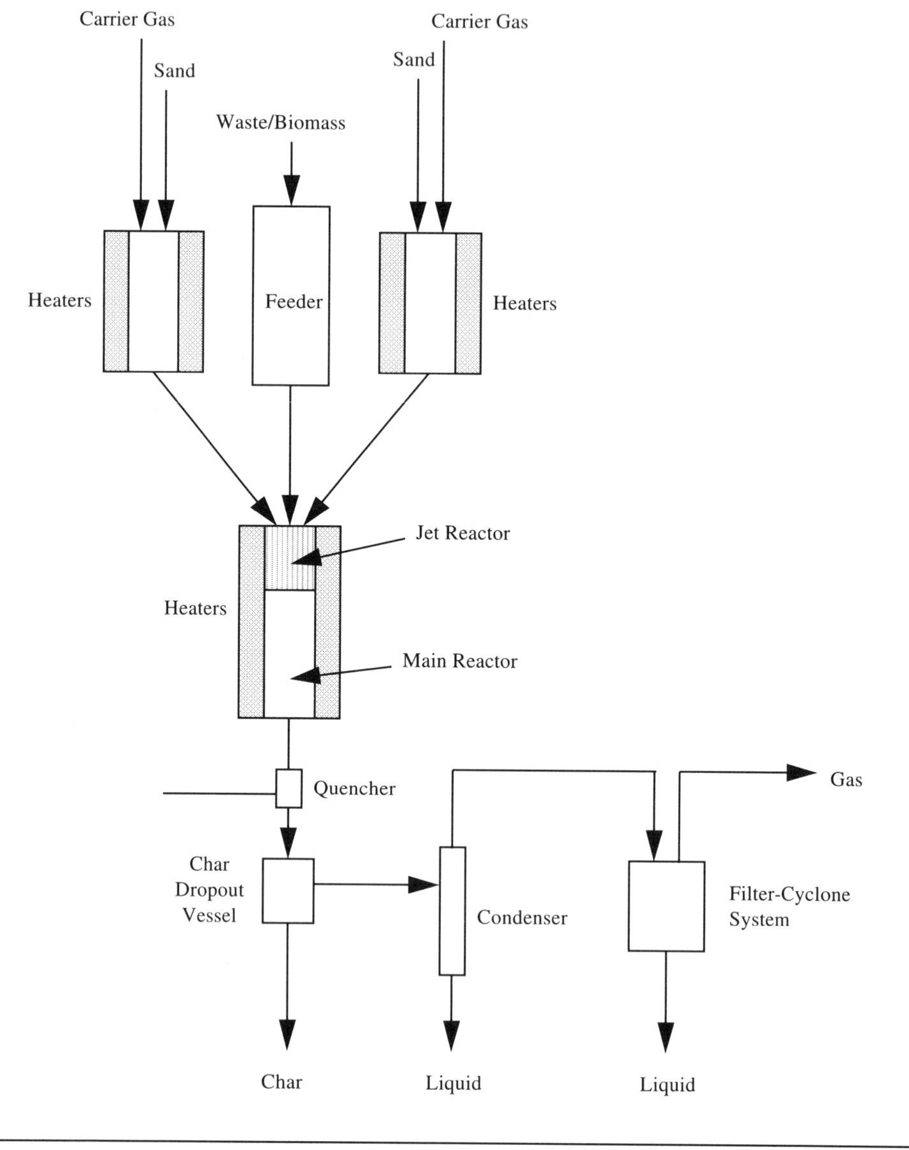

Box 7.1
(continued)

rapidly quenched to produce a liquid product. Temperature ranges from 400 to 950 °C are possible and residence times from 50 to 1500 ms. Maximum feedstock throughputs are proposed in the range of 25 tonne/day. High liquid yields up to 67% of the original feedstock have been recorded depending on the feedstock material. The solid char is removed in a drop-out vessel, and a condenser and filter/cyclone system is used to separate the liquids from the gaseous products.

Source: Bridgwater A.V. and Evans G.D., An assessment of thermochemical conversion systems for processing biomass and refuse. Energy Technology Support Unit, Harwell. Report ETSU/T1/00207/REP, Department of Trade and Industry, 1993. Crown copyright is reproduced with the permission of the Controller of Her Majesty's Stationery Office.

produce a gas product, ash and a tar product. Partial combustion occurs to produce heat, and the reaction proceeds exothermically to produce a low to medium calorific value fuel gas. The operating temperatures are relatively high compared with pyrolysis, at 800–1100 °C with air gasification, and 1000–1400 °C with oxygen. Calorific values of the product gas are low for air gasification, in the region of 4–6 MJ/m^3, and medium, about 10–15 MJ/m^3, for oxygen gasification (Bridgwater and Evans 1993). Steam gasification is endothermic for the main char–steam reaction and consequently steam is usually added as a supplement to oxygen gasification to control the temperature. Steam gasification under pressure, however, is exothermic, and steam gasification at pressures up to 20 bar and temperatures of between 700 and 900 °C produces a fuel gas of medium calorific value, approximately 15–20 MJ/m^3 (Bridgwater and Evans 1993; Rampling 1993) The product calorific values can be compared to natural gas at about 37 MJ/m^3.

The principal reactions occurring during gasification of waste in air are (Francis and Peters 1980; Rampling 1993)

$C + O_2 \Rightarrow CO_2$ Oxidation – exothermic
$C + CO_2 \Rightarrow 2CO$ Boudouard reaction – endothermic

Overall this is

$2C + O_2 \Rightarrow 2CO$ Exothermic

In steam gasification the reactions are

$C + H_2O \Rightarrow CO + H_2$ Carbon–steam reaction – endothermic
$C + 2H_2O \Rightarrow CO_2 + 2H_2$ Carbon–steam reaction – endothermic
$CO + H_2O \Rightarrow CO_2 + H_2$ Water–gas shift reaction – exothermic
$C + 2H_2 \Rightarrow CH_4$ Hydrogenation – exothermic

Table 7.4 The main types of waste gasifier reactor systems

Updraft gasification
Air flows up from the base of the reactor with the waste flowing down counter-current to the air flow. Gasification takes place in a slowly moving 'fixed' bed. Because the moisture, tar and gases generated do not pass through a hot bed of char there is less thermal breakdown of the tars and heavy hydrocarbons, and therefore the product gas is relatively high in tar. The tars may be condensed and recycled to increase thermal breakdown of the tars.

Downdraft gasification
The air and the waste flow co-currently down the reactor. Gasification takes place in a slowly moving 'fixed' bed. There is an increased level of thermal breakdown of the tars and heavy hydrocarbons as they are drawn through the high temperature oxidation zone, producing increased concentrations of hydrogen and light hydrocarbons. The air/steam or oxygen is introduced just above a 'throat' or narrow section in the reactor, which influences the degree of tar cracking.

Fluidised bed gasification
Waste is fed into the fluidised bed at high temperature. The fluidised bed may be a bubbling bed where the solids are retained in the bed through the gasification process. Alternatively, circulating beds may be used with high fluidising velocities; the solids are elutriated, separated and recycled to the reactor in a high solids/gas ratio, resulting in increased reaction. Twin fluidised-bed reactors may be used where the first bed is used to gasify the waste, and the char is passed to a separation unit and then to a second fluidised bed where combustion of the char occurs to provide heat for the gasifier reactor.

Source: Rampling T.W.A., Downdraft fixed bed gasification. Paper W93058. Presented to the Department of Trade and Industry, Renewable Energy Programme Advanced Workshop, Natural Resources Institute, Chatham, December, 1993.

In high pressure steam gasification, additional reactions include

$$CO + 3H_2 \Rightarrow CH_4 + H_2O \quad \text{Hydrogenation – exothermic}$$
$$CO_2 + 4H_2 \Rightarrow CH_4 + 2H_2O \quad \text{Hydrogenation – exothermic}$$

In practice there is usually some moisture present with the air, which produces some hydrogen. In addition, the heating of the waste produces pyrolytic reactions and methane, and higher molecular weight hydrocarbons are formed. When air is used the non-combustible nitrogen in the air inevitably reduces the calorific value of the product gas by dilution. Therefore, the major components of the product gas from waste gasification are carbon monoxide, carbon dioxide, hydrogen and methane, and where air gasification is used, nitrogen will also occur as a major component.

For wastes and biomass the development of gasification technologies has been via air or oxygen/steam gasification (Bridgwater and Evans 1993). Table 7.4 shows examples of the systems used for waste gasification (Rampling 1993). The characteristics of the gasifier system, the waste composition and operational conditions can give rise to tars, hydrocarbon gases and char; these are products of the incomplete gasification of the waste. The characteristics of the gasifier have most influence on the quality of the product gas, for example, down-draft gasifiers have

Table 7.5 *Product gas characteristics from different gasifier types*

Gasifier type	Calorific value of the product gas (MJ/m^3)	Gas quality[2]	Efficiency (%)
Downdraft–air	4.0–6.0	****	70–90
Downdraft–O_2	9–11	****	60–80[1]
Updraft–air	4.0–6.0	***	75–95
Updraft–O_2	8–14	***	65–85[1]
Fluidised bed–air	4–6	***	70–90
Fluidised bed–O_2	8–14	***	60–75[1]
Fluidised bed–steam	12–18	***	70–80
Circulating fluidised bed–air	5–6.5	**	75–95
Circulating fluidised bed–O_2	10–13	***	70–80[1]
Twin fluid bed	13–20	***	65–75
Cross flow–air	4.0–6.0	*	75–95
Horizontal moving bed–air	4.0–6.0	**	60–70
Rotary kiln–air	4.0–6.0	**	70–85
Multiple hearth	4.0–6.0	**	60–80

[1] Oxygen system efficiencies include a notional energy used for oxygen production.
[2] Gas quality is a relative assessment in terms of tars and particulates in raw gas: *, worst; *****, best.

Source: Bridgwater A.V. and Evans G.D., An assessment of thermochemical conversion systems for processing biomass and refuse. Energy Technology Support Unit, Harwell. Report ETSU/T1/00207/REP, Department of Trade and Industry, 1993. Crown copyright is reproduced with the permission of Her Majesty's Stationery Office.

all the products of gasification passing through a high temperature zone with high turbulence. This arrangement results in a high conversion of the pyrolysis intermediates and a gas with a low tar content, whereas the up-draft gasifier produces gas which is hot, and when it passes up through the down-flowing waste it produces pyrolysis reactions and a higher concentration of tar in the final product gas. Fluidised-bed reactors produce intermediate pyrolysis tar products which are passed out of the fluid bed into the freeboard by the fluidising gas. As the tars pass up through the hot freeboard of the fluidised bed some thermal cracking of the tars to gases may occur, but an overall gas tar content similar to that from an up-draft gasifier is usually found (Rampling 1993). Table 7.5 shows the characteristics of gasification product gas from different gasifier types (Bridgwater and Evans 1993). Box 7.2 shows an example of a circulating waste gasification system (Bridgwater and Evans 1993).

Utilisation of the gaseous product is often by direct combustion in a boiler or furnace. The heat energy is used for process heat or to produce steam for electricity generation. However, the raw gas will contain tar, char and hydrocarbon gases, and therefore the boiler or furnace burner system must be able to tolerate these contaminants and not be susceptible to fouling or clogging. In addition, gasification of heterogeneous waste such as municipal solid waste produces a gas which can vary in composition, and consequently the burner system of the boiler or furnace should be able to handle a range of gas compositions and calorific values. The advantages

Box 7.2
The Lurgi Circulating Fluidised Bed Waste Gasifier

The Lurgi circulating fluidised-bed gasifier has been used for the gasification of processed municipal solid waste in the form of refuse derived fuel, wood bark and waste wood, coal and petroleum coke. The full throughput capacity of the system is approximately 6500 kg/hour and uses air as the reactant gas, although oxygen or steam can be used. The product gas is of low calorific value, between 4 and 6 MJ/m^3, and has been used as fuel for firing a lime kiln. The figure shows a schematic diagram of the commercial system used to gasify wood bark. The wet bark is dried to approximately 10% moisture content in a rotary drier utilising the derived gases from the gasifier as fuel to provide the heat. The dried bark is fed to the gasifier at an operating temperature of 800 °C and atmospheric pressure. Solids removed from the gasifier are captured by the cyclones and are returned to the fluidised bed. Ash formed at the base of the gasifier is cooled and removed for disposal. The product gas is then used directly to fuel a lime kiln using a multifuel burner. The product gas composition for wood bark gasification is: hydrogen, 20.2%; carbon monoxide, 19.6%; carbon dioxide, 13.5%; hydrocarbons, 3.8%; nitrogen, 42.9%. Nitrogen content is high due to the use of air as the source of oxygen, and consequently the product gas has a low calorific value.

Source: Bridgwater A.V. and Evans G.D., An assessment of thermochemical conversion systems for processing biomass and refuse. Lurgi Energie- und Umwelttechnik GmbH Circulating Fluid Bed Gasifier, Energy Technology Support Unit, Harwell. Report ETSU/T1/00207/REP, Department of Trade and Industry, 1993. Crown copyright is reproduced with the permission of the Controller of Her Majesty's Stationery Office.

of direct combustion systems are that the gas does not have to be cleaned to any great extent before combustion, and the gases are used hot, maintaining the sensible heat in the system.

Where the utilisation of the product gas is into gas turbines or internal combustion engines to generate power or electricity, then the gas has to be cleaned to a higher specification than in direct combustion systems. Piping of the gas to the combustion unit requires that it be cooled and cleaned before utilisation to prevent pipe corrosion and deposition of tars and water. Removal of particulate material is by cyclones and bag filters, and tar removal is by secondary cracking at high temperature or catalyst cracking at lower temperatures. Gas turbines have been suggested as a suitable utilisation system for electricity generation, particularly for pressurised waste gasifiers. However, the fuel gas specifications for gas turbines are very stringent (Bridgwater and Evans 1993).

Some modern developments in thermochemical processing of waste have utilised both pyrolysis and gasification. Whiting (Warmer Bulletin 53, 1997) has reviewed several combined systems which are at or near the commercial scale. These include the combined pyrolysis–combustion system of Siemens, Germany, two-stage pyrolysis by TSK, Japan, the pyrolysis–vitrification system of Proler, USA, and the combined pyrolysis–gasification system of the Noell process, Germany, and the Thermoselect process in Italy.

Box 7.3 shows the combined pyrolysis–gasification Noell process developed in Germany (Leipnitz 1995).

Box 7.2
(continued)

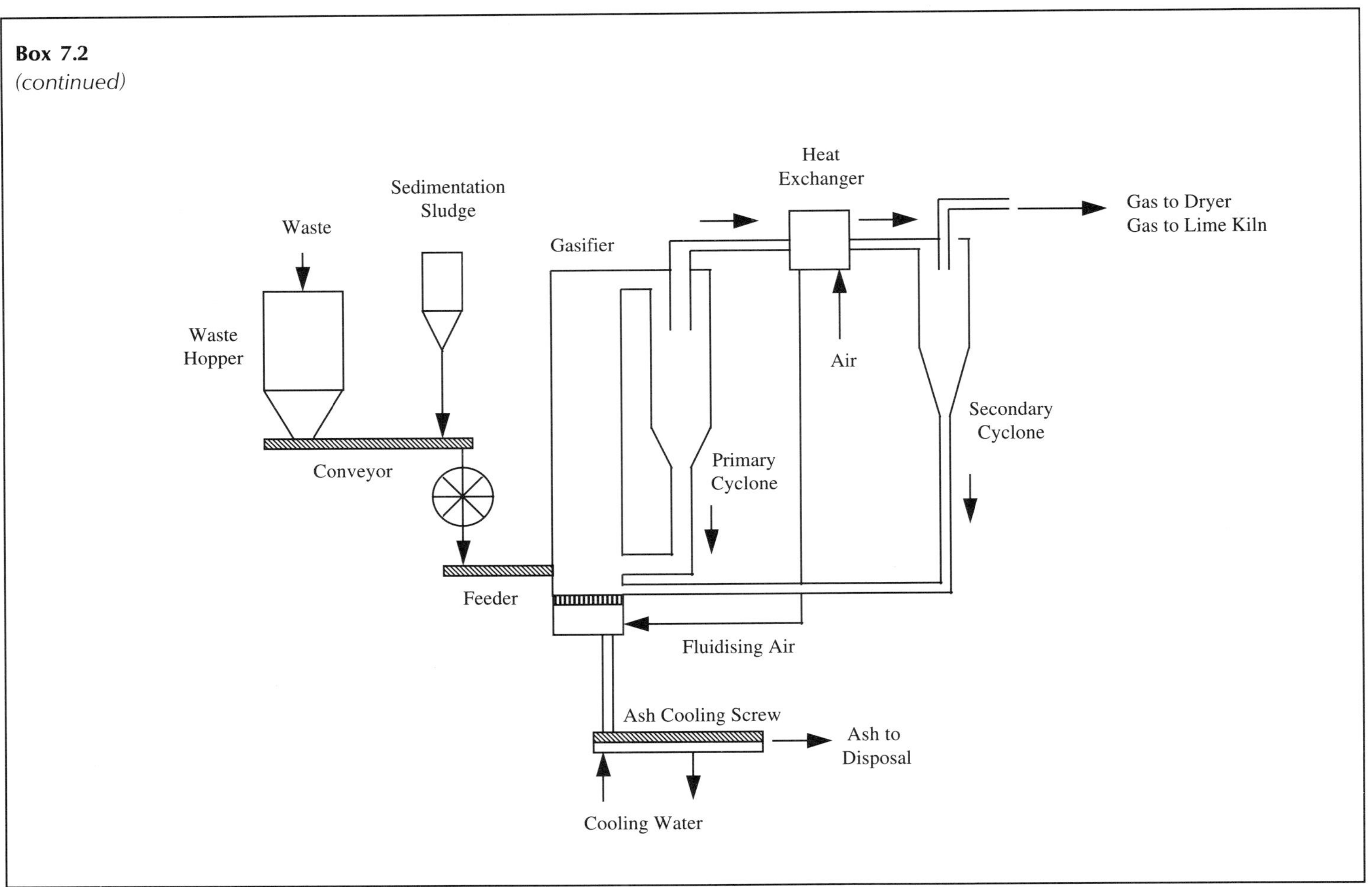

Box 7.3
The Noell Waste Treatment Process

The Noell waste treatment process is based on a combination of pyrolysis and entrained flow gasification. The system is designed to treat domestic waste, sewage sludge, hazardous waste and biomass with throughputs of up to 100 000 tonnes per year. The figure shows a schematic diagram of the system. The pyrolysis section consists of an indirectly heated, gas-fired rotary kiln operated under an inert gas atmosphere, the gas being derived from the waste treatment process. Shredded waste is fed into the pyrolysis reactor at approximately 550 °C and solids retention times in the kiln are about 1 hour. The char product is separated, and ferrous and non-ferrous metals are separated from the char. The char is then ground and passed to the entrained flow gasifier. The pyrolysis gases and oil, water and dust carry-over are quenched. The condensed oil, dust carry-over and gas plus the ground char are passed to the gasifier. The gasifier is an entrained flow type where an inert solid material of particle size <1 mm and high loading of about 350 kg/m^3 of solids is fed with the pyrolysis products and oxygen into a burner operating at sub-stoichiometric conditions. High temperatures of the order of 1400 °C are produced in the gasifier. The gasifier reaction under partial oxygenation conditions, i.e. sub-stoichiometric, generates a gas composed of over 80% carbon monoxide and hydrogen. Any solid inert material is converted to an ash slag because of the high temperatures involved, quenched and granulated. The resultant gas is cooled, scrubbed and utilised for energy recovery. The high gasifier temperatures completely destroy toxic hydrocarbon compounds, and because the operating conditions are reducing the de-novo synthesis of dioxins and furans is eliminated, thus reducing the costs of gas clean-up.

Source: Leipnitz Y., Noell Abfall Und Energietechnik GMBH, Germany. In: The Future of European Thermal Waste Treatment. Librairie des Enterprises, Conference, 7/8 September 1995, Paris.

7.4 Composting

Composting is the aerobic rather than anaerobic biological degradation of biodegradable organic waste such as garden and food waste. Composting is a relatively fast biodegradation process, taking typically about 4–6 weeks to reach a stabilised product. Small-size composting has been practised for many years at the individual household level. The UK Government is seeking to expand the practice to try to reach a target of 40% of domestic properties with a garden to carry out home composting by the year 2000 (Making Waste Work 1995). In addition, large scale composting schemes are being developed where organic waste is collected from parks, household garden waste is collected from civic amenity sites, and garden and food waste are collected directly from households in separated kerbside collections. The organic waste collected is composted at large central facilities. There is a separate UK Government target to recycle or compost 25% of household waste by the year 2000, and such schemes help the target to be met. However, currently less than 0.5% of collected household waste is composted (Making Waste Work 1995). The degraded product is a stabilized product which is added to soil to improve its structure, especially for clay soils, acts as a fertilizer to improve the nutrient content, acts as a mulch, and is used to retain moisture in the soil.

Box 7.3
(continued)

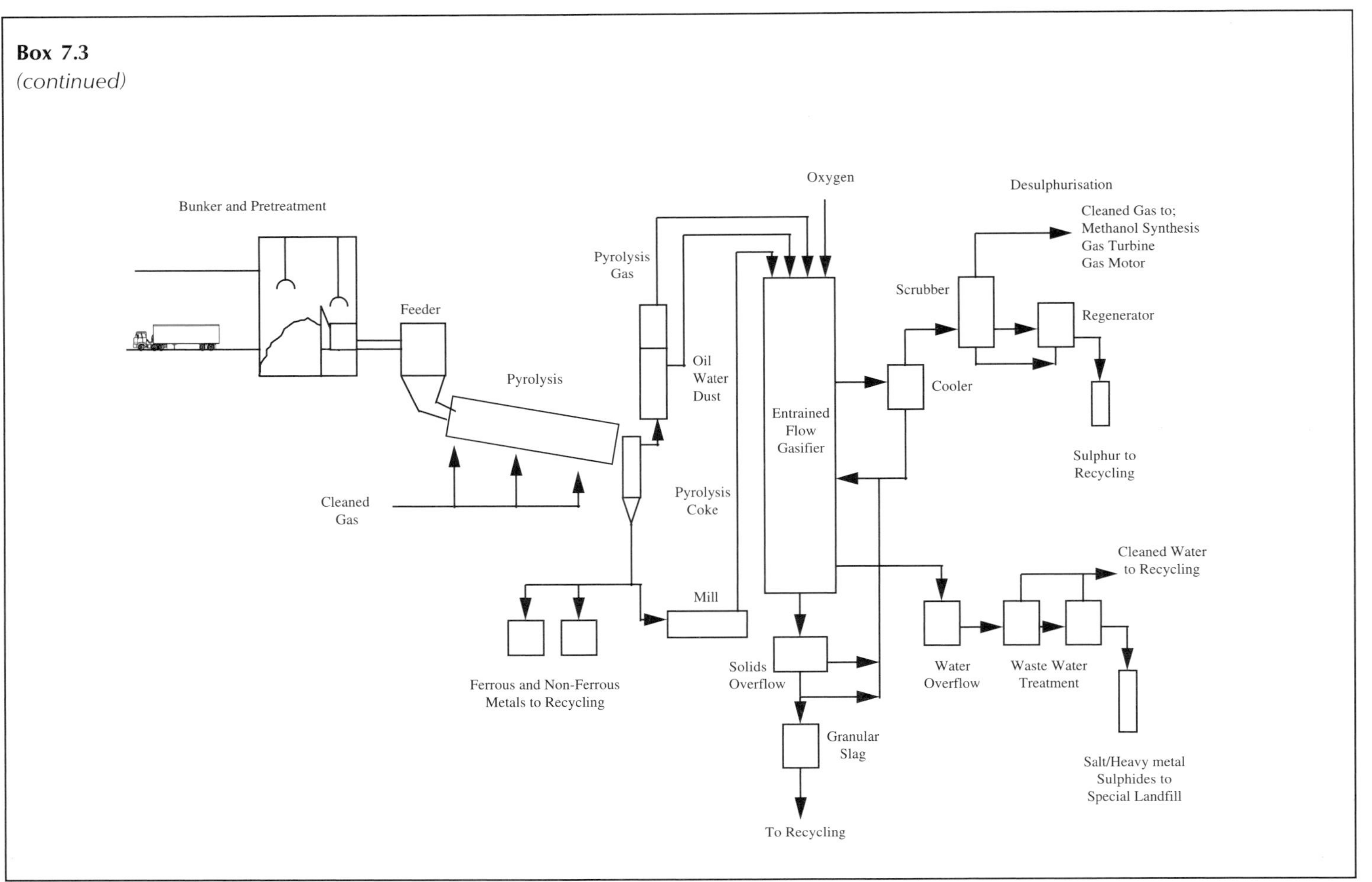

Table 7.6 *Average heavy metal concentrations found in various composted wastes (p.p.m.)*

Heavy metal	EU eco-label standard	Mixed refuse compost	Green waste compost	Source-segregated household waste compost
Zinc	300	1510	214	290
Lead	140	513	87	87
Copper	75	274	37	47
Mercury	1.0	2.4	0.4	0.4

Source: Walker M., Spreading the muck message. The Waste Manager, February, 19–21, 1996.

Composting removes a large part of the organic biodegradable waste from the waste stream. Consequently, the biodegradation process operating in landfills will be reduced. This has the result that less landfill gas and leachate will be produced, which aids in the operation and management of landfill but is clearly a disadvantage for landfills designed as bioreactors to generate landfill gas for energy recovery.

The organic waste used for large scale composting schemes can be variable, and may also contain contaminants due to incorrect sorting. This in turn may lead to a variable product composition, and quality may be difficult to guarantee for the end user. Acceptance of the product in the market place relies on a wide range of criteria, including the price, quality and consistency of the product, and guarantees that the product will be free of contaminants such as heavy metals, glass and other inert materials, and also free from plant and animal pathogens. The impact on consumer confidence which would occur if a contaminant such as glass or a hypodermic syringe were found in composted municipal solid waste would be extremely adverse (Border 1995). Table 7.6 shows the concentration of heavy metals found in compost derived from various wastes compared with the European Union standard (Walker 1996).

The compostable fraction of municipal solid waste, including putrescibles, paper and cardboard, can comprise up to 60% (Waste Management Paper 28, 1992). The composting process is aerobic and consequently relies on a plentiful supply of oxygen. Regular aeration is required to maintain aerobic conditions, and this is achieved by regularly turning the composting waste or by air injection. Biodegradation of the waste by micro-organisms is an exothermic process, and temperatures in the compost pile can reach up to 70 °C (Warmer Bulletin 29, 1991). Box 7.4 shows the biological processes operating during composting (de Bertoldi et al 1983).

A typical aerobic composting system for organic waste includes a pre-processing stage, the aerobic biodegradation stage and a maturation stage. Pre-processing of 'green waste', i.e., garden waste from parks and civic amenity sites, may only require shredding or pulverisation. There are several 'green waste' composting facilities in the UK, and it is estimated that over 250 000 tonnes of such waste is composted each year (Border 1995). Source-segregated organic waste, segregated by the householder and collected in kerbside or bring systems, would require a greater degree of pre-processing to remove contaminants and poorly segregated wastes. Mixed municipal solid waste would require separation of the component waste on the scale

Box 7.4
The Biological Processes Operating During Composting

The table below shows the organic fraction composition of MSW. The simple carbon compounds, such as soluble sugars and organic acids, are easily metabolised and mineralised by heterotrophic and heterogeneous micro-organisms. High metabolic activity and exothermic processes produce an increased temperature in the compost heap which, because of the low thermal conductivity, cannot dissipate the heat and consequently temperatures rise. This rise in temperature allows only the thermophilic micro-organisms to be active. Cellulose, pectin, starch and lignin are degraded later by fungi and actinomycetes; the decrease in temperature and also moisture and pH increases the activity of the fungi and actinomycete micro-organisms. Cellulose decomposition is intense throughout the process, but particularly during the final stages, mostly through degradation by eumycete micro-organisms. The degradation of lignin is restricted to the basidiomycete group of fungi. The nitrogen content is decreased via ammonia formation and volatilisation during the composting process. However, the loss of carbon dioxide and water from carbon- and hydrogen-containing constituents in the waste results in an overall decrease in the C:N ratio, representing a relative (to C) increase in nitrogen overall. In addition, the nitrogen content later slightly increases by nitrogen fixation from the micro-organisms.

The six main factors influencing composting are listed below.

1. Suitable oxygen content to maintain aerobic conditions: a minimum oxygen content in the compost of 18% is recommended.
2. Temperature: maximum micro-organism activity is observed in the temperature range 30–35 °C.
3. Moisture content: below a minimum 40% moisture content biodegradation is significantly reduced; high moisture contents are also to be avoided since the moisture occupies intraparticle spaces and thereby produces anaerobic conditions.
4. pH range of the waste material: optimal composting is achieved in the pH range 5.5–8. Bacteria prefer a near neutral pH, whereas fungi develop better in a slightly acidic environment.
5. C:N ratio of the waste material: optimal C:N ratio in the starting waste material is about 25, higher values resulting in a slow rate of decomposition, and lower ratios resulting in nitrogen loss. The organic fraction of MSW has a C:N ratio between 26 and 45 and for raw sewage sludge it is 7–12.
6. Size range of waste material: shredding of the starting waste material increases the surface area and results in enhanced rates of composting.

Constituent	%
Volatile matter	70–90
Protein	2–8
Lipids	5–10
Total sugar	5
Cellulose	35–55
Starch	2–8
Lignin	3–8
Phosphorus	0.4–0.7
Potassium	0.7–1.7
Crude fibre	35–40

Source: de Bertoldi M., Vallini G. and Pera A., The biology of composting: A review. Waste Management and Research, 1, 157–176, 1983.

of a materials recovery facility to remove the inert materials such as glass, ferrous and non-ferrous metals etc. The separated organic fraction may be shredded or pulverised to give a size range of between 1 and 10 cm^2 depending on the type of waste.

The composting stage involves the biodegradation of the sample under aerobic conditions, and therefore requires aeration of the waste. Aeration can be achieved by several methods, and as these increase in sophistication and control of the process, they also increase in cost. In the 'windrow' method, the waste is piled into elongated conical heaps about 2 m high, 3–4 m in width and 50 m in length and turned periodically by mechanical turning (Warmer Bulletin 29, 1991; Diaz et al 1993). Turning rates vary from one turn per day in the early stages of composting to one turn every 5 days towards the end of the process. The windrows are usually placed on a gravel bed to aid the collection of any leachate that may be formed. Windrows would normally be arranged in rows, and the mechanical turning vehicle moves up and down each row turning the waste, thereby periodically fully aerating the pile.

Forced aeration systems involve air being blown or sucked through the pile of composting waste by a fan. The compost pile is located on an aeration block and remains undisturbed. Air is distributed via a perforated pipe covered with a porous base material, which is usually finished compost which acts as a filter and even distributor of the air. The waste pile for composting is constructed over the filter and perforated pipe. Typical forced-aeration systems have pile heights of 2–3 m, are 2–6 m in width and up to 30 m in length (Diaz et al 1993). The air is passed through the pile either continuously or periodically. If the air is drawn down through the pile, the odours from the compost are contained in the system, allowing for control and treatment if required. Where the air is blown up through the compost pile, this serves to transfer the heat from the inner pile to the outer regions. Figure 7.3 shows a schematic diagram of the windrow and forced aeration systems (Warmer Bulletin 29, 1991). Windrows and forced-aeration systems are most common in the UK, with more sophisticated systems of composting being developed throughout Europe and the USA (Warmer Bulletin 29, 1991).

In more sophisticated systems the waste is processed in mechanical systems which allow closer control of temperature, moisture, aeration and waste mixing rates (Diaz et al 1993). For example, in drum/tunnel systems the waste is placed in a long rotating drum/tunnel. Forced aeration is employed by passing air through the rotating drum while the waste is continuously stirred and tumbled. Figure 7.4a shows a typical rotating drum composter (Diaz et al 1993). Residence times in the drum are typically only a few days, and the waste would require a subsequent maturation stage to reach completion of the composting process. Composting reactors are similarly more sophisticated composting systems. For example, a reactor system is shown in Figure 7.4b (Diaz et al 1993). The waste is stirred by a series of augers, which are perforated and allow air to be blown into the composting waste pile. The augers are located on a stirring arm.

The compost arising from the above processes may be stabilised but not ready for use, since some of the intermediate products from the biodegradation processes acting on the large organic waste compounds may be toxic to plant matter. The

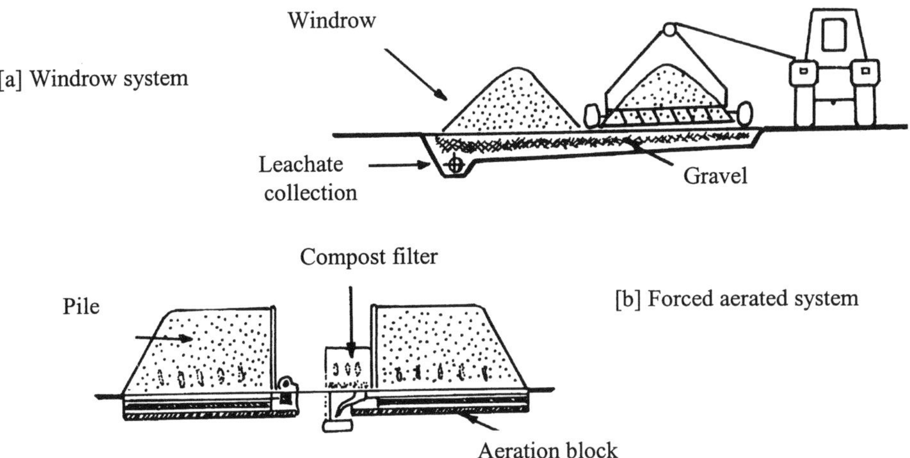

Figure 7.3 Schematic diagram of a windrow system and forced-air system for aeration of compost. Source: Warmer Bulletin 29, Compost. Warmer Fact Sheet. The World Resource Foundation, Tonbridge, 1991. Reproduced by permission from the World Resource Foundation

maturation stage involves further biodegradation of these intermediate compounds and may take several weeks for completion (Warmer Bulletin 29, 1991).

The final stages of composting would be processes to remove uncomposted materials and contaminants such as glass, plastics and metals. The compost would then be size-reduced and screened.

The composting process is a biodegradation process and leads to the formation not only of compost, but also other products which may require control and treatment. Leachate may form in cases of high moisture content. The leachate will have many of the properties and composition of leachate generated in the early stages of landfill. The leachate is allowed to collect in channels and discharged to a sewer or treated on site depending on the level of leachate generated. Gaseous emissions from the composting process consist mainly of volatile organic compounds. The emissions are often malodorous and potentially toxic. Table 7.7 shows some gaseous emissions from the composting of municipal solid organic waste using a variety of composting systems from windrows and forced aeration, to reactor systems (Eitzer 1995). The results show that whilst a wide range of emissions were detected they were below permissible workplace exposure limits. However, to limit the nuisance value of odours escaping from the composting facility and for better control of the composting process, the system may be enclosed in a covered building, further adding to costs.

Increasing sophistication and control of the composting process inevitably lead to increased costs. An estimation of the costs involved in the various types of composting systems are shown in Table 7.8 (Walker 1996). In addition, the collection systems for the organic waste add to the costs of the composting process. Consequently, in the

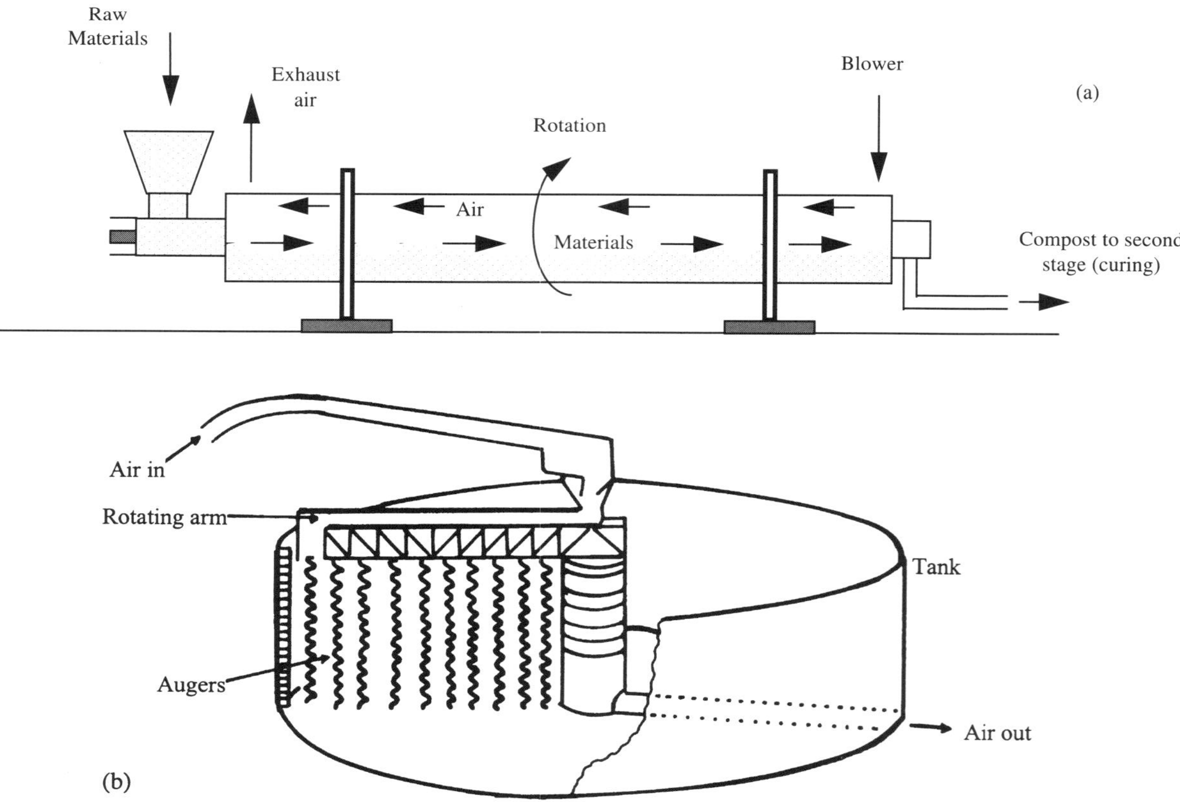

Figure 7.4 *Schematic diagram of a rotating drum system and a reactor system for composting. Source: Reprinted with permission from Diaz L.F., Savage G.M., Eggerth L.L. and Golueke C.G., Composting and Recycling of Municipal Solid Waste. Lewis, Boca Raton, FL, 1993. Copyright Lewis Publishers, an imprint of CRC Press, Boca Raton, Florida*

Table 7.7 *Emissions of selected volatile organic chemicals from municipal solid waste composting facilities*

Component	Emission ($\mu g/m^3$)	Limit value[1] ($\mu g/m^3$)
Trichlorofluoromethane	915 000	5 620 000
Acetone	166 000	1 800 000
Carbon disulphide	150	31 000
Methylene chloride	260	174 000
1,1-Dichloroethane	1	400 000
2-Butanone	320 000	590 000
Chloroform	54	49 000
1,1,1-Trichloroethane	15 000	1 900 000
Carbon tetrachloride	290	31 000
1,2-Dichloroethane	2	40 000
Benzene	700	32 000
Trichloroethene	1 300	270 000
2-Hexanone	6 600	20 000
Toluene	66 000	188 000
Tetrachloroethene	5 600	339 000
4-Methyl-2-pentanone	16 000	205 000
Chlorobenzene	29	46 000
Ethylbenzene	178 000	434 000
m,o-Xylene	15 000	434 000[2]
p-Xylene	6 900	434 000[2]
Styrene	6 100	213 000
Isopropyl benzene	370	246 000
n-Propyl benzene	1200	–
4-Chlorotoluene	240	–
1,3,5-Trimethylbenzene	2 200	123 000[2]
1,2,4-Trimethylbenzene	1000	123 000[2]
sec-Butylbenzene	220	–
1,3-Dichlorobenzene	2	–
1,4-Dichlorobenzene	90	451 000
p-Isopropyl toluene	4 800	–
1,2-Dichlorobenzene	1	150 000
n-Butylbenzene	210	–
1,2,4-Trichlorobenzene	9	123 000[2]
Naphthalene	1 400	52 000
Hexachlorobutadiene	4	210
1,2,3-Trichlorobenzene	6	–

[1] American Conference of Governmental Industrial Hygienists, threshold limit values – time weighted average for workplace air.
[2] Value for the sum of all the isomers of each compound.

Source: Reprinted with permission from Eitzer B.D., Emissions of volatile organic chemicals from municipal solid waste composting facilities. Environmental Science and Technology, 29, 896–902, 1995. Copyright 1995 American Chemical Society.

Table 7.8 *Typical gate fees for various composting systems*

Composting system	Typical gate fee[1] (£/tonne)
Open air, turned windrow systems	15–20
Open air, forced aeration systems	17–22
Covered building, forced aeration systems	20–30
Simple tunnel systems (non-turning)	30–35
Complex tunnel systems (with turning)	35–40
Enclosed hall (negative pressure)	35–45
Reactor systems	45–60

[1] Assuming 15 000 tonnes per annum.

Source: Walker M., Spreading the muck message. The Waste Manager, February, 19–21, 1996.

Table 7.9 *Potential markets for compost derived from waste*

Market	Potential annual market (million m^3)
Retail horticulture	0.5–1.0
Professional horticulture	0.5–1.0
Landscaping and land reclamation	1.0–2.0
Agriculture	200 +

Source: Border D., Taking a green product to market. The Waste Manager, March, 12–15, 1995.

UK the development of large scale composting has been through the relatively simple and cheap use of 'green waste' deposited by householders at civic amenity sites and from local authority parks and gardens.

Whilst there is great interest in composting organic waste from domestic sources, parks and garden waste deposited at civic amenity sites, there are other major sources of organic waste suitable for composting. For example, agricultural waste, sewage sludge, forestry waste and food waste (Border 1995). In terms of marketability, the retail and professional horticultural retail market and the landscaping and land restoration markets for compost produced from waste are important and significant (Table 7.9, Border 1995). The main competitor to waste-derived compost in the horticultural market is peat. There is a big potential market for the large scale use of waste-derived compost via land reclamation and land restoration as a substitute for top soil and as a means of improving the chemical and physical properties of soils. By far the largest potential market for compost derived from waste has been identified as the agricultural market (Table 7.9, Border 1995). However, penetration of waste-derived compost into the agricultural market has proved difficult. This is due in part to the high specifications and performance requirements of the agricultural end-user and the resistance to a new product.

Whilst composting of waste is currently only developed on a small scale in the UK, some countries have a well-developed waste composting industry. For example, approximately 95 composting facilities operate in France, processing 1.5 million tonnes of mixed domestic waste each year to produce 650 000 tonnes of compost (Warmer Bulletin 29, 1991). In the USA, nine composting facilities produce over 2.5 million tonnes of compost per year, with a further 79 plants in various stages of planning.

Portugal has one of the highest rates of municipal solid waste composting, with a composting rate of 13.5% representing approximately 450 000 tonnes of waste composted each year (ENDS 1996). Spain composts approximately 1.5 million tonnes of waste, representing 10.9% of total municipal solid waste, and Denmark and Belgium have composting rates of over 8%. By contrast, the UK has a composting rate of less than 0.5% and Japan about 0.1% (ENDS 1996).

7.5 Anaerobic Digestion

The anaerobic degradation processes found in landfills which lead to the formation of methane and carbon dioxide from organic waste are utilised in anaerobic digestion but in an enclosed, controlled reactor. The better control of the process means that all of the gas is collected for utilisation unlike landfills where collection efficiencies are relatively low at 50% or less. In addition, the solid residue arising from anaerobic digestion can be cured and used as a fertiliser. The main aim of the process is to produce a product gas rich in methane, which can be used to provide a fuel or act as a chemical feedstock. Anaerobic digestion has been used to treat sewage sludge and agricultural wastes for several years. In addition, the process is now being developed for municipal solid wastes and industrial wastes. Figure 7.5 shows the main steps in the anaerobic digestion of municipal solid waste (IEA Bioenergy 1996). The biodegradable fraction of the waste requires separation from the other components. Source separation, mechanical separation and hand-sorting may be used. Pre-treatment involves removing contaminants, homogenising the waste, and protecting the down-stream processes. The main anaerobic digestion generates gas consisting of methane and carbon dioxide. The methane is combusted to produce energy. The anaerobic process also serves to stabilise the waste and disinfect and deodorise. Post-treatment involves removal of further contaminants such as glass and plastics, and then further stabilisation takes place through the composting of the residue (IEA Bioenergy 1996).

The organic components of waste can be classified into broad biological groups, represented by proteins, carbohydrates and lipids or fats. Carbohydrates are by far the major component of biodegradable wastes and include cellulose, starch and sugars. Proteins are large complex organic materials composed of hundreds or thousands of amino acids groups. Lipids or fats are materials containing fatty acids. The degradation of the organic components takes place largely by biological processes, but also involve interrelated physical and chemical processes (Wheatley 1990, Diaz et al 1993).

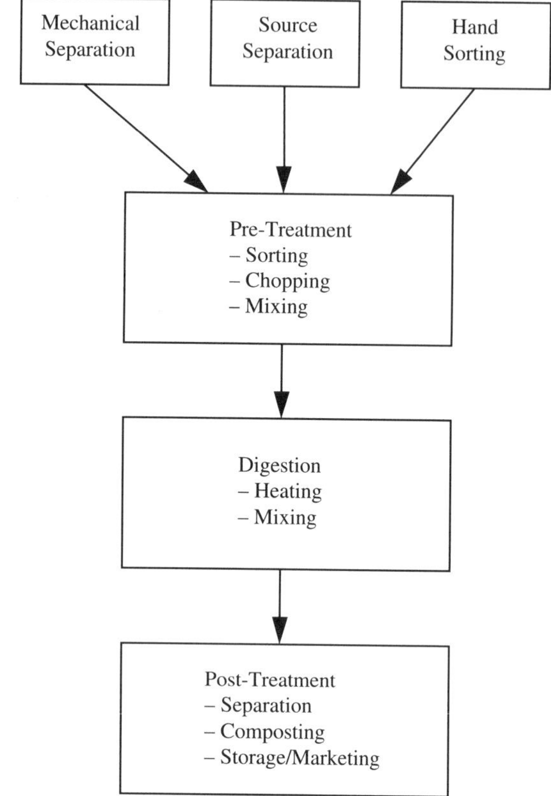

Figure 7.5 *The main steps in the anaerobic digestion of municipal solid waste. Source: Adapted from: IEA Bioenergy, Biogas from municipal solid waste: Overview of systems and markets for anaerobic digestion of MSW. Energy Recovery from MSW. Task Anaerobic Digestion Activity, National Renewable Energy Laboratory, Harwell, 1996*

The initial stages of the decomposition involve the hydrolysis and fermentation of the cellulosic, protein and lipid compounds in the waste by micro-organisms, the facultative anaerobes which can tolerate reduced oxygen conditions. Carbohydrates, proteins and lipids are hydrolysed to sugars, which are then further decomposed to carbon dioxide, hydrogen, ammonia and organic acids. Proteins decompose via deaminisation to form ammonia and also carboxylic acids and carbon dioxide. Gas concentrations may rise to levels of up to 80% carbon dioxide and 20% hydrogen. The second stage of anaerobic digestion is the acid stage, where organic acids formed in the hydrolysis and fermentation stage are converted by acetogen micro-organisms to acetic acid, acetic acid derivatives, carbon dioxide and hydrogen. Other organisms convert carbohydrates directly to acetic acid in the presence of carbon dioxide and hydrogen. Hydrogen and carbon dioxide levels then begin to decrease. The final stage of anaerobic digestion is the main methane gas forming stage. Low hydrogen levels promote the methane-generating micro-organisms, the methanogens, which generate methane and carbon dioxide from the organic acids

Table 7.10 *Types of anaerobic digestion systems*

Dry continuous digestion
The waste is fed continuously to a digestion reactor with a digestate dry matter content of 20–40%. Both completely mixed and plug-flow systems are available, with plug-flow systems relying on external recycling of a proportion of the outgoing digested waste to be mixed with the incoming waste feedstock to initiate digestion.

Dry batch digestion
The waste is fed to the reactor with digested material from another reactor. The reactor is then sealed and left to digest naturally. Leachate derived from the biodegradation process is collected from the bottom of the reactor and recirculated to maintain a uniform moisture content and redistribute nutrients and micro-organisms. When digestion is complete the reactor is opened, unloaded and refilled to start the batch process again.

Leach-bed process
The process is similar to the dry batch process but the leachate derived from the biodegradation of the waste is exchanged between established and new batches of waste to facilitate start-up of the biodegradation process. After methanogenesis has become established in the waste, the reactor is uncoupled and reconnected to fresh solid waste in a second reactor.

Wet continuous digestion
The waste is slurried with a large proportion of water to provide a dilute (10% dry solids) waste feedstock that can be fed to a conventional completely mixed digester. Effective removal of glass and stones is required to prevent accumulation in the bottom of the reactor. When used for municipal solid waste alone, filter pressing of the wet digestate to recover liquor to recycle for feed preparation is required to avoid generating an excessive volume of diluted digestate for disposal. Alternatively, the process can be used for co-digestion with dilute wastes such as sewage sludge.

Multi-stage wet digestion
There are also a range of multi-stage wet digestion processes where municipal solid waste is slurried with water or recycled liquor and fermented by hydrolytic and fermentative micro-organisms to release volatile fatty acids, which are then converted to gas in a specialist high-rate industrial anaerobic digester.

Source: IEA Bioenergy, Biogas from municipal solid waste: overview of systems and markets for anaerobic digestion of MSW. Energy Recovery from MSW. Task Anaerobic Digestion Activity, National Renewable Energy Laboratory, Harwell, 1996.

and their derivatives generated in the earlier stages. There are two classes of micro-organisms which are active in the methanogenic stage, the mesophilic bacteria which are active in the temperature range 30–35 °C, and the thermophilic bacteria active in the range 45–65 °C. The methanogenesis stage is the main gas generation stage of anaerobic digestion, with the gas composition generated at approximately 60% methane and 40% carbon dioxide. Ideal conditions for the methanogenic micro-organisms are a pH range from 6.8 to 7.5 (Wheatley 1990; Diaz et al 1993; IEA Bioenergy 1996).

Anaerobic digestion takes place in an enclosed, closely controlled reactor. There are a range of systems available, as shown in Table 7.10. Biodegradation takes place in a slurry of waste and micro-organisms. The yield of gas from anaerobic digestion depends on the composition and biodegradability of the waste. The rate of decomposition of the waste depends on the micro-organism population and temperature. The slurry consists largely of water. For sewage sludge, agricultural wastes and some industrial wastes it might have over 90% moisture content, and for municipal solid waste it may have about 60% moisture content. Lower moisture contents are preferred since they reduce the liquid effluents from the plant (Wheatley 1990; IEA Bioenergy 1996).

Operational parameters for the reactor would be controlled to give anaerobic conditions and maximum gas yield. For maximum gas production, a temperature in the range 30–35 °C for mesophilic bacteria and 45–65 °C for thermophilic bacteria would be used. Whilst higher gas production is found at the higher temperature range where the thermophilic bacteria are active, this should be balanced against the costs of the higher energy input. In the UK, sewage sludge digestion mainly operates in the thermophilic temperature range 30–35 °C. Depending on the type of waste, some form of shredding may be required to increase the surface area for reaction, whereas for sewage sludge, agricultural slurries and industrial liquids/sludges no pre-processing size reduction will be required. However, other factors such as the pH of the waste, the nutrient content and the C:N ratio are important factors which require control. For example, a high C:N ratio produces a high acid content and low methane production. Municipal solid waste can have a high C:N ratio of above 50, whereas sewage sludge has a C:N ratio below 10. Therefore the co-digestion of municipal solid waste and sewage sludge has been suggested. The optimum pH in the reactor would be 7, although high gas production rates are observed between pH 6.8 and 7.5. Low pH, that is acidic conditions, inhibit the activity of the methanogenic micro-organisms, however, a certain level of organic acids are required as nutrients for the methanogens (Wheatley 1990; Diaz et al 1993).

Mixing of the waste slurry within the digester is important in maintaining a high rate of anaerobic biodegradation and consequently a high production of gas. The mixing process serves to disperse the incoming waste within the actively digesting sludge, thereby improving contact with the micro-organisms and replacing the previously degraded products with fresh nutrients in the form of the waste. Mixing also reduces any stratification in the digester. For example, different temperatures may result in differential biological activity or sedimentation, where gravity separation results in a liquid top layer with low biodegradation activity, an active biodegradation layer and a lower degraded product sludge layer. The methods of mixing include internal mechanical mixing using rotating paddles, thus recirculating the product gas through the digestor, or recirculation of the slurry (Wheatley 1990).

The methanogenic micro-organisms are the centre of the process and their continued biodegradable activity is essential to the continued operation of the digester. Consequently, the pH range, temperature, nutrient level, C:N ratio, mixing etc. must be controlled to maintain maximum activity. The initiation of the whole process is by introduction of the waste and a starter sludge of micro-organisms, for example, taken from an already operating digester or anaerobically biodegrading organic waste from other sources (Wheatley 1990; Diaz et al 1993; IEA Bioenergy 1996).

Gas production rates from anaerobic digestion depend on the starting waste materials and the operational characteristics. Table 7.11 shows the production of

Table 7.11 *Production of gas from the anaerobic digestion of various wastes*

Waste material	Gas production (m^3/kg dry solids)	Temperature (°C)	Methane content (%)	Retention time (days)
Cattle manure	0.20–0.33	11.1–31.1	–	–
Poultry manure	0.31–0.56	32.6–50.6	58–60	9–30
Pig manure	0.49–0.76	32.6–32.9	58–61	10–15
Sheep manure	0.37–0.61	–	64	20
Sugarbeet leaves	0.5	–	–	11–20
Municipal refuse[1]	0.31–0.35	35–40	55–60	15–30

[1]US refuse, estimated yield per kilogram of organic solids.

Source: Reprinted with permission from Diaz L.F., Savage G.M., Eggerth L.L. and Golueke C.G., Composting and Recycling of Municipal Solid Waste. Lewis, Boca Raton, FL, 1993. Copyright Lewis Publishers, an imprint of CRC Press, Boca Raton, Florida.

gas, represented by methane and carbon dioxide, in relation to different types of waste and operational characteristics (Diaz et al 1993). Other gases present in low concentrations would be hydrogen, hydrogen sulphide, nitrogen and hydrocarbon gases at trace levels. Anaerobic digestion produces a gas which is similar in composition to landfill gas, and the potential uses are similar to those of landfill gas. The gas has a calorific value of between 20 and 25 MJ/m^3 (IEA Bioenergy 1996). The gas may be used directly as replacement fuels for kilns, boilers and furnaces located close to the anaerobic digestion site. If the gas is to be used in power generation a greater degree of gas clean-up is required, for example, to remove corrosive trace gases, moisture and vapours from the gas stream. Such possible pre-treatments may include further filtration, gas chilling to condense certain constituents, absorption and adsorption systems to scrub the gases and other gas clean-up systems such as membranes and molecular sieves to remove trace contaminants (IEA Bioenergy 1996).

Box 7.5 describes the Valorga, France, anaerobic digestion plant for municipal solid waste (Cayrol et al 1990).

The end products from anaerobic digestion depend on the type of digestion process. The main product from the dry digestion of waste is a solid residue or digestate which can be matured into a compost product. Small quantities of surplus liquor are also produced which are generally similar to a dilute digested sewage sludge. Products from the wet digestion process are generally similar to a concentrated digested sewage sludge. They can be spread directly onto farmland or dewatered to provide separate liquid fertiliser and solid compost products.

As a waste treatment and disposal method, anaerobic digestion is not a major option in the UK, with only a small number of operational plants, for example, Irvine which handles 300 tonnes/year of mixed waste, and plants at Stockbridge which handle 260 tonnes/year of manure, fuel crops and mixed waste (IEA Bioenergy 1996). However, several other plants are at the planning or construction stage, which will raise the throughput of waste treatment via anaerobic digestion to over 250 000 tonnes/year. Other countries, notably Denmark, Germany, Italy, the

Box 7.5
The VALORGA Process Anaerobic Digestion Plant for Municipal Solid Waste

The VALORGA company of France has developed an anaerobic digestion plant for the treatment of municipal solid waste. The commercial plant at Amiens, north of Paris, was built in 1988 and designed to treat 109 000 tonnes of municipal solid waste per year, which includes 104 000 tonnes/year domestic waste and 5000 tonnes/year of industrial effluent waste. The mixed waste is delivered to the plant and a grab transfers the waste to a crusher. The large combustible components in the waste are screened through a trommel, and in addition, ferrous metals are removed by a magnetic belt and glass and other inert materials are removed in a densimetric sorting apparatus. The large combustible materials are combusted in a two-stage pyrolytic/combustion unit and energy is recovered via a steam boiler. The sorted organic material from the pre-processing stage is passed to the anaerobic digestion stage (methanisation unit). The waste is digested in three reactors of 2400 m^3 capacity each. The waste is mixed with liquid effluent from the digestor and passed into the reactor as a slurry of between 30 and 35% solids content. Operation of the reactor is under mesophilic (35–40 °C) or thermophilic (45–60 °C) conditions. The product gas produced contains between 55 and 60% methane. The gas is decarbonised to remove carbon dioxide, desulphurised to remove traces of hydrogen sulphide (3000–4000 p.p.m.), dried and compressed, and then sold for direct use or to the national gas company of France, Gaz de France, for injection into the public network. The residual material from the digester is pressed to reduce the moisture content to about 40% and is used as a soil conditioner. The liquid from the press is recirculated with the incoming organic fraction of the waste back into the digester. The plant is estimated to produce about 6550 m^3 product gas/day, representing a yield of about 131 m^3 of gas per tonne of sorted waste.

Source: Cayrol F., Peillex J.-P. and Desbois S., Industrial application of the VALORGA process for the treatment of household wastes of the city of Amiens. In: Biomass for Energy and Industry. 5th EC Conference, Grassi, G., Gosse G. and dos Santos G. (Eds.) Elsevier Applied Science, London, 1990.

Netherlands, Sweden and Switzerland, have fully developed anaerobic digestion plants for handling waste. For example, Denmark treats manure, organic industrial waste and biowaste totalling 1.1 million tonnes/year, Germany treats sewage sludge, manure, biowaste and organic industrial waste totalling 360 000 tonnes/year, and Italy handles manure, mixed waste, organic industrial waste and manure totalling 660 000 tonnes/year (IEA Bioenergy 1996).

Box 7.5
(continued)

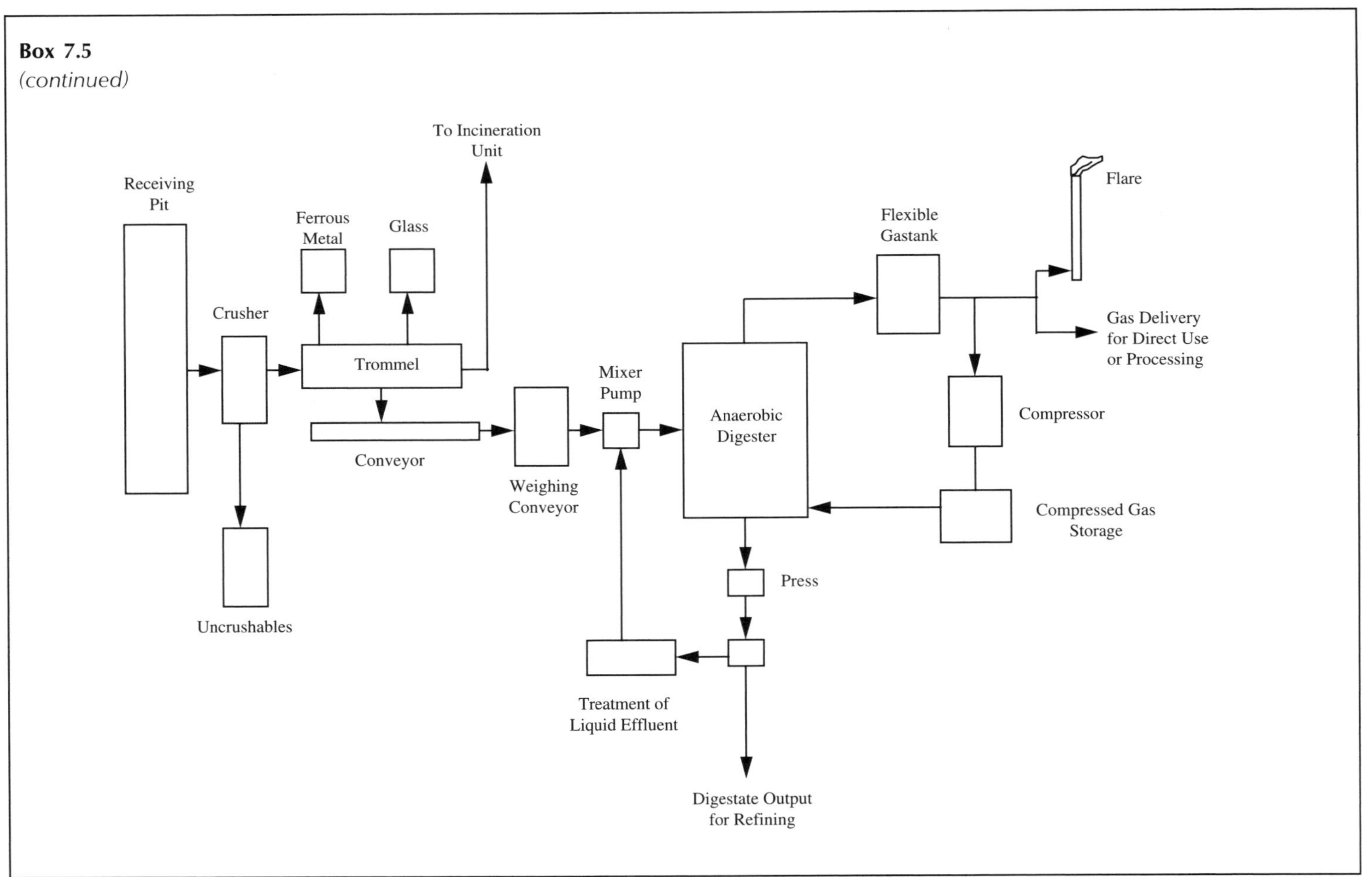

Bibliography

IEA Bioenergy, 1996. Biogas from municipal solid waste: overview of systems and markets for anaerobic digestion of MSW. Energy Recovery from MSW. Task Anaerobic Digestion Activity, National Renewable Energy Laboratory, Harwell.
Border D. 1995. Taking a green product to market. The Waste Manager, March, 12–15.
Bridgwater A.V. and Bridge S.A. 1991. A review of biomass pyrolysis and pyrolysis technologies. In: A.V. Bridgwater and G. Grassi (Eds.) Biomass Pyrolysis Liquids: Upgrading and Utilisation. Elsevier Applied Science, London.
Bridgwater A.V. and Evans G.D. 1993. An assessment of thermochemical conversion systems for processing biomass and refuse. Energy Technology Support Unit, Harwell, Report ETSU/T1/00207/REP, Department of Trade and Industry.
Cayrol F., Peillex J.-P. and Desbois S. 1990. Industrial application of the VALORGA process for the treatment of household wastes of the city of Amiens. In: Grassi G., Gosse G. and dos Santos G. (Eds.) Biomass for Energy and Industry. 5th EC Conference. Elsevier Applied Science, London.
de Bertoldi M., Vallini G. and Pera A. 1983. The biology of composting: A review. Waste Management and Research, 1, 157–176.
Desbene P.-L., Essayegh M., Desmazieres B. and Basselier J.-J. 1991. Contribution to the analytical study of biomass pyrolysis oils. In: A.V. Bridgwater and G. Grassi (Eds.) Biomass Pyrolysis Liquids: Upgrading and Utilisation. Elsevier Applied Science, London.
Diaz L.F., Savage G.M., Eggerth L.L. and Golueke C.G. 1993. Composting and Recycling of Municipal Solid Waste. Lewis, Boca Raton, FL.
ENDS, 1996. Report 255. Environmental Data Services Ltd., Bowling Green Lane, London.
Eitzer B.D. 1995. Emissions of volatile organic chemicals from municipal solid waste composting facilities. Environmental Science and Technology, 29, 896–902.
Francis W. and Peters M.C. 1980. Fuels and Fuel Technology. Pergamon, Oxford.
Graham R.G., Bergougnou M.A. and Overend R.P. 1984. Fast pyrolysis of biomass. Journal of Analytical and Applied Pyrolysis, 6, 95–135.
Horne P.A. and Williams P.T. 1996. Influence of temperature on the products from the flash pyrolysis of biomass. Fuel, 75, 1051–1059.
Kaminsky W. and Sinn H. 1980. Pyrolysis of plastic waste and scrap tyres using a fluidised bed process. In: Jones J.L. and Radding S.B. (Eds.) Thermal Conversion of Solid Wastes and Biomass. ACS Symposium Series 130, American Chemical Society, Washington, DC.
Lee M. 1995. Feedstock recycling: New plastic for old. Chemistry in Britain, July, 515–516.
Leipnitz Y. 1995. Combined pyrolysis and gasification. Noell Abfall Und Energietechnik GMBH, Germany. In: The Future of European Thermal Waste Treatment. Librairie des Enterprises, Conference 7/8 September, Paris.
Making Waste Work, 1995. Department of the Environment and Welsh Office, HMSO, London.
Menachem L. and Goklestein I. 1984. International Fibre Science, Vol. 11. Wood Structure and Composition. Marcel Dekker, New York.
Pakdel H., Roy C., Aubin H., Jean G. and Coulombe S. 1991. Formation of dl-limonene in used tyre vacuum pyrolysis oils. Environmental Science and Technology, 25, 1646–1649.
Pober K.W. and Bauer H.F. 1977. The nature of pyrolytic oil from municipal solid waste. In: Anderson L.L. and Tillman D.A. (Eds.) Fuels from Waste. Academic Press, London.
Rampling T.W. and Hickey T.J. 1988. The laboratory characterisation of refuse derived fuel. Warren Spring Laboratory Report, LR643, Department of Trade and Industry, HMSO, London.
Rampling T.W.A. 1993. Downdraft fixed bed gasification, Paper W93058. Presented to the Department of Trade and Industry, Renewable Energy Programme Advanced Workshop, Natural Resources Institute, Chatham.
Rick F. and Vix U. 1991. Product standards for use as fuel in industrial firing plants. In:

Bridgewater A.V. and Grassi G. (Eds.) Biomass Pyrolysis Liquids: Upgrading and Utilisation. Elsevier Applied Science, London.
Roy C. 1994. State of the art of thermolysis techniques in depollution. Technological Days in Wallonia. International Meeting Thermolysis, a Technology for Recycling and Depollution, 24–25 March, ISSeP, Liege.
Stoikos T. 1991. Upgrading of biomass pyrolysis liquids to high value chemicals and fuel additives. In: Bridgwater A.V. and Grassi G. (Eds.) Biomass Pyrolysis Liquids: Upgrading and Utilisation. Elsevier Applied Science, London.
Walker M. 1996. Spreading the muck message. The Waste Manager, February, 19–21.
Warmer Bulletin 29, 1991. Compost. Warmer Fact Sheet. The World Resource Foundation, Tonbridge.
Warmer Bulletin 53, 1997. Whiting K. Thermal processing of solid waste. Journal of the World Resource Foundation, Tonbridge.
Waste Management Paper 28, 1992. Recycling. Department of the Environment, HMSO, London.
Wheatley A. 1990. Anaerobic Digestion: A Waste Treatment Technology. Elsevier Applied Science, London.
Williams P.T. and Besler S. 1992. The pyrolysis of municipal solid waste. Journal of the Institute of Energy, 65, 192–200.
Williams P.T., and Besler S. 1996. Influence of temperature and heating rate on the slow pyrolysis of biomass. Renewable Energy, 7, 233–250.
Williams P.T. and Horne P.A. 1994. Characterisation of oils from the fluidised bed pyrolysis of biomass with zeolite catalyst upgrading. Biomass and Bioenergy, 7, 223–236.
Williams E.A and Williams P.T. 1997. The pyrolysis of individual plastics and a plastic mixture in a fixed-bed reactor. Journal of Chemical Technology and Biotechnology, 70, 9–20.
Williams P.T., Besler S., Taylor D.T. and Bottrill R.P. 1995. Pyrolysis of automotive tyre waste. Journal of the Institute of Energy, 68, 11–21.

8

Integrated Waste Management

Summary

This concluding chapter discusses the integration of treatment and disposal options described in the previous chapters to introduce the concept of 'integrated waste management'. The different approaches to integrated waste management are described.

8.1 Integrated Waste Management

The treatment and disposal of waste has developed from its early beginnings of mere dumping to a sophisticated range of options including re-use, recycling, incineration with energy recovery, advanced landfill design and engineering, and a range of alternative technologies, including pyrolysis, gasification, composting and anaerobic digestion. The further development of the industry is towards integration of the various options to produce an environmentally and economically sustainable waste management system.

Integrated waste management has been defined as the integration of waste streams, collection and treatment methods, environmental benefit, economic optimisation and societal acceptability into a practical system for any region (Warmer Bulletin 49, 1996). Integrated waste management implies the use of a range of different treatment and disposal options, including the areas covered in this book, i.e., waste reduction, re-use and recycling, landfill, incineration, and alternative options such as pyrolysis, gasification, composting and anaerobic digestion. However, integration also implies that no one option of treatment and disposal is better than another, and each option has a role to play, but that the overall waste management system chosen should be the best environmentally and economically sustainable one for a particular region (Figure 8.1, Warmer Bulletin 49, 1996).

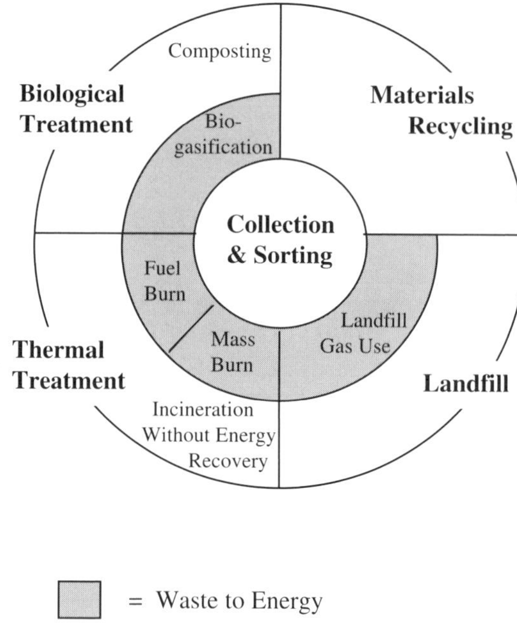

Figure 8.1 Elements of an integrated waste management system. Source: Warmer Bulletin 49, White P., So what is integrated waste management. Journal of the World Resource Foundation, Tonbridge, 1996. Reproduced by permission from the World Resource Foundation

Environmental sustainability means that the options, and the integration of those options should produce a waste management system that reduces the overall environmental impacts of waste management, including energy consumption, pollution of land, air and water and loss of amenity (White et al 1995; Warmer Bulletin 49, 1996). Economic sustainability means that the overall costs of the waste management system should operate at a cost-level acceptable to all areas of the community, including householders, businesses, institutions and government (White et al 1995; Warmer Bulletin 49, 1996). In assessing the most environmentally and economically sustainable system, the local existing waste management infrastructure such as availability of landfill sites, existing incinerators, the types of waste to be managed, waste tonnages generated etc. should all be considered.

Figure 8.1 (Warmer Bulletin 49, 1996) shows that at the centre of an integrated waste management system is the collection and sorting of the waste, since this influences the treatment and disposal options of that waste, for example, recycling, composting, use for energy recovery etc. (White et al 1995). Materials recycling allows the useable materials in the waste to be removed, e.g., paper, glass, metals etc., in a materials recycling facility. The residual waste may then be processed as refuse derived fuel or combusted in an incinerator to recover energy. The waste may be landfilled to produce landfill gas, and energy recovered from combustion of the derived gas. Biological treatment of the waste via anaerobic digestion to produce a combustible gas, or treatment to produce compost, may also be options. In most cases the treatment options require landfill as a final disposal route for the residual product. An integrated waste management system would include one or all of the above options (White et al 1995).

Integrated waste management may also be interpreted as integration in terms of the management of wastes from different sources, such as commercial, household and industrial, or else in terms of different materials, such as metals, paper and putrescible wastes, or of waste from different product areas, such as packaging waste, white goods etc. (Warmer Bulletin 49, 1996). In a truly integrated waste management system, wastes such as demolition products, sewage sludge, and hazardous, agricultural, industrial and household wastes would all be included in the waste management system. However, such diverse wastes are covered by different authorities, are subject to different legislation and arise in different amounts, and are therefore more difficult to integrate than, for example, 'municipal solid waste'.

Tchobanoglous et al (1993) define integrated waste management in terms of the integration of six functional elements.

1. Waste generation – Assessment of waste generation and evaluation of waste reduction.
2. Waste handling and separation, storage and processing at source – Involves the activities associated with the management of wastes until they are placed in storage containers for collection. This may include source separation of household waste into recyclable and non-recyclable materials. Provision for suitable storage for the wastes, which may encompass a wide variety of different types, is also part of this element. Processing includes such processes as compaction, or composting of putrescible materials.
3. Collection – This element of the waste management system covers the collection and transport of the waste to the location where the collection vehicle is emptied. This location may be, for example, a materials recycling facility, a waste transfer station or a landfill disposal site.
4. Separation, processing and transformation of solid waste – The recovery of separated materials, the separation and processing of waste components and transformation of wastes are elements which occur primarily in locations away from the source of waste generation. This category includes waste treatment at materials recycling facilities, activities at waste transfer stations, anaerobic digestion, composting and incineration with energy recovery.
5. Transfer and transport – This element involves the transfer of wastes from the smaller collection vehicles to the larger transport equipment and the subsequent transport of the wastes, usually over long distances, to a processing or disposal site. The transfer usually takes place at a waste transfer station.
6. Disposal – Final disposal is usually landfill or landspreading, i.e., the disposal of waste directly from source to a landfill site, and the disposal of residual materials from materials recycling facilities, residue from waste incineration, residues from composting or anaerobic digestion etc. to the final disposal in landfill.

The inter-relationships of the six functional elements of an integrated solid waste management system are shown in Figure 8.2.

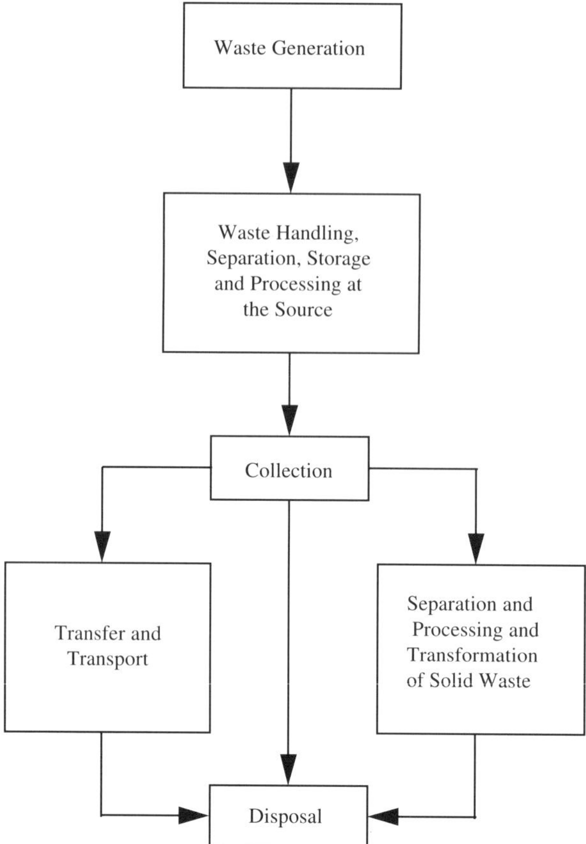

Figure 8.2 *Simplified diagram showing the inter-relationships between the functional elements in a solid waste management system. Source: Tchobanoglous G., Theisen H. and Vigil S., Integrated Solid Waste Management: Engineering Principles and Management Issues. McGraw-Hill, New York, 1993. Figure reproduced with permission of The McGraw-Hill Companies, New York*

Integrated waste management, as described by Tchobanoglous et al (1993), involves an evaluation of the use of the functional elements, and the effectiveness and economy of all the interfaces and connections to produce an integrated waste management system. They define integrated waste management as the selection and application of suitable techniques, technologies and management programmes to achieve specific waste management objectives and goals.

An example of an integrated waste management system is described in Box 8.1 (Riley 1996).

Whilst integrated waste management is a development now and in the near future, the far future of waste treatment and disposal is under active research. A recent United States of America NASA (National Aeronautics and Space Administration) symposium was dedicated to Waste Processing in Space (Golub and Wydeven 1991).

Box 8.1
Project INTEGRA, Hampshire

Project INTEGRA is an integrated waste management plan for Hampshire operated by Hampshire Waste Services. The project involves waste minimisation, recycling, composting, anaerobic digestion, incineration with energy recovery, and landfill with energy recovery from landfill gas. In terms of waste management, Hampshire is divided into three regions, north, south east and south west. The total waste arisings from the population of about 1.6 million is 600 000 tonnes per year, which is predicted to rise to 750 000 per year by the year 2010. The project aims to recycle 25% of the waste, with a future target option of 40% recycling. Recycling will include materials recycling, composting, anaerobic digestion and energy from waste. Currently, household waste recycling is carried out at three materials recycling facilities, which together recover approximately 13 000 tonnes/year of glass, plastics, paper and card, aluminium and steel. The Waste Collection Authorities use different recycling collection systems throughout the county, including twin bins, split vehicles, bags, banks, trailers, boxes etc. The materials recycling facilities have the capabilities to handle all these types of input. There is a plan to develop four further materials recovery facilities throughout the county. Composting is to be carried out on a large scale of an approximate throughput of 50 000 tonnes/year. Anaerobic digestion of the biodegradable waste, segregated from the waste stream after the removal of dry recyclable materials, will be in large digestors and will generate methane gas for energy recovery. Waste incineration is also to be part of the waste management package, with three medium sized mass-burn incinerators located throughout the county, with throughputs of 90 000 tonnes in the north region, 165 000 tonnes in the south east region and 105 000 tonnes in the south west region. Landfill of waste with energy recovery by landfill gas utilisation is also an option, with several sites throughout the county. The projected throughputs of waste into the various sectors, at a target of 25% waste recycling and the future option of a 40% target, is shown below for an estimated waste generation of 750 000 tonnes/year by the year 2010.

Type of recycling	25% recycling target		40% recycling target	
	Tonnes/year (1000s)	%	Tonnes/year (1000s)	%
Dry recycling	123	16	235	31
Composting	50	7	50	7
Anaerobic digestion	15	2	15	2
Energy from waste	360	48	360	48
Landfill	202	27	90	12
Total	750	100%	750	100%

Source: Riley K., Personal communication and Project INTEGRA data sheet, Hampshire Waste Services, Otterbourne, Hampshire, 1996.

Long-duration space missions demand an efficient and cost-effective waste management system. Such a system would involve waste minimisation and extensive recycling of sewage, liquid wastes, putrescible materials etc. Physico-chemical and/or biological processes are required for the recovery of the constituents for life support, and also to minimise launch costs and re-supply requirements (Golub and Wydeven 1991). The aim is to get as close as possible to a closed loop waste management system. The NASA symposium papers cover a range of waste management issues in space, for example, waste management for Space Station Freedom, a lunar base and a manned mission to Mars.

Bibliography

Golub M.A. and Wydeven T. 1991. Waste management in space – preface. Waste Management and Research, 9, 323.

Riley, K. 1996. Personal communication, Hampshire Waste Services, Otterbourne.

Tchobanoglous G., Theisen H. and Vigil S. 1993. Integrated Solid Waste Management: Engineering Principles and Management Issues. McGraw-Hill, New York.

Warmer Bulletin 49, 1996. White P. So what is integrated waste management? Journal of the World Resource Foundation, Tonbridge.

White P., Franke M. and Hindle P. 1995. Integrated Solid Waste Management: A Lifecycle Inventory. Blackie Academic and Professional, London.

Index